Aircraft, Engines and Airmen:

A Selective Review of the Periodical Literature, 1930-1969

by

August Hanniball

The Scarecrow Press, Inc.
Metuchen, N.J. 1972

ISBN 0-8108-0430-1

Library of Congress Catalog Card Number 70-171927

To my wife

Alice

I affectionately dedicate

this book

Contents

Legend of Symbols Used in Main Text

Part One: Aircraft

TEXTUAL SYMBOLS (Only one symbol is used--that which best describes the total article.)

1 General description, history, design, development

2 Testing, trials, first flights, performance

3 Construction, production, servicing, maintenance, modernization

4 Operational data, handling, piloting aspects, in-flight characteristics, record flights

5 Color scheme description

6( ) System description (with system in parenthesis)

ILLUSTRATIVE SYMBOLS

C Cockpit, flight deck, instrument panel

C* Same, with legend

D Drawing or painting

D* Same, in color

P Photograph of superior quality or detail

P* Same, in color

Pr Profile with outstanding information

Pr* Same, in color

Pr† Same, with color legend

S Schematic, cutaway or diagram of aircraft or system

S* Same, with legend

For an example (with explanatory notes) of a typical entry in Part One using the above symbols (and other abbreviations elsewhere explained) see pages 2ff.

Part Two: Engines

TEXTUAL SYMBOLS (Only one symbol is used--that which best describes the total article.)

1 General description, history, design, development

2 Testing, performance

3 Construction, production, servicing, maintenance

4( ) System description (with system in parenthesis)

ILLUSTRATIVE SYMBOLS

P Photograph of superior quality or detail

P* Same, in color

S Schematic, cutaway or diagram of engine or system

S* Same, with legend

The sample format (noted above) for Part One, on pages 2ff, may also serve as a guide to the format of Part Two.

List of Periodicals Indexed

Periodicals which ceased publication prior to December 1969 are indicated by an asterisk following the title. A list of periodical title abbreviations used in main text follows the full citation list below.

1. Aero Album -- Fallbrook, Calif. Quarterly. Vol. 1, Spring 1968 to Vol. 8, Winter 1969.

2. Aero Digest* -- Washington, D.C. Monthly. Vol. 16 No. 1, January 1930 to Vol. 73 No. 6, December 1956.

3. Aeronautics* -- London. Monthly. Vol. 1, August 1939 to Vol. 45 No. 5, March 1962. Previous title: Popular Flying (See #44).

4. Aeronautics* -- Chicago. Monthly. Vol. 5 No. 3, September 1929 to Vol. 6 No. 6, June 1930. Previous and succeeding title: Popular Aviation (See #43).

5. Aeroplane* -- London. Weekly. Vol. 38 No. 1, January 1, 1930 to Vol. 116 No. 2974, October 16, 1968. Title variants included are: Aeroplane and Astronautics; Aeroplane and Commercial Aviation News; and Aeroplane, the International Air Transport Journal. Merged with Flight International (See #35).

6. Aerospace Historian -- Washington, D.C. Quarterly. Vol. 12 No. 4, October 1965 to Vol. 16 No. 4, Winter 1969. Previous title: Airpower Historian (See #20).

7. Aerospace International -- Washington, D.C. Bimonthly. Vol. 2 No. 11, November 1966 to Vol. 5 No. 6, November/December 1969. Previous title: Air Force/Space Digest International (See #11).

8. Air BP -- London. No. 1, Autumn 1956 to No. 46, 1969. Published by International Aviation Service of British Petroleum Co., Ltd.

9. Air Classics -- Canoga Park, Calif. Irregular, then bimonthly. Vol. 1 to Vol. 6 No. 2, December 1969.

10. Air Force -- Washington, D.C. Monthly. Vol. 25 No. 7, December 1942 to Vol. 52 No. 12, December 1969.
Includes title variant: Air Force and Space Digest.
Previous title: Air Forces News Letter (See #12).

11. Air Force/Space Digest International* -- Washington, D.C. Monthly. Vol. 1, January 1965 to Vol. 2 No. 10, October 1966.
Became: Aerospace International (See #7).

12. Air Forces News Letter* -- Washington, D.C. Monthly. Vol. 24 No. 15, September 1941 to Vol. 25 No. 6, September 1942.
Suspended in December 1941 and January 1942.
Previous title: Air Corps News Letter.
Became: Air Force (See #10).

13. Air News* -- Chicago. Monthly. Vol. 1, May 1941 to Vol. 10 No. 5, May 1946.
Includes title variant: Air News with Air Tech.
Became: Skyways (See #51).

14. Air Pictorial -- London. Monthly. Vol. 12 No. 4, April 1950 to Vol. 31 No. 12, December 1969.
Includes title variant: Air Pictorial and Air Reserve Gazette.
Previous title: Air Reserve Gazette.

15. Air Pilot and Technician*-- New York. Monthly. Vol. 29 No. 6, June 1943 to Vol. 33 No. 3, March 1945.
Previous title: Sportsman Pilot.
Merged with: Aero Digest (See #2).

16. Air Power* -- London. Quarterly. Vol. 1, Autumn 1953 to Vol. 8 No. 1, Autumn 1960.
Previous title: Royal Air Force Quarterly (See #47).
Became: Royal Air Forces Quarterly (See #48).

17. Air Progress -- New York. Irregular, then monthly. Vol. 1, 1938 to Vol. 25 No. 6, December 1969.

18. Air Tech* -- Chicago. Monthly. Vol. 1, October 1942 to Vol. 7 No. 2, August 1945.
Merged with Air News (See #13).

19. Aircraft Engineering -- London. Monthly. Vol. 1, March 1929 to Vol. 41 No. 12, December 1969.

20. Airpower Historian -- Montgomery, Ala. Quarterly. Vol. 1, September 1954 to Vol. 12 No. 3, 1965.
Includes title variant: Air Power Historian.
Became: Aerospace Historian (See #6).

21. American Aviation Historical Society Journal -- Brooklyn, N.Y. Quarterly. Vol. 1, Spring 1956 to Vol. 14 No. 4, Winter 1969.

22. American Helicopter* -- New York. Monthly. Vol. 1, December 1945 to Vol. 54 No. 7, April 1960.

23. Army Aviation Digest* -- Ft. Rucker, Ala. Monthly. Vol. 1, February 1955 to Vol. 3 No. 2, February 1957.
Became: United States Army Aviation Digest (See #54).

24. Aviation* -- New York. Weekly until Vol. 29 No. 1, July 1930, then monthly. Vol. 28 No. 1, 1930 to Vol. 46 No. 6, June 1947.
Became: Aviation Week (See #27).

25. Aviation Age* -- New York. Monthly. Vol. 14 No. 3, September 1950 to Vol. 30 No. 3, September 1958.
Includes title variant: Aviation Age formerly Aviation Operations.
Former title: Aviation Operations (See #26).
Became: Space/Aeronautics (See #52).

26. Aviation Operations* -- New York. Monthly. Vol. 11 No. 4, March 1949 to Vol. 14 No. 2, August 1950.
Previous title: Aviation Maintenance and Operations and Aviation Service Magazine.
Became: Aviation Age (See #25).

27. Aviation Week -- New York. Weekly. Vol. 47 No. 1, July 7, 1947 to Vol. 91 No. 26, Mid-December 1969.

Includes title variant: Aviation Week and Space Technology.
Previous title: Aviation (See #24).

28. Bee-Hive -- Hartford, Conn. Irregular until January 1946, then quarterly. Vol. 1, January 1927 to Vol. 44 No. 4, Fall 1969.
Includes title variant: United Aircraft Quarterly Beehive.

29. Boeing Magazine -- Seattle, Wash. Monthly. Vol. 14 No. 7, July 1944 to Vol. 39 No. 12, December 1969.
Previous title: Boeing News.

30. Bristol Siddeley Journal* -- London. Quarterly. Vol. 1, Autumn 1959 to Vol. 8 No. 1, 1967.
Became: Rolls Royce Journal.

31. Canadian Aviation -- Toronto. Monthly. Vol. 1, June 1928 to Vol. 42 No. 2, December 1969.

32. Canadian Aviation Historical Society. Journal -- Toronto. Quarterly, Vol. 1, January 1963 to Vol. 7 No. 4, Winter 1969.
Vol. 1 No. 1 had the title The Early Bird Enthusiast.

33. Commercial Aviation* -- Montreal and Toronto. Monthly. Vol. 1, February 1939 to Vol. 8 No. 1, January 1946.
Title variants included are: Commercial Aviation and Aircraft Production; Commercial Aviation, Production, Maintenance.
Became: Aircraft and Airport.

34. Cross and Cockade -- Santa Ana, Calif. Quarterly. Vol. 1, Spring 1960 to Vol. 10 No. 4, Winter 1969.

35. Flight -- London. Weekly. Vol. 22 No. 1097, January 3, 1930 to Vol. 96 No. 3172, December 25, 1969.
Includes Flight International and other minor title variants.

36. Flying -- Chicago. Monthly. Vol. 27 No. 2, August 1940 to Vol. 85 No. 6, December 1969.

Title variants included are: Flying and Popular Aviation, and Flying, including Industrial Aviation. Previous title: Popular Aviation (See #43).

37. Flying Review International -- London. Monthly. Vol. 19 No. 1, October 1963 to Vol. 25 No. 8, December 1969.
Previous title: Royal Air Force Flying Review (See #46).

38. Historic Aviation -- Kansas City, Kan. Monthly. Vol. 1, September 1968 to Vol. 2 No. 12, December 1969.

39. Interavia (English Edition) -- Geneva. Monthly. Vol. 1, April 1946 to Vol. 24 No. 12, December 1969.

40. Lockheed Horizons -- Burbank, Calif. Quarterly. No. 1, Spring 1965 to No. 7, January 1969.

41. Naval Aviation News -- Washington, D.C. Monthly. October 1943 to December 1969. Volume numbers irregular.
Previous title: BUAER News Letter.

42. Pegasus* -- New York. Monthly. Vol. 1, January 1943 to Vol. 27 No. 1, January 1958.

43. Popular Aviation* -- Chicago. Monthly. Vol. 1, August 1927 to Vol. 26 No. 7, July 1940.
Between September 1929 and June 1930 title was Aeronautics (See #4).
Became: Flying and Popular Aviation (See #36).

44. Popular Flying* -- London. Monthly. Vol. 1, April 1932 to Vol. 8 No. 4, July 1939.
Became: Aeronautics (See #3).

45. Roundel* -- Ottawa. Monthly. Vol. 1, November 1948 to Vol. 17 No. 5, June 1965.
Became: Sentinel, Canadian Forces (See #49).

46. Royal Air Force Flying Review* -- London. Monthly. Vol. 8 No. 6, March 1953 to Vol. 18 No. 12, September 1963.
Previous title: Royal Air Force Review.
Became: Flying Review International (See #37).

[Regular Edition used. Add one month to citation when using International Edition between April 1958 and June 1961.]

47. Royal Air Force Quarterly* -- London. Quarterly. Vol. 1, January 1930 to Vol. 20 No. 2, April 1949. New Series: Vol. 1, January 1950 to Vol. 5 No. 3, July 1953.
Includes variant title: Royal Air Force Quarterly and Empire Air Forces Journal.
Became: Air Power (See #16).

48. Royal Air Forces Quarterly -- London. Quarterly. Vol. 1, February 1961 to Vol. 9 No. 4, Winter 1969.
Previous title: Air Power (See #16).

49. Sentinel, Canadian Forces -- Ottawa. Ten issues per year. Vol. 1, November 1965 to Vol. 5 No. 10, November-December 1969.
Previous title: Roundel (See #45).

50. Skyline -- Inglewood and El Segundo, Calif. Quarterly. Vol. 1, 1940 to Vol. 6 No. 3, May-June 1945 and Vol. 7 No. 1, May 1949 to Vol. 27 No. 4, 1969.

51. Skyways* -- New York. Monthly. Vol. 1, November 1942 to Vol. 15 No. 7, July 1956.
Includes title variant: Flying Sportsman and Skyways, June to October 1947.
Became: Skyways for Business.

52. Space/Aeronautics -- New York. Monthly. Vol. 30 No. 4, October 1958 to Vol. 52 No. 7, December 1969.
Previous title: Aviation Age (See #25).

53. United States Air Services* -- New York and Washington, D.C. Monthly. Vol. 15 No. 1, January 1930 to Vol. 41 No. 12, December 1956.

54. United States Army Aviation Digest -- Ft. Rucker, Ala. Monthly. Vol. 3 No. 3, March 1957 to Vol. 15 No. 12, December 1969.
Previous title: Army Aviation Digest (See #23).

55. Western Aerospace* -- Los Angeles. Monthly.

Vol. 42 No. 5, May 1962 to Vol. 44 No. 6, June 1964.
Previous title: Western Aviation (See #56).

56. Western Aviation* -- Los Angeles. Monthly. Vol. 31 No. 7, July 1951 to Vol. 42 No. 4, April 1962.
Includes title variant: Western Aviation, Missiles and Space.
Previous title: Western Flying (See #57).
Became: Western Aerospace (See #55).

57. Western Flying* -- Los Angeles. Monthly. Vol. 7 No. 1, January 1930 to Vol. 31 No. 6, June 1951.
Became: Western Aviation (See #56).

List of Periodical Title Abbreviations

Aero Album -- Aero Album

Aero Digest -- Aero Digest

Aeronautics -- Aeronautics. London

Aeron (US) -- Aeronautics. Chicago

Aeroplane -- Aeroplane

Aerosp Hist -- Aerospace Historian

Aerosp Intl -- Aerospace International

Air BP -- Air BP

Air Cl -- Air Classics

Air Force -- Air Force (and Space Digest)

AF/Sp D Intl -- Air Force/Space Digest International

AF News Ltr -- Air Forces News Letter

Air News -- Air News

Air Pict -- Air Pictorial

Air Pilot -- Air Pilot and Technician

Air Power -- Air Power

Air Prog -- Air Progress

Air Tech -- Air Tech

Acft Eng -- Aircraft Engineering

Airpwr Hist -- Airpower Historian

AAHSJ -- American Aviation Historical Society Journal

Am Helicop -- American Helicopter

Army Avn D -- Army Aviation Digest

Aviation -- Aviation

Avn Age -- Aviation Age

Avn Opns -- Aviation Operations

Avn Wk -- Aviation Week

Bee-Hive -- Bee-Hive

Boeing Mag -- Boeing Magazine

Bris Sid Jnl -- Bristol Siddeley Journal

Can Avn -- Canadian Aviation

CAHS -- Canadian Aviation Historical Society. Journal

Comm Avn -- Commercial Aviation

Cr & Cockade -- Cross and Cockade

Flight -- Flight (International)

Flying -- Flying

Fly Rev Intl -- Flying Review International

Hist Avn -- Historic Aviation

Interavia -- Interavia

Lock Hor -- Lockheed Horizons

Nav Avn News -- Naval Aviation News

Pegasus -- Pegasus

Pop Avn -- Popular Aviation

Pop Fly -- Popular Flying

Roundel -- Roundel

RAF Fly Rev -- Royal Air Force Flying Review

RAF Qtr -- Royal Air Force Quarterly

RAFs Qtr -- Royal Air Forces Quarterly

Sentinel -- (Canadian Forces) Sentinel

Skyline -- Skyline

Skyways -- Skyways

Sp/Aeron -- Space/Aeronautics

US Air Svcs -- United States Air Services

US Army Avn -- United States Army Aviation Digest

West Aerosp -- Western Aerospace

West Avn -- Western Aviation

West Fly -- Western Flying

Key to Abbreviations of Terminology

The abbreviations listed below include only those which are not standard or which may cause difficulty. Standard abbreviations, such as those used for the months of the year, are omitted, as are those whose meaning is self-evident.

Airlines which are better known by their acronyms (e.g. KLM, PAA, TWA) are not included.

Acft	Aircraft
AF	Air Force
Ag	Agricultural
Arm	Armament
ATO	Assisted Take Off
Aut	Autumn
Auto	Automatic
Aux	Auxiliary
AVG	American Volunteer Group
Bomb	Bombing
CBI	China-Burma-India (Theater of action in World War II)
CG	Commanding General
Cntrl	Control
CO	Commanding Officer
Comm	Communication
Cond	Conditioning
Czech	Czechoslovakia
DFC	Distinguished Flying Cross
Div	Division
ECM	Electronic Countermeasures
Elec	Electrical
Elecrn	Electronic
Equip	Equipment
Escad	Escadrille
Ext	Extinguisher
Flt	Flight
(fp)	Full Page

Ftr	Fighter
GB	Great Britain
GHQ	General Headquarters
Gp	Group
Grnd	Ground
Hydr	Hydraulic
ILS	Instrument Landing System
Injec	Injection
Inst	Instruments
JG	Jagdgeschwader
JS	Jagdstaffel
Lndg	Landing
Lube	Lubrication
Mnfld	Manifold
Mk	Mark
NAAS	Naval Auxiliary Air Station
NAS	Naval Air Station
Nav	Navigation
OTU	Operational Training Unit
Por(s)	Portrait(s)
POW	Prisoner(s) of War
Press	Pressurization
Prop	Propellers
Proto	Prototype
Pur	Pursuit
RAAF	Royal Australian Air Force
RAF	Royal Air Force
RATO	Rocket Assisted Take Off
RCAF	Royal Canadian Air Force
Retract	Retraction
RFC	Royal Flying Corps
RNAS	Royal Naval Air Service
RNZAF	Royal New Zealand Air Force
SAC	Strategic Air Command
SHAPE	Supreme Headquarters, Allied Powers Europe
Spg	Spring
Sq	Squadron
STOL	Short Take Off and Landing
Sum	Summer
Supp	Supplement
Suppr	Suppression
Sync	Synchronization

TAC	Tactical Air Command
Tech	Technical
Temp	Temperature
USAAC	United States Army Air Corps
USAAF	United States Army Air Force
USAF	United States Air Force
USAFE	United States Air Force, Europe
USMC	United States Marine Corps
VC	Victoria Cross
V/STOL	Vertical/Short Take Off and Landing
w	With
Win	Winter
WW I	World War I
WW II	World War II

Preface

In an era when rapid access to information is receiving great attention, it is an anachronism that millions of aeronautical devotees, both interested dilettantes and serious students, are not better able to locate some excellent written and illustrative material in the periodical literature of aeronautics.

The Library Index to Military Periodicals, produced by the Air University at Maxwell Air Force Base, Alabama, does list a considerable amount of aeronautical material. For the aeronautics buff, however, it not only encompasses far more than just his field of interest, its military focus results in the exclusion of many excellent aviation magazines. In addition, this Air University index began in 1949, so many of the contemporary accounts of aviation's glorious past are not cited in its pages.

Another important index is the United States National Advisory Committee for Aeronautics (NACA) Bibliography of Aeronautics, an out-growth of Paul Brockett's pioneering work of the same title. The two works include between them the years 1909 through 1932. Neither publication is confined to periodicals, however, and the indexing does not measure up to modern standards.

The aim of this work, then, is to fill the sixteen year void between the NACA and Air University indexes and to bring together in one volume forty years of significant aeronautical writing and illustrative effort from periodical literature.

The study is divided into five major parts: aircraft, aircraft engines, biography of airmen, air forces of the world, and a miscellany. It was necessary to fix certain limitations before embarking on the formidable task of reviewing the fifty-seven periodicals that are included:

a. The review begins in 1930. If this work is well-received, a review of the periodical literature prior to

1930 will be made.

b. Publications selected for review are limited to the English language, and were chosen primarily for their detailed treatment of aeronautical subjects or for outstanding illustrations. Local availability of these publications is not a significant problem today when modern copying methods and interlibrary loan procedures make the contents available to a wide audience.

c. Technical journals for the scientist and engineer contain articles not often understood by the layman. As a result, they have been omitted from consideration.

d. Spacecraft, lighter-than-air and soaring craft, gliders (except troop-carriers) and most business, executive, private, and home-built aircraft have not been cited. Their inclusion would have necessitated a review of many additional magazines which are devoted only to that specialized aspect of aviation of which they are a part. There is a fine distinction between certain categories of aircraft, however, that makes the decision for inclusion or exclusion difficult. A small transport feederliner, for example, can also be a business or executive aircraft. If the former role is appreciable, it will be included here.

e. To preserve conciseness of format, full bibliographic citations have been omitted. A number symbol has been used to indicate primary textual content and one or more letter symbols denote illustrative material.

f. Most articles are accompanied by photographs. To indicate all of these by symbol serves no useful purpose, unless the photograph is artistically exceptional or provides clarity in detail. Similarly, three-view or side-view drawings (herein termed profiles) abound and only those which appear to be outstanding or which are in color are cited.

Every precaution has been taken to insure that names and designations are correct and that periodical citations are accurate. Nevertheless, human frailty makes it inevitable that mistakes will occur when thousands of aircraft types and variants are included. The author can only beg forgiveness and solicit corrections for future editions and supplements.

It is, of course, impossible to verify the authenticity of each article. Accuracy in text and drawing can only be assumed. Deletions were made when written material or illustrations were known to be in error, but there is little doubt that others have escaped observation. Such faults are acknowledged and hopefully they are infrequent.

A review of the "List of Periodicals Indexed" will reveal that some titles are indexed for a few years prior to 1930, the standard point of departure. These periodicals had their inception so near 1930 that it seemed wiser to include the few additional earlier years than to omit them arbitrarily. This insures that such titles are indexed in their entirety rather than in part.

In conclusion, a word to the serious aircraft modeler who has often impressed the author with his search for authenticity. An appreciation of your needs has guided this work. To obtain the best rewards you are urged to read the introductory material to each section carefully.

For an example of the format of entries in Part One, with detailed explanations of each item within an entry, see pages 2ff. The format of entries in Part Two is similar.

The author wishes to express his thanks to all those whose cooperation helped to make this volume possible, especially to his former colleagues at the United States Air Force Academy Library, and to Major Frederick Kiley, Department of English, for reviewing material and providing many helpful suggestions.

Thanks are also due to the staff of the Air University Library, Maxwell Air Force Base, Montgomery, Alabama for their considerable assistance.

In a work of this magnitude there are always a few items that are difficult to obtain. Mr. Doug MacRitchie of the Canadian Aviation Historical Society was most kind for unhesitatingly providing early issues of that society's Journal for review. Similarly, acknowledgment is due Mrs. F. H. Aubrey, Aeronautical Library, National Science Library, Ottawa, Canada, and Mary Edith Fry, Technology Department, Seattle Public Library, Washington, for furnishing review copies of Commercial Aviation.

Indispensable aid in typing, proofreading, and criticism was provided by the author's wife. Without her assistance this book would never have reached print.

August Hanniball

University of Utah Libraries
Salt Lake City, Utah
January 1971

Part One

AIRCRAFT

Introduction to Part One

The articles indexed in this section have been chosen for information about the aircraft per se--not for congressional investigations or parliamentary debate concerning their worth, the problems they would bring when they appear on the scene, or disasters in which they have played a part.

Format

An example with explanatory notes best illustrates the format of this section:

ILYUSHIN[1] Russia [2]
Fly Rev Intl, D '66:268-270 (Pt 1) [3]
Ja '67:315-316 (Pt 2)

1704.[4] IL-18 [5] "Moskva" [6] (Moscow) [7] [Coot] [8]
"Moskva" is name for 1957 prototype only.[9]

Air BP, No. 39:8 [10](Tabso-)[11]			Pr* [13]
Air Force, N '60:46-48	1 [12]		
Air Prog, N/D '65:17 (-D)	1		
Flight, Ap 25, '58:560	6	(Cockpit)	C
Jl 25, '58:110-111	1		S*

1. Primary arrangement is alphabetical by designer or manufacturer. Names containing prefixes are entered under the prefix, e.g. De Havilland. Abbreviations are used when they are better known than the full name. In such cases the full name will follow in parentheses.

2. The country in which the aircraft designer or manufacturer has produced his aircraft follows the name.

3. General or historical articles about the manufacturing firm follow directly after the name and country. A distinction is made between this type of article and a biographical sketch of the founder or designer who has given his name to the business. The latter will be cited in Part Three, "Biography."

If the article is limited in time or subject, a note will indicate such restrictions. Of course, the date that the article appeared will also provide an insight into limitations.

4. An identification number is assigned each aircraft for indexing.

5. A primary identifier is listed first for each aircraft cited. This can be:

   a. An official or common name.
   b. A letter or letter/number combination.
   c. A number.

   The arrangement of aircraft under any given designer or manufacturer will follow the above order. For example, Lockheed "Sirius" precedes Lockheed A-11, which precedes Lockheed 212. The choice of a primary identifier, whether names, letters, or numbers, is solely the author's. Usually the best known element was selected, although successive manufacturers' designations were sometimes chosen to preserve chronological order in design.

   United States military aircraft which never went beyond the experimental or prototype stage, as evidenced by the symbols X or Y preceding the basic mission symbol, are placed alphabetically as though the X or Y did not exist. This procedure tends to keep aircraft of similar categories together and allows the reader with incomplete information to locate his entry more rapidly. As an example, the Bell XP-77 is listed in the pursuit category along with the P-39, P-59, P-63 and XP-83--not with the X-1, X-2, X-5 and others that were conceived solely as experimental aircraft from their inception.

6. Other identifiers, if any, follow the primary identifier. No special effort has been made to produce a complete list of descriptors. It is enough to know Curtiss "Kingbird," for example.

7. The direct translation of a foreign aircraft name is placed in parentheses.

8. Some Russian and communist satellite nation aircraft,

as well as aircraft of the Japanese Empire in World War II, have been assigned code names. These appear in brackets.

9. Notes of interest are located directly under the identifying elements. It must be remembered that this is an index, not a reference work on aircraft; therefore, no effort has been made to be comprehensive. Notes have been furnished primarily to guide the researcher or to point out an important fact.

10. The periodical citation consists of three elements:

    a. Title abbreviation: A key is provided on page xv.
    b. Date or volume/number: When dates are not continuously provided for or are entirely absent, a number or volume and number may be found instead. An example is the Air BP citation above.
    c. Pagination: Page numbers always follow a colon. Intervening pages of advertising are sometimes ignored if a truer insight to the comprehensiveness of the article is achieved.

11. If the written or illustrative material of an article is concerned only (or primarily) with a specific aspect of an aircraft, that specific quality is indicated within parentheses following pagination. By this means articles limited to a model or variant, a famous named aircraft, the aircraft of an ace or a particular airline can all be indicated. In the example, the Air BP citation is a profile in color of an Ilyushin 18 in the markings of the Bulgarian airline "Tabso." Similarly, the Air Progress article is concerned only with the D model of the Il-18.

    Variants from the basic military mission for which an aircraft was designed can also be identified through parenthetical treatment. The (R-/-C) accompanying some citations of the Douglas B-66 "Destroyer" indicates that the article or illustration is concerned with the reconnaissance version--the RB-66C. The hyphens indicate position relative to the basic symbol or name.

    Mark numbers, those model designations favored particularly by the British, have been designated by Roman or Arabic numerals. Generally, they are cited as they appear in the article itself.

Multi-part articles are often an indication of in-depth treatment. To make such series evident, a second pair of parentheses may occur: e.g. (Pt 1), (Pt 2).

When parentheses are absent, one of two possibilities exists:

a. Discussion or portrayal involves all or a number of different variants--too many to be specific.
b. Particularity is not indicated in text or drawing. This occurs most frequently when a new aircraft appears on the scene and a specific designation has not yet been assigned. It is, of course, impractical to attempt individual verification whenever this occurs.

12. The meaning of the six textual symbols is provided on page vi. Only one is used--that which best summarizes the entire content of the article. Accordingly, some ingenuity must be employed at times to locate precise information. If performance figures are desired and no "2" symbols are in evidence, one should next check citations accompanied by symbol "1"--general description. Performance figures are often provided in these broader articles. Ideally, one should check all references before abandoning the search.

13. The legend for the five basic illustrative symbols is also located on page vi. Asterisks are added to these to denote color or description.

The broad term "schematic" requires explanation. Ranging from highly detailed cutaway views of an entire aircraft to a simple line drawing of a component part, the criteria for inclusion is relevance. If the illustration adds appreciably to an understanding of the text or is noteworthy in itself, it will be cited.

As stated in the preface, photographs are included only when judged outstanding. Though more subjective, this approach will provide greater usefulness than citing every photograph that occurs. Some excellent photographs appear on the front covers of periodicals. Nevertheless, because library copies of these magazines will most often be used, front cover citations are normally not included. Library property stamps are invariably placed here and extraneous written material often crowds in upon what would otherwise be a fine picture.

Addendum

Some outstanding articles have been written that are not specific enough to warrant inclusion in the main portion of this section. Their excellence seemed so pronounced that an addendum was added where these broader topics could be cited.

In no sense is this addendum to be considered comprehensive. It simply includes those articles which appear to the author to have considerable merit.

Part One

AIRCRAFT

ABRAMS			USA
1.	"Explorer"		
	Acft Eng, D '39:415-416	1	
	Aviation, Ja '38:41	1	
	Can Avn, N '38:14-15	1	
	Flight, O 20, '38:347	1	
	Pop Avn, F '38:29-30+	1	
	RAF Fly Rev, Jl '57:40	1	
	West Fly, D '37:21-22	1	
A. D. (Air Department of the Admiralty)			GB
2.	"Flying Boat"		
	Built by Pemberton-Billing Ltd., 1917-18		
	Air Pict, N '63:350-355	1	
A. E. (Designed by Royal Aircraft Factory)			GB
3.	3 "Farnborough Ram"		
	Air Pict, Jl '61:202-206+	1	
AEC (Aircraft Engineering Corp.)			USA
4.	"Ace"		
	Air Pict, D '56:429	1	
AEG (Allgemeine Elektrizitäts Gesellschaft)			Germany
5.	G. IV		
	Aeronautics, F'46:81		C*
	Cr & Cockade, Sum '69:178	1	P
	Pop Fly, Ja '36:541	1	D
6.	J-1		
	Cr & Cockade, Spg '69:55-62	1	P, Pr, S
	63-65	4	

AERFER			Italy
7.	"Ariete" (Battering Ram)		
	RAF Fly Rev, D '59:24-25	1	P*
8.	"Sagittario 2" (Sagittarius 2)		
	Avn Age, Ag '57:78-81	1	S
	Avn Wk, My 13, '57:60-63	1	
	Flight, Mr 15, '57:343	1	
	RAF Fly Rev, O '57:30-33	1	Pr, S*
AERFER/MACCHI			Italy
9.	AM-3		
	Avn Wk, Je 2, '69:356-358(-c)	1	C, P
	Fly Rev Intl, N '67:958-959+	1	P*, S
AERIAL DISTRIBUTERS			USA
	See DISTRIBUTER WING		
AERIAL EXPERIMENT ASSOCIATION			Canada
	Hist Avn, Jl '69:11-21		
	Pop Avn, Ap '36:241-242		
	See also CURTISS		
10.	"Silver Dart"		
	First aircraft flown in British Empire, 1909.		
	Can Avn, Ja '37:6-7+	1	
	S '66:68	1	D*
	Ap '69:137	1	D*
	CAHS, Win '64:100-102 (Replica-)	4	
	Hist Avn, Jl '69:20	1	P
	Interavia, No. 2, '59:173	1	
	Pop Fly, S '37:335-336+	1	
	Roundel, Ap '59:9-10 (Replica-)	1	
	Jl/Ag '59:22-29	1	
AERMACCHI			Italy
	See MACCHI		
AERO			Czech
11.	A300		
	Flight, Ja 19, '39:61, 65	1	
	Fly Rev Intl, N '63:47	1	

AERO (cont.)

12.	Ae-50		
	RAF Fly Rev, My '63:41	1	
13.	L-39		
	Fly Rev Intl, F '69:31	1	

AERO COMMANDER USA

14.	A-9 "Ag Commander"		
	Air Prog, Mr '68:52-53+	4	C, P
15.	520		
	Flying, Ag '51:30-31+	4	
	My '53:28-29+	1	
	Skyways, O '51:20-22+	4	C
	N '52:10-13+	4	
	O '55:18	6 (Cabin Temp)	S*
16.	560		
	Flight, Ap 20, '61:537-538 (-F)	4	C
	Flying, Mr '56:32+ (-A)	4	
	My '61:86 (-F)	1	
17.	600 "Snow Commander"		
	Can Avn, Ag '66:11-12	1	
18.	680		
	Air Pict, Ja '57:18-19 (-S)		Pr
	19-20 (-A&-S)	1	C
	Flying, O '59:34-36+ (-E)	4	
	D '60:84 (-F)	4	
19.	720 "Alti-Cruiser"		
	Flying, Mr '58:54	1	

AERO SPACELINES USA

See ON MARK/AERO SPACELINES

AEROCAR USA

20.	"Aerocar"		
	Can Avn, Ag '50:28+ (-I)	1	
	Flying, S '57:38-39+ (-II)	4	
	D '69:53-55+ (-III)	4	
	West Avn, Ja '57:24 (-II)	1	

AEROCENTRE France

21. NC 211 "Cormoran" (Cormorant)
Avn Wk, O 11, '48:44 1
Interavia, No. 1, '47:44-46 1 C
RAF Fly Rev, N '62:39 1

22. NC 1070
Acft Eng, Ap '48:114-115 1

23. NC 2001 "Abeille" (Bee)
Interavia, No. 4, '47:31-34 1 C

24. NC 3020 & 3021 "Belphegor"
Flight, Je 12, '47:554 1
Fly Rev Intl, Jl '67:748 1
Interavia, No. 6, '47:37-40 1 C

AERONAUTICAL PRODUCTS USA

25. A-3 Helicopter
Am Helicop, Mr '46:31 1
Aviation, Je '45:122-130 1 S*

AERONCA USA

26. "Chief"
Flight, S 8, '38:209-210 4
Flying, Ap '68:54-59+ 1 S
Skyways, Ap '45:34-41+ 1 S*
West Fly, S '38:19-20 (-50-C) 1

27. KC "Scout"
Aviation, N '37:40 1
Can Avn, Ap '38:21 1
West Fly, D '37:34 1

28. L-16 "Grasshopper"
Flying, N '52:30+ 4

29. L-58B "Defender"
Flight, My 27, '43:554b 1

AEROTECHNICA Spain

30. AC-13
Am Helicop, Jl '55:11+ 1

AFA (Aktiengesellschaft für Flugzeugunternehmungen Altenrhein) Swiss

31. P-16
Earlier development: FFA P-16
Later development: AFU AJ-7
Avn Wk, My 30, '60:66-75 (-III) 1

AFU (Same as AFA; change in initials only) Swiss

32. AJ-7
Earlier versions: AFA P-16 and FFA P-16
Avn Wk, My 17, '65:111-115 1

AGO (Aerowerke Gustav Otto) Germany

33. Ao-192 "Kurier"
Fly Rev Intl, Ja '68:53 1
RAF Fly Rev, Jl/Ag '59:9 1

34. C.I
Pop Avn, Jl '35:27+ 1

35. C.II
RAF Fly Rev, O '60:11 1

36. C.IV
Cr & Cockade, Sum '69:173 1 P

AGUSTA Italy
Fly Rev Intl, F '65:13-15 (Helicopters)
Interavia, No. 6, '58:615-616

37. A 101 G
Interavia, No. 8, '64:1126-1127 1

38. A 105
Interavia, No. 8, '64:1127 1

39. A 106
Fly Rev Intl, Jl '69:30 1

AGUSTA/BELL Italy/USA

40. AB 102
Avn Wk, Je 27, '60:86 1

AGUSTA/BELL (cont.)

41. AB 204
Fly Rev Intl, D '69:53 (-B:Austria) Pr*

AIBA Japan

42. "Tsubami VI"
Pop Avn, S '36:20 1

AICHI

43. B7A1 "Ryusei" (Shooting Star) [Grace]
RAF Fly Rev, S '59:61 1

44. D3A1-2 Type 99
RAF Fly Rev, Mr '57:54+ 1

45. D4Y1-4 "Suisei" (Comet) [Judy]
RAF Fly Rev, O '57:37-38 1

46. M6A1 "Seiran" (Mountain Haze): Float version
"Nansan" (Southern Mountain): Wheel version
RAF Fly Rev, Mr '56:11 1
F '61:39 1

AIR & SPACE USA

47. 18A
Flying, Ap '65:40-41+ 4

AIR PARTS NZ
See FLETCHER

AIRCO (Aircraft Manufacturing Company) GB
See DE HAVILLAND

AIRCRAFT BUILDERS USA

48. "Student Prince"
Aviation, F 1, '30:210 1

AIRCRAFT INTERNATIONAL USA
See STEWARD-DAVIS

AIRMARK GB

49. AM 21
Fly Rev Intl, D '69:31 1

AIRPLANE DEVELOPMENT CORPORATION USA
See VULTEE

AIRSPEED GB
Flight, F 23, '51:217-228 (Acft list and photos)

50.	"Ambassador" A.S. 57			
	See also "Elizabethan"			
	Aeronautics, Je'48:44-47	1		P
	N '48:44			P*
	Aeroplane, O 26, '45:481-487	2		
	O 11, '46:416-420	1		S*
	Jl 25, '47:99-103	1		S*
	O 21, '49:559-562	4		C
	Jl 14, '50:54-57	6	(Elec.)	
	Air Pict, Ag '51:231	1		
	Acft Eng, F '49:42-48	1		S
	Ag '52:216-220	6	(Anti-ice)	
	Avn Wk, Ja 26, '48:21-25	1		
	Can Avn, S '47:30-31	1		
	Comm Avn, N '45:72	1		
	Flight, O 25, '45:440-443	1		
	Ja 30, '45:107-109	1		S*
	Ap 10, '47:307-311	1		S*
	Jl 10, '47:40	1		
	Jl 24, '47:75-78	6	(Controls)	S
	Ag 28, '47:231	1		
	O 23, '47:481	4		
	D 4, '47:628-629	1		
	O 20, '49:523-527	4		C
	D 15, '49:759	1		
	F 23, '51:229-230	1		
	Interavia, No. 5, '47:36-39	1		
51.	"Consul" A.S. 65			
	Acft Eng, O '49:325-327	1		S*
	Flight, Ap 11, '46:357	1		C
52.	"Courier" A.S. 5			
	Aeroplane, Mr 22, '33:513-516	1		S
	N 15, '33:854	4		
	Can Avn, D '34:27 (Canadian-)	1		

AIRSPEED (cont.)

52.	"Courier" (cont.)		
	Flight, O 20, '32:982	1	
	Mr 23, '33:266-269	1	Pr, S
	Ap 20, '33:379-380	1	
	D 14, '33:1253-1254	4	
	Pop Fly, Mr '34:627-628	4	
53.	"Elizabethan" A.S. 57		
	Special "Ambassador" for British European Airways (BEA).		
	Aeroplane, N 14, '52:661-664	4	C
	Flight, F 23, '51:231-234	1	C, S
	My 2, '52:536-537	4	
	Jl 11, '52:36-38	4	
54.	"Envoy" A.S. 6		
	Aeroplane, Jl 11, '34:50-54	1	C*, S*
	Ja 9, '35:40-42	1	C
	Mr 18, '36:361-362 (-II)	4	
	Jl 1, '36:15-16 (Converted-)	1	
	Mr 3, '44:247	1	
	Acft Eng, S '34:244	1	
	Flight, Jl 12, '34:712-713	1	C*, S
	Ja 10, '35:30-31 ("Overseas"-)	1	C
	Mr 19, '36:300-301 (-II)	4	
	Pop Fly, N '36:399 (-II)	1	
	S '37:338 ("Kings Flight"-)		P
	S '38:284-286 ("Kings Flight"-)	4	P
55.	"Ferry" A.S. 4		
	Aeroplane, Ap 13, '32:652-656	1	S
	Flight, Ap 15, '32:317-320	1	Pr, S
56.	"Fleet Shadower" A.S. 39		
	RAF Fly Rev, My '59:9	1	
57.	"Horsa" A.S. 58		
	Troop glider.		
	Aeroplane, My 26, '44:584-589	1	C*, S
	Can Avn, S '43:52-53+	1	C*, S*
	Flight, Jl 15, '43:64-67	1	S*
	S 30, '43:365	1	
	S 14, '44:286a-b	3	

AIRSPEED (cont.)

58.	"Oxford" A.S. 40		
	Aeroplane, Jl 28, '37:111-114	1	C*, S*
	Ag 3, '38:137		P
	Ja 25, '46:112-114	4	C*
	Air BP, No. 28:24-25	4	C
	Aviation, F '41:132		S*
	Comm Avn, Jl '39:24	1	
	Flight, Jl 1, '37:26-27	1	S*
	Je 16, '38:603		C*
	Je 30, '38:628-632	1	C
	D 24, '42:688b	1	
	Ja 21, '43:60-61	1	
	My 27, '43:549-551	4	C*
	O 18, '45:421-422	4	
	My 8, '69:Inset	1	P*
59.	"Queen Wasp" A.S. 30		
	Pop Avn, Ja '38:55		P
60.	"Viceroy" A.S. 8		
	Aeroplane, O 3, '34:393-394	1	S*

AISA (Aeronautica Industrial S.A.) Spain

61.	AVD-12C		
	Flight, N 16, '56:785	1	
	RAF Fly Rev, O '56:30-33	1	Pr, S*
62.	I-115 or E.9		
	Interavia, No. 10, '53:586-587	1	P, S*

ALBATROS Germany

Hist Avn, Je '69:28-30 (C. Series)
RAF Fly Rev, Mr '59:16-19+ (D.I-VII)

63.	B.I		
	Air Prog, O '65:23-27	1	Pr
64.	B.II		
	Aeronautics, F '46:78		C*
	Avn Wk, My 27, '63:58		D*
	Flying, Jl '60:53		D
65.	C.III		
	Cr & Cockade, Aut '62:257		Pr

ALBATROS (cont.)

66.	C.V		
	Cr & Cockade, Spg '69:66-90	1	Pr, S*
	Sum '69:168 (Mohr's-)	1	P
67.	C.VII		
	Avn Wk, Ag 31, '64:54-55	1	D*
68.	C.X		
	Cr & Cockade, Sum '69:170 (Wollenhaupt's-)	1	P
69.	D.I		
	Air Cl, Ap '69:29	1	D
70.	D.III		
	Air Cl, Ap '69:30	1	D
	Avn Wk, O 21, '63:36		D*
	Can Avn, Ap '68:138	1	D*
	Cr & Cockade, Aut '64:289 (Austrian-)		C
	Sum '69:169 (Simon's-)	1	P
	Pop Fly, Je '33:129	1	D
71.	D.V		
	Cr & Cockade, Sum '62:162-167 (-A)	1	S
	Aut '62:246 (-A:JS4)		P
	Spg '64:83-84 (-A)		P
	Aut '64:227-232 (v. Hipple, v. Schleich, Dilthey & other units)	5	Pr
	Sum '69:172 (Clausnitzer's-)	1	P
	:174 (Bertraub's-)	1	P
	:176 (-A:Sakowsky's & Waldhausen's)	1	P
	:177 (Wackwitz's-)	1	P
	:182 (- A:Hohmuth's)	1	P
	:185 (-A)	1	P
	Flying, F'61:77		D
	Fly Rev Intl, N '64:52-53 (Dilthey, Loerzer, v. Schleich, Jasta Boelke-)	1	Pr*
	N '68:53 (-A:Restored)		Pr*
	RAF Fly Rev, Mr '59:16 (-A:Dilthey's)		Pr*

ALBATROS (cont.)

72.	D.XI		
	Cr & Cockade, Win '63:378-381	1	P, Pr

ALLIED			USA
73.	XLRA-1 Glider		
	Fly Rev Intl, D '63:61	1	
AMBROSINI			Italy
74.	"Sagittario"		
	Flight, Ap 24, '53:508-509	1	
75.	"Supersette" S. 7 Super		
	Aeroplane, Ag 8, '52:188-191	4	C, P
	Flight, Ag 24, '51:224-227	4	C
	Fly Rev Intl, My '64:63	1	
76.	P. 512		
	Flight, N 11, '48:570	1	S
ALCOR			USA
77.	C-6-1 "Junior Transport"		
	Aero Digest, My '38:62+	1	
	Aviation, My '38:36-37	1	C
	Flight, My 12, '38:466a	1	
	Pop Avn, Jl '38:41	1	
AMERICAN AERONAUTICAL CORPORATION			USA
	See BUDD		
AMERICAN EAGLE			USA
	Hist Avn, N '69:11-26		
78.	A-1		
	Pop Avn, Mr '28:29		P
79.	A-430		
	US Air Svcs, Mr '30:54	1	

AMERICAN HELICOPTER USA

80.	XA-5 "Top Sergeant"		
	Am Helicop, Mr '49:6-7+	1	
	Avn Wk, F 14, '49:22-26	1	
	Flight, Mr 23, '50:371	1	
	Interavia, Ap '49:220-221	1	
81.	XA-6 "Buck Private"		
	Am Helicop, Ap '51:8+	1	
	Interavia, No. 8, '52:457	1	
	Skyways, Jl '51:56	1	
82.	XH-26 "Jet Jeep"		
	Am Helicop, S '52:8+	1	
	Avn Age, Ja '54:30-31	1	C, S
	Avn Wk, O 13, '52:21-28	1	
	West Avn, O '52:53	1	

AMERICAN HOPPI-COPTERS USA
See HOPPI-COPTER

AMERICAN JET INDUSTRIES USA

83.	T-610 "Super Pinto"		
	Earlier version: TEMCO TT-1 "Pinto"		
	Air Prog, Mr '69:47-51+	4	C
	Fly Rev Intl, F '69:32	1	

AMIOT France

84.	110S		
	Pop Avn, Ap '34:222	1	
85.	142M and 143M		
	Fly Rev Intl, Je '65:49-51	1	Pr*
	RAF Fly Rev, My '59:11	1	
	RAF Qtr, Ja '35:16-17+	1	S*
86.	350 to 356		
	RAF Fly Rev, Ag '56:15	1	
87.	370		
	RAF Fly Rev, Ag '56:15	1	

ANBO Lithuania

88. IV
Flight, Ag 23, '34:884 1

ANSALDO Italy

89. S.V.A. 5 "Primo" (First)
Avn Wk, N 27, '61:34-35 1 D*
Cr & Cockade, Win '69:345-349 1 C*
Pop Fly, N '34:420 1 D

90. S.V.A. 10
Fly Rev Intl, O '68:53 (Uruguay-) Pr*

ANT (A. N. Tupolev) Russia

91. 5
Military designation: I-4
Air Pict, Ag '64:251 (I-4) 1
Fly Rev Intl, S '64:45+ 1

92. 6
Military designation: TB-3
RAF Fly Rev, Ap '57:49-50 1
Jl '62:43+ (TB-3) 1

93. 9
Aero Digest, My '30:142 1

94. 14
Acft Eng, Jl '42:197 1

95. 20
Aeroplane, Ja 10, '34:48 ("Maxim Gorky") 1
Air Pict, Ap '66:151 1
Airpwr Hist, Ap '62:80-83 ("Maxim Gorky") 1
Pop Avn, Mr '35:180 1
RAF Fly Rev, Je '58:8+ 1

96. 25
Aero Digest, Ag '37:32-34+ 1
Air Prog, Ag/S '64:74-75 1 Pr†
D '68:42-43 (Moscow-Calif.-) Pr*

ANT (cont.)

96. 25 (cont.)
AAHSJ, Sum '68:103-111 (Moscow-Calif.-) 4
Aviation, Ag '37:50-53 1
Flying, S '56:43+ ("Moscow") 4
Jl '62:38-39+ (Moscow-Calif.-) 4
US Air Svcs, Jl '37:11-12+ (Moscow-Calif.-) 4

97. 40
See TUPOLEV SB-

98. 44
Fly Rev Intl, N '68:71 1

ANTOINETTE France

99. Monoplane
Flight, My 14, '42:476-477 1
Pop Avn, My '37:36 (Latham's-) 4

ANTONOV Russia

100. A-7
RAF Fly Rev, Mr '59:50 1

101. An-2 "Kolchosnik" [Colt]
Aeroplane, S 7, '56:385-388 4 C*
D 5, '58:828-829 1
Air Pict, Ap '63:108-109 1 Pr
Flight, Jl 13, '56:97 1 S
Fly Rev Intl, My '66:540 1
RAF Fly Rev, Ag '56:22-25 1 Pr, S*
F '59:8-9 1 P*
US Army Avn, My '59:17 1

102. An-8 [Camp]
Air Force, Jl '57:82-83 1
Fly Rev Intl, Mr '68:164 1

103. An-10 "Ukrainia" [Cat]
Air Force, N '57:84-86 1
Avn Wk, O 14, '57:43-45 1
Je 23, '58:32 1
D 28, '59:18-19 1

ANTONOV (cont.)

103.	An-10 (cont.)			
	Flight, Jl 25, '58:112 (-A)	1		
	F 27, '59:289-292	1		S
	N 27, '59:635 (-A)	1		
	My 6, '60:628 (-A)			Pr
	Fly Rev Intl, My '66:543 (-A)	1		
104.	An-12 [Cub]			
	Cargo version of An-10A			
	Air Prog, N/D '65:22-23	1		C
	Avn Wk, Mr 26, '62:37	1		
	Ag 2, '65:52-53	1		
	My 22, '67:66	6	(Skiis)	
105.	An-14 "Pchelka" (Little Bee) [Clod]			
	Air Pict, O '63:324-325	1		Pr
	Avn Wk, Ap 21, '58:43	1		
	Ag 18, '58:39	1		S
	Jl 31, '67:65	1		
	Fly Rev Intl, Ja '66:298	1		
	My '66:545	1		
106.	An-11 "Antheus" [Cock]			
	Aeroplane, Je 24, '65:16-19	1		
	Air Prog, N/D '65:10-13	1		
	Avn Wk, Je 28, '65:32-33	1		
	Jl 12, '65:94-97	1		
	Ag 14, '67:52-53	4		
	:58	6	(Cargo)	
	S 15, '69:94	2		
	Fly Rev Intl, My '66:544	1		
107.	An-24 [Coke]			
	Aeroplane, Je 21, '67:4-5	1		
	Air BP, No. 43:9 (-V:LOT)			Pr*
	Air Pict, F '63:44-45	1		Pr
	Air Prog, N/D '65:26 (-B)	1		C
	Avn Wk, My 28, '62:52-53	1		S*
	O 7, '63:126-127	1		
	Ag 16, '65:54	1		
	Jl 10, '67:54- 64 (-TV&-RV)	1		
	Je 30, '69:68-69 (-RT&-TV)	1		
	Fly Rev Intl, Je '66:597-598+	1		C, S*, Pr
	S '69:26-28	1		
	Interavia, No. 2, '67:223 (-B)	3		

ARADO Germany

RAF Fly Rev, S '55:26+ (Some WW II types)

108.	Ar-68		
	Fly Rev Intl, F '66:379, 381	1	P
	:380 (-E&-F)		Pr*
	RAF Fly Rev, Ag '60:9+	1	
109.	Ar-79		
	Acft Eng, O '38:321-322	1	
110.	Ar-80		
	Fly Rev Intl, F '67:396-397	1	
111.	Ar-81		
	RAF Fly Rev, Mr '61:9	1	
112.	Ar-95		
	Later development: Ar-196		
	Acft Eng, N '40:325-326+	1	P
	Fly Rev Intl, Je '67:678-679	1	
113.	Ar-96		
	Later version: Ar-396		
	Aero Digest, S '40:111-112	1	
	Flying, D '40:44 (-B)	1	
	Fly Rev Intl, S '69:51 (-B:Hungary)		Pr*
114.	Ar-196		
	Development of Ar-95		
	Air Prog, Je/Jl '63:48 (-A-3)		Pr†
	Fly Rev Intl, Je '67:690 (-A-1&-3)		Pr*
	RAF Fly Rev, N '57:49-50	1	P
115.	Ar-197		
	Fly Rev Intl, D '65:253 (Graf Zeppelin's-)	1	
	RAF Fly Rev, S '61:40	1	
116.	Ar-198 "Flying Aquarium"		
	RAF Fly Rev, O '61	1	
117.	Ar-232		
	Aeroplane, S 7, '45:271	1	
	Fly Rev Intl, Jl '66:708-709	1	
	RAF Fly Rev, Mr '56:9 (-A&-B)	1	
	Mr '59:37-38	1	

ARADO (cont.)

118.	Ar-234 "Blitz" (Lightning)		
	World's first production jet bomber.		
	Air News, D '45:23	1	
	RAF Fly Rev, My '54:39	1	
	Ap '56:31-34	1	S*
	:30	5	Pr
119.	Ar-240		
	RAF Fly Rev, Ag '58:37-38	1	
120.	Ar-396		
	Earlier version: Ar-96		
	RAF Fly Rev, Jl '63:51+	1	

ARMSTRONG WHITWORTH GB

Interavia, Ag '49:464-466

121.	"Albemarle" A.W. 41		
	Aeroplane, Je 30, '44:728-731	1	S*
	Air BP, No. 39:26-29	4	C
	Air Tech, Ap '45:60-62+	1	S
	Aviation, Ap '44:180-181	1	
	Flight, Ja 27, '44:87-91	1	
	Ap 4, '46:340a-341	4	
	RAF Fly Rev, N '57:42+	1	
122.	"Apollo" A.W. 55		
	Aeronautics, Mr '50:42-45	1	C, P, P*
	Aeroplane, Ag 6, '48:157-158	1	
	Ap 22, '49:446-450	1	S*
	S 9, '49:354-355	4	
	O 26, '51:556		S*
	Flight, Ag 21, '47:187	1	
	Ag 26, '48:235-240	1	S*
	Ap 21, '49:445	2	
	S 15, '49:336-338	4	C
123.	"Argosy" A.W. 650		
	Later development: HAWKER-SIDDELEY "Argosy"		
	Aeronautics, Mr '59:34-43	1	
	Jl '61:42-43	1	
	Aeroplane, Ja 18, '57:77	1	
	Ja 17, '58:70-71	1	C
	My 30, '58:742-744	1	
	N 21, '58:748-758	1	S*

ARMSTRONG WHITWORTH (cont.)

123.	"Argosy" (cont.)		
	Air Pict, Ap '59:117-119	1	Pr
	Avn Wk, D 8, '58:92-105	3	
	Ag 17, '59:46-47	1	C
	F 13, '61:97-99	3	
	F 12, '62:54-67	4	C
	Flight, Ja 18, '57:66-67	1	
	Mr 15, '57:326-327	1	
	N 1, '57:718-720	1	S*
	My 30, '58:752-753	1	C
	Jl 25, '58:113-115	1	S*
	O 17, '58:619-632	1	C*, S*
	N 28, '58:847	3	
	Ja 16, '59:89	2	
	F 6, '59:196	6 (Tail de-ice)	S*
	F 27, '59:293	1	Pr
	S 4, '59:113		P
	Ag 5, '60:201-203	4	C
	Interavia, No. 2, '57:145	1	
	No. 2, '59:163-165	1	C
	Sp/Aeron, S '59:59-61	3	S*
124.	"Argosy" A.W. 660		
	Later development: HAWKER-SIDDELEY "Argosy"		
	Aeronautics, Ja '62:32-34	4	
	Aeroplane, Mr 3, '61:229-234 (-C.1)	1	S*
	Avn Wk, F 13, '61:97-99	3	
	Flight, Ag 21, '59:41	1	C
	O 7, '60:586+	1	
	F 10, '61:181-185	1	S*
	S 28, '61:510-511 (-C.1)	1	P
	Interavia, No. 2, '59:163-165	1	C
	No. 1, '62:67-73	1	
	RAF Fly Rev, O '61:59 (-C.1)	1	P*
	RAFs Qtr, Spg '62:6 (-C.1)		P
	Sp/Aeron, S '59:59-61	3	S*
125.	"Atalanta" A.W. 15		
	Aeroplane, Jl 6, '32:22-28	1	S*
	Acft Eng, Ag '32:200-202	1	P
	Flight, Jl 8, '32:619-623 (Pt 1)	1	Pr, S
	Jl 15, '32:661-665 (Pt 2)	3	S
	Pop Fly, Ap '33:21		C

ARMSTRONG WHITWORTH (cont.)

126.	"Atlas"		
	CAHS, Sum '67:60		S*
	Flight, O 16, '31:1033-1036 (Pt 1)	1	S
	O 23, '31:1057-1061 (Pt 2)	1	Pr, S
	RAF Fly Rev, Jl/Ag '59:39	1	
	RAF Qtr, Jl '32:Plate 7		P
127.	"Awana"		
	Air Pict, Jl '66:263	1	
128.	"Ensign" A.W. 27		
	Aeroplane, Mr. 31, '37:377-381	1	S*
	Air Pict, F '68:60-61 ("Enterprise")	4	
	Aviation, Je '37:32-33	1	S
	Can Avn, N '38:18-19	1	
	Flight, Ap 1, '37:316-322	1	C, S*
	S 15, '38:224-225	1	C, S
	Mr 11, '43:256b	1	
	F 15, '57:203-206 (Pt 1)	1	
	F 22, '57:247-250 (Pt 2)	1	C
	Pop Avn, F '38:23-24	1	
	O '38:39		P
	US Air Svcs, Jl '38:30	1	
129.	"Meteor"		
	Original development: GLOSTER "Meteor"		
	Aeronautics, Mr '51:36-37 (-NF.11)		P
	S '52:141 (-NF.11)		P
	Aeroplane, S 7, '51:283 (-NF.11)		P
	Flight, My 30, '58:744 (-TT.20)	1	
130.	"Scimitar" A.W. 35		
	Acft Eng, Ag '34:213-214	1	
	Fly Rev Intl, O '68:75-76	1	
131.	"Sea Hawk"		
	Earlier development: HAWKER "Sea Hawk"		
	Air Pict, Ap '59:127-130	1	S*
132.	"Siskin"		
	Air Pict, Mr '57:90-91 (-IIIA:41 Sq)	5	Pr
	:91+	1	
	Ap '67:127 (-V)	1	
	Can Avn, Ag '66:44 (-IIIA)	1	D*
	CAHS, Sum '63:32 (-IIIA:Canadian)	1	

ARMSTRONG WHITWORTH (cont.)

132.	"Siskin" (cont.)		
	Fly Rev Intl, My '68:265 (-IIA:43 Sq)		Pr*
	RAF Fly Rev, Mr '60:61 (-III, -IIIA & -IIIB)	1	
	RAF Qtr, Ap '38:Supp (-IIIA)	1	D
133.	"Whitley" A.W. 38		
	Aeroplane, Ag 11, '37:168-172 (-I)	1	S*
	Air BP, No. 29:6-8	4	C
	Air Pict, D '56:412-414 (Pt 1)	1	
	Ja '57:13-15 (Pt 2)	1	
	Flight, O 21, '37:396-402 (-I to -III)	1	S
	My 26, '38:528-529		S*
	Mr 21, '40:264a-d	4	
	Ja 2, '41:13-16	1	
	D 3, '42:608b	1	
	Ap 4, '46:340a-341	4	
	Fly Rev Intl, My '68:268 (-V:102 Sq)		Pr*
	Pop Avn, F '37:18	1	
	D '37:25		P
134.	A.W. 16		
	Aeroplane, O 14, '31:925-928	1	S
	Flight, O 16, '31:1033-1036 (Pt 1)	1	S
	O 23, '31:1057-1061 (Pt 2)	1	Pr, S
	Fly Rev Intl, O '68:75	1	
135.	A.W. 19		
	Air Pict, Ja '63:20-21	1	Pr
136.	A.W. 23		
	Flight, O 17, '35:Supp:b-c	1	P
137.	A.W. 27		
	Flight, Ja 7, '37:8-10	3	S
138.	A.W. 29		
	Air Pict, O '58:360-361	1	
	RAF Fly Rev, Jl '63:55	1	
139.	A.W. 52		
	(-G) is glider version.		
	Aeronautics, Ap '47:48-49	1	

ARMSTRONG WHITWORTH (cont.)

139.	A.W. 52 (cont.)		
	Aeroplane, My 10, '46:547-551	1	C*, S
	D 20, '46:748-753	3	
	D 27, '46:773-777	3	S*
	Ja 16, '48:75-77		S*
	Acft Eng, My '49:144-148+	1	
	Aviation, Ag '46:63-64 (-G)	1	
	F '47:53-54	1	
	Flight, My 9, '46:463-468 (-G)	1	S*
	D 19, '46:674a-679	1	S*
	N 20, '47:567	2	S*
	D 25, '47:724-725	2	
	Ja 15, '48:69-73	1	C*
	Mr 31, '49:369 (Nene & Derwent-)	2	
140.	F.K. 3 "Little Ack"		
	Air Pict, Jl '55:203	1	
141.	F.K. 8 "Big Ack"		
	Air Pict, N '60:387-389	1	
	Avn Wk, Ja 7, '63:36-37	1	D*
	Pop Fly, Ag '33:261	1	D
142.	F.K. 10		
	Air Pict, Ja '59:11	1	
	RAF Fly Rev, Mr '60:58	1	

ARMY AIRCRAFT FACTORY — Sweden

142a.	S. 21		
	Original design: PHÖNIX 122 and 222		
	Fly Rev Intl, Ja '66:318	1	

ARROW — GB

143.	"Active"		
	Aeroplane, Jl 15, '31:180-181	1	
	Flight, Jl 17, '31:705		P
	Jl 24, '31:727-730	1	Pr
144.	"Sport"		
	Flying, Jl '68:52-57	4	C, P*

ARSENAL			France
145.	VB. 10 Earlier designation: VG. 30 Acft Eng, F '47:43-45+	1	S
146.	VG. 30 RAF Fly Rev, Je '59:11	1	
147.	VG. 33 RAF Fly Rev, Je '59:11	1	
148.	VG. 39 Fly Rev Intl, Mr '65:43	1	
149.	VG. 70 and 90 Air Pict, My '51:147, 152 (-90)	1	P
	RAF Fly Rev, Mr '57:57	1	
ARSENAL-DELANNE			France
150.	10-C.2 RAF Fly Rev, S '56:31	1	
ARUP			USA
151.	"Flying-Wing" Pop Avn, Ag '33:84-86	1	
	Ag '35:92	1	
ATLANTA			USA
152.	Commercial Monoplane (1931) Flight, Ja 2, '31:10-11	1	
AUSTER See also earlier TAYLORCRAFT			GB
153.	"Auster" Aeroplane, F 22, '46:221-222 (J.1)	4	C
	Air Pict, D '51:360-363	1	
	F '52:48 (T.7)	1	P
154.	A.O.P. (Air Observation Post) Aeroplane, Je 11, '54:756-760 (-9)	1	S*
	D 2, '55:856-857 (-9)	4	C

AUSTER (cont.)

154.	A.O.P. (cont.)		
	Flight, Jl 8, '48:46 (-7)	1	C
	Je 11, '54:761-765 (-9)	1	S*
	D 10, '54:835 (-9)	2	
	D 2, '55:832-834 (-9)	4	
155.	B. 4 Ambulance/Freighter		
	Aeroplane, D 7, '51:728-733	1	C, S*
	Ag 22, '52:239-241	4	C
	Air Pict, Ja '52:24	1	
	Avn Wk, N 19, '51:31-33	1	
	Can Avn, Ja '53:50+	1	
	My '53:66	1	
	Flight, O 5, '51:456-457	1	
	Ap 4, '52:370-372	4	C
	Jl 3, '53:28	4	
156.	B. 8 "Agricola"		
	Aeroplane, Ja 13, '56:51-55	1	S*
	O 26, '56:625-628	4	C
	Avn Wk, D 26, '55:45	1	
	Flight, D 16, '55:890-891	1	
	Ja 13, '56:47-51	1	C, S*
	O 19, '56:646-647	4	C
	O 26, '56:687	1	P
157.	J. 1U "Workmaster"		
	Air Pict, Ag '58:269	1	P
	Avn Wk, Ap 7, '58:95-97	1	
158.	J. 5 "Aiglet"		
	Flight, Ap 18, '52:456-457	4	C

AUSTIN-BALL GB

159.	A.F.B.-1		
	Air Pict, S '57:319	1	
	N '57:401	1	
	D '57:439	1	
	F '58:69	1	
	Hist Avn, Mr '69:14-15	1	

AUSTRALIAN GOVERNMENT FACTORIES Australia

160.	"Pika"		
	Flight, Ap 25, '52:512	1	

AUSTRIAN-AVIATIK — Austria

161. D.I "Berg Scout"
Air Pict, Mr '56:78 — 1
Cr & Cockade, Aut '60:14-27 — 1 — Pr, S
Win '62:366 (Vienna Tech. Museum's-) — 5 — Pr

AUTOGIRO — GB
Flying, Ja '60:34-35+

162. C. 19
Flight, Ja 17, '30:120-121 (-II) — 1
My 9, '30:501-506 (-IIa&-III) — 1 — Pr, S
N 20, '31:1145-1148 (-IV) — 1
Pop Avn, Ag '32:91-92+ — 4

163. C. 24
Flight, N 20, '31:1145-1148 — 1

164. C. 30
Later development: AVRO 671
Aeroplane, My 23, '34:845-846 — 4
Ag 1, '34:141-143 (-P) — 1 — S*
Flight, My 4, '33:425-428 — 4
N 9, '33:1115-1116 (-P) — 1 — P
N 16, '33:1157 (-P) — 1
Ag 2, '34:785-792 — 1 — Pr, S
Ag 9, '34:813-816 — 4 — C*, S*

165. C. 40
Flight, Je 8, '39:580-582 — 1

AVIA — Czech

166. B. 33
Original design: ILYUSHIN IL-10 [Beast]
Fly Rev Intl, Ap '66:500 — 1

167. B. 35 and 135
Fly Rev Intl, Ap '68:229 — 1
RAF Fly Rev, Ja '56:11 (B. 35) — 1

168. B. 534
Air Prog, Ag/S '64:78 (-IV) — Pr†
RAF Fly Rev, My '60:40 (Slovak A.F. Sq) — Pr*
:41 — 1
Ap '56:9+ — 1

AUTOGIRO (cont.)

169. S. 34
Pop Fly, My '38:76 P

170. S. 199 "Mezec" (Mule)
Development of MESSERSCHMITT Bf-109
Air Pict, Mr '60:99-100 1
My '67:173 1
Fly Rev Intl, Ag '67:809-811 1 Pr*
RAF Fly Rev, Ap '61:47+ (Israel-) 1
Ap '63:34-36 1 Pr*

171. 14
Original design: ILYUSHIN IL-14
Avn Wk, Jl 8, '57:75-76 1
Flight, Je 14, '57:795-796 4

172. 51
Flight, Ja 18, '34:54-56 1
Pop Avn, My '34:290+ 1

173. 158
Fly Rev Intl, Jl '65:59 1

AVIAN Canada

174. 2/180 "Gyroplane"
Air BP, No. 27:26-29 1 P*
Avn Wk, N 16, '59:99-100 1 S
D 7, '59:156 1 S
Can Avn, D '61:28+ 1
My '66:17 1
Flight, D 18, '59:766 1
Flying, Ap '64:44-45+ 1

AVIATIK Germany

175. G.
Von Mises project
Cr & Cockade, Win '62:295-309 1 Pr
Win '63:352-353 1 P

AVIATION TRADERS GB

176. "Accountant"
Aeroplane, Jl 12, '57:48-52 1 C, S*
Jl 19, '57:78-79 1 C
Flight, S 25, '53:434-435 1
O 21, '55:645-646 1 S*
Jl 6, '56:36-38 (-I&-II) 1 S*
Jl 12, '57:47-52 1 C, S*
N 1, '57:715-717 1 S*
Interavia, No. 10, '55:777-778 1 C
No. 6, '57:594-596 1
RAF Fly Rev, S '63:39 1

177. "Carvair"
Aeroplane, Je 29, '61:749-752 1 S*
F 8, '62:148-150 2
Ap 8, '65:34-36 1
Air BP, No. 30:9 (Aer Lingus-) Pr*
Avn Wk, S 4, '61:46-47 4
S 23, '63:114-117 4
Can Avn, Je '65:37-39 1
Flight, Je 29, '61:890-891 2

178. "Merchantman"
Converted VICKERS "Vanguard" 953
Flight, My 22, '69:839-840 3

179. "Super-Carvair"
"Carvair" with Rolls-Royce "Dart" engines
Flight, D 12, '63:953 1

AVIONS FAIREY Belgium
Flight, My 2, '52:527-530
See also FAIREY, Great Britain

AVRO GB
AAHSJ, Win '67:270-272 (Early aircraft)
Comm Avn, Ap '44:36-49
Ag '45:26-28
Flight, Ja 13, '38:30-33 and Supp a-d
Interavia, Jl '49:400-402

180. "Viper"
CAHS, Fall '64:85 Pr

AVRO (cont.)

181.	F & G (1912)		
	Flight, Je 18, '42:609-610	1	
182.	504		
	Aeronautics, Jl '41:32 (-K)		C*
	Aeroplane, Ja 30, '42:137	1	
	Air BP, No. 12:20-23	1	
	AAHSJ, Win '61:Supp (-K)		P*
	Avn Wk, S 2, '63:50-51	1	D*
	Can Avn, Jl '64:45 (-K)	1	D*
	CAHS, Fall '68:63-67 (-K)	3	
	Flight, Jl 9, '54:41-44 (Pt 1)	1	
	Jl 16, '54:83-88 (Pt 2)	1	
	O 23, '69:12	4	
	:17-19 (-K)	1	C, S*
	Flying, D '53:54	1	
	Ag '61:65	1	D
	Fly Rev Intl, O '67:942 (-K)		P*
	Hist Avn, Ap '69:12-13	1	
	Pop Fly, S '38:310-312 (-J)	4	
	RAF Qtr, Jl '37:Supp (-K)	1	D
	Jl '39:Plate (-K)		D
183.	531 "Spider"		
	Air Pict, Mr '57:88	1	
184.	539		
	Air Pict, D '64:394-416	1	Pr
185.	571 "Bison"		
	Air Pict, N '67:410-411	1	
	RAF Qtr, Ap '38:Supp (-II)	1	D
186.	594 "Avian"		
	Aeroplane, My 7, '30:850-852	4	
	Ag 29, '52:286-287	4	
	Acft Eng, Je '29:130-133 (-IV)	1	Pr
	Flight, O 31, '30:1183-1187 ("Southern Cross Jr.") (Pt 1)	1	Pr
	N 7, '30:1214-1217 ("Southern Cross Jr.") (Pt 2)	2	
	Pop Avn, Ap '28:21-22	4	

AVRO (cont.)

187.	618 "Ten"		
	Original design: FOKKER F.VII/3m. Variant: AVRO 642		
	Air BP, No. 13:2-4 ("Southern Cloud")	1	
	Flight, Ap 9, '64:585 ("Southern Cloud")	1	
188.	619 "Five"		
	Aeroplane, N 22, '33:896	4	
189.	621 "Tutor"		
	Aeroplane, Ap 23, '30:758-760	1	
	Jl 1, '31:49		P
	N 1, '33:772	4	
	Can Avn, S '30:25-26	1	
	Flight, My 2, '30:475-482	1	P, Pr, S
	Ap 27, '33:391-395	3	
190.	626 Trainer		
	AVRO 637 is a variant		
	Aeroplane, Ja 21, '31:106c-112	1	
	Jl 1, '31:42		P
	:50-55	1	S
	N 29, '33:936	4	
	Flight, Jl 3, '31:xxxv-xxxviii	1	
	N 9, '33:1128	1	
191.	627 "Mailplane"		
	Aeroplane, Ag 12, '31:426-434	1	S*
	Flight, Ag 14, '31:802-807	1	Pr, S
192.	631 "Cadet"		
	Aeroplane, Mr 23, '32:512-517	1	S
	:517-518	4	
	Flight, Mr 25, '32:253-257	1	Pr, S
193.	636		
	Aeroplane, Jl 31, '35:149-150	1	
	:151	4	
	Flight, Ag 1, '35:122-124	1	C
194.	637		
	AVRO 626 variant		
	Flight, N 16, '33:1142-1143	1	

AVRO (cont.)

195.	638 "Club Cadet"		
	Aeroplane, Ag 23, '33:350	4	
	Flight, Jl 13, '33:702-705	1	S
	Pop Fly, S '34:296-297	4	
196.	639 "Cabin Cadet"		
	Flight, Jl 13, '33:702-705	1	S
197.	640 "Cadet 3-Seater"		
	Flight, Jl 13, '33:702-705	1	C, S
198.	641 "Commodore"		
	Aeroplane, My 30, '34:868-872	1	C*, S*
	Acft Eng, S '34:242	1	
	Flight, My 31, '34:533-538	1	C*, Pr, S
	Pop Avn, S '35:159	1	
199.	642 "Eighteen"		
	AVRO 618 variant		
	Aeroplane, Ap 4, '34:610-613	1	S
	Acft Eng, S '34:243	1	
	Flight, Ap 5, '34:319-323	1	Pr
200.	643 "Cadet"		
	Aeroplane, Ap 15, '36:466-467	4	
	Flight, Jl 13, '33:702-705	1	S
	F 6, '36:144-145 (-II)	1	
201.	652 "Avatar," "Avalon," and "Ava"		
	Aeroplane, Mr 13, '35:304-306	4	
	Flight, Mr 7, '35:240-242	1	C, S
202.	652A "Anson," "Annie"		
	See also 652A "Nineteen" variant		
	Aeroplane, Ja 29, '36:127-130	1	S*
	N 17, '44:557-560	1	
	My 30, '52:653	1	
	Air BP, No. 31:22-23 (-I)	4	C
	Aviation, N '43:226 (-II&-V)	1	
	Can Avn, Mr '40:10-12	3	
	F '41:22-24	4	
	Je '41:45-47	6	(Engine Mnfld)
	Je '42:42-44 (-II)	4	C
	Mr '43:52-53+ (-II&-V)	1	
	O '43:53-55 (-V)	4	C
	:111-120 (-V)	3	

AVRO (cont.)

202. 652A (cont.)

Comm Avn, N '41:112 6 (Fuel Tank)
Ag '42:44-54 (-II) 1
:68 6 (Props)
N '42:142 6 (Control)
S '43:38-41 (Federal-/-V) 3
N '43:69-72+ 1
Flight, O 24, '35:442 1
Ja 30, '36:116c&d-119 1 S*
N 9, '44:503 (Civil-) 1 C
Jl 17, '47:58-59 (Converted V) 1 C
Je 6, '52:682-685 (-I to -22) 1
Ja 9, '69:Inset 1 P*
Fly Rev Intl, Je '69:73 (-I:Greece) Pr*
Pop Avn, Ap '38:33 P
RAF Fly Rev, My '61:36 1
S '61:27 4
F '63:18-19 4
Roundel, My '54:49 4
Jl/Ag '62:33 (-II:Canada) 4 P

203. 652A "Nineteen"
Variant of 652A "Anson"
Aeroplane, Ja 17, '47:71-73 4 S*
Flight, Ag 9, '45:144-145 1

204. 671 "Rota"
Original design: AUTOGIRO C. 30
RAF Qtr, Ap '35:Supp 1 P
RAF Fly Rev, Ja '63:25-26+ 4 D

205. 679 "Manchester"
Aeroplane, N 14, '41:540-541 1
N 21, '41:564-565 (-I) 1
Je 12, '42:674-675 (-I) 1
Air Pict, D '52:362-365 1
Can Avn, O '42:44-45 1
Flight, N 13, '41:339-341 1
N 20, '41:359-361 4
Je 4, '42:555-557 1
S 17, '42:308b 1
RAF Fly Rev, Jl '58:6 1

AVRO (cont.)

206.	683 "Lancaster"			
	Aeronautics, Ap '45:43-46	1		S
	Aeroplane, Jl 10, '42:42-43			D
	Ag 14, '42:178-181	3		C
	:190-191			S*
	O 23, '42:479-482 (-I)			S*
	D 17, '43:698-699 (-II)			P
	Air News, O '43:26-27+	4		
	Air Pict, D '52:362-365	1		
	Air Prog, F '43:58-59			S*
	Ag '43:58-59			S*
	Air Tech, N '42:12-13+	1		S*
	Aviation, F '43:155			S*
	F '44:145			S*
	Jl '46:97			S*
	Can Avn, O '42:35-37+	1		
	Je '43:78 (-I)			S*
	Ag '43:57-70+	3		
	My '45:58-59+	6	(Lndg gear)	S*
	F '64:30	1		
	Comm Avn, S '42:24-32	1		S*
	Jl '43:46-58+	1		C, S*
	Ap '44:44-49	1		C
	:53-54	6	(Tires)	
	Flight, Jl 9, '42:32-33	1		P
	Ag 13, '42:174-179	1		S*
	Ag 20, '42:202a	1		
	N 4, '43:503 (Civil-)	1		
	D 9, '43:634-637 (-I&-II)	1		S*
	F 17, '49:200 (Python-)	1		
	O 26, '56:674-675	4		
	Je 3, '65:874-875	4		
	Ap 17, '69:639-640	1		P*
	Flying, Mr '43:41 (-I&-II)	1		
	My '44:59			P*
	Fly Rev Intl, My '68:269 (-I:106 Sq)			Pr*
	Roundel, Ja/F '57:23-25	1		
	N '62:33 (-X:RCAF)	4		P
	Mr '64:10-14	1		
	RAF Fly Rev, F '54:27-30	4		
	D '58:45	5		
	N '60:46 (RAF Scampton's-)	1		
	Skyways, Je '43:30-33	3		
	D '44:18-23+	1		S*

AVRO (cont.)

207.	685 "York"		
	Aeroplane, N 12, '43:556-557 (-I)		P
	Ja 28, '44:105-106	1	
	Ag 18, '44:181-185 (-I)	1	C*, S*
	Air Pict, D '52:362-365	1	
	Can Avn, O '44:72-73+ (-I)	1	C*, S*
	Comm Avn, N '44:36-40	1	S*
	Flight, N 11, '43:532-533	1	
	D 2, '43:615	1	
	Ja 20, '44:74-75	1	
	Je 29, '44:694-695	1	
	Ag 17, '44:176-180	1	S*
	F 22, '45:204a-d (Duke of Glouster's-)	1	S
	Ja 17, '46:70	1	
	O 1, '47:385-388	1	S
208.	688 "Tudor I"		
	Aero Digest, S '46:80-81 (-I)	6	(Cabin Press) S*
	Aeronautics, Ja '46:52, 54 (-I)	1	
	Aeroplane, Ag 17, '45:185-191 (-I)	1	D*
	F 27, '48:253-255 (-IV)	4	
	Aviation, Jl '46:70-71 (-I)	6	(Cabin Press) S*
	Avn Wk, Ap 11, '49:33	1	
	Can Avn, D '45:88-90	1	
	Comm Avn, Ag '45:30-34	1	
	Flight, Je 28, '45:690-695 (-I)	1	S*
	My 16, '46:483-486 (-I)	2	
	Mr 6, '47:188-192a	1	
	Mr 11, '48:275-277 (-IV)	4	C*
	:285-287	1	
	D 2, '48:660-661 (-VIII)	1	
	Mr 24, '49:353 (-VIII)	2	
	Mr 31, '49:381 (-VIII)	6	(Cabin Press)
	F 26, '54:231-232 (Aviation Traders'-)	1	
	Fly Rev Intl, F '67:397 (-8)	1	
	Interavia, No. 3, '46:37-40	6	(Cabin Press) S*
209.	689 "Tudor 2"		
	Aeroplane, Je 7, '46:666-673	1	S*
	F 26, '54:240-242	1	
	Flight, Mr 28, '46:316a-318	1	S*
	Interavia, No. 1, '46:23-25	1	

AVRO (cont.)

210.	691 "Lancastrian"			
	Aeroplane, F 2, '45:140-141			P
	Ag 22, '47:217-219	1		S*
	Air News, Je '45:56-59	1		
	Air Pict, D '52:362-365	1		
	Flight, F 1, '45:118-121	1		
	S 26, '46:336-338 (Nene-)	1		
	Mr 13, '47:216a-217 (Nene-)	6	(Engine)	S*
211.	694 "Lincoln"			
	Aeronautics, Ja '46:42-45	1		P
	Aeroplane, Ja 18, '46:77-82	1		S*
	:83-84	6	(Turret)	S*
	Flight, O 4, '45:375	6	(Turret)	
	N 1, '45:467-468 (-II)	1		
	Ja 17, '46:57-62	1		S*
	:62a-64	6	(Turret)	S*
	Fly Rev Intl, My '68:268 (-I:57 Sq)			Pr*
	RAF Fly Rev, S '60:25	1		
212.	696 "Shackleton"			
	Aeroplane, D 9, '49:788-792	1		C, S*
	Air Pict, F '69:34-39	1		
	Avn Wk, Ap 25, '49:15-16	1		
	Flight, My 18, '50:611-617	1		C, S*
	Ja 29, '54:125-128 (M.R.2)	4		
	Ja 13, '56:52-53 (M.R.3)	1		P
	N 27, '69:843-848 (M.R.3:Phase 3)	4		
	Roundel, Ag '49:48	1		
	RAF Qtr, Ap '52:142 (M.R.I)			P
213.	698 "Vulcan"			
	Projected transport: 722 "Atlantic"			
	Aeronautics, N '52:42-45	1		
	D '53:40-43			P
	F '55:28-31			P
	Aeroplane, S 19, '52:420-421			P
	S 27, '57:487-490	1		
	Ja 31, '58:143-152	3		S*
	Air Pict, N '59:405-407+ (-B.2)	1		Pr
	Avn Age, O '53:24-25	1		
	Avn Wk, My 4, '53:28-34	1		
	O 14, '57:68-80	4		

AVRO (cont.)

213.	698 (cont.)		
	Bris Sid Jnl, Sum '61:86-91 (83 Sq-)	4	
	Flight, S 19, '52:385-387	1	P
	S 27, '57:502-505	1	
	D 13, '57:923-927	3	S*
	Ja 31, '58:142-150	1	S*
	Jl 18, '58:66-67	4	C*
	S 4, '59:115 (-B.2)		P
	N 30, '61:840-841	4	
	Fly Rev Intl, My '68:269 (-B.2)		Pr*
	Interavia, No. 11, '52:628-629	1	P
	No. 9, '60:1058-1064	4	P
	RAF Fly Rev, Jl '53:13-15	1	
	Ap '55:19-20+	2	
	O '56:48-49 (-B.1)	1	P*
	D '57:28-30	1	
	D '58:13-15+	4	
214.	701 "Athena"		
	Aeronautics, Ja '49:32-33		P
	Aeroplane, Ja 30, '48:132-134	1	C, S*
	Ag 19, '49:230-233 (-II)	1	S*
	N 18, '49:676-678 (-II)	4	
	Avn Wk, My 3, '48:21-26 (-I)	1	C
	Flight, Ap 8, '48:383-388 (-I)	1	S*
	Je 24, '48:698-699 (-I)	4	
	Ag 18, '40:197-199 (-II)	1	
	Interavia, O '48:549-551 (-I&-II)	1	
215.	706 "Ashton"		
	Aeroplane, N 17, '50:422-428	1	S*
	O 26, '51:557		S*
	Flight, D 7, '50:524-530	1	C, S*
216.	707		
	Aeroplane, Ag 3, '51:125 (-A)		P
	Ag 17, '51:178-179 (-B)	1	
	Avn Wk, O 9, '50:20-21 (-B)	1	
	S 3, '51:27 (-A)	1	
	Flight, Ag 10, '51:160-161 (-A&-B)	1	
	:172-174	2	
	Ag 17, '51:196-197 (-B)	1	
	RAF Qtr, Jl '51:234 (-B)		P
	Ap '52:174 (-A)		P

AVRO (cont.)

217. 722 "Atlantic"
Projected 698 "Vulcan" transport
Avn Age, Ag '53:27 1

218. 730
Projected supersonic bomber
Fly Rev Intl, D '66:276-277 1

219. 748
Later versions: See HAWKER-SIDDELEY 748 and "Andover"

Aeronautics, D '59:30-33	1	
Aeroplane, S 4, '59:115		S*
S 11, '59:177-179	1	C
F 12, '60:185-189	3	S*
Jl 8, '60:39-46 (-1)	1	S*
F 22, '62:197-200	4	C
Avn Wk, S 14, '59:74-83	1	
Can Avn, Je '60:59	1	
Jl '62:18-19	1	
Flight, My 8, '59:636-637	1	S
S 4, '59:106	1	C
Je 3, '60:757-764	1	C, S*
Jl 29, '60:167	1	C
Ap 27, '61:566-567	1	
F 1, '62:157 (-MF)	1	
F 8, '62:217	2	
Mr 1, '62:328-330 (-1)	4	C
Jl 19, '62:81-82 (-MF)	1	S
Interavia, No. 4, '62:495 (-MF)	1	
Sp/Aeron, Jl '61:71+	6 (Fail-safe)	S

AVRO CANADA Canada

220. C-102 "Jetliner"

Aero Digest, N '49:24-25+	1	
My '50:42-43+	3	
Aeroplane, Ja 21, '49:72-74	1	
S 30, '49:480		P
Acft Eng, Ag '50:228-237	1	C, S*
Avn Wk, N 1, '48:18-21 (X-)	1	
Ag 22, '49:36 (X-)	1	
O 3, '49:15-18 (X-)	1	

AVRO CANADA (cont.)

220.	C-102 (cont.)		
	Can Avn, Ag '48:20-21	3	
	Ja '49:22-23	1	
	O '49:18-21+	1	C
	F '50:25+	4	C
	Flight, S 29, '49:434-435	1	
	Ja 5, '50:13-15+	1	
	Jl 13, '50:78-79	6 (Cabin Press)	S*
	Interavia, No. 5, '50:261-266	1	S*
	Roundel, Mr '49:29-30 (X-)	1	
	N '49:49		P
221.	CF-100 "Canuck"		
	Aeroplane, Ag 19, '55:268-273	1	S
	Ap 12, '57:505-509	4	P
	Air Pict, Ja '53:10-11	1	
	Air Power, Win '53:202		P
	Avn Wk, Jl 5, '54:28-30	1	
	Can Avn, O '53:38-44 (-IV)	3	
	Ja '55:89 (-III:conversion)	1	
	F '57:50-51	4	C
	CAHS, Fall '66:86		S
	Spg '65:24-25 (Canadian-)	5	
	Flight, Ap 13, '50:455	4	
	O 23, '53:567-568	4	
	S '51:18-19+	1	
	Interavia, No. 3, '54:178	1	
	Roundel, Je '54:30-33	4	
	RAF Fly Rev, F '54:37-38	1	
	S '55:39+	2	
	O '55:11	6 (Armament)	
	Ja '56:30-33 (-IV)	1	Pr, S*
	Ja '57:36-37	4	
	Jl '60:24	1	
	:25 (-V:Belgium)		Pr*
222.	CF-105 "Arrow"		
	Aeroplane, O 18, '57:572		S*
	:585-588	1	
	Ap 4, '58:483	4	
	O 10, '58:566-568	1	S
	Air Pict, F '65:50-53	1	S
	Air Power, Sum '58:318-319	1	P
	Avn Wk, Je 14, '54:11	1	
	O 21, '57:50-59	1	
	N 25, '57:75-76	3	
	Ja 13, '58:108-109	3	

AVRO CANADA (cont.)

222. CF-105 (cont.)

Can Avn, O '57:50-52	1		
N '57:62-63	1		
D '57:44-45+	6	(Grnd Support)	
F '58:48	4		
:53-55+	6	(Lndg Gear)	
Mr '58:30-31+	6	(Hydr)	
My '58:38-39			P
Flight, O 18, '57:604-605	1		
O 25, '57:647-654	1		S*
O 10, '58:573	1		
Fly Rev Intl, F '64:62+	1		
Interavia, No. 12, '57:1261-1263	1		
RAF Fly Rev, F '59:18-21 (-I)	1		Pr, S*
Sp/Aeron, My '59:57-59	1		S

B.A. (British Aircraft Manufacturing Company) GB

223. "Double Eagle"

Aeroplane, Je 24, '36:797		S*
Ap 28, '37:496		S*
Flight, Ap 29, '37:422-423	1	S*

224. "Swallow"

Flight, My 16, '35:522-524 (-II)	1	C*, S

BAADE East Germany

225. 152

Later development: VEB 152

Avn Wk, Mr 23, '59:58-61	1
Fly Rev Intl, Mr '69:70-71	1

BAC (British Aircraft Corporation) GB

226. "Jet Provost"

Civil version: See BAC 145

Earlier version: HUNTING PERCIVAL "Jet Provost"

Air Force, Mr '69:116 (-T.5)	1	
Flight, Jl 6, '67:30-31	4	
Ag 24, '67:301-306 (-T.5)	1	S*

BAC (cont.)

227.	"Lightning"		
	Earlier version: ENGLISH ELECTRIC "Lightning"		
	Flight, Mr 21, '68:408	1	P*
	S 5, '68:371-378 (F. 53)	1	C*, S*
	Fly Rev Intl, My '68:272 (F. 6, 5 Sq)		Pr*
228.	TSR-2		
	Air Pict, D '63:393-395	1	Pr
	N '64:347-348	1	
	Ap '65:122-123	4	
	Air Prog, Ja '65:40		Pr
	Acft Eng, D '63:358-362+	1	
	N '64:338-353+	1	
	Avn Wk, N 4, '63:30-32	1	
	O 5, '64:22-23	1	
	Bris Sid Jnl, Win '63-64:30-34	1	
	Flight, Ja 9, '59:54	1	
	O 31, '63:710-711+	1	S*
	Ap 9, '64:569-578	1	D, S*
	Ag 27, '64:332+	6 (Engine Cntrl)	
	O 1, '64:576-577	2	
	O 8, '64:637-640	2	
	Ag 5, '65:218-220	4	
	Fly Rev Intl, Ja '64:6	1	
	F '64:26-27	1	S*
	O '64:35-37+	1	
	Je '65:3-4, 13-16	1	
	Interavia, No. 12, '63:1873	1	
229.	145		
	RAF version: BAC "Jet Provost"		
	Fly Rev Intl, Mr '65:13-16+	1	S*
230.	111 "One-Eleven"		
	Aeroplane, Ag 29, '63:10-18	2	
	:23-40	1	
	O 22, '64:20	6 (Anti-Stall)	
	F 25, '65:Supp:3-5	1	
	:6-14	4	
	:15-18	3	
	Mr 2, '67:22-23	1	
	:24-30	4	
	F 7, '68:34-35 (-500)	1	
	S 18, '68:24 (-400)	1	
	:27 (-500)	1	

BAC (cont.)

230. 111 (cont.)

Air BP,	No. 33:24 (Braniff-)			Pr*
Air Pict,	S '63:283-285	1		Pr
Acft Eng,	Je '61:175	1		
	My '63:123, 127-147, 149-154	1		C*, S*
	:124-126	2		
	:148	6	(Engine)	
	D '63:348-352	6	(Landing)	S*
	O '64:298-303	2		
	N '64:359	3		S*
	Mr '66:18-29	6	(Interior)	S*
	Jl '66:28-29	6	(Interior)	
	Mr '67:21-31 (-500 Series)	1		S*
	:32-34 (-500 Series)	4		
Avn Wk,	My 15, '61:42-43	1		
	Je 18, '62:79-85	3		
	Ag 19, '63:42-45	1		
	S 7, '64:38-41	3		
	O 25, '65:145-147	4		
	N 1, '65:55-57 (-400)	4		
	Ja 23, '67:75-81	6	(Hydraulic)	
Can Avn,	Mr '63:12-13	1		
Flight,	My 18, '61:673-674	1		
	Ag 10, '61:182-186	1		
	My 23, '63:737-738 (-200, 300, 400)	1		
	Jl 11, '63:44-56 (-200, 300, 400)	1		C, S*
	Mr 19, '64:413-417	2		S*
	Je 18, '64:1003-1004	1		
	Ag 27, '64:324-326	1		
	D 3, '64:955-957	4		C
	:958-961	3		
	:962-964	1		
	Ja 14, '65:45-46	6	(Auto Landing)	
	F 25, '65:284-288	2		
	Ap 7, '66:554-557 (-401)	4		
	F 2, '67:155-156 (-500)	1		
	Jl 6, '67:6 (-500)	2		
	F 22, '68:256	1		P*
	Mr 14, '68:364-365 (-500)	1		
	Ag 22, '68:293-295	6	(Navigation)	
	N 7, '68:741-746 (-500)	1		
	:747-750 (-500)	4		C
Fly Rev Intl,	O '63:33-37	1		Pr, S*
	Mr '65:25-31	1		
	N '68:38-41 (BEA's-)	4		

BAC (cont.)

230. 111 (cont.)

Interavia, No. 9, '61:1212-1214	1		
No. 4, '62:436-437	1		
No. 5, '63:656-657	3		
No. 7, '63:1068-1072 (-200, 300, 400)	1		S*
No. 9, '64:1185-1187	4		
:1188-1193, :1198-1200	1		
:1193-1194	6	(Engine)	
:1195-1196	6	(Flt Controls)	
:1196-1197	4		C*
No. 2, '66:172-173	4		C

231. 167 "Strikemaster"
Development of "Jet Provost"

Flight, Ag 24, '67:301, 304-306	1	C
Ap 17, '69:640-641	1	
Fly Rev Intl, Ag '69:48-51	1	C, P*
Interavia, No. 12, '69:1918-1920	1	P

232. 211 "Two-Eleven"

Flight, Ag 24, '67:284	1	

233. 221

Avn Wk, My 11, '64:37	1	
Flight, Jl 23, '64:133-138	1	S*
Interavia, No. 7, '64:986	1	

234. 311 "Three-Eleven"

Aeroplane, Jl 17, '68:34	1	
Acft Eng, N '68:22+	1	
Avn Wk, N 25, '68:27	1	
Flight, N 14, '68:777-780	1	
Ag 14, '69:229-230	2	
Fly Rev Intl, Ja '69:23-25	1	
Interavia, No. 1, '69:51-53	1	S*

BAC/SUD GB/France

235. "Concorde"

Aero Album, Sum '69:28-31	1	
Aeroplane, Ap 21, '66:20-21	6	(Flt Controls)
Ap 12, '67:Supp:3-26	1	
:28-36	6	(Engine)
S 18, '68:17	1	

BAC/SUD (cont.)

235. "Concorde" (cont.)

Aerosp Intl,	S/O '67:20-21	1		
Air Pict,	N '67:393-396	1		
	Ap '69:125-127	3		S*
Acft Eng,	N '63:326-331	3		
	F '64:40-42	6	(Engine)	S*
	Ap '64:111-119	1		S*
	My '64:125	1		
	Jl '65:218-224	1		S
	Mr '66:16-17	6	(Engine)	
	Ap '66:38-40	2		S*
	Ja '67:10-12	6	(Oxygen)	S
	Ap '67:6-9	6	(Engine)	
	My '67:40-42	6	(Auto Flight)	
	Mr '68:7-30+	1		S*
	:31-38	6	(Engine)	
	Ja '69:7-11	2		S*
	:12-16+	1		
	:20-22	6	(Engine)	S
	:27-28	6	(Fuel)	S*
	:29-40	6	(Equip)	S*
	F '69:9-10	1		
	Je '69:21-25+	6	(Auto Landing)	
Avn Wk,	S 17, '62:34-36	1		
	Je 10, '63:56-57	1		
	F 24, '64:40-43	1		
	My 11, '64:36	1		
	S 7, '64:61-69	1		
	Je 14, '65:100-105	1		
	Ap 4, '66:40-43	3		
	My 29, '67:194-205	1		
	:218-223	6	(Flight Inst)	
	Jl 3, '67:40-43	1		C
	D 18, '67:24-27	1		
	Ja 1, '68:31-33	3		
	Ag 12, '68:72-75	6	(Auto Flt Cntrl)	
	S 16, '68:55-69	2		
	S 23, '68:30-31	2		
	N 25, '68:73-76	6	(Engine Cntrl)	
	Mr 10, '69:283-290	2		
	Ap 14, '69:45	2		
	Ap 21, '69:32-34	2		
	Je 23, '69:39-45	4		
	Ag 25, '69:30-32	2		
	N 24, '69:35	4		

BAC/SUD (cont.)

235.	"Concorde" (cont.)				
	Bris Sid Jnl,	Win '62-63:30-35	1		
		:36-41	6	(Engine)	
		Vol. 6, No. 3, '65: 10-14	6	(Engine)	
		Vol. 6, No. 4, '65:1-6	1		
	Can Avn,	Jl '63:8-9+	1		
		Ag '63:14-15+	6	(Engine)	
		Je '69:24-25	1		
		:26-29	4		C
	Flight,	O 24, '63:679-681	1		C, S*
		N 21, '63:824-825	6	(Engine)	
		D 12, '63:950-952	1		
		F 20, '64:271-273+	1		
		Ap 23, '64:649-654	1		
		:654-656	6	(Engine)	
		My 7, '64:744-745	1		
		S 24, '64:548-550	6	(Nav)	C*
		Ja 21, '65:99	6	(Furnishing)	
		Mr 4, '65:317-328	1		C*, S*
		Ja 13, '66:54-55	6	(Engine)	
		F 17, '66:262-263	6	(Fuel)	
		Je 9, '66:977-978	6	(Nav)	C
		Mr 2, '67:310-311	1		
		My 18, '67:799-816	1		C*, S*
		D 7, '67:953-966	1		C
		D 14, '67:976-977, 987-988	2		
		D 21, '67:1035-1037	1		
		My 30, '68:819-820	1		
		N 28, '68:885-886	1		
		Mr 6, '69:352-353	2		
		Mr 6, '69:Supp:1-11	1		
		:12-15	2		
		:16-26	6	(Engine)	S
		:31-34	6	(All)	
		:39-41			S*
		:42-43	6	(Hydr & Brake)	
		:44-51	6	(Control)	C
		:52-63	2		
		:64-68	4		
		:71-76	3		S*
		Ap 17, '69:614-615	4		
		:630+	1		
		Jl 24, '69:119-120	4		
		Ag 28, '69:313-314	1		
		D 4, '69:866	2		

BAC/SUD (cont.)

235. "Concorde" (cont.)

	Fly Rev Intl,	D '63:17-21	1		
		My '64:18-19	3		
		Jl '65:5-8	1		C
		Mr '67:412-414+	1		C
		Ap '68:179-182	1		
		:183-184+	6	(Engine)	
		:191-196	4		
		:197-201	3		S*
		:202-204	6	(Electronics)	
		:205-206	1		
		My '69:22-23	2		
	Interavia,	No. 1, '63:29-30	1		
		No. 8, '63:1185-1187	1		
		No. 3, '64:321-325	1		
		:325-326	6	(Engine)	
		No. 6, '64:780	1		
		No. 7, '64:985-987	1		
		No. 6, '65:807-809	1		
		No. 8, '65:1249-1250	1		
		No. 1, '66:48-51	3		
		:63-65	6	(Air Cond)	
		No. 5, '66:773-775	3		S*
		No. 8, '66:1167-1168	1		
		:1169-1171	4		C, S*
		No. 8, '67:1238-1239	1		
		No. 2, '68:149-161	1		C*, S*
		:162-164	6	(Flt, Nav, Comm)	S*
		:165-167	4		
		No. 1, '69:173	2		
		No. 4, '69:395-396	2		
		No. 9, '69:1520-1524	2		C
		No. 11, '69:1824-1825	6	(Engine Cntrl)	
	Sp/Aeron,	Ap '64:84-87	1		

BACHEM Germany

236. BP-20 and BA-349 "Natter" (Viper) Piloted bomb

Air Power,	Aut '54:3+	1	
Aviation,	Ja '46:77-80	1	
Interavia,	No. 1, '47:41-43	1	
RAF Fly Rev,	N '56:19-21+	1	S*

BADDECK Canada

See CANADIAN AERODROME

BARLING USA

237.	NBL-1 "Barling Bomber"		
	Air Prog, F/Mr '63:65-67	1	Pr
	AAHSJ, Fall '61:176		C
	Hist Avn, N '68:2	1	
	US Air Svcs, Ap '55:7	1	

BRITISH AERIAL TRANSPORT COMPANY, LTD. (BAT) GB
Air Pict, Ap '60:126-128+ (With aircraft photos)

238.	"Bantam" F.K. 22 and 23		
	Air Pict, My '59:164	1	

B.E. (Designed by Royal Aircraft Factory) GB

239.	2		
	Air BP, No. 12:15 (-C)		D*
	Cr & Cockade, Spg '69:Cover		P
	Fall '69:246 (-D)		C
	Flight, Ap 2, '54:393-397 (Pt 1)	1	
	Ap 16, '54:478-482 (Pt 2)	1	
	Fly Rev Intl, Ap '69:52 (-E:Norway)		Pr*
	Pop Fly, Je '34:136 (-E)	1	D
	Roundel, Mr '57:31 (-E)		P
	RAF Qtr, Jl '37:Supp (-C, -E)	1	D
	Ja '39:Plate		D
	Ap '39:196-200	1	
	:201-202 (-C)	1	D
	:203-204 (-E)	1	
240.	8		
	Air Pict, N '62:354-357	1	
241.	9		
	Air Pict, Mr '63:80-83	1	
242.	12		
	Air Pict, N '56:401	1	
	Air Prog, O '65:58		Pr†
	RAF Qtr, Ap '39:205-206	1	
	Jl '39:Plate		D

BEACHEY USA

243.	"Little Looper"		
	AAHSJ, Fall '65:183-185	1	Pr

BEACHEY (cont.)

244.	80-H. P. Monoplane (1915)			
	AAHSJ, Spg '64:57	1		Pr†

BEAGLE GB

245.	A. O. P. (Aerial Observation Post)			
	Air Pict, Je '63:187-189 (-II)	4		C
	Flight, Je 13, '63:931-933 (-II)	4		C
246.	B. 125 "Bulldog"			
	Fly Rev Intl, S '69:30-31	1		
247.	B. 206 "Basset"			
	Aeroplane, S 18, '68:40 (-S)	1		
	Air Pict, Mr '66:105	1		
	Acft Eng, Ag '66:8-18, 31-33 (-S)	1		S*
	:19-26 (-S)	6	(Engine)	S*
	:22-23 (-S)			S*
	:27-29, 34 (-S)	2		
	:35-36 (-S)	3		
	Flight, Ag 24, '61:255-262 (Prototype-)	1		C*, S*
	O 3, '63:575-580	1		S*
	D 22, '66:1063-1068 (-S)	1		C*, S*
	:1068 (-S)	4		
	Flying, Mr '64:42-43+	4		
	Fly Rev Intl, My '64:34-39	1		C, Pr, S*
248.	B. 218			
	Smaller B. 206 "Basset"			
	Acft Eng, Ja '63:4-6	1		
	Flight, Ag 23, '62:273-279	1		C*, S*

BEARDMORE GB

249.	"Inflexible"			
	RAF Fly Rev, S '60:18+	1		
250.	S. B. 3D			
	Naval folding wing SOPWITH "Pup"			
	Flight, Ja 1, '54:8-12	1		

BEECH USA

Bee-Hive, Sum '65:22-27
Fly Rev Intl, Ja '65:27-28 (Twin-engine aircraft)
Nav Avn News, N '57:18-19 (US Naval aircraft)
Skyways, Mr '45:20-22+ (WW II trainers)
West Fly, My '34:14-16

251.	"King Air" 90-Series		
	Aeroplane, N 8, '67:18	1	
	Acft Eng, N '64:360-361	1	
	Flight, Jl 2, '64:14-15+	4	C
	N 19, '64:863-866	4	C
	Mr 19, '64:423-424	4	
	Flying, Jl '64:36	1	
	N '67:42-46	4	C, P
	Interavia, No. 3, '69:302-304	1	
252.	"King Air" 100-Series		
	Can Avn, O '69:35	1	
	N '69:30-31	1	
	Flight, Je 12, '69:983	1	
	D 25, "69:991		C
	Fly Rev Intl, Ag '69:30	1	
	Interavia, No. 9, '69:1543-1545	1	C, P
253.	XA-38 "Destroyer" or "Grizzly"		
	Air Cl, O 68:6-12+	1	P
	Air News, F '46:44-45	1	
	Air Pict, My '62:149-150	1	
	Comm Avn, D '45:78-80	1	
	RAF Fly Rev, D '57:52	1	
254.	AT-7 "Navigator"		
	Aero Digest, Mr '46:82-85+	1	C, S*
255.	AT-10 "Wichita"		
	Aviation, Je '42:154		D*
	Comm Avn, Je '42:36		D*
	Flight, D 9, '43:642b	1	
256.	AT-11 "Kansas"		
	Air Tech, Ja '43:28		P
	Aviation, Mr '43:167		S
	Flight, O 21, '43:450	1	

BEECH (cont.)

257. C-45 "Expeditor"
Improved design: PAC "Tradewind"
US Navy version: SNB "Navigator"
See also F-2 and 18
Flight, Ag 8, '40:102 1

258. F-2
Military photo version of C-45
Flying, S '41:105 P

259. L-23 "Seminole"
Later designation: U-8
Army Avn D, Jl '55:4-12 (-B: Photo) 1 C
N '55:5-7 4

260. MS-760
See MORANE-SAULNIER 760 "Paris"

261. PD-183
Counter-insurgency (COIN) aircraft
Avn Wk, Je 1, '64:19-20 1

262. SNB "Navigator"
See also C-45
Nav Avn News, S '49:18-19 3

263. T-34 "Mentor"
Commercial version: BEECH 45
Aero Digest, Ag '54:30-36 (-A) 1 S*
Air Prog, Ag '66:84 P
S '66:6-9+ 4 C
:10-11 1 Pr
Avn Wk, Jl 16, '51:12 1
Flight, D 22, '49:800-801 1
Mr 2, '50:296-297 1
Nav Avn News, Ag '54:8-9 1
Jl '62:36-37 4
RAF Fly Rev, N '56:34-36 1 Pr, S*
N '57:7 (-B) 5
Skyways, O '50:36 (Y-) 1
West Avn, Jl '54:20 1

264. T-36
Avn Age, O '52:38-39 (-A) 6 (Cabin Press) S*
Flight, Ja 4, '52:20 1

BEECH (cont.)

265. U-8

Earlier designation: L-23

US Army Avn, O '63:34-40	6 (Lndg gear)	S*
Ja '68:18-23	4	

266. U-21

US Army Avn, Je '67:64-65	1	P
O '67:65 (-A)	4	

267. 17 "Staggerwing" or "Traveller" (British)

Aero Digest, D '32:52-53 (-R)	1	
Ap '34:61 (-L)	1	
Air Cl, Mr/Ap '65:24-28	1	P
Air Prog, Je/Jl '64:27-32		Pr
Ag '67:84		P*
Je '68:52-53+ (A-/-F)	1	P
N '69:32-35+	4	C
AAHSJ, Sum '63:121-125	1	D
Win '67:296-298		P
Aviation, D '32:488-489	1	
Can Avn, My '69:20 (D-/-S:Canada)	1	
Flight, My 17, '34:495-496 (-J, -R)	1	
Flying, O '47:86 (D-/-S)	1	
Mr '64:36-37+	1	C
O '66:70-75	4	
Pop Avn, D '36:22 (C-/-B)	1	
RAF Fly Rev, Ja '61:44 (C-&D-)	1	

268. 18

"Super" version: E 18S

See also C-45, F-2 and SNB

Air BP, No. 10:28 (Super-)		P*
Aviation, Ja '37:40-41	1	
Jl '40:55		S*
Ag '40:64-65+ (-F2)	1	S*
Avn Wk, Ap 14, '69:88-90 (Modified	6 (Wing flap)	
Can Avn, S '39:20 (-S)	1	
D '56:59-60 (Super-)	4	C
Flight, Mr 18, '37:288 (M.-)	1	C
Ag 8, '40:102	1	
Flying, Jl '65:28-33	1	P*
Pop Avn, Je '37:27-28	1	
Skyline, Ja '46:34-41+ (D-/-S)	1	C, P*, S*
Skyways, F '55:8-9+ (Super-)	4	C
West Fly, Ja '37:21+	1	

BEECH (cont.)

269.	34 "Twin Quad"			
	Avn Wk, N 10, '47:13	1		
	Can Avn, S '47:67	1		
	Flight, O 27, '49:559-560	6	(Engine)	S*
	Interavia, No. 5, '47:49	1		
	RAF Fly Rev, Je '63:45+	1		
	Skyways, N '48:29	4		
	West Fly, O '49:28+	2		
270.	36			
	Air Prog, O '68:61+	4		
271.	45			
	Military version: T-34			
	Avn Wk, D 20, '48:16	1		
	Can Avn, Ja '49:54	1		
	Skyways, F '49:58	1		
272.	73 "Jet Mentor"			
	Avn Age, Jl '56:38-39	1		C
273.	79 and 89			
	Fly Rev Intl, N '68:30-31	1		
274.	99			
	Aeroplane, F 28, '68:24	1		
	Avn Wk, Ap 15, '68:115	3		
	S 2, '68:36-52	1		C
	Flight, My 18, '67:791	1		
	Ag 28, '69:317-320 (-A)	4		C
	S 25, '69:487 (-A)	2		
	Interavia, No. 12, '68:1552-1553	1		C

BEECHCRAFT USA

See BEECH

BELL USA

Flying, My '54:23-25 (Aircraft genealogical chart)

Nav Avn News, Jl '58:22-23 (US Naval aircraft)

275.	AH-1G "Hueycobra"			
	Army version			
	See also 209 "Hueycobra"			
	Avn Wk, Jl 31, '67:49	1		
	My 20, '68:26-28, 64-76	4		P

BELL (cont.)

275.	AH-1G (cont.)			
	Flight, D 15, '66:1046	1		
	Jl 27, '67:154-156	4		C
	Fly Rev Intl, D '68:28	1		
	Interavia, No. 2, '66:170-171	1		C
	No. 10, '67:1560-1561	1		C
	No. 5, '68:539			P
	US Army Avn, Ag '68:64	6	(Armament)	
	Ag '69:10-11	4		
276.	AH-1J "Sea Cobra"			
	Marine version			
	See also 209 "Hueycobra"			
	Avn Wk, O 27, '69:20-21	1		P
	Can Avn, N '69:43	1		P
	Fly Rev Intl, D '68:28	1		
277.	D 188A			
	V/STOL aircraft			
	Avn Wk, D 5, '60:54-69	1		
	Interavia, No. 1, '61:80	1		D
	RAF Fly Rev, D '61:57+	1		
278.	D 216			
	Am Helicop, Mr '56:10	1		D
279.	XFL-1 "Airabonita"			
	Shipboard version of P-39			
	RAF Fly Rev, D '56:9+	1		
280.	X and YFM-1 "Airacuda"			
	Aero Digest, Ag '37:25	1		
	D '37:20			P
	Aviation, S '37:49	1		
	D '40:78			S*
	Pop Avn, N '37:33	1		Pr
	RAF Fly Rev, Je '56:11	1		
	Jl '63:53+	1		
	US Air Svcs, F '38:27+	1		P
	West Fly, Ag '37:34	1		
281.	H-12			
	See also XR-12			
	Avn Wk, N 6, '50:17 (Y-/-B)	1		

BELL (cont.)

282.	H-13 "Sioux"			
	Later designation: OH-13			
	Commercial version: BELL 47			
	Aerosp Hist, Sum '69:52 (-J:Eisenhower's)	4		
	Am Helicop, Je '57:14+ (-J:Eisenhower's)	1		C
	Army Avn D, My '55:25-27	4		
	Avn Wk, Ap 9, '62:111-113 (-K)	1		
	US Army Avn, Ap '59:24+	4		
	S '60:10-11 (-E)	6	(Armament)	C
	West Avn, Ja '52:17 (-D)	4		
	West Fly, Ja '51:13 (-D)	1		
283.	H-40			
	Re-designated HU-1A			
	Commercial version: BELL 204			
	Am Helicop, D '55:10+ (X-)	1		
	Army Avn D, F '56:15-18 (X-)	1		
	Avn Wk, D 5, '55:18 (X-)	1		
	Mr 18, '57:68-69 (X-)	1		C
	S 30, '57:61-62 (X-)	2		
	Ag 18, '58:94-95 (Y-)	3		
	US Army Avn, My '57:5-10 (X-)	2		
	West Avn, Ja '56:12 (X-)	1		
284.	HSL-1			
	Am Helicop, F '51:10 (X-)	1		
	Ap '54:8	6	(Auto Pilot)	P
	Jl '54:13	2		
	Avn Wk, Mr 29, '54:32-33	3		
	Flight, Je 4, '54:728	1		
285.	HU-1 "Iroquois" or "Huey"			
	Earlier designation: H-40			
	Later designation: UH-1			
	Commercial version: BELL 204			
	Avn Wk, D 8, '58:76-91	4		C
	My 16, '60:132-133 (-A)	3		
	N 14, '60:59 (-A)			C*
	Ja 15, '62:66-67 (-D)	1		
	Fly Rev Intl, D '68:25	1		
	RAFs Qtr, Vol. 1, No. 3:180			P

BELL (cont.)

285. HU-1 (cont.)

US Army Avn,	Je '59:11-15	1	C
	D '59:17 (-A)	6 (Suspension)	
	Je '60:25-27 (-A)	4	
	S '60:20-22	3	
	D '60:8 (-D)	1	
	S '61:16-17	2	
	Mr '62:4-5 (-B&-D)	1	
	Ap '62:8-10	2	
	Ja '64:12-13	4	

286. HUL-1 "Ranger"
US Navy helicopter
Commercial designation: BELL 47

Avn Wk, F 13, '61:109 (-M)	2	

287. L-39
Swept-wing P-63

Avn Wk, Mr 8, '48:30	1	
Nav Avn News, Jl '46:16	1	
RAF Fly Rev, My '59:11	1	

288. OH-4A
US Army helicopter
Commercial designation: BELL 206

Avn Wk, S 9, '63:86	1	
F 17, '64:104-111	1	C
Ap 6, '64:68-74	6 (Armament)	
Flight, Ag 6, '64:210-212	1	
US Army Avn, Ag '64:6-7+	1	C

289. OH-13 "Sioux"
US Army version of BELL 47G

US Army Avn, S '64:40 (-T)	1	
F '68:14-19	4	
O '68:17	6 (Door Gun)	

290. OH-58A "Kiowa"
Commercial designation: 206A "Jetranger"

US Army Avn, N '69:26-27	1	

291. P-39 "Airacobra" or "Caribou" (Early RAF)
See XFL-1 for US Navy version

Aero Digest, My '41:72-75	3	
Mr '43:256-257		S

BELL (cont.)

291.	P-39 (cont.)			
	Aeroplane,	O 24, '41:448-449	1	
		O 31, '41:476-479 (-I)	1	S*
	Air BP,	No. 18:14-17	4	C, P
	Air Pilot,	Ag '43:16+ (Russian-)	4	
	Air Prog,	Jl '42:42-43		S*
		Je/Jl '63:49 (-D)		Pr†
		S '67:34-35	1	S*
	Air Tech,	N '42:30 (-C)		P
		F '43:22-25+	1	
		Mr '43:29		P
		Ag '44:17-23+	1	C*, S*
	Acft Eng,	D '41:343-344	1	P
	AAHSJ,	Fall '63:185-198 ("Cobra II")	1	Pr†
	Aviation,	Mr '40:40-41+	1	
		Mr '41:36-37+ (Pt 1)	1	
		Ap '41:46-47+ (Pt 2)	1	S
		Jl '41:78-79+	1	
		:84-85+	2	
		F '42:116-117		S*
		My '43:126-155	1	C, S*
		Ja '44:199		S*
		F '44:140-143		S*
	Can Avn,	Mr '40:14+	1	P
		Ap '41:27-30+	1	
		D '43:164 (T-)	1	
	Comm Avn,	My '40:21-22+	1	P
		Ag '41:78-80	1	
		:86-94+	3	
		:96	6 (Armament)	
		:98	2	
	Flight,	My 22, '41:364	1	
		O 23, '41:281-283	4	
		Je 25, '42:Supp:a	1	
	Flying,	N '42:61	1	
		S '43:103-104+	3	
		N '43:116	3	S
		Mr '44:49		C
	Fly Rev Intl,	Je '65:52 (-D, L, N, Q)		Pr*
	Pop Avn,	Je '40:53 (X-)	1	
	RAF Fly Rev,	Ja '63:49	1	
	US Air Svcs,	Mr '40:38	1	
	West Fly,	Jl '43:52+ (Russian-)	4	
		O '43:84-85+	3	
		D '43:124 (T-)	1	

BELL (cont.)

292.	P-59 "Airacomet"		
	Later jet version: XP-83		
	Air Prog, '56/'57 Annual:90-91 (-A)		Pr, S*
	Ap/My '63:77		C, S*
	AAHSJ, Fall '66:155-178 (X-/-A)	1	C, P, Pr
	Comm Avn, N '44:54-58 (-A)	1	
	Flying, N '44:29+	4	
	Mr '57:44-46	1	
	Fly Rev Intl, F '64:62 (X-/-A)	1	
	S '68:99-102	1	Pr*
	RAF Fly Rev, Ag '58:11+	1	
	Skyways, S '45:55 (-A)		P*
	West Avn, Mr '56:31 (Last-)	1	
	West Fly, N '44:51	1	
293.	P-63 "Kingcobra"		
	Swept-wing version: L-39		
	Air Prog, Ap/My '63:74		C
	Air Tech, O '44:26		P
	N '44:26-27	1	
	Can Avn, Ap '45:170 (R-)	1	
	RAF Fly Rev, Jl '55:9	1	
	S '61:22 (-C:Russia)		Pr*
	:23	1	
	Skyways, My '45:32-38+	1	P*, S*
	West Fly, S '44:120	1	
	O '44:74-75+	3	
294.	XP-77		
	Air Force, F '45:37		P
	Air Pict, Ap '60:131-132+	1	
	Can Avn, Mr '45:121	1	
	Comm Avn, Ap '45:54+	1	
	RAF Fly Rev, Mr '59:50	1	
295.	XP-83		
	Basic design: P-59		
	US Air Svcs, D '45:34	1	
	West Fly, D '45:98	1	
296.	XR-12		
	See also H-12		
	Avn Wk, Ag 25, '47:15	1	
	West Fly, Je '44:54+	1	

BELL (cont.)

297. UH-1B, C, E, F, L "Iroquois" or "Huey"
Earlier designation: HU-1 "Iroquois"
Civil version: BELL 204
Avn Wk, D 10, '62:52-57 (-B) 2
Ja 21, '63:32 (-B) 1
My 11, '64:32 D*
My 20, '68:69-76 (-B) 4
Fly Rev Intl, D '68:25-26 1
US Army Avn, My '65:16-18 4
N '65:24-26 2
My '67:36-46 4
Jl '69:18-20 4
Ja '68:61 (-B) 6 (Voice Warning) S*

298. UH-1D and H "Iroquois" or "Huey"
Earlier designation: HU-1 "Iroquois"
Civil designation: BELL 205
Avn Wk, My 17, '65:119 (-D) 2
Je 13, '66:54-59 (-H) 4 C
Ap 1, '68:79-82 (-D) 6 (Fire Suppr) S*
Fly Rev Intl, D '68:26-27 (-D&-H) 1
Interavia, No. 9, '63:1331-1334 (-D) 1 S*
US Army Avn, My '66:13-15 (-H) 1 C*
Mr '67:26-31 (-D) 6 (Engine Filter)

299. UH-1M "Iroquois"
UH-1 Night-fighting version
Avn Wk, Ag 25, '69:23-24 6 (Night Vision)

300. UH-1N "Iroquois"
For USAF
Canadian version: BELL 212
Fly Rev Intl, D '68:29 1

301. VTOL Test Bed (1955)
Am Helicop, Ja '55:9+ 1
Avn Wk, F 7, '55:16 1
Flight, F 11, '55:186 1 S*
Skyways, Mr '55:43 1

302. X-1
Earlier designation: XS-1
Avn Wk, Ja 1, '51:22-24 1
Ap 25, '55:42 1
Ag 15, '55:17 (-A) 2

BELL (cont.)

302.	X-1 (cont.)		
	Flight, Mr 20, '47:231-232	6 (Engine)	S
	Ja 18, '51:77	4	
	Pegasus, Je '52:1-15	1	S*
	N '55:2-3	1	S*
	RAF Fly Rev, S '55:57-58+	2	
	West Avn, D '54:33 (-A&-B)	1	
303.	X-2		
	Air Force, Mr '57:41-47	4	
	Am Helicop, S '55:14	1	
	Avn Wk, Ag 15, '55:15	1	
	Ag 6, '56:454	2	
	S 24, '56:29	2	
	N 5, '56:26-27	2	
	D 23, '57:65-68	2	
	Nav Avn News, O '55:35	1	
	Pegasus, N '55:3	1	
	West Avn, S '55:16	1	
304.	X-5		
	Aero Digest, Ag '53:28-46	1	S*
	Am Helicop, Je '51:11+	1	P
	Avn Wk, Je 18, '51:16-17	1	
	Interavia, No. 7, '51:392	1	
	No. 8, '52:446	1	
	No. 5, '62:617-619	1	
	RAF Fly Rev, F '54:5	1	
	Skyways, S '51:25+	1	
	West Avn, Jl '51:9	1	
305.	X-14A		
	Air Power, Win '57/'58:98		P
	Am Helicop, Ag '57:11	1	
	Avn Wk, Jl 8, '57:50-54	1	
	Fly Rev Intl, N '63:49	1	
	Interavia, No. 8, '57:817	1	
306.	X-22A		
	Air Prog, S '67:41-43+	1	S*
	Avn Wk, My 24, '65:40-45	1	
	Ag 2, '65:84-85	1	
	Ap 11, '66:31	1	
	Ag 15, '66:31	2	
	My 15, '67:32-33	2	

BELL (cont.)

306.	X-22A (cont.)			
	Flight, Ag 8, '63:205-206	1		
	My 27, '65:816	1		
	Mr 23, '67:445-449	1		S*
	Fly Rev Intl, Ag '65:13-15	1		C
	Interavia, No. 5, '63:658-660	1		
	Nav Avn News, O '62:18-19	1		D
	Je '65:29	1		
	Ap '68:16	2		
307.	XS-1			
	Later designation: X-1			
	Am Helicop, Mr '47:10-11+	1		
	Avn Wk, D 22, '47:9-10	1		
	Jl 26, '48:22-27	1		
	Flying, N '47:52-53+	1		C
	Interavia, No. 8/9, '46:74-75	1		P
	Skyways, F '48:21+	1		
	West Fly, Ja '47:18-19	1		
308.	XV-3 "Convertiplane"			
	Am Helicop, Ja '55:6+	1		
	D '58:7	1		
	Army Avn D, My '55:18-20	1		
	Avn Wk, F 14, '55:15	1		P
	F 21, '55:17-18	1		
	F 20, '56:37	1		
	D 29, '58:24-25	1		
	O 5, '59:124-129	4		C
	US Air Svcs, S '55:19	1		
309.	47			
	Military versions: H-13, OH-13 "Sioux," and HUL-1			
	Aeroplane, Ja 9, '48:42-47	1		C*, S*
	N 18, '55:770-773 (-G)	4		
	Am Helicop, My '47:12-13+ (-B)	4		S*
	O '48:14+ (-D)	6	(Dusting)	S
	O '52:8+ (-D1)	4		
	O '54:12 (-H)	1		
	S '56:13+ (-G-2)	4		
	O '56:15 (-J)	1		
	Army Avn D, Mr '55:15-18 (-J)	1		

BELL (cont.)

309. 47 (cont.)

Avn Age, S '53:30-31 (-G)	2	
N '53:30-31 (-G)	1	S*
Ag '55:40-43 (-G)	2	
O '56:64-69 (-J)	1	C
Avn Wk, Ja 5, '53:56-63 (-D)	4	
Ap 27, '53:18 (-G)	1	
N 8, '54:15 (-G&-H)	1	C
O 8, '56:109-113 (-J)	1	
N 16, '59:79-83 (-G-3)	1	
Can Avn, My '47:27-29+	4	C*
Flight, S 25, '47:347-348 (-B)	1	
My 6, '48:500-502 (Irvin-Bell-/-B)	1	C*, S*
O 31, '52:562-566	4	
N 26, '54:777 (-G1&-H)	1	C
Flying, F '47:43-50 (-B)	4	C
Ap '59:52-53+ (-J)	4	
Je '61:42-43+ (-G-3B)	4	
Interavia, No. 3, '51:139-141	4	
West Avn, S '56:29 (-J)	1	
West Fly, My '46:38	1	

310. 204

Military designations: H-40, HU-1, and all UH-'s

Avn Wk, S 12, '60:146-147 (-B)	1	C
Ja 21, '63:100 (-B)	1	
Ap 29, '63:84-93 (-B)	4	C
Flying, D '60:38-39+ (-B)	1	
Jl '63:42-43+ (-B)	4	
Fly Rev Intl, D '68:25-26	1	

311. 205

Military designations: UH-1D and H, HU-1, and H-40

Flight, O 19, '67:8 (-A)	1	
Flying, O '69:60-64 (-A)	4	C
Fly Rev Intl, D '68:26	1	

312. 206A "Jetranger"

Army version: OH-58A "Kiowa"

Avn Wk, O 18, '65:25	1	
Ja 24, '66:99-102	1	
Flying, My '66:102-107	1	P*
F '67:34-37	4	

BELL (cont.)

313. 207 "Sioux Scout"
Avn Wk, S 23, '63:30-32 1
N 4, '63:66-67 1

314. 209 "Hueycobra"
Military designations: AH-1G and J
Air Prog, Jl '66:52-53 1 C
Avn Wk, S 20, '65:30-32 1
O 18, '65:50 1 C
Flight, S 30, '65:591 1
Fly Rev Intl, D '68:27-28 1

315. 211 "Huey Tug"
Avn Wk, S 9, '68:55-58 1
Fly Rev Intl, D '68:26 1

316. 212
For Canada
USAF version: UH-1N
Can Avn, D '68:30-31 1
Flying, Ja '69:81 1
Fly Rev Intl, D '68:28-29 1

BELLANCA USA
AAHSJ, Spr '65:42-46 (History-making aircraft)
Pop Avn, Jl '35:11

317. "Airbus"
US Army designation: C-27
Aero Digest, S '30:98-100 1
Can Avn, S '30:22-23 1
Flight, Ja 2, '31:16-17 1
Pop Avn, N '30:34 1
Ap '34:230-231+ (C-27A) 1 Pr
US Air Svcs, N '30:32 1 P

318. "Aircruiser"
Aero Digest, Jl '39:63-64 1
Can Avn, N '38:15 1
My '69:26-27+ (Canadian-) 1

BELLANCA (cont.)

319.	"Pacemaker"		
	Aero Digest, Ag '35:53 (-Series VII)	1	
	CAHS, Win '64:103-104 (-CF-ATN)	4	
	:132 (RCAF-)	1	Pr
	Pop Avn, Ja '31:39-40	4	
	Ag '31:17	4	
320.	"Senior Skyrocket"		
	Aviation, Jl '34:228	1	
	Bee-Hive, Je '34:4-6+	1	C
	Flight, Ag 16, '34:846	1	
	West Fly, Je '34:22-23	1	
321.	"Skyrocket"		
	Can Avn, Mr '46:35+	1	
322.	"Tandem"		
	Aero Digest, N '30:72-76	1	C
	Pop Avn, O '30:20 ("Blue Streak")	1	C
	US Air Svcs, Ag '30:44-45 ("Blue Streak")	1	P
323.	J-300 Special		
	AAHSJ, Spr '65:44 ("Liberty")		C, P
324.	28 "Flash" or "Swope"		
	Air Pict, My '56:178 (-70:"Irish Swope")	1	
	O '56:367 (-70&-90)	1	
	F '57:67 (-70:"Irish Swope")	4	
	Aviation, D '36:33-34 (-70:"Irish Swope")	1	
	Flight, N 5, '36:465 (Mollison's "Dorothy"-90)	4	
	Fly Rev Intl, D '67:1068-1069 (-70&-90)	1	
	Pop Avn, O '37:33 (-90)	1	
	Ja '38:33 (-90)	1	
	RAF Fly Rev, O '59:11 (-90)	1	
325.	77-140 and 320 (1935 Bomber)		
	Aero Digest, O '34:40-41	1	
	S '35:40	1	
	Air Cl, F '69:54-61	1	P, Pr
	Aviation, F '35:43-45	1	

BELLANCA (cont.)

325. 77-140 and 320 (cont.)

	Flight, F 28, '35:229-230	1	
	RAF Fly Rev, Ap '63:45	1	

BENDIX USA

326.	J "Whirlaway"		
	Am Helicop, O '47:20-21	1	
	Avn Wk, My 17, '48:21-24	1	

BENNETT USA

327.	Plywood Monoplane		
	Aviation, S '38:40	1	
	West Fly, S '38:21	1	

BENNETT AVIATION New Zealand

328.	"Airtruck"		
	RAF Fly Rev, Ja '62:49	1	

BENSEN USA

329.	B-4 "Sky-Scooter"		
	Am Helicop, Je '55:9+	1	P
330.	B-7 "Gyro-Copter"		
	Am Helicop, F '56:8	1	P
	My '56:6-7+ (-M)	1	
	S '57:9	4	
331.	B-9 "Little Zipster"		
	Am Helicop, Ap '60:13+	1	P

BEREZNYAK-ISAEV Russia

332.	BI Rocket Interceptor		
	Fly Rev Intl, F '66:363-364	1	
	Je '69:79-81	1	P, Pr*

BERIEV Russia

Fly Rev Intl, My '67:563-564 (Pt 1) (Seaplane designs)
Je '67:681-683 (Pt 2) (Seaplane designs)

BERIEV (cont.)

333.	Be-2 or MBR-2 [Mote]		
	Air Pict, Ja '67:25	1	
	RAF Fly Rev, Je '58:6+	1	
334.	Be-4		
	RAF Fly Rev, Je '58:6+	1	
335.	Be-10 [Mallow]		
	Air Pict, Je '62:174-175	1	Pr
	Flying, Je '63:24-25+	1	
	Fly Rev Intl, N '64:18-20+	1	Pr
336.	Be-12 "Chaika" (Seagull) [Mail]		
	Avn Wk, Ag 12, '68:29		P
	Fly Rev Intl, O '67:921	1	P
337.	Be-30 [Cuff]		
	Aeroplane, N 1, '67:25	1	
	Avn Wk, N 6, '67:30-31	1	S
	Je 23, '69:51	1	C
	Flight, S 28, '67:515	1	

BERLIN — Germany

338.	B-9		
	Prone pilot aircraft		
	RAF Fly Rev, N '60:9	1	

BERLIN DOMAN — USA

339.	BD-19		
	Fly Rev Intl, Ap '69:31	1	

BERLINER-JOYCE — USA

340.	Two-Place Monoplane (1930)		
	Flight, F 28, '30:254	1	
341.	XFJ-1		
	Fly Rev Intl, Mr '64:61	1	
342.	XOJ-1		
	Later development: GENERAL AVIATION OJ-2		
	US Air Svcs, N '32:36 (With floats)		P

BERNARD France

343. 52-C. 1
Flight, Mr 22, '34:286 1

344. 80
Long-distance monoplane
Flight, Ap 17, '31:333 ("Tango") 4

345. 190
Interavia, O '49:599-602 ("Oiseau Canari") 4

BFW (Bayerische Flugzeugwerke) Germany
See also MESSERSCHMITT and UDET

346. M-18 and M-20
Acft Eng, Ap '29:76 (-20) 1
RAF Fly Rev, N '62:37 1

347. M-23
Flight, Mr 21, '30:317-318 1

348. M-35
Aeroplane, My 9, '34:782-783 1

BIRD WING USA
Hist Avn, O '69:4-7, 30-31

B/J USA
See BERLINER-JOYCE and GENERAL AVIATION

BLACKBURN GB
Comm Avn, D '44:122-134
Flight, My 25, '39:533-544
My 1, '59:602-613 (Aircraft types with photos)

349. "Baby"
See SOPWITH "Baby"

350. "Beverley" B. 101
Commercial version: "Universal Freighter"
Aeronautics, S '53:162-163 P
Ap '57:42-45 P
Aeroplane, Ap 8, '55:446-450 3
Ag 26, '55:307-310 4 C
Air Pict, My '68:172-174 1

BLACKBURN (cont.)

350.	"Beverley" B. 101 (cont.)		
	Air Power, Aut '57:18		P
	Can Avn, Je '55:116 (-Tanker)	1	
	N '55:44-45	4	C
	Flight, F 19, '54:205-207	1	
	F 4, '55:145-148 (Proto-)	1	C, S
	D 21, '67:1048	1	
351.	"Blackburd"		
	Pop Fly, O '37:390-391 (-Skids)	2	P
352.	"Blackburn" R. 1		
	Aeronautics, Je '55:52		P
353.	"Blackfish"		
	BLACKBURN-built FAIREY "Swordfish"		
	Comm Avn, D '44:136-140	1	P
354.	"Bluebird" L. 1		
	Later development: B. 2		
	Aeroplane, My 21, '30:935-956	1	C, S
	:972-974	4	
	Acft Eng, S '29:219-222 (-IV)	1	S
	Pop Avn, My '33:308 (-IV)	1	
355.	"Botha" B. 26		
	Aero Digest, Ap '41:155-156	1	
	Air BP, No 33:6-9	4	C
	Air Pict, Ag '66:299	1	
	Aviation, My '41:67 (-I)	1	
	Flight, Ja 30, '41:99 (-I)	1	
	O 2, '41:214a-d	1	
	D 17, '42:660b	1	
	Ag 1, '46:110-111	4	
	RAF Fly Rev, My '58:6	1	
	F '63:17	4	
356.	"Buccaneer" B. 103		
	Earlier development: "Buccaneer" N.A. 39		
	Later development: HAWKER-SIDDELEY "Buccaneer"		
	Aeronautics, Mr '62:24-25	1	P
	Air Prog, Ja '65:36 (-S. 1)		Pr
	Flight, Ja 20, '61:87-89	2	
	F 1, '62:165-168	4	
	Je 28, '62:1003	4	
	Ap 4, '63:467-478 (-S. 1)	1	S*

BLACKBURN (cont.)

357. "Buccaneer" N.A. 39
Later development: "Buccaneer" B. 103
Aeroplane, My 9, '58:653-656 1
Air Pict, N '58:376-377+ 1
Avn Wk, S 21, '59:104-107 1
Ag 7, '61:81-82 1
Flight, My 9, '58:627-631 1
N 14, '58:767 1 Pr
My 1, '59:602-603 1
RAF Fly Rev, D '58:21-23+ 4
Ap '60:24-25 1 P*

358. "Dart"
RAF Qtr, Ap '38:Supp 1 D

359. "Firebrand" B. 37
Aeronautics, Je '46:32-35 1 P
Aeroplane, S 28, '45:357-361 (-IV) 1 S*
S 12, '47:366 (-V) P*
N 19, '48:646-648 (-V) 4
Air Pict, S '52:275 (-Up to VA) 1
D '65:443 (-V) 1
Aviation, D '45:188+ (-IV) 1
Comm Avn, O '45:84-90 1
Flight, S 27, '45:338-341 (-IV) (Pt 1) 1 S*
O 4, '45:367-369 (-IV) (Pt 2) 1 S
F 13, '47:179+ 1

360. "Firecrest" B. 48
Development of B. 37 "Firebrand"
Other designations: Y.A. 1 and S. 28/43
Flight, My 29, '47:487-492 (S. 28/43) 1 S*
Ag 21, '47:207 (S. 28/43) 1
Mr 3, '49:255-257 (Y.A. 1) 4 S*
Fly Rev Intl, D '63:57 1
Interavia, No. 8, '47:23-25 (S. 28/43) 1 P

361. "H.S.T. (High Speed Transport) 10" B. 9
Aeroplane, My 6, '36:551 S*
Flight, N 29, '34:1281-1282 1

362. "Iris" R.B. 1
Flight, F 5, '32:xii D
Ap 8, '32:299 (-V) 1
My 27, '32:459 (-V) 1 P
RAF Qtr, Ja '30:64 (-III) P

BLACKBURN (cont.)

363.	"Kangaroo" R.T. 1		
	Aeroplane, F 26, '43:257	1	
	Air Pict, D '61:385-390	1	C
364.	"Lincock" F. 2		
	Can Avn, Ap '30:27	1	
	Flight, Je 27, '30:709		P
	Fly Rev Intl, Ja '64:61 (-I to -III)	1	
	RAF Fly Rev, Jl '60:11	1	
365.	"Perth" R.B. 3A		
	Acft Eng, N '33:256-258	1	
	Flight, O 12, '33:1017-1019	1	
366.	"Ripon" T. 5		
	Flight, F 28, '30:257		P
	O 17, '30:1129 (-III)		P
367.	"Roc" B. 25		
	"Skua" variant		
	Flight, F 25, '43:202a	1	
	RAF Fly Rev, S '56:27-28	1	
368.	"Seagrave" or "Seagrave Meteor"		
	Air Pict, F '55:44-45	1	Pr
369.	"Shark" B. 6		
	Earlier version: B. 3		
	CAHS, Sum '64:33 (RCAF-)	1	
	Comm Avn, Jl '39:24 (-III)	1	
	Flight, My 17, '34:490	1	P
	D 13, '34:1318-1322	1	Pr, S
	Mr 7, '35:253 (Float-)	1	S
	S 26, '35:347	6 (Torpedo Gear)	S
	Mr 12, '36:282-283 (Float-)	1	
	RAF Fly Rev, Je '59:6+	1	
	Ap '62:32	4	
	RAF Qtr, Ap '35:Supp	1	
370.	"Skua" B. 24		
	See also "Roc"		
	Aeroplane, Ag 9, '39:192-195	1	S*
	Flight, Ag 10, '39:125-131 (-I)	1	S*
	Pop Avn, My '38:63	1	
	Ag '39:13		P

BLACKBURN (cont.)

370.	"Skua" (cont.)		
	RAF Fly Rev, S '56:27-28	1	
	D '61:35	4	
371.	"Sprat"		
	Air Pict, Je '57:197	1	
372.	"Sydney" R. B. 2		
	Aero Digest, N '31:64	1	
	Aeroplane, Ag 27, '30:502	1	
	D 3, '30:1312		P
	Acft Eng, Ag '30:201	1	
	Flight, S 5, '30:986-987	1	
	Pop Avn, F '31:44	1	
373.	"Universal Freighter" B. 101		
	Earlier designation: GENERAL AIRCRAFT "Universal Freighter"		
	RAF version: "Beverley"		
	Aero Digest, S '50:44-45+	1	
	Aeroplane, Mr 24, '50:343-351	1	S
	Jl 7, '50:22-23		S*
	N 23, '51:665-671	1	S
	F 1, '52:131-133	4	C
	Avn Age, Je '52:42	1	
	Avn Wk, Je 27, '49:26-28	1	
	Jl 3, '50:26-27	1	
	D 31, '51:20-23	1	
	Flight, Mr 23, '50:372-378	1	C, S*
	Ag 24, '51:212-213 (-II)	1	
	Ag 22, '52:218-219 (-II)	1	S*
	Jl 6, '56:66-67	1	
374.	B. 2		
	"Bluebird" successor		
	Aeroplane, S 6, '33:430-432	4	
375.	B. 3 or M1/30 and 30A		
	Later development: "Shark"		
	Flight, Mr 9, '33:217		P
	RAF Fly Rev, Ja '57:15	1	
376.	B. 7		
	Flight, My 2, '35:476-477	1	

BLACKBURN (cont.)

377. B. 20
Retractible-hull flying boat
Aeroplane, Je 22, '45:714-715 1
Air Pict, Ag '57:273 1
Aviation, O '45:176+ 1
Can Avn, Ag '45:94-96 1
Flight, Ap 18, '46:383-384 6 (Hydraulic) S*
RAF Fly Rev, D '56:9 1

378. B. 44
Design for retractible-hull flying boat--not built
RAF Fly Rev, D '56:9 1

379. C.A. 15C Civil Biplane (1932)
Aero Digest, Ap '33:57 1
Acft Eng, O '32:253-254 (Pt 1) 1
N '32:287-288 (Pt 2) 1
Flight, O 27, '32:1004-1007 1 Pr
RAF Qtr, Jl '32:Plate 14 P

380. C.A. 15C Civil Monoplane (1932)
Acft Eng, O '32:253-254 (Pt 1) 1
N '32:287-288 (Pt 2) 1
Flight, O 27, '32:1004-1007 1 Pr

381. F. 3
Air Ministry specification F. 7/30--never flown
RAF Fly Rev, Ap '59:11 1

BLERIOT France

382. I to XII
Aeroplane, Ja 11, '52:53 (-XI) 4
Air Prog, '52 Annual:78-79
(-XI:Cross-Channel) 1 S*
Acft Eng, Jl '35:161-163 (-I to -XI) 1
AAHSJ, Win '61:Supp (-IID) P*
Sum '63:97-106, 111 1
:106-110 (-XI) 4 Pr
Win '64:288-293 1 P
CAHS, Sum '65:46+ (-XI:Canadian) 1
Flight, Jl 21, '49:71-73 (-XI:Cross-Channel) 4
Flying, N '64:49-53 4
Pop Avn, O '36:17-18+ (-XI:Cross-Channel) 4

BLERIOT (cont.)

382. I to XII (cont.)
RAF Fly Rev, Je '59:21-23 (-XI:Cross-Channel) 4
RAF Qtr, Ja '39:83-88, Plate 4 D

383. XXVII (1911)
Aeronautics, F '46:75 C*

384. 110
Long distance monoplane
Flight, Mr 13, '31:219 1
Ap 29, '32:371 4
Ag 10, '33:805 (Codos & Rossi) 4
Ag 24, '33:843 (Codos & Rossi) 4

385. 127M
Pop Avn, Ja '31:36 1

386. 290
Pop Avn, O '32:234 1

387. 5190 Flying Boat
Aeroplane, Ag 16, '33:309-310 1
Flight, Ag 5, '32:734-735 1
Ag 17, '33:833 1

BLERIOT-SPAD France

388. 91-7 Racer
Pop Avn, S '32:149 1

BLOCH France

389. M.B. 130 and 131 "Guynemer"
RAF Fly Rev, Ja '58:38 1

390. M.B. 135
RAF Fly Rev, F '57:9 1

391. M.B. 150
RAF Fly Rev, Je '57:43 1

392. M.B. 151
Air Pict, Je '63:173 1
RAF Fly Rev, Je '57:43 1

BLOCH (cont.)

393.	M.B. 152		
	Air Pict, Je '63:173	1	
	Air Prog, O '65:59		Pr†
	Flight, My 23, '40:474e-f	1	D
	Fly Rev Intl, Je '68:352, 354		Pr*
	RAF Fly Rev, Je '57:43	1	
394.	M.B. 155		
	Development of M.B. 152		
	Fly Rev Intl, Mr '64:61	1	
	Je '68:352, 354		Pr*
	RAF Fly Rev, Je '57:43	1	
395.	M.B. 157		
	Fly Rev Intl, Je '68:354		Pr*
	RAF Fly Rev, Je '57:43	1	
396.	M.B. 162		
	RAF Fly Rev, F '57:9	1	
397.	M.B. 175		
	Air Pict, Mr '52:75	1	
398.	M.B. 220		
	Pop Avn, N '37:24+	1	

BLOHM & VOSS — Germany

Air Pict, O '63:312-316 (Pt 1) (WW II Projects)
N '63:368-370 (Pt 2) (WW II Projects)
D '63:404-406 (Pt 3) (WW II Projects)

399.	Bv 40		
	Experimental glider-fighter		
	RAF Fly Rev, Jl '60:30 (-V6)		Pr†
	:31-34	1	C*, S*
400.	Bv 138		
	Earlier designation: Ha 138		
	RAF Fly Rev, Jl '57:39-40	1	
401.	Bv 141		
	Air Prog, Je '43:20-23	1	D*
	Air Tech, N '42:25+	1	
	Flight, Jl 16, '42:64-66	1	

BLOHM & VOSS (cont.)

401.	Bv 141 (cont.)		
	Fly Rev Intl, Ag '64:52-55 (Pt 1)	1	Pr*
	S '64:40-41 (Pt 2)	1	
	RAF Fly Rev, Je '55:9	1	
	D '58:41-42	1	
402.	Bv 144		
	RAF Fly Rev, Ap '62:41	1	
403.	Bv 155		
	RAF Fly Rev, Ag '55:9 (-B&-C)	1	
	Mr '60:53-54	1	
404.	Bv 222 "Wiking" (Viking)		
	Aero Album, Sum '69:22-27	1	Pr†
	Aeroplane, Ag 30, '46:246-250	1	C, S*
	Flight, D 10, '42:633	1	
	N 4, '43:502	1	
	RAF Fly Rev, Ag '55:9-10	1	
	Ja '57:41-42	1	
405.	Bv 238		
	Flying scale model: See FGP FG-227		
	RAF Fly Rev, Ag '57:36+	1	
	My '60:45	1	
406.	Bv 250		
	RAF Fly Rev, Ag '57:36+	1	
407.	Ha 137		
	Flight, S 28, '39:270d-f	1	S
	RAF Fly Rev, Ap '61:51	1	
408.	Ha 138		
	Later designation: Bv 138		
	Flight, S 4, '41:131	1	
	N 19, '42:552a	1	
409.	Ha 139		
	Acft Eng, My '37:136	1	
	Ag '41:218-220	1	
	Bee-Hive, S '37:5	4	
	Can Avn, Ag '40:44	1	
	Pop Avn, D '37:17 ("Nordmeer")	4	
	RAF Fly Rev, F '63:47	1	

BLOHM & VOSS (cont.)

410. Ha 140
Flight, O 6, '38:296 1
RAF Fly Rev, O '57:9 1
S '62:43 1

411. P 188
RAF Fly Rev, Ag '57:39 1

412. P 194.01
Flight, Ja 10, '46:29-31 1

413. P 211
Fly Rev Intl, Jl '67:749 1

414. P 214 and 215
RAF Fly Rev, My '58:44 1

BOEING USA

Air News, Ag '45:54-58
Air Pict, Ap '52:104-108 (Bomber history)
Air Prog, Sum '61:12-46 (Pt 1)(Photographic history)
Fall '61:14-36 (Pt 2)(Photographic history)
AAHSJ, O/D '57:213-223 (Stearman and Boeing aircraft list)
Bee-Hive, Sum '66:16-23
Boeing Mag, Jl '46:3-18 (30th Anniversary)
Flying, D '53:23-25 (Aircraft genealogical chart)
Interavia, No. 2, '67:201-206 (1967 activities)
US Air Svcs, Jl '41:16-19+ (25th Anniversary)
Ag '56:8-10 (40th Anniversary)
West Fly, Mr '34:16-18
Jl '41:22-25+ (25th Anniversary)

415. "Clipper" Model 314
Aero Digest, D '38:44-45+ 2
F '41:72+ 3
Aeroplane, F 6, '42:163-165 (-A) 1 C
Air Cl, Je '69:10-25+ 1 C, P
AAHSJ, Win '61:266 C
Sum '68:93-102 1 P
Aviation, Jl '38:21+ 1
Mr '43:241+ 6 (Fire Ext.)
Jl '44:175-177+ 3

BOEING (cont.)

415.	"Clipper" (cont.)		
	Boeing Mag, Je '46:8-9+	4	
	O '46:10-11 (BOAC's-)	4	
	D '64:12-13	4	
	Je '68:3-6 (Pt 1)	1	
	Jl '68:3-6 (Pt 2)	1	P
	D '68:8-9	1	
	Can Avn, Mr '38:22-23	1	
	D '41:52	4	
	Comm Avn, My '39:44	1	
	Ag '42:92-96	4	
	Flight, Jl 21, '38:67-68	1	
	Ap 13, '39:370-373+	4	C
	F 5, '42:111 (-A)	1	
	D 23, '43:694 (-A)	1	
	Flying, O '41:44-46+	4	
	Ag '46:50-51	1	
	Jl '56:25-26+	1	
	Pop Avn, My '38:20 ("South Seas")	1	
	Mr '39:17	1	C
	O '39:16-18+	4	
	Pop Fly, My '37:64-67	4	
	Ag '38:13 ("Atlantic Clipper")	1	
	Ja '39:506-507+	1	
	RAF Fly Rev, Mr '63:45+	1	
	S '63:39+ (Correction)	1	
	US Air Svcs, Je '37:25	1	
	S '37:14		S*
	Mr '38:29 ("South Seas")	1	
	Jl '38:13	4	
	O '38:18+	2	
	D '38:21-22+	1	
	My '39:24	6 (Control Panel	C
	West Fly, S '37:22	1	
	Jl '38:18-20+	1	
	O '38:20-21	2	
	D '38:18-19	2	
416.	"Monomail" Model 200		
	Aviation, O '30:232-240	1	
	US Air Svcs, Ag '30:47	1	P
	N '30:46-47	1	P
417.	XAT-15 "Crewmaker" Model X-120		
	Flight, Ag 13, '42:186	1	P
	US Air Svcs, Jl '42:30		P

BOEING (cont.)

418.	B & W Model 1		
	AAHSJ, Spr '66:38-46	1	Pr
	Boeing Mag, Mr '62:13-14	1	
	Jl '66:4-7	1	P*
	Fly Rev Intl, Ap '67:517-518 (Replica-)	1	P*
	D '68:54-55 (Replica-)		P*
419.	B-1 Model 6		
	Boeing Mag, Jl '51:12-15	1	
	Pop Avn, My '40:56	1	
	US Air Svcs, Jl '39:34	1	
420.	Y1B-9 Model 214		
	Pop Avn, D '31:81	1	
	RAF Fly Rev, Ag '63:41	1	
	US Air Svcs, N '31:32		P
421.	XB-15 Model 294		
	Air Pict, Ap '69:145	1	
	Air Prog, '39 & '40 Annual:51		P
	Bee-Hive, My '37:8-9	1	
	F/Mr '39:1		P
	Pop Avn, Ag '37:18	1	
	Jl '38:42-43	1	P
	US Air Svcs, My '37:18-19	1	D
	My '38:16		P
422.	B-17 "Flying Fortress"		
	Prototype: BOEING 299		
	USA-RAF Equivalents: C=I, E=IIA, F=II, G=III		
	Aeronautics, Jl '45:32-35	1	P
	Aeroplane, Mr 29, '40:422-423	3	
	Ag 22, '41:198-199	1	
	Ja 29, '43:125-129	4	C
	F 4, '44:123-125	1	
	Aerosp Hist, Sum '69:21 (Y-)	4	
	Air BP, No. 24:13-15 (-E)	4	C
	Air Cl, D '68:14-18 (Names)		P
	Ap '69:36-40 (Names)		P
	D '69:19 (-F:63 Sq "Talisman")		Pr*
	:19 (-G:234 Sq "Just Nothing")		Pr*
	Air Force, Ap '47:19-23	1	C
	Ag '69:81-88 (Japanese-)	1	
	AF News Ltr, Je '42:29-31	1	

BOEING (cont.)

422. B-17 (cont.)

Air Pict,	D '64:400-402 (Pt 1)	4	
	Ja '65:17-19 (Pt 2)	4	
Air Pilot,	S '43:20		D*
Air Prog,	Ap/My '63:68 (-G)		C
	F/Mr '65:44-47	1	D
	D '68:42-43 (-G)		Pr*
Air Tech,	O '42:32 (-E)		P
	F '44:15-22+	1	S*
AAHSJ,	Spg '64:42-43+ (PB-1)	1	
Aviation,	Ja '43:106-111+	3	
	Je '43:189, 191		S*
	Jl '43:124-145+	3	
	F '44:136-139		S*
	Ja '45:121-144	1	S*
Boeing Mag,	S '46:12-14 (-G:"Two Kind Words")	4	
	My '51:13 (S-)	4	
	Ap '57:8-11	1	
Flight,	Ag 21, '41:106 (-I)	4	
	Ag 20, '42:202b (-II)	1	
	:203-204	4	
	S 17, '42:311-312	4	
	D 16, '43:662-663	1	
	My 4, '44:473-477+ (-G)	1	S*
	N 1, '45:481	4	
	Je 20, '63:971-972	4	
Flying,	Ag '43:138	3	
	Mr '44:49		C
	Ag '44:58-59	6 (Bail-out)	S*
	D '44:50-52+	2	
	D '65:100 (-F)	1	D
Interavia,	D '48:698 (Wright "Typhoon" test bed)		P
Pegasus,	N '57:6-9 (Drones)	1	C
Pop Avn,	Ag '37:18 (Y-)	1	
	D '37:28	4	
	Ag '38:10-12+ (So. Amer. flight)	4	
	N '39:25 (-B)		P
	My '40:29 (-B)	3	
Roundel,	Ja/F '62:18-20 (Canadian-)	4	
RAF Fly Rev,	Ap '61:49 (German-)	1	
Skyways,	Mr '43:24-29+ ("Susy-Q")	4	
	O '43:18-21+	1	S*

BOEING (cont.)

422. B-17 (cont.)

US Air Svcs,	Mr '38:12-14 (So. Amer. flight)	4	
	O '38:11-13+ (Columbian flight)	4	
	Mr '39:34	4	
	O '41:12 (-E)	1	P
	N '43:34 (-A to -G)	1	
	Ap '44:11, 38 ("Hells Angels")	4	P
	Je '44:39 (-G)	6 (Guns)	
	Ap '47:11-12+	1	
	N '48:25 ("Typhoon" test bed)	1	
	Ap '54:4 ("Typhoon" test bed)		P
West Fly,	My '43:44-46	4	
	O '43:80-81+	3	

423. B-29 "Superfortress" and "Washington" (British)
Photo-reconnaisance version: F-13
Tanker version: KB-29

Aero Digest,	Jl 1, '44:70-75+	1	
	Ja 1, '45:79-81+	6 (Armament)	
	Je 1, '45:70-71+	6 (Bomb Door)	S*
	Ag 15, '45:61-66	1	S*
	O 1, '45:51-53	1	S*
Aeroplane,	Ag 11, '44:162	1	
	Mr 30, '45:370-372	1	
Air Force,	Jl '44:4-7	1	
	O '44:26-27	6 (Cabin Press)	
	Ap '45:6-8+	3	
Air News,	Mr '45:28-29	6 (Armament)	S*
Air Power,	Sum '54:388 (K-)		P
Air Tech,	F '45:26-27+	6 (Armament)	
Aviation,	Jl '44:110-111+	1	
	O '44:197-201		S*
	F '45:187-189		S
	My '45:193		S*
	S '45:133-137 (Pt 1)	6 (Armament)	S
	O '45:152-153 (Pt 2)	6 (Armament)	S
Bee-Hive,	Jl/Ag '44:12-13	6 (Props)	
Boeing Mag,	Jl, '44:3-5+	1	
	Ag '44:3-4+	6 (Cabin Press)	
	S '44:8-9+	6 (Wing)	
	:10-11	3	
	O '44:10-11	3	

BOEING (cont.)
423. B-29 (cont.)
Boeing Mag, N '44:12 6 (Armament)
Ap '45:10-11+ ("Eddie Allen") 4
Je '45:8-9 3
Ap '46:8-9 1
Je '46:6-7 4
O '46:3-5 ("Pacusan Dreamboat") 4
Can Avn, Jl '44:54-55+ 1
O '44:68+ 6 (Armament)
Comm Avn, Ag '44:120 6 (Engine)
N '44:44+ 6 (Armament)
Mr '45:82-83+ 1
Je '45:76-80 (Incl. F-13A) 1
Flight, Jl 6, '44:19-20 1
Ag 10, '44:143-145+ 1
F 1, '45:115 1
Jl 12, '45:40-42 1 S*
F 7, '46:140a-142 1
F 9, '50:182-183 1 S*
Ag 17, '50:202-203 1
Flying, Ag '44:48+ 1
Ap '48:28-29+ 1
RAF Fly Rev, Ag '60:9 (British-) 1
Skyways, S '44:30-33+ 1
Ap '45:22-25+ 4
Ag '45:34-41+ 1 S*
D '50:20-21 3
US Air Svcs, Jl '44:20-26+ 1
N '45:9 4
Ja '46:26 ("Pacusan Dreamboat") 4
O '46:24-25 ("Pacusan Dreamboat") 4
West Fly, Jl '44:44-45+ 1
O '44:70-71+ 3
F '45:42-43 1 C
My '45:51+ 6 (Fire Control) S*
Ag '45:60 3

424. XB-38
LOCKHEED conversion of B-17E
Aero Album, Spg '69:14-17 1 P

BOEING (cont.)

425.	XB and YB-40			
	B-17F escort bomber			
	XB= BOEING-built			
	YB= VEGA-built			
	Air Pict, Je '61:172-173+ (Y-)	1		
	RAF Fly Rev, Mr '60:61 (Y-&X-)	1		
	My '60:50 (Y-)	1		
426.	XB-44			
	Boeing Mag, Mr '46:9	1		
427.	B-47 "Stratojet" Model 450			
	Reconnaisance version: RB-47			
	Weather version: WB-47			
	Aero Digest, Ag '48:26-27 (X-)	2		
	N '50:22-25+	1		
	Aeroplane, Jl 20, '51:72-74	1		
	Jl 10, '53:49-53	1		
	D 11, '53:790-791	4		
	Ja 18, '57:80-83	4		
	Air Force, Jl '50:28-31	1		
	Ap '56:90-96	4		
	AAHSJ, Fall '68:204-207	1		
	Avn Age, F '53:24-26	3		
	Je '53:151-157	1		
	Avn Wk, S 22, '47:14-15 (X-)	1		
	Ap 30, '51:25-31	4		
	Ag 31, '53:21 (R-/-B&E)	1		
	O 18, '54:16	1		
	Mr 7, '55:18 (X-/-D)			P
	Boeing Mag, O '47:3-5 (X-)	1		
	N '47:3-4 (X-)	2		
	Ja '48:3-5 (X-)	2		
	:6-7+ (X-)	4		
	Ag '48:8-9 (X-)	1		P
	Mr '50:7-9+	3		
	My '50:6-7+	6	(Skin)	
	F '51:3-7	2		
	My '51:3-5	4		
	N '51:3-5	3		
	:15	6	(Brake Chute)	
	Ag '52:3-5+	3		
	My '54:10-11	6	(ATO)	
	Je '54:12-13 (R-/-E)	1		
	S '54:12-13	6	(Canopy)	

BOEING (cont.)

427.	B-47 (cont.)				
	Boeing Mag,	O '54:6-7	3		
		Ja '55:3-5 (-E)	4		
		S '55:11	6	(Canopy)	
		D '55:6-7 (R-/-E)	3		
		Ag '56:6-7	1		
		Je '57:14-15	6	(ATO Rack)	
		Mr '63:8-9 (W-/-E)	1		
		Ap '66:15	1		
		Jl '69:9			P
		S '69:14-15 (W-)	1		
	Flight,	Ap 7, '49:401 (X-)	1		
		Ag 20, '54:239-243 (-E)	4		C
		Ja 25, '57:115-118	4		
	Flying,	Je '48:19 (X-)	1		
		Ag '49:15-17+	1		
		Ja '51:18-19+	3		
		My '55:45+	6	(ATO)	
	Fly Rev Intl,	Je '68:356-357 (X-/-D)	1		
	Interavia,	F '48:74-77 (X-)	1		
		No. 6, '52:329 (-B)	1		
		No. 1, '53:32-34	4		
	RAF Fly Rev,	Ag '53:29-31	4		
		Mr '55:30-33 (To -E)	1		Pr, S*
	RAF Qtr,	Ap '52:138 (-A)			P
	US Air Svcs,	Ja '48:22-23 (X-)	1		
		Je '49:26 (X-)	6	(Engine)	
		F '51:11	2		
		D '51:10-14	1		P
		S '55:22 (X-/-D)	1		
	West Avn,	Ja '52:8	3		
	West Fly,	O '49:16-17 (X-)	6	(Yaw Damper)	S*
		Mr '51:14 (-B)	2		
428.	B-50 Model 345				
	Trainer version: TB-50				
	Reconnaisance version: RB-50				
	Air Force,	Ag '55:46-52	4		
	Air Prog,	'52 Annual:46-47 (-D)			S*
	Avn Age,	Ag '51:28-29 (-D)	1		S*
	Avn Wk,	Jl 14, '47:13	1		
		Ap 23, '51:17 (T-/-D)	1		
	Boeing Mag,	N '46:3-4+	1		
		Jl '47:3-4	2		
		My '48:8-9			P
		N '49:6-9 (-D)	1		P
		Ja '50:6-7 (R-)	1		

BOEING (cont.)

428. B-50 (cont.)
Flight, N 17, '49:655 (-A to -D) 1
Flying, Ap '48:28-29+ 1
Interavia, No. 7, '47:24-25 1
US Air Svcs, Jl '47:12-13 1
Ag '47:7-8 1
Mr '49:21 ("Lucky Lady II") 4
N '49:16 (-D) 1 P
West Fly, N '47:14-15 6 (Anti-ice)

429. B-52 "Stratofortress" Model 464
Air Force, Ap '56:99-107 1
Air Power, Win '56/'57:90 (-C) P
Air Prog, D '68:50-53+ 4
Avn Wk, Ag 8, '52:12-17 1
Mr 22, '54:13 (X-&-A) 1
Mr 19, '56:26-30 4
Ag 20, '56:79-80 6 (Armament)
Je 24, '57:81-88 6 (Electrical)
Ap 13, '59:126-127 3
N 29, '65:16-21 4
F 7, '66:48-56 6 (ECM)
My 1, '67:55-59 4
Bee-Hive, Sum '52:4-5 (Y-) 2
Boeing Mag, Ap '52:3-5 (Y-) 2
Jl '52:3-5+ 1
Mr '53:10-11 2
N '53:3-5 2
Jl '54:3-5+ 3
O '55:3-6 2
Ja '56:10-11 1
F '56:10-11 2
Ja '57:8-9 2
F '57:3-5 (Round-the-World) 4
My/Je '58:6-7 (-G) 1
Ja '59:10-11 (-G) 1
Jl '59:10-11 (-G) 6 (Wing)
N '59:12-13 (-G) 2
Mr '60:7 (-H) 1 D
Je '60:6-7 (-H) 1
F '61:8-9 (-H) 1
My '61:7 (-H) 6 (Starter)
Ag '62:13 1
Ap '64:12-13 (-F&-H) 2
S '64:12-13 6 (Electronics)

BOEING (cont.)

429. B-52 (cont.)

Boeing Mag,	Ja '65:7 (-H)	6 (Rockets)	
	Mr '67:6-7 (-H)	2	
	Ag '69:12-13 (-H)	2	
Flight,	D 5, '52:702-704	1	
	Mr 4, '55:278-279	3	
	F 8, '57:187	6 (Tail Gun)	
	N 15, '57:771-778	1	S*
	Ja 1, '60:15 (-G)	1	Pr
	Ap 29, '60:596-597 (To -H)	1	
Interavia,	No. 10, '52:576-577 (X-)	1	
	No. 10, '53:585 (X-)		P
	No. 11, '55:856-857	2	
RAF Fly Rev,	N '55:52-53 (To R-/-B)	1	P*
	Je '57:13 (-A to -D)	1	
	O '62:32-33+ (-H)	1	P*
US Air Svcs,	My '56:10 (-C)		P
West Avn,	Ag '55:36	6 (Lndg Gear)	
	Je '56:26	6 (Flight Cntrl)	
	S '56:11	6 (Fire Cntrl)	

430. C-97 "Stratocruiser" and "Stratofreighter" Model 367
Tanker version: KC-97
Commercial version: BOEING 377

Aeroplane,	Mr 2, '45:249-250	1	
Air Force,	D '46:54-55	4	C
Air Pict,	My '50:145 (-A)		P
Acft Eng,	N '48:335-336	6 (Air Duct & Anti-ice)	
Aviation,	Ag '45:170-171	4	C
Boeing Mag,	Ja '45:10-11	1	
	F '45:13+	1	
	Je '45:14	1	
	S '45:1-7	1	
	O '45:8-9+	1	C
	Ap '46:3-4+	4	C
	Je '47:4-5 (Hospital-)	1	
	Je '51:6-7 (K-/-A)	3	
	D '51:6-9	1	
	N '53:12-14	1	
	N '65:12-13 (K-)	4	
	Jl '67:15 (K-/-L)	4	
Flight,	Ja 12, '50:48-49	1	
	F 26, '54:235-236	1	
	O 8, '54:544-546 (K-/-G)	4	

BOEING (cont.)

430.	C-97 (cont.)				
	US Air Svcs,	F '45:14-16+	1		
		F '48:20 (X-&Y-/-A)	1		
		O '49:25 (-A)	1		
		Jl '52:16 (K-/-F)	1		
		D '54:11+	1		
		Ap '55:10-11 (K-)	1		
	West Avn,	Ja '52:14	1		
	West Fly,	F '45:52+	1		
		O '50:19-21+	1		
431.	C-135 "Stratolifter," "Stratotanker" (KC-135) Model 717				
	Air Force,	N '69:66-69 (-L)	4		
	Air Power,	Win '56/'57:98 (K-:"City of Renton")			P
	Avn Wk,	Jl 23, '56:34 (K-/-A)	1		
		Jl 18, '60:127-129	1		
		Ap 9, '62:98-100 (-B)	1		
	Boeing Mag,	My '56:8-9 (K-)	3		
		Ag '57:6-7 (K-)	2		
		S '57:3-5	1		
		N/D '57:6-7 (K-)	2		
		Ja/F '58:14-15 (K-)	2		
		Ag '58:10-11	6	(Cargo)	S*
		Ag '60:10-11 (K-)	1		
		Ap '61:3-5	1		
		Ap '61:8-9 (-B)	1		
		Ja '65:3-5 (R-/-A&-B)	1		D
		Ja '67:3-5	1		
		D '68:4-5	1		
	Flight,	My 28, '54:697-699+ (Proto-)	1		
		F 1, '62:169-173	4		
	Interavia,	No. 3, '62:300-301 (-A&-B)	1		C
	RAF Fly Rev,	N '61:42-43 (-A)	1		P*
		Ap '62:28-31 (K-/-A)	1		Pr, S*
432.	VC-137C USAF BOEING 707 for US President				
	Boeing Mag,	O '62:8-9	1		P
		Ap '67:6-7	4		
433.	F2B-1 Model 69				
	AAHSJ,	Spg '65:55			P

BOEING (cont.)

434.	F4B		
	US Army: P-12		
	Air Prog, Spg '58:14-22	1	C, P
	AAHSJ, Win '61:265 (-4)		C
	Sum '64:151 (-4)		D
	Spg '67:50-51 (-3)		C
	Flight, Je 14, '34:593-594 (-4)	1	
	Fly Rev Intl, Ag '66:770 (-1A, -3, -4)		Pr*
	Ap '68:230 (-1:Restored)		P*
	RAF Fly Rev, O '61:63+	1	
	US Air Svcs, O '32:38 (-4)		P
	Je '34:22 (-4)		P
435.	XF8B-1 Model 400		
	Air Pict, N '67:417	1	
	Boeing Mag, N '45:3-4+	1	
	D '45:15	2	
436.	GA-1 or GA-X Model 10		
	Flying, F '55:56	1	
	RAF Fly Rev, Mr '58:6	1	
437.	XL-15 "Scout" Model 451		
	Air Force, Je '48:22-23	1	
	Am Helicop, Ja '48:4	1	
	Boeing Mag, F '47:3-5	1	
	S '47:8-9	1	
	Skyways, Jl '48:27	1	
	US Air Svcs, F '47:13	1	
438.	NB-2		
	Aero Album, Fall '68:37	1	P
439.	P-12		
	US Navy: F4B		
	See also 100 F		
	Aero Album, Win '68:9-11 (First-)	1	P
	Aero Digest, Ap '33:12		P
	Aerosp Hist, Win '68:49	1	P
	Air Prog, Spg '58:14-23	1	P
	AAHSJ, Win '61:265 (-E)		C
	Flying, Ag '64:54-56+	4	
	Fly Rev Intl, Ag '66:772 (-D&-E)		Pr*
	Pop Avn, S '30:14 (-C)	1	
	Ap '33:227 (-E)	1	Pr†, S*

BOEING (cont.)

439. P-12 (cont.)
RAF Fly Rev, My '57:55 1
US Air Svcs, F '30:62 4
D '31:54 (-C) 1 P

440. P-26 "Peashooter"
Export version: BOEING 281
Aero Album, Spg '68:32-38 1 P
Air Cl, Mr/Ap '65:18-21+ 1 P
Air Pict, D '57:426-427 (-A:95 Pur Sq) 5 Pr
:426 1
D '58:430-431 (USAF Museum's-) 1
Air Prog, '55/'56 Annual:74-76 (-A) Pr, S*
Jl '65:52-56 (-A) 1 C*, Pr†, S*
Bee-Hive, Jl '32:5 1
Fly Rev Intl, My '65:49-52 1 Pr*
Pop Avn, Ja '34:26-27 (X-) 1 Pr
N '37:25 4 P
RAF Fly Rev, O '55:9-10 1
Ap '59:30-33 1
:31 (-A:34 Pur Sq) Pr*
West Fly, Ag '34:37 (-A) 1

441. YP-29 Model 264
Pop Avn, My '37:22 1

442. XPBB-1 "Sea Ranger" Model 344
Boeing Mag, Jl '42:2-3 2
My '69:10-11 1 P
Flight, O 1, '42:370 1
D 23, '43:695 1
RAF Fly Rev, N '56:9+ 1

443. PT-13 and PT-17
See STEARMAN PT-13 and PT-17

444. PW-9 Model 15
AAHSJ, Sum '62:110-114 1 Pr

445. SST 733 Design
Acft Eng, O '65:300-304 1 S*
Avn Wk, My 4, '64:36-43 1 C
Boeing Mag, O '66:7-9 1
My '67:10-11 3
Can Avn, Je '64:30-33 1
Interavia, No. 6, '64:828-829 1
No. 8, '66:1160-1162 1

BOEING (cont.)

446.	SST 969 and 2707 Designs			
	Aeroplane, O 6, '66:26-27	1		S*
	F 2, '67:4-7	1		S*
	Ag 30, '67:12-14	3		
	D 6, '67:27	1		
	Aerosp Intl, S/O '67:16-18	1		
	Ja/F '69:34-35 (-300)	1		
	Air Force, Jl '66:74-89	1		
	F '67:41-45	1		S*
	AF/Sp D Intl, Jl '66:6-15	1		
	Acft Eng, D '66:29-34	1		S*
	O '69:12-16	1		
	Avn Wk, Je 13, '66:43	1		
	Je 20, '66:34-35	1		
	Ag 15, '66:36-39 (Pt 1)	1		
	Ag 22, '66:30-33 (Pt 2)	1		
	O 3, '66:34-37	1		
	O 17, '66:39-45	1		
	N 13, '67:91-95	6	(Antennas)	
	D 4, '67:34-36	1		
	:109-119	6	(Wing Pivot)	S*
	O 28, '68:41-43 (-300)	1		
	N 18, '68:36-37 (-300)	1		
	D 9, '68:16-20 (-300)	1		S*
	D 16, '68:41-43 (-300)	3		S*
	:45-61 (-300)	1		S*
	Boeing Mag, Ja '68:9	1		D*
	Flight, Je 23, '66:1035-1036	1		S*
	O 6, '66:577	1		
	Ja 12, '67:46	1		
	D 7, '67:931	1		
	O 31, '68:687	1		
	Ja 16, '69:104-108 (-300)(Pt 1)	1		S*
	Ja 23, '69:133-136 (-300)(Pt 2)	1		
	Fly Rev Intl, Mr '67:415+	1		S*
	Mr '69:22-23 (-300)	1		
	Interavia, No. 12, '66:1904-1906	1		S*
	No. 2, '69:168-170 (-300)	1		C, S*
	No. 12, '69:1911-1913	1		
	Sp/Aeron, Jl '67:74-88	1		
	D '67:76-82	6	(Life Support)	S*
447.	40 "Mailplane"			
	Aero Digest, Ja '31:41 (-B.4)			C
	AAHSJ, Fall '61:203 (-B)			D
	:179 (-A)			C

BOEING (cont.)

447.	40 (cont.)			
	Flying, S '63:13 (-A)	1		D
	Pegasus, Ap '46:9 (-B.4)			P
	US Air Svcs, Ja '31:29 (-B.4)			C
448.	80A			
	AAHSJ, Win '61:262			C
	Aviation, My 10, '30:943-947	1		C
449.	100F			
	US Army equivalent: P-12F			
	US Air Svcs, Mr '32:13	1		P
450.	203			
	Boeing Mag, Ag '69:6	1		P
451.	204			
	US Air Svcs, Ja '30:18	1		P
452.	247			
	Aero Digest, Ap '33:50-51	1		
	D '35:40 (-D)			C
	Air Cl, Ag '69:12-13, 66, 68 (-D:"Adaptable Annie")	1		P
	Air Pict, Jl '62:194-197	1		
	Air Prog, Sum '61:66-69	1		
	Acft Eng, F '35:44 (-D)	1		
	AAHSJ, Win '64:239-271	1		C, P, Pr
	Aviation, Ap '33:124-126	1		
	Je '35:19-21 (-D)	1		
	Bee-Hive, Je '32:6-7	1		
	Ap '33:8-10	1		
	Je '33:7-8	6	(Props)	
	Boeing Mag, Ag '53:12-13	1		
	Can Avn, O '58:35-36 (-D)	1		
	Flight, Ag 12, '32:756	1		
	Je 15, '33:563-565	1		C
	O 25, '34:1121-1122	1		
	N 8, '34:1180-1181 (-D)	1		
	Flying, Ja '54:36+ (-D:"Adaptable Annie")	1		
	Pop Avn, Je '33:362-364	1		C
	Jl '33:19, 34			C*, P
	N '33:323			S*
	D '33:392-394+	1		
	Pop Fly, Mr '34:616, 624-626+	4		P

BOEING (cont.)

452.	247 (cont.)		
	RAF Fly Rev, Mr '62:45+	1	
	US Air Svcs, Ap '33:12-15	1	
	My '33:20		C
	Je '33:12	1	P
	Mr '34:28		C*
	Ap '34:21	2	
	N '34:27 (-D)	1	
	N '53:22 (-D:"Adaptable Annie")	1	
	West Fly, Ap '33:18-19	1	
	N '34:14+ (-D:England-Australia)	4	
453.	281		
	Export version of P-26A		
	Aero Digest, Mr '35:38	1	
	Bee-Hive, Ap '35:9	1	
	Flight, Ap 11, '35:393	1	
454.	299		
	Prototype of BOEING B-17		
	Pop Avn, O '35:219+	1	
	D '35:358	2	
	US Air Svcs, Ag '35:18-19	1	
	O '35:16	4	P
	D '35:38 (Crash-)	4	
455.	307 "Stratoliner"		
	Aero Digest, Ap '38:21-23	1	C
	F '39:67-68	2	
	Air BP, No. 14:9 (-B:TWA)		Pr*
	Air Pict, F '65:58-60	1	
	AAHSJ, Spg '63:23-29	1	Pr
	Aviation, My '40:46-49+	1	C, S
	Mr '45:148-151	1	
	Boeing Mag, S '44:6-7+	1	
	Ap '45:17	1	
	Ja '68:3-5	1	
	D '68:8-9	1	
	Can Avn, Ap '38:6-7+	1	S
	F '39:10	1	
	Comm Avn, Ap '40:15-18	1	
	Flight, Ap 7, '38:343-344	1	
	Jl 4, '40:6-6a+	1	

BOEING (cont.)

455.	307 (cont.)		
	Flying, D '42:30-32+	1	
	My '53:61 (-B)	4	
	Pop Fly, S '38:294		C
	RAF Fly Rev, O '59:53	1	
	Mr '60:58	1	
	US Air Svcs, Je '37:25	1	
	Ap '38:16-18+	1	C
	Jl '39:15	2	
	Ap '40:12-14	1	C
	My '40:14-16+ (-B)	1	
	West Fly, My '38:18-20	6 (Cabin Press)	
	F '39:20	1	
	My '40:14-16+	6 (Cabin Press)	
456.	377 "Stratocruiser"		
	Civil version		
	Military version: C-97		
	Aero Digest, D 15, '45:56-57+	1	
	F '49:38-43+	1	C, S*
	Jl '50:36-37+	6 (Electrical)	
	Aeronautics, Mr '52:26-32	1	
	Aeroplane, Ja 30, '48:128-131	1	S*
	F 27, '48:248-251	1	
	Ap 15, '49:424-426	1	C
	AAHSJ, Win '69:217-228	1	P, S*
	Aviation, N '45:176-177+	1	
	Avn Opns, Je '50:26	6 (Engine)	S
	Avn Wk, D 13, '48:29-33	1	
	Ja 10, '49:22-29	1	
	Boeing Mag, N '44:3-7	1	
	Je '46:12-13	6 (Rudder Boost)	
	O '46:6-7+	4	C
	Je '47:6-7+	6 (Press)	
	:10-11+	6 (Electrical)	
	Ag '47:8-9		P
	Je '48:7-9+	6 (Engine)	
	Ag '48:8-9		P
	F '49:1-20	1	S
	Je '49:3-5+	4	
	:8-11	1	P
	O '49:6-9	4	P
	Can Avn, Ja '46:30-31+	1	C
	D '49:18-19+	1	
	Comm Avn, Ja '45:42-44+	1	
	Ja '46:29-32	1	

BOEING (cont.)

456. 377 (cont.)

Flight, Ja 4, '45:10-12	1		
Ja 27, '49:105-108	1		S*
Ap 14, '49:438-439	1		C
D 15, '49:774-777	4		
Je 14, '57:812-813	1		
Interavia, No. 3, '46:21-26	1		C, P
No. 8, '47:32	6	(Electrical)	
Mr '49:155-160	1		P
US Air Svcs, D '44:20-22	1		
O '45:10-12	1		
S '46:36 (United-)	1		
Ag '48:15	6	(Engine Analyzer)	
Mr '50:21	1		
West Fly, D '44:56-57	1		
N '45:44-45+	1		
Ag '47:14-15	3		
N '47:14-15	6	(Anti-ice)	S*
Jl '48:15-16	2		

457. 377 "Stratofreighter"
377 cargo version

Aero Digest, S '46:71-73+	1		
Aviation, O '46:81	1		
Boeing Mag, Ag '46:3-5	1		
Can Avn, N '46:48-50	1		
US Air Svcs, N '50:11	1		

458. 417
Project cancelled while under construction

Boeing Mag, My '46:3-5	1		S*
S '46:6-7	1		

459. 707 "Stratoliner"

Aeronautics, N '54:27			P
Aeroplane, Ja 15, '60:72-77	1		S*
Air BP, No. 14:9 (-131:TWA)			Pr*
No. 27:21 (-437:Air-India)			Pr*
No. 32:17 (-138B:Qantas)			Pr*
Air Pict, N '54:333+	1		
Je '56:202-203	1		Pr
N '60:371 (-120)			S*
Avn Age, My '57:34-41	6	(Flight Cntrl)	
Ag '57:26-33	3		
S '57:48-49	6	(Flight Cntrl)	S*

BOEING (cont.)

459.	707 (cont.)				
	Avn Wk,	Je 29, '53:12-13	1		
		Mr 8, '54:14-15	1		
		My 24, '54:16-17, 38	1		
		Jl 26, '54:14-16	2		
		Ag 2, '54:63-67	1		
		Ag 9, '54:30-42	1		
		D 6, '54:87-95	6	(Hydraulic)	
		N 21, '55:21-25	4		C
		N 28, '55:29-35	1		
		F 27, '56:37-38	2		
		Je 4, '56:80-83	1		
		Je 25, '56:63-65	4		
		Mr 18, '57:38-39	4		
		Ja 20, '58:48-67	1		
		S 1, '58:28-31 (-120)	1		
		O 6, '58:70-79 (-120)(Pt 1)	4		C
		O 13, '58:79-97 (-120)(Pt 2)	4		
		N 9, '59:42-45	1		
		Ag 13, '62:40-43 (-320B)	1		
		Mr 8, '65:38-45	6	(Auto Landing)	
	Bee-Hive,	Fall '68:15-17 (-121: "Mayflower")	4		
	Boeing Mag,	Mr/Ap '54:3-7 (-Proto)	1		
		Je '54:3-9 (-Proto)	1		
		Ag '54:3-5 (-Proto)	2		
		O '54:3-5+ (-Proto)	1		
		N '54:3-5 (-Proto)	2		
		F '55:3-5+	1		
		Ap '55:8-9 (-Proto)			P*
		S '55:8-10	1		
		Je '56:6-12	6	(Interior)	P*
		Ag '56:3-5	1		
		S '56:8-9 (-120)			S*
		My '57:6-7	4		
		O '57:6-7 (-Proto)	1		
		Ja/F '58:6-9	2		P*
		My/Je '58:14-15	2		
		O '58:2-13	1		
		My '59:8-10 (-320)	1		P*
		Jl '59:3-5 (-Proto)	1		
		O '59:3-5 (-Proto)	1		
		My '62:10-11 (-320B)	2		
		Jl '64:3-7 (-Proto)	1		
	Can Avn,	Ag '58:28-29	4		
		S '61:33 (-Proto)	1		

BOEING (cont.)
459. 707 (cont.)
Flight, Ag 7, '53:166-167 (Provisional-) 1
Ja 27, '56:115-118 1
Je 15, '56:764-765 (-Proto) (Pt 1) 4 C
Je 22, '56:799-803 (-Proto) (Pt 2) 4
Jl 6, '56:11-14 1 S*
N 1, '57:677-680 1 S*
Jl 25, '58:144-147 (-120 to -720) 1 S*
S 19, '58:480+ (-120) 1
O 31, '58:682-684 (-121) 4
Mr 27, '59:437 (-123:American) 1 Pr
N 20, '59:584-585 (-420) S*
Ja 15, '60:75-76 (-436) 1
Je 3, '65:860-861 (-820) 1
Je 10, '65:897-898 (-820) 1
Interavia, No. 4, '54:237 (-Proto) 1
No. 9, '54:607-609 (-Proto) 1 P
No. 4, '57:305-307 (-Proto) 4
No. 5, '58:400-403 6 (Navigation) C
No. 1, '67:84-85 3
Skyways, Ja '56:8-9+ 1
US Air Svcs, Mr '54:4 (-Proto) P
Mr '55:21-22 1
F '56:19 (-120 to -320) 1
O '56:22 2
West Avn, Mr '56:11 2

460. 720
Air BP, No. 22:15 (-United) Pr*
Avn Wk, Ag 15, '60:56-61 4
Mr 20, '61:67-75 (-B) 4
Mr 8, '65:38-45 6 (Auto Landing)
Bee-Hive, Spg '60:1 1 P
Boeing Mag, My '59:5 D*
D '59:6-7 1
Flying, N '60:26-28+ 4 C
Fly Rev Intl, Ap '66:450-451 1 Pr

461. 727
Aeroplane, D 16, '60:798 1
N 30, '61:689 1 S*
N 29, '62:26 1
F 21, '63:9-10 6 (Flight Cntrl)
F 24, '66:12-14 2
Jl 5, '67:16-19 (-200) 1
My 1, '68:30 (-200) 4
Jl 24, '68:17, 20-23 4

BOEING (cont.)

461.	727 (cont.)				
	Air BP,	No. 31:9 (Lufthansa-)			Pr*
		No. 41:9 (Pan Am-)			Pr*
	Air Prog,	O/N '63:46-47	1		S*
	Acft Eng,	Jl '66:30-31	6	(Interior)	
	Avn Wk,	D 12, '60:40-43	1		
		Mr 6, '61:60-65	1		
		D 11, '61:40-41	2		
		N 26, '62:42	1		
		D 10, '62:60-77	1		
		N 4, '63:50-65	1		C
		N 23, '64:43-46	6	(Hydraulic)	
		F 15, '65:40-41 (-M)	1		
		Ap 12, '65:42-43	4		
		Ag 7, '67:45 (-200)	1		
		S 11, '67:40-43 (-200)	1		
		My 27, '68:42-47 (-QC)	4		
	Bee-Hive,	Spg '63:34			P
		Ja '64:18-21	1		
	Boeing Mag,	Ja '61:3-5, 8-9	1		
		N '61:12-13	1		
		D '61:8-9	1		S*
		O '62:6-7	3		
		D '62:8-9	1		
		Mr '63:3-5	2		
		My '63:6-9	2		
		S '64:11 (-C)	1		
		Je '65:3-5	1		
		O '65:8-9 (-200)	1		
		Ap '68:8-11 (-QC)	1		
	Flight,	D 16, '60:938+, 948	1		
		D 30, '60:1010-1012	1		
		D 7, '61:891	1		S*
		My 9, '63:674-688	1		S*
		D 5, '63:922-923+	1		
		Je 18, '64:1029-1034+	4		C
		Ag 19, '65:282 (-200)	1		
		Mr 7, '68:328	1		P*
	Flying,	D '63:32-33+	4		
	Fly Rev Intl,	Ap '64:44 (-C)	1		
		S '64:26-31	1		Pr, S*
	Interavia,	No. 4, '62:428-429	1		S*
		No. 9, '62:1122-1123 (Mockups)	1		
		No. 4, '63:446-450	1		
		:461-462	6	(Auto Pilot)	

BOEING (cont.)

462. 737
Aeroplane, Mr 4, '65:23-24 1
Ja 3, '68:22-23 4
Air BP, No. 44:17 (-200:NAC) Pr*
Air Pict, My '68:162-163 (-130) 4
Air Prog, Jl '67:84 P
Avn Wk, Mr 1, '65:26-28 1
Ap 12, '65:41 (-100 to -222) 1
Ap 17, '67:40-41 2
S 18, '67:56-70 4 C
Mr 18, '68:225 P*
Mr 3, '69:28-29 (Improved-) 1
My 19, '69:40-43+ (Short Field-) 1
S 29, '69:36-37 4
O 27, '69:40-41+ (Advanced-) 1
Bee-Hive, Sum '68:26-29 6 (Cabin Press)
Boeing Mag, Ap '65:6-7 1
Je '65:3-5 1
N '66:12-13 3
Jl '67:8-9 6 (Lndg Gear)
Can Avn, My '65:37 1
Mr '67:16-17+ 1
Flight, D 10, '64:995-996 1
Mr 4, '65:314 1
Ap 22, '65:628 1
Jl 1, '65:8-10 (-100&-200) 1
F 3, '66:181-188 1 C, S*
My 11, '67:758-759 2
Ja 4, '68:6 (-100&-200) 1
Jl 18, '68:88-89 (Britannia-) 4
N 13, '69:732-734 (Improved-) 1
Fly Rev Intl, My '66:515-517+ 1 C, S
Je '67:617 1
Je '69:64-66+ 1 P*, Pr
Jl '69:20-21 (Improved-) 1
Interavia, No. 4, '65:505 1
No. 5, '65:637 1
No. 7, '65:1142 (-100&-200) 1
No. 10, '66:1533-1534 1
No. 3, '67:374-376 3 C

BOEING (cont.)

463. 747			
Aero Album, Win '69:20-23	1		C, Pr
Aeroplane, D 27, '67:20-22	1		
Ap 3, '68:23	1		
My 22, '68:26	1		
Je 12, '68:4-5	3		
Jl 17, '68:30-32	3		
Ag 7, '68:13-16	1		
Aerosp Intl, S/O '67:22-23	1		
Air Force, N '68:68-70	1		
AF/Sp D Intl, S '66:16-24	1		
O '66:40-41	1		
Avn Wk, F 21, '66:37-39	1		
Mr 21, '66:42-43	1		
Ap 18, '66:41	1		
Ap 25, '66:40-47	1		
O 31, '66:205-209	6	(Avionics)	
D 12, '66:43	1		
Mr 20, '67:45-48	1		
N 20, '67:35-74	1		
:79-85	6	(Engine)	
:89-94	6	(Navigation)	
:99-144	3		
:147-157	2		
Ap 29, '68:57-58	3		
Jl 15, '68:26-27	3		
O 7, '68:38-43	1		
O 28, '68:116-117	3		
D 2, '68:25 (-B)	1		
F 17, '69:26-31	2		P
:81-82+	6	(Entertainment)	
Ap 28, '69:39-49	3		
S 1, '69:24-27	1		
O 20, '69:36-43	4		
D 1, '69:33	6	(Evacuation)	
Boeing Mag, Ap '67:12-13	6	(Nose)	
My '67:15	2		
N '67:12-13	3		
:15	6	(Flaps)	
N '68:3-5	1		
Mr '69:8-9	2		P
Can Avn, N '68:16-17	1		

BOEING (cont.)

463. 747 (cont.)

Flight,	Ja 6, '66:8-9	1	
	Ap 21, '66:655-656	1	
	Ja 11, '68:52-55	2	S*
	Mr 7, '68:344	6 (Auto Landing)	
	O 17, '68:597-598	1	
	D 12, '68:978-997	1	C*, S*
	D 19, '68:1027-1034	3	S*
	S 4, '69:389	6 (Navigation)	C
Fly Rev Intl,	Jl '66:683-684	1	
	Ag '68:425-426	1	
	Ja '69:46-50	1	
	F '69:27-28 (-B)	1	
Interavia,	No. 5, '66:666-667	1	
	No. 6, '66:935	3	
	No. 8, '66:1198	1	
	No. 4, '67:489-490 (Freighter)	1	S*
	No. 7, '68:853-865	4	
	No. 12, '68:1499-1501	1	
	No. 10, '69:1661-1664 (Pan Am-)	4	C

464. 767

Avn Wk, F 24, '69:26-29 (-100 to -300) 1

BOEING-CANADA Canada

465. "Totem"

Flight,	S 23, '32:890-892	1	C

BOEING-VERTOL USA

466. CH-46 "Sea Knight" Model 107

US Navy designation
Earlier designation: HRB-1
US Marine designation: UH-46
Civil version: 107

Avn Wk,	O 22, '62:30-32 (-A)	1
	N 5, '62:108-109 (-A)	1
Boeing Mag,	Ag '65:6-7	4
Flight,	D 12, '68:975-976	1
Nav Avn News,	Ja '65:22 (-A)	1

BOEING-VERTOL (cont.)

467. CH-47 "Chinook" Model 114
US Army designation
Earlier designation: HC-1
Civil version: 107

Aeroplane, S 29, '66:16-18 (-A)	4		
Aerosp Intl, Mr '67:24-26 (-A)	6	(Armament)	S
Avn Wk, N 29, '65:64-65 (-A)	1		
Ap 17, '67:84-85 (-B)	1		
Boeing Mag, O '68:3-5 (-A to -C)	4		
Flight, Je 1, '67:922 (-A&-B)	1		
D 12, '68:975-976 (-A&-B)	1		
Fly Rev Intl, F '64:34-38+ (-A)	1		Pr, S*
Interavia, No. 4, '64:514-517	1		C, S*
US Army Avn, Ap '64:18-20	1		
Ap '65:12-15	4		
S '65:24-25	4		
Jl '66:26-29 (-A)	6	(Armament)	
Ag '66:26-28	4		
Ag '67:12-13 (-B)	1		
D '67:22-27 (-A, -B, -C)	1		
Ja '68:60 (-A)	6	(Voice Warning	S*
Jl '68:64 (-B&-C)	6	(Airfoil)	

468. CH-113 "Labrador" and "Voyageur" Model 107
Canadian version of Model 107

Can Avn, F '63:16-17	1		
Roundel, My '63:14-15	1		

469. H-21 "Work Horse"
Earlier design: PIASECKI H-21
Civil designation: BOEING-VERTOL 44

Am Helicop, Ag '56:10 (-C)	6	(Engine)	
S '56:12+ (-C)	4		
Avn Age, Ap '56:36-37	1		S*
Avn Wk, Jl 30, '56:35	1		
S 17, '56:28-31	4		
Mr 25, '57:64-65 (-B)	4		

470. HC-1B "Chinook"
Later designation: CH-47

Aeroplane, F 22, '62:203			C, P
Am Helicop, Ap '59:12 (Y-)	1		
Ap '60:14+ (Y-)	1		

BOEING-VERTOL (cont.)

470. HC-1B (cont.)

Avn Wk,	Jl 7, '58:78-84 (Y-)	1	
	F 15, '60:120-135 (Y-)	1	C, S*
	Ja 1, '62:78-79	1	C
	Jl 16, '62:31	1	
Boeing Mag,	N '61:3-5	1	
	My '62:6-7	2	
Interavia,	No. 7, '62:866-868	1	
US Army Avn,	Je '59:37 (Y-)	1	
	Ag '62:3-8	1	

471. HRB-1 "Sea Knight" Model 107
Later designation: CH-46

Boeing Mag,	Ja '62:3-5	1	
Nav Avn News,	Ap '61:33	1	

472. UH-46 "Sea Knight" Model 107
US Marine designation
US Navy designation: CH-46

Nav Avn News,	S '64:28 (-A)	1	

473. 44
Military designation: H-21

Air BP,	No. 13:14 (-B:N.Y. Airways)		Pr*
Am Helicop,	My '57:8-9+ (-A to -C)	1	
Avn Age,	Je '57:204-211	1	
Avn Wk,	Ap 15, '57:33-34	1	
	F 17, '58:30-32	2	
Flight,	My 10, '57:646 (-A to -C)	1	S
	Je 21, '57:844-846	4	

474. 76

Avn Wk,	N 24, '58:80-88	1	
	O 24, '60:56-65	1	
Boeing Mag,	N '69:5	1	
Nav Avn News,	S '58:12	1	

475. 107
US Navy and Marine designations:
CH-46 and UH-46
US Army designation: CH-47
RCAF designation: CH-113

Aeroplane,	D 7, '61:725-727 (-II)	1	
Am Helicop,	Jl '58:10+	1	
	Ap '60:7+	4	

BOEING-VERTOL (cont.)

475. 107 (cont.)

	Avn Wk, Jl 7, '58:78-84	1	
	My 18, '59:42-43 (-A)	1	
	Mr 6, '61:52-57	1	
	O 30, '61:36-38	4	
	Boeing Mag, Ag '60:3-5	1	
	Flying, Ap '61:26-27+	4	
	Interavia, No. 1, '62:93-94	4	
476.	147		
	Avn Wk, Mr 8, '65:46-51	1	

BOK (Biuro Opytnykh Konstruktsii) Russia

477.	5 (1937)		
	Fly Rev Intl, Mr '66:420	1	

BOLKHOVITINOV Russia

478.	DB-A (1934)		
	Fly Rev Intl, N '67:1004	1	
479.	S (1940)		
	Fly Rev Intl, Je '68:357	1	

BÖLKOW Germany

	Boeing Mag, Ap '65:3-5		
	Interavia, No. 9, '65:1435-1439		
480.	Bo. 46		
	Interavia, No. 7, '64:1020-1021	1	
481.	Bo. 105		
	Air Force, Ap '69:30-31	1	
	Avn Wk, Je 26, '67:80-89	1	S
	Ap 8, '68:63	1	
	Flight, My 18, '67:794	1	
	Interavia, No. 7, '64:1020-1021	1	S*
	No. 6, '67:888	1	

BORDELAISE France

482.	A.B. 20		
	Flight, F 26, '32:170	1	

BORDELAISE (cont.)

483.	D.B. 70			
	Aero Digest, Ja '32:59	1		
	Acft Eng, Ja '30:19	1		
	Flight, F 21, '30:225-227	1		

BORGWARD Germany

484.	"Kolibri I"			
	Avn Wk, Je 15, '59:105-107	1		

BOULTON PAUL GB

Flight, Jl 8, '55:39-53 (Forty year history)
O 7, '65:619-625 (Fifty year history)

485.	P. 12 "Bodmin"			
	Aeroplane, Ja 1, '43:26	1		
486.	P. 32 "Superstrand"			
	Flight, Je 17, '32:533	1		
	S 12, '35:294-295	1		
487.	P. 29 "Sidestrand"			
	Aeroplane, Je 17, '30:1142	1		
	Flight, O 31, '30:1192 (-III)	6	(Engine)	
	Je 19, '31:549 (-III)	1		
	RAF Fly Rev, Mr '58:11	1		
	RAF Qtr, Jl '32:Plate 11			P
488.	P. 64 "Mailplane" or "Mail Carrier"			
	Aeroplane, Ap 5, '33:593-602	1		S*
	Je 21, '33:1142-1143	1		
	Flight, Ap 6, '33:316-322	1		Pr, S
	Je 22, '33:607-609	1		
	RAF Fly Rev, F '62:43	1		
489.	P. 71A Commercial Biplane			
	Aeroplane, Ja 30, '35:129-132	1		S*
	Flight, Ja 31, '35:118-123	1		Pr, S
490.	P. 75 "Overstrand"			
	Air Pict, O '66:376-379	1		Pr
	Flight, Je 27, '35:700-701			S*
	O 22, '36:414a	6	(Turret)	
	Pop Fly, Ag '36:253	1		
	RAF Fly Rev, Mr '58:11	1		
	RAF Qtr, Ap '35:Supp	1		P

BOULTON PAUL (cont.)

491.	P. 82 "Defiant"		
	Aero Digest, Ap '42:144+	1	
	Aeroplane, F 13, '42:188-189 (-I)	3	
	Ag 31, '45:246-247 (-III)	4	
	Air Pict, Ag '61:238-242 (Pt 1)	1	Pr†
	S '61:286-290+ (Pt 2)	1	Pr†
	Acft Eng, Ap '42:98-102	1	
	Aviation, S '42:201+	1	
	Flight, F 12, '42:132-135	1	
	F 25, '43:202b	1	
	S 2, '43:250 (-Target tug)	1	
	Jl 18, '46:60-61	4	
	Fly Rev Intl, My '68:272 (-II:125 Sq)		Pr*
	RAF Fly Rev, O '56:15+	1	
	Jl '63:25	4	
	S '63:17-19+	4	
492.	P. 92		
	A half-scale model was built		
	Flight, Je 13, '46:594	1	
	RAF Fly Rev, Mr '62:45	1	
493.	P. 108 "Balliol"		
	Aeroplane, O 24, '47:553-554	1	C, S
	Jl 2, '48:16-20 (-I)	4	S*
	Mr 3, '50:255		P
	N 9, '51:610-611		P
	F 6, '53:174-175		P
	F 13, '53:204-205	4	
	Air Pict, O '50:298 (-II)	1	
	Avn Age, Mr '55:32-33	1	S
	Avn Wk, Ag 2, '48:20-24	1	
	Flight, Ag 22, '46:185	1	
	Ag 7, '47:129	1	
	Je 24, '48:679-680	4	
	Jl 1, '48:13-18 (-I)	1	S*
	Jl 8, '48:50-51 (-I)	1	
	F 3, '49:123-125 (-I)	4	
	My 26, '49:623-624 (-II)	1	
	F 23, '50:246-248 (-II)	1	
	Interavia, O '48:549-551 (-I&-II)	1	
494.	P. 111		
	Air Pict, Ja '51:11		P
	Avn Wk, N 27, '50:19	1	
	RAF Fly Rev, Jl/Ag '59:11	1	

BOULTON PAUL (cont.)

495.	P. 119		
	Aeroplane, Ag 31, '51:241	1	
	Avn Wk, N 12, '51:33	1	
	Flight, Ag 31, '51:245	1	
496.	P. 120		
	Modification of P. 111		
	RAF Fly Rev, Jl/Ag '59:11	1	

BRANTLY USA

497.	B 1		
	Aero Digest, S '47:62-63+	1	
498.	B 2		
	Aeroplane, N 29, '62:30	1	
	Avn Wk, My 16, '60:111-121	4	C
	Can Avn, Mr '61:28-29	4	
	Flying, Ja '61:52-53+	4	C
	Jl '62:28-29+	4	C
	D '65:40-44 (-B)	4	
499.	305		
	Avn Wk, My 31, '65:88-91	1	
	Flying, D '65:40-44	4	
	F '67:34-35	1	P*

BREDA Italy

500.	B.P. 471		
	Avn Age, D '51:34-35	1	C, S*
501.	19		
	Flight, F 13, '31:135	1	
502.	25		
	Flight, Ja 5, '33:15	1	P
503.	27		
	Operational with one Chinese Air Force squadron		
	Flight, S 6, '34:925	1	
	RAF Fly Rev, O '52:6+	1	
504.	28		
	Flight, Je 14, '34:594	1	

BREDA (cont.)

505.	32		
	Acft Eng, S '32:231-232	1	
	Flight, Mr 25, '32:258-260 (Pt 1)	1	C
	Ap 1, '32:286-287 (Pt 2)	3	
506.	33		
	Flight, Ag 19, '32:784	1	
507.	64		
	Flight, N 21, '35:545	1	
508.	65		
	Flight, Jl 4, '40:12-14	1	S
	Pop Avn, F '39:35		P
	RAF Fly Rev, My '58:11	1	
509.	75		
	RAF Fly Rev, Ja '62:47	1	
510.	88 "Lince" (Lynx)		
	Aeroplane, O 26, '39:519	1	
	Flight, Jl 4, '40:12-14	1	D, S
	Ag 12, '43:177	1	
	RAF Fly Rev, N '60:11	1	

BREDA-ZAPPATA Italy

511.	B. Z. 308		
	Aeroplane, Jl 9, '48:54-56	1	
	Air Pict, Ag '68:297	1	
	Interavia, D '48:670-673	1	P

BREGUET France

512.	14		
	Avn Wk, Mr 13, '61:224-225 (-A2)		D*
	Cr & Cockade, Spg '62:15-16	4	
	Aut '66:232-236	1	P
	Fly Rev Intl, O '68:51 (-T bis)		P
	Je '69:72 (-A2:Greece)		Pr*
	Pop Avn, Mr '32:169-170		Pr

BREGUET (cont.)

513.	19 "Grand Raid"		
	Air Prog, D '68:42 ("Question Mark")		Pr*
	Flight, S 12, '30:1015-1017 ("Question Mark")	1	C
	Fly Rev Intl, Je '69:72 (-B2: Greece)		Pr*
	Pop Avn, D '30:39 ("Question Mark")	1	
514.	27 "Tout Acier"		
	See also 413		
	Acft Eng, Jl '29:181	1	
515.	41		
	Flight, F 22, '34:179-180	1	
	RAF Qtr, Ja '35:18-19	1	S*
516.	76 "Deux-Ponts"		
	See also 761, 763, and 765		
	Avn Wk, Ag 27, '51:30	1	
	Jl 7, '52:21-28	1	
517.	96 "Vultur"		
	Avn Wk, Ag 27, '51:30	1	
518.	121 "Jaguar"		
	Avn Wk, Jl 12, '65:65	1	
519.	124		
	Interavia, No. 12, '65:1954-1955	1	
520.	270		
	Acft Eng, S '30:219-222	1	
	Pop Fly, D '30:20	1	
521.	330		
	Flight, F 5, '32:107 (Codos & Robida)	4	
522.	410		
	Acft Eng, My '32:115-116	1	
523.	413 "Tout Acier"		
	See also 27		
	Acft Eng, Ap '35:86-87	3	
524.	482		
	RAF Fly Rev, S '58:20+	1	

BREGUET (cont.)

525. 521 "Bizerte"
Developed from SHORT "Calcutta"
Acft Eng, Jl '34:183-184 1
RAF Fly Rev, Je '62:47 1

526. 530 "Saïgon"
Flight, O 18, '34:1102 1

527. 670 "Breguet-Wibault"
Flight, Ap 25, '35:443 1

528. 690 Series
Flight, My 2, '40:400c-401 (690) 1 D
Fly Rev Intl, Jl '67:740 (691, 693, 695) Pr*
RAF Fly Rev, S '59:19 (690-698) 1
My '63:33, 35-36 (Series) 1
:34 (691AB2) Pr*

529. 761 "Deux-Ponts"
See also 76 and 763
Interavia, No. 9, '47:36-38 1 C

530. 763 "Provence" and "Deux-Ponts"
See also 76 and 761
Aeroplane, Je 25, '54:826-827 1
Air Pict, F '52:48 1 P
S '54:287 4
Flight, Ap 10, '53:458-461 4

531. 940 "Integral"
Aeroplane, Je 6, '58:793-796 1 S*
Air Pict, Je '59:221-224 1 S*
Air Prog, Fall '59:8 S*
Flight, My 30, '58:730 1
Jl 18, '58:84 Pr

532. 941
US licensee: MCDONNELL DOUGLAS 188
Aeroplane, Ag 5, '65:15-17 1
Air Pict, Je '59:221-224 1
O '63:332-334 4
Air Prog, Ag/S '64:46-47 1 C

BREGUET (cont.)

532.	941 (cont.)		
	Avn Wk, Je 12, '61:103-107	1	
	Jl 15, '63:73-80	1	
	Jl 29, '63:42-44	4	
	Flight, My 30, '58:730	1	
	Fly Rev Intl, N '63:2		P*
	:23-26	1	Pr, S*
	D '69:38-39	4	
	Interavia, No. 2, '60:221-222	1	S*
	No. 11, '63:1756-1760	1	C*
	Sp/Aeron, F '61:51-54	1	S*
533.	942		
	Avn Wk, Je 12, '61:103-107	1	
534.	944		
	Avn Wk, F 8, '65:77-78	1	
535.	960 "Vultur"		
	Later development: 1050 "Alize"		
	Flight, My 15, '53:606-608	1	
536.	1001 "Taon" (Gnat)		
	Avn Age, D '57:26-33	1	S
	Avn Wk, O 28, '57:50-55	1	S
	RAF Fly Rev, N '57:30-33+	1	Pr, S*
537.	1003 "Taon" (Gnat)		
	RAF Fly Rev, S '58:40-42	1	
538.	1050 "Alize" (Trade-wind)		
	Development of 960 "Vultur"		
	Aeroplane, Mr 22, '57:418	1	S*
	Je 27, '58:898-901	1	C*, S*
	Air Pict, My '59:160-163	1	Pr
	Avn Wk, Mr 25, '57:37	1	P
	RAF Fly Rev, Ja '58:26-29	1	Pr, S*
539.	1100		
	Aeroplane, Je 14, '57:853	1	
	RAF Fly Rev, Ag '57:16		Pr*

BREGUET (cont.)

540. 1150 "Atlantic"
Aeroplane, My 18, '61:553-554 (-P) 1
Air Pict, Ja '62:22-23 1
Avn Wk, N 27, '61:52-53 1
Je 10, '63:72-77 1
Je 14, '65:240-244 1
Flight, N 9, '61:723-724 1
Flying, Je '63:24-25+ 1
Fly Rev Intl, Jl '65:29-37 1 Pr
Interavia, No. 12, '61:1684-1685 1
No. 1, '69:63-67 1 C, P, S*
RAF Fly Rev, Ap '60:19-20 1
F '63:26-29+ 1 Pr, S*
RAFs Qtr, Vol. 3, No. 3:166 P

BREGUET/BAC Anglo-French

541. "Jaguar"
See SEPECAT "Jaguar"

BREGUET-DORAND France

542. "Gyroplane"
Flight, D 19, '35:Supp:c-d 2
Pop Avn, My '37:25 1

BREGUET-RICHET France

543. "Gyroplane"
Air Pict, Mr '67:97 1

BREWSTER USA

544. XA-32
Air Pict, My '62:149 1
Fly Rev Intl, Ap '66:500-501 1
RAF Fly Rev, Ja '61:40 1

545. F2A-Series
See also 139, 239, 339, and 439
AAHSJ, Fall '69:153-158 4

BREWSTER (cont.)

546. F2A-1
See also 239

Flight, Ag 8, '40:102a (X-)	1	
Fly Rev Intl, Ap '65:52		Pr*
RAF Fly Rev, O '58:11+	1	
Ja '59:6+	1	
Ap '60:16-17	1	

547. F2A-2 "Buffalo" or "Peanut Special"
See also 339

Aero Digest, S '41:185-186	1	
Air Prog, N '66:48		Pr†
Acft Eng, Ja '41:15		P
AAHSJ, Fall '69:218		Pr†
Aviation, Ag '41:106+	1	
Comm Avn, D '41:42+	1	
Flight, F 18, '43:176b	1	
RAF Fly Rev, O '58:11+	1	
Ap '60:16 (Netherland Indies AAC)		Pr*
:17-18	1	
West Fly, Ag '41:71		P*

548. F2A-3 "Buffalo" or "Peanut Special"
See also 439

Fly Rev Intl, Ap '65:52		Pr*
RAF Fly Rev, O '58:11+	1	
Ap '60:16-17	1	

549. SB2A "Buccaneer"
See also 340 "Bermuda"

Air BP, No. 40:16 (-4)		C
Air Prog, My '43:43 (-1)		P
Air Tech, Mr '43:28 (-2)		P
Flight, Ja 7, '42:12b	1	
Je 18, '42:622 (-1)	1	
Ag 22, '46:192a-b	4	
Fly Rev Intl, Mr '66:420	1	
RAF Fly Rev, Jl '58:42	1	

550. XSBA-1

Pop Avn, S '37:22	1	
West Fly, D '37:34	1	

BREWSTER (cont.)

551. 138
Aero Digest, O '38:71-72 1

552. 139, 239, 339, and 439 "Buffalo"
See also F2A
Fly Rev Intl, O '63:39-40 (Luukkanen's-)4
Ap '65:52 (-239&-339 D&E) Pr*
D '68:51 (-239:Finnish) Pr*
RAF Fly Rev, S '55:22+ 1
US Air Svcs, Ag '41:34+ (-339) 4

553. 340 "Bermuda"
See also SB2A "Buccaneer"
Fly Rev Intl, Mr '66:420 1
RAF Fly Rev, Jl '58:42 1

BRISTOL GB
Aero Album, Spg '69:33-38 (Early racers)
Air Pict, Je '50:180-181 (40th Anniversary)
Am Helicop, S '55:6-7+ (Helicopter history)
Comm Avn, D '44:96-106
Flight, Mr 2, '39:212-218
F 15, '40:149-152
F 16, '50:213-232 (Photos of aircraft and engines)
F 12, '60:205-212 (50th Anniversary)
Pop Avn, F '38:38-42+ (History and aircraft types)
RAF Fly Rev, F '60:20-25 (50th Anniversary - Many photos)
My '60:9+ (50th Anniversary - Many photos)

554. "Scout"
Air Pict, Mr '60:89+ (-F) 1
Avn Wk, My 27, '63:58-59 1 D*
Flight, S 26, '58:525-528 (Pt 1) 1
O 3, '58:544-546 (Pt 2) 1
O 17, '58:644-645 (Pt 3) 1
O 24, '58:670-671 (Pt 4) 1
O 31, '58:701-702 (Pt 5) 1
RAF Qtr, Ap '39:Plate (-D) D

BRISTOL (cont.)

555.	14 to 17 "Fighter" or "Brisfit"		
	Air Pict, D '68:462-464	1	C*
	AAHSJ, Fall '60:202-208	1	
	Win '61:Supp		P*
	Spg '67:80		D
	Avn Wk, Ja 25, '60:84-85	1	D*
	Flight, Je 27, '30:704		P
	N 7, '52:587-591	1	
	Je 6, '69:856b	1	P*
	O 23, '69:36		P*
	Flying, Mr '61:53 (McKeever's-)	1	D
	Fly Rev Intl, My '68:265 (139 Sq)		Pr*
	Pop Avn, F '34:97	1	
	Pop Fly, Je '32:160-161		D
	O '32:383	1	D
	RAF Qtr, Ap '37:Supp	1	D
	Ap '39:Plate		D
556.	20 M. 1 Monoplane Scout		
	Aero Album, Spg '69:38 (-D)	1	P
	Air Pict, Jl '57:237 (-A to -D)	1	
	Pop Fly, D '33:492 (-C)	1	D
	RAF Qtr, Jl '39:Plate		D
557.	24 and 25 "Braemar"		
	Flight, Mr 5, '42:209	1	
	RAF Fly Rev, My '59:42 (-I&-II)	1	
558.	26 and 33 "Pullman" Development of BRISTOL "Braemar"		
	Flight, Mr 5, '42:209	1	
	RAF Fly Rev, My '59:42	1	
559.	32 "Bullet"		
	Aero Album, Spg '68:21		P
	Spg '69:34-35	1	P
	Pop Avn, Je '38:58 ("Pick-a-back"-)	1	
560.	37 "Tramp"		
	RAF Fly Rev, My '59:42	1	
561.	72 "Racer"		
	Aero Album, Spg '69:36-37	1	P
	Flight, Mr 26, '42:286-287	1	
	RAF Fly Rev, Mr '58:6	1	

BRISTOL (cont.)

562. 73 "Taxiplane"
Air Pict, Ja '64:25 1

563. 105 "Bulldog"
Aeronautics, Jl '41:34 C*
Air Prog, Ap '65:2 P
:20 (-IIA) Pr†
Bris Sid Jnl, Aut '61:28 (Restored-) 1 P
Flight, Je 27, '30:696 P
Fly Rev Intl, D '64:19 P
My '68:265 (-IIA:3 Sq) Pr*
S '68:62 (Denmark-) Pr*
D '68:50 (-IVA:Finland) Pr*
Pop Fly, S '33:320 D
F '38:614 (Sweden-) P
RAF Fly Rev, Ag '60:11+ (All) 1
Ja '61:44 (-I) 1
Ja '62:22 1
:23 (-IIA:56 Sq Rebuilt) P*
RAF Qtr, Jl '32:Plate 3 (-IIIA) P

564. 107 "Bullpup"
Flight, D 19, '30:1462-1463 1

565. 118 General Purpose Biplane
Aero Digest, Ja '32:66-67 1
Aeroplane, D 2, '31:1280-1282 1
Flight, D 4, '31:1185-1187 1 Pr

566. 120
RAF Qtr, Jl '32:Plate 9 P

567. 123 Biplane Fighter
Air Ministry specification: F./30
Air Pict, Je '58:206-208 1
RAF Fly Rev, Je '59:9+ 1

568. 130 "Bombay"
Aeroplane, Ag 2, '39:157-160 1 S*
Air Pict, Jl '64:210-213 1 Pr
Flight, Ag 3, '39:103-106 1 C
Pop Avn, F '38:43 P
Pop Fly, Jl '39:154-155 1

BRISTOL (cont.)

569. 133 Monoplane Fighter
Air Ministry specification: F.7/30
Air Pict, Je '58:206-208 1
Flight, F 28, '35:221 1
RAF Fly Rev, Je '59:11 1

570. 138A High Altitude Monoplane
Acft Eng, D '36:327-330 1
Flight, O 1, '36:338-340+ 1 S*
O 8, '36:360a-b 4
Pop Fly, N '36:391 4
RAF Qtr, Ap '37:173-177 1 C*

571. 142M "Blenheim"
Canadian-built: FAIRCHILD, LTD. "Bolingbroke"
Aero Album, Win '69:17 (Finland-) P
Aeronautics, F '40:18-19 S*
Aeroplane, Je 16, '37:734 1 S*
Jl 30, '43:128-129 P
Ag 27, '43:239-245 1 P
Air BP, No. 15:2-5 4
:4 (-I) C
Air Pict, Jl '52:201-203 1
Comm Avn, Jl '39:25 1
Flight, Je 24, '37:621-625 4 C, S*
My 19, '38:502a-f 3
D 10, '42:632a (-IV&-IVF) 1
Fly Rev Intl, Je '66:636-639 (-I) 1 Pr*
My '68:268 (-I:90 Sq) Pr*
Je '69:73 (-IV:Greece) Pr*
Pop Avn, O '37:29 1
Pop Fly, N '37:461 1
RAF Fly Rev, My '54:22 1
US Air Svcs, O '40:21 1

572. 148 Army Co-operation Monoplane
Air Ministry specification: A.39/34
RAF Fly Rev, My '58:11+ 1

573. 152 "Beaufort"
Aeroplane, F 12, '43:184-185 (-I) S*
Air Pict, Jl '52:201-203 1
Air Prog, Je '43:14 D
Acft Eng, Jl '41:188-189 1 P
S '41:253-254 1 P

BRISTOL (cont.)

573.	152 (cont.)		
	Can Avn, Ag '41:29+	1	
	Flight, Je 12, '41:402a-d	1	
	N 5, '42:498a (-II)	1	
	Ag 1, '46:110-111	4	
	Flying, S '42:114		P*
	Roundel, Mr '65:33 (RCAF-)	1	P
	RAF Fly Rev, F '56:13+	1	
574.	156 "Beaufighter"		
	Aero Album, Fall '69:30-32 (RAAF-)	4	P
	Aero Digest, D '43:172-173		S
	Aeroplane, Ag 8, '41:139-140	1	
	Ag 22, '41:208-209	3	
	S 12, '41:282-283 (-I)		S*
	Jl 31, '42:128-129		C*, S*
	S 3, '43:270-271		P
	Air BP, No. 16:13-16 (-VI)	4	C, P
	No. 41:16-19	4	C, P
	AF News Ltr, O '41:25-26+	1	
	Air News, Ap '43:29		D
	Air Pict, Ag '52:232-235	1	
	Air Prog, Jl '42:46-47		S*
	Je '43:15		D
	Jl '65:50-51 (-I)	1	S*
	Air Tech, S '43:26-28	3	S*
	D '44:26-27+	1	S
	Acft Eng, S '41:255-256	1	P
	Aviation, S '41:93+	1	
	D '41:121		S*
	Ja '42:98-99		S
	Mr '44:117-137		C, S*
	Can Avn, N '41:22-23+	1	S*
	CAHS, Win '65:112	1	
	Flight, Ag 14, '41:90-90d	1	S
	S 24, '42:336a (-I)	1	
	S 2, '43:253-255 (Torpedo-)	1	
	N 2, '44:468-471 (Rocket-)	4	
	Je 14, '45:634-637	4	
	Flying, N '41:21	1	
	Fly Rev Intl, My '68:272 (-IF:604 Sq)		Pr*
	RAF Fly Rev, Je '55:9 (-II)	1	
	N '60:57	4	
	West Fly, Mr '43:56+	1	

BRISTOL (cont.)

575.	163 "Buckingham"			
	Aeroplane, D 14, '45:689-691 (-I)	1		
	Air Pict, Ag '52:232-235	1		
	Mr '60:72-73	1		
	Flight, D 13, '45:623-625+ (-I)	1		S
	RAF Fly Rev, F '59:42	1		
576.	164 "Brigand"			
	Aeroplane, D 14, '45:689-693 (-I)	1		S*
	My 20, '49:578-581 (-III)	4		C
	Air Pict, Ag '52:232-235	1		
	Mr '60:74-78	1		S*
	Flight, D 13, '45:623-625+	1		S
	Ap 29, '48:467 (-B.1)	1		
577.	166 "Buckmaster"			
	Air Pict, Ag '52:232-235	1		
	Mr '60:78	1		
	Flight, O 18, '45:420-421	1		
	RAF Fly Rev, F '59:42	1		
578.	167 "Brabazon"			
	Aero Digest, N '49:17+	1		
	Aeronautics, N '49:43-74	1		
	Aeroplane, Jl 26, '46:100-105	1		S*
	S 23, '49:425-441	1		S*
	O 26, '51:557			S*
	Acft Eng, O '49:322-324	6	(Rear Spar)	
	My '51:124-133	6	(Engine)	S
	Am Helicop, S '49:9+	1		
	Avn Wk, Je 27, '49:20-22	1		
	Ag 15, '49:21-29	6	(Engine)	
	Can Avn, O '49:24+	1		
	Comm Avn, Ag '45:106-107	1		
	Flight, D 26, '46:693-698	6	(Engine & & Props)	S*
	Mr 27, '47:262-266 (Pt 1)	3		S
	Ap 3, '47:288-290 (Pt 2)	3		S
	Jl 24, '47:84a-d	1		S*
	O 16, '47:430-432	1		
	Ja 6, '49:11-13	1		
	F 10, '49:175-177	6	(Ground Equip)	
	S 8, '49:277-278	2		
	S 29, '49:407-432	1		C, S*
	Ag 3, '51:148-150	1		
	Flying, Mr '64:44-45+	1		

BRISTOL (cont.)

578.	167 (cont.)		
	Interavia, No. 10, '47:44-46	6 (Prop Drive)	S
	US Air Svcs, Mr '50:22	6 (Press)	
579.	170 "Wayfarer" and "Freighter"		
	Aero Digest, Ap '46:54-55+	1	
	Je '52:40-41		P
	Aeroplane, Mr 8, '46:288-291	1	
	Mr 15, '46:314-319	1	C, S*
	F 20, '48:220-222	1	
	:222-224	4	
	S 7, '51:287		P
	O 2, '53:483-485 (-32)	4	C
	Am Helicop, My '46:31	1	
	Avn Wk, D 31, '51:20-23	1	
	N 3, '52:38 (-32)	1	
	O 19, '53:76-81	4	
	Ap 5, '54:66-79	4	
	Can Avn, Ap '46:38-39	1	
	Mr '53:36-37+	4	
	Flight, N 23, '44:555-556	1	
	Mr 7, '46:235-240	1	C, S
	My 16, '46:488-490	4	C
	O 17, '46:411-414	4	C
	F 19, '48:205-208	1	S*
	My 27, '48:586-587 (Military-)	1	
	N 24, '49:689	1	
580.	171 "Sycamore" Helicopter		
	Aeroplane, O 31, '47:583-587	1	S*
	D 17, '54:887-888		S*
	Jl 1, '55:13-18	4	C, S*
	Air Pict, Ap '51:103 (-III)	1	
	Ja '69:4-8	1	
	Acft Eng, Mr '51:67-71 (-I to -III)	1	S*
	Am Helicop, Ap '47:17-18	1	
	Ag '48:9-10+	1	
	N '51:12-13+ (-III)	1	S*
	Mr '56:4	1	
	Avn Wk, D 29, '47:22-27	1	S*
	My 27, '52:38 (-10)	1	S*
	Can Avn, Mr '53:110 (-50)	1	
	Flight, O 30, '47:494a-496	1	S*
	F 10, '49:166-170	6 (Engine)	S*
	Jl 4, '52:17-19	4	C*, S
	Ja 16, '53:58-59 (-1 to -3 & -50)	1	

BRISTOL (cont.)

580. 171 (cont.)

Flight, Ja 23, '53:104-105 (-50)	1	S*
O 23, '53:559-560 (-4)	4	

581. 173 Helicopter

Aeronautics, Mr '53:34-35	1	P
Ap '55:51 (-3)	1	
Aeroplane, Mr 16, '51:321-328	1	S*
Air Pict, Jl '51:221	1	
Ja '54:29	2	
Am Helicop, Mr '51:14-15	1	
Ja '52:13	1	
F '54:12 (-1)	2	
Ap '56:6-7+	2	C
D '56:20	2	
Mr '57:10-13	1	
Avn Age, O '55:32-33	1	S*
Avn Wk, My 7, '51:21-25	1	S*
Flight, Mr 16, '51:311-319	1	C*, S*
S 28, '51:416	1	
Ja 11, '52:33-34	4	
Jl 4, '52:16-17	1	
Ja 16, '53:58-59 (-2&-3)	1	
Ja 23, '53:104-105	1	S
F 25, '55:234-235 (-1 to -3)	1	

582. 175 "Britannia"

Canadian development: CANADAIR CL-28

Aeronautics, Jl '54:60-77	1	S
Ag '57:26-32	1	
Aeroplane, F 29, '52:249-250	1	
Je 6, '52:698-700	1	
Ag 22, '52:243-254	1	S*
F 17, '56:188-191	4	C
F 1, '57:141-179	1	C, P*, S*
Air BP, No. 5:19 (-102:BOAC)		Pr*
No. 23:16 (-310:Britain United)		Pr*
Air Pict, Ap '52:100-101	2	
Jl '55:208-209 (-100)		S*
D '69:442-447	1	
Avn Age, Ap '55:32-33	1	
N '55:178-185	4	
F '57:30-35	1	S*

BRISTOL (cont.)

582. 175 (cont.)

Avn Wk, Mr 24, '52:21-29 1
S 1, '52:30-33 2
O 25, '54:13-15 1
F 27, '56:53 C
My 28, '56:44-55 1 C
Bris Sid Jnl, Win '60/'61:44-49
(-102&-312) 6 (Engine)
Can Avn, Ap '53:64-67 1 P*
D '55:30-31 (-300LR) 1 C
Mr '56:38-42 4
N '56:42-43 4 C
Flight, Ja 25, '52:86-88 1 C
F 29, '52:236-237 2
Mr 7, '52:268-271 (Proto-) 1
Ag 8, '52:149-150 2
Ag 22, '52:209-214 1 S*
F 6, '53:158-159 (Proto-) 1
Ja 29, '54:119 (Proto-) 4
Ag 13, '54:215-216 (-250&-300) 1
O 15, '54:572-573+ 1
Ja 21, '55:68-69 (Proto-) 1
F 11, '55:173-174 3
Ag 12, '55:223-232 1 S*
N 11, '55:734-735+ (-100&-300) 1
Ja 6, '56:4 4
Mr 30, '56:353-354 4
Jl 6, '56:23-26 1 S*
F 1, '57:133-138, 143-147 1
:139-142 4
:147-148 6 (Engine)
:150-153 6 (Flt Deck) C*
N 1, '57:685-688 (-310) 1 S*
Ja 31, '58:135-137 (-102) 1
Jl 25, '58:116-118 (-310&-320) 1 S*
D 19, '58:951-952 (Pt 1) 2
D 26, '58:983-984 (Pt 2) 2
Ag 14, '59:7-8 (RAF-) 1 S
Interavia, No. 2, '53:78-80 1
West Avn, F '56:38 (-100) 1

583. 188 Hypersonic Research Aircraft

Aeroplane, Ag 19, '60:216-219 3
My 11, '61:517-521 1 S*
Air Prog, Je/Jl '63:50-51 Pr†

BRISTOL (cont.)

583.	188 (cont.)			
	Avn Wk, O 10, '60:98-99	1		
	Je 12, '61:93-100	3		
	Ap 23, '62:29	1		
	Ap 30, '62:58-63	1		
	Flight, D 11, '59:707-708	1		
	My 3, '62:693-703	1		S*
	Interavia, No. 1, '60:102	1		
	No. 9, '60:1159	3		
	No. 6, '61:804-805	1		
	RAF Fly Rev, F '60:13-14	1		
	Jl '62:15-16+	1		
584.	192 Helicopter			
	Derivative of 173			
	Aeroplane, Ag 1, '58:167-171	1		C, S
	Am Helicop, Ag '56:7	6	(Engine)	
	N '57:10-11	1		
	Avn Wk, Ag 18, '58:64-70	6	(Engine)	S
	Flight, Ja 18, '57:69-71	1		S*
	Ag 1, '58:169-173	1		C, S*
585.	221			
	FAIREY F.D. 2 with ogee wing			
	Avn Wk, N 11, '63:64	1		

BRISTOL AERONAUTICAL USA

586.	XLRQ-1 Transport Glider		
	Fly Rev Intl, D '63:61	1	

BRITTEN-NORMAN GB

587.	BN-2 "Islander"		
	Aeroplane, Ag 23, '67:30-31	4	
	S 13, '67:20-28	1	
	S 18, '68:39 (-A)	1	
	Air Pict, Mr '68:97-100	4	C*
	Air Prog, Ap '68:12, 42-43		C, S*
	Avn Wk, S 5, '66:111-112	3	C
	Jl 1, '68:76-78	1	C
	Can Avn, Je '68:14-15+	4	
	Flight, Je 24, '65:1017-1021+	1	C*, S*
	S 30, '65:575-576	2	
	Mr 30, '67:484-485	3	
	Ag 24, '67:295-297	1	S*
	:298-300	4	

BRITTEN-NORMAN (cont.)

587. BN-2 (cont.)

Flying, Ap '68:74-77	1	
Fly Rev Intl, Ja '68:27-31	1	C
Interavia, No. 3, '66:316	1	S
No. 1, '68:44-46	1	C

BROWN USA

Air Prog, O/N '64:37-40+ (With aircraft photos)

BÜCKER Germany

Air Prog, Ag/S '63:60-69+ (With aircraft photos)

588. Bü-131 "Jungmann" (Young Man, Cadet)

Aeroplane, My 9, '34:781-782	1	
Flying, Ap '69:68-71	4	

589. Bü-133 "Jungmeister" (Young Master)

Air Prog, Ag/S '63:60-69+	1	P, Pr†
Flying, Ap '68:90-94+	1	C, P*
Fly Rev Intl, N '64:57	1	

590. Bü-181 "Bestmann" (Best Man)

Aeroplane, N 16, '56:717-718 (-D)	4	
Flight, Jl 29, '43:118b	1	
Fly Rev Intl, Ja '65:51+	1	

BUDD USA

591. BB-1

Original design: SAVOIA-MARCHETTI S. 31

Air Pict, Ap '64:119	1	
Acft Eng, F '32:38+	1	
Flight, D 8, '32:1169-1170	1	
Ja 26, '33:76	1	
US Air Svcs, D '31:48	1	

592. RB-1 "Conestoga"

Aero Digest, My 15, '44:75-77+	1	
Je 1, '44:80-83	3	S*
Je 15, '44:92-95		C*, S*
Air Pilot, Je '44:24-25	1	
AAHSJ, Spg '67:54-61	1	C
Aviation, Je '44:186-187	1	
Comm Avn, Jl '44:74-76+	1	
Flight, Jl 20, '44:74-75	1	
Fly Rev Intl, Jl '68:413	1	

BUDD (cont.)

592.	RB-1 (cont.)		
	Nav Avn News, Je 15, '44:38	1	
	RAF Fly Rev, Ja '61:44	1	
	West Fly, Je '44:50-51+	1	

BUHL USA

Air Prog, F '68:56-59+

593.	CA-1 "Airster"		
	Aviation, O '30:232-240	1	

BURGESS COMPANY, LTD. GB

AAHSJ, Sum '65:79-87 (Pt 1)
Win '65:241-249 (Pt 2)
Win '66:270-278 (Pt 3)
Win '67:273-280 (Pt 4)
Spg '68:37-42 (Pt 5)
Pop Avn, Ja '38:44-46+

BURNELLI USA

594.	UB-14		
	Aero Digest, Ag '35:38+	1	
	Air News, Mr '42:23-27	1	S*
	Can Avn, F '44:72+	1	
	Flight, N 21, '35:544-545 (-A)	1	
	O 22, '36:422	1	
	Pop Fly, N '36:404-405+	1	
	US Air Svcs, N '44:48	1	
	West Fly, D '44:156	1	
595.	UB-20		
	Aero Digest, S '30:108-110	1	
	RAF Fly Rev, Mr '61:11	1	

BUTLER USA

596.	"Blackhawk"		
	Hist Avn, Ag '69:4-7	1	P

CAC (Central Aircraft Company) GB

597.	C/F 2A "Centaur 2A"		
	Air Pict, F '57:44	1	

CAMS (Chantiers Aéro-Maritimes de la Seine) France

598. 55 Flying Boat
Aero Album, Fall '68:31 (3d Exploration Sq) P

599. 58/2 Flying Boat
Flight, Ap 13, '33:351-352 1

CANADAIR Canada
Interavia, No. 9, '51:506-508 (Activities in 1951)
Roundel, D '58:8-11

600. C-4 "Argonaut"
See also DC-4M
Aero Digest, Je '48:26-27+ 1
Aeroplane, F 27, '53:262-265 4 C

601. CC-106 "Yukon"
Commercial designation: CL-44
Roundel, Ja/F '60:4-8 1
O '61:18-21 2 C
RAF Fly Rev, F '60:18-19 1
Mr '61:34-37 1 Pr, S*

602. CC-109
Commercial designation: 540
Roundel, Ja/F '60:4-8 1

603. CF-104
RCAF LOCKHEED F-104G
See also CL-90
Air Pict, N '61:341-342 4
Can Avn, F '60:20-21 1
Ag '61:24-27 6 (Navigation)
Mr '62:8-12 3
D '62:29-30 4
Fly Rev Intl, Ag '64:11-13+ 1 S*
Roundel, O '59:4-7 (-111) 1
My '61:33 3
S '61:5-7 4
Ap '63:2-7 6 (Ejection Seat)
Sentinel, Mr '68:18-19 4

CANADAIR (cont.)

604. CL-13 "Sabre"
USA version: NORTH AMERICAN F-86 "Sabre"
Australian version: COMMONWEALTH
CA-27 "Sabre"

Air Pict, F '54:40 (-3)	6 (Wing)		
Can Avn, F '49:26+	1		
F '50:18-20+	3		
S '50:22-23	1		
My '53:38 (-D&-F)	1		
N '58:54-55	1		
Flight, Ja 30, '53:136-146	1		C*, S*
:146-147	4		
Fly Rev Intl, O '64:94 (Acrobatic teams)			Pr*
Je '69:73 (-E:Greece Aero Team)			Pr*
Roundel, D '59:4 ("Golden Hawks"-)			P
RAF Fly Rev, Mr '56:44-45	1		
Ap '57:16 (-E:Italy)			Pr*
Mr '59:8-10	1		
Sentinel, F '69:14-16	1		

605. CL-21

Flight, Je 13, '52:704	1		
Interavia, No. 9, '52:490	1		

606. CL-28 "Argus"
Adapted from BRISTOL "Britannia"
RCAF designation: CP-107

Avn Wk, Je 7, '54:48-58	3		
D 31, '56:50-55	6	(Hydraulic)	
Ja 21, '57:85-86	6	(Hydraulic)	
F 11, '57:26-28	1		
Ap 22, '57:28-29	1		
Jl 29, '57:50-54	2		
Can Avn, Mr '55:21-23+	1		
F '57:58-61+	1		
:62-63+	6	(Hydraulic)	
Ag '58:48-49			C
S '59:32-35	4		C
:36-38	6	(Navigation)	S*
Flight, Ap 1, '55:409	1		
Ja 25, '57:105-106	1		
O 4, '57:543-549	1		S*
N 27, '59:633-634	1		
Interavia, No. 3, '63:328-329	1		
Nav Avn News, Je '60:20			P

CANADAIR (cont.)

606. CL-28 (cont.)

Roundel,	My '58:2-5+	1	
RAF Fly Rev,	N '59:14-15	1	P*
	My '60:24-27 (Mk. 2)	1	Pr, S*
Sentinel,	My '68:28-31	4	C

607. CL-30 "Silver Star"
Canadian-built LOCKHEED T-33

Fly Rev Intl,	Ja '68:50 ("Golden Hawks," "Red Knights," No. 1 Air Div)		Pr*

608. CL-41 "Tutor"

Aeroplane,	S 7, '61:332-334	4	C
Air Prog,	Fall '60:74-77		C, S*
	Ag/S '65:34-35 (-A&-G)	1	
Avn Wk,	N 25, '57:81-82	1	
	N 23, '59:79-90	1	C
	D 11, '61:54-63	4	C
	Mr 25, '63:61-67	1	C
	Ja 4, '65:66-67	1	
Can Avn,	D '57:46-47	1	
	F '60:40-41+	1	S*
	Jl '61:20-24	4	S*
	F '65:30+	4	
	F '66:20-21	4	
Flight,	Ja 29, '60:142-143	1	S*
	S 7, '61:391-392 (-A)	4	C
Flying,	N '61:50-51+	4	C
Interavia,	No. 11, '61:1518-1519 (-R)	1	C*, S*
	No. 6, '63:854-857 (-A, -G, -R, -K)	1	C*, S*
Nav Avn News,	My '65:13 (-G)	1	
Roundel,	D '61:3	1	
	Ja/F '63:12-15 (-A)	1	S*
	My '64:16-17	3	
	S '64:10-13	2	C
RAF Fly Rev,	Ja '60:26-29	1	Pr, S*
	D '61:42-43	1	
	Jl '63:26-28+ (-A&-R)	1	S*

609. CL-44
Military designation: CC-106 "Yukon"
Conversion: CONROY CL-44-0 "Airlifter"

Aeroplane,	N 27, '59:537-541	1	S*
Air BP,	No. 25:17 (Flying Tiger Line-)		Pr*
	No. 42:8 (Loftleider-)		Pr*

CANADAIR (cont.)

609. CL-44 (cont.)

	Air Prog, Fall '60:72-73 (-D4)		S*
	Avn Wk, Ap 22, '57:28-29	1	
	N 11, '57:48-51	1	
	N 2, '59:52-74 (-D-4)	1	
	Mr 13, '61:271 (-D)		P*
	N 20, '61:72-83 (-D)	4	
	D 20, '65:42-45 (-J)	1	
	Can Avn, D '57:76-77 (-C)	1	
	Je '59:73-74	1	
	O '59:22-27+	1	S*
	Mr '61:8-11	4	C*
	D '62:33	4	
	Flight, Jl 25, '58:127 (-D)	1	
	D 12, '58:913-916 (-D)	1	S*
	N 20, '59:588-589 (-D-4)		S*
	Ag 26, '60:298 (-6)	1	Pr
	F 17, '61:207-208	4	
	Flying, O '61:54-55+	4	C
	Interavia, No. 8, '60:951-955 (-D-4)	1	S*
	No. 1, '62:74-75	4	
	No. 12, '62:1619-1621	4	
	West Avn, O '61:36	1	

610. CL-66 "Cosmopolitan"

	Can Avn, Mr '58:22-24+	1	

611. CL-84 "Dynavert"

Military version: CX-84

	Avn Wk, N 25, '63:70-77	1	
	My 11, '64:84	1	
	D 21, '64:20-21	1	
	Je 30, '69:19-20	2	
	Can Avn, F '64:22-23	1	S*
	F '65:33	1	
	Mr '66:25	1	P
	Ag '67:26-27	1	
	Flight, F 21, '63:264	1	S
	D 5, '63:920-921+	1	S*
	S 30, '65:582-583	1	
	Fly Rev Intl, N '64:24, 26-27	1	S*
	My '66:513-514	1	
	My '69:26-28 (-1&-1C)	1	
	Interavia, No. 2, '64:233-235	6 (Wing)	S*
	No. 7, '65:1127	1	
	No. 11, '65:1710-1711	1	S*

CANADAIR (cont.)

611. CL-84 (cont.)

Interavia, No. 1, '66:53-54	2	
Sentinel, Je '66:28-29	1	

612. CL-90

RCAF designation: CF-104

USAF designation: LOCKHEED F-104G

RAF Fly Rev, S '61:34-35	1	P*

613. CL-215

Aeroplane, F 16, '67:24-25	1	
Air Prog, F '68:42-43		S*
Avn Wk, Ag 16, '65:96-101	1	
Jl 22, '68:71	2	
S 22, '69:64-70+	4	C
Can Avn, Je '66:24-25	1	
Ag '67:27	1	
Ag '68:13-16	1	S*
Flight, F 22, '68:269-272	1	C
Fly Rev Intl, My '66:567	1	
Mr '68:127-128+	1	S*
N '68:56-57	1	
Interavia, No. 5, '66:748-749	4	
No. 1, '68:50-51	4	

614. CX-84

Civil predecessor: CL-84 "Dynavert"

Avn Wk, Ap 14, '69:98	2	
Flight, Jl 31, '69:182	1	
Interavia, No. 7, '69:906-908	1	S*
Sentinel, Ap '69:10-13	1	

615. DC-4M "Four" or "North Star"

See also C-4 "Argonaut"

Aeroplane, Jl 30, '48:132-134	1	S*
Mr 4, '49:246-249	1	S*
Can Avn, Ja '47:40-42+	4	
:43-48+	1	
:49+	6 (Electrical)	C
:50-51	6 (Engine)	S*
D '47:40-41+ (-1&-2)	1	C
O '50:20+	6 (Press)	S*
Flight, Je 26, '47:609-610 (-1&-4)	1	
Ap 7, '49:391-393 (BOAC-)	1	C
F 29, '52:242-243	6 (Engine)	S
Interavia, Jl '48:374-375	1	

CANADAIR (cont.)

615. DC-4M (cont.)
Roundel, Jl/Ag '51:38-42 ("Ice Wagon") 2
Ja/F '60:9 4
Sentinel, Ja/F '66:20-23 4
Skyways, D '51:43 C

616. 540
Development of CONVAIR 440 "Metropolitan"
RCAF designation: CC-109
Air Prog, Sum '61:7 S*
Avn Wk, Ap 20, '59:103-111 1
Can Avn, Mr '59:63 4
Flying, Je '59:43+ 1
Interavia, No. 10, '59:1246 2

CANADIAN AERODROME Canada

617. "Baddeck 1" and "Baddeck 2"
Roundel, Jl/Ag '59:22-29 (-1)(Pt 1) 1
S '59:12-16 (-1&-2) (Pt 2) 1

CANADIAN CAR AND FOUNDRY Canada

618. "Goblin I" G-23
Original design: GRUMMAN FF-1
Can Avn, D '66:18-19 1 C
CAHS, Win '65:118, 120 1 Pr†
Comm Avn, Jl '39:28 1
Fly Rev Intl, O '66:150-151 1 Pr*
Roundel, D '64:33 1 P
RAF Fly Rev, F '60:11+ 1

619. "Loadmaster" CBY-3
Aero Digest, F '46:50-51+ 1
AAHSJ, Fall '69:165 (Last-) 1
Can Avn, S '45:50-52+ 1
D '45:75 P*
Comm Avn, S '45:56+ 1
Fly Rev Intl, Ap '65:43-44 1
RAF Fly Rev, Mr '61:11 1

620. "Maple Leaf Trainer"
Can Avn, Ag '40:20-21 (-II) 1
Comm Avn, D '39:32 (Proto-) 1

CANADIAN CAR AND FOUNDRY (cont.)

621. FDB-1

	Aero Digest, Je '39:89-90	1	
	Acft Eng, Mr '40:69-72	1	S*
	Can Avn, My '39:33	1	
	Comm Avn, Jl '39:28	1	
	RAF Fly Rev, S '59:19	1	

CANADIAN VICKERS — Canada

Comm Avn, Ag '39:10-13

622. "Vancouver"

	Flight, F 6, '31:114-115 (-II)	1	
	Roundel, N '54:22	4	

623. "Vedette"

	AAHSJ, Win' 63:303 (-Va)		D
	Pop Fly, O '34:380	1	D

624. "Viking"

	CAHS, Spg '67:18-19, 30 (-IV)	1	Pr
	Sum '67:57 (-IV)	1	

CANAMERICAN — Canada

625. SG VI-E

	Am Helicop, Jl '53:6-8+	1	S*
	Ja '54:6-9+	2	

CANARY — Germany

See BÜCKER

CANT (Cantieri Riuniti Dell'Adriatico) — Italy

RAF Fly Rev, D '56:61-62 (Some WW II aircraft types)

626. Z-501 "Gabbiano" (Seagull) and "Mammaiuto" (Mamma, Help!)

	Acft Eng, F '35:45	1	
	Flight, Jl 18, '40:45-48	1	
	US Air Svcs, D '34:35	1	

627. Z-506 "Airone" (Heron)

	Acft Eng, S '37:239	1	
	Flight, Jl 18, '40:45-48	1	
	RAF Fly Rev, F '61:37	1	

CANT (cont.)

628.	Z-1007 "Alcione" (Kingfisher)		
	Flight, Je 20, '40:541	1	
	D 24, '42:688A (-bis)	1	
	RAF Fly Rev, My '55:10	1	
629.	Z-1012		
	Pop Avn, F '39:62	1	
630.	Z-1018 "Leone" (Lion)		
	RAF Fly Rev, Ag '61:45	1	P

CAO France

See SNCAO

CAPELIS USA

631.	XC-12		
	Fly Rev Intl, Ap '67:515-516	1	
	West Fly, Ag '33:20	1	

CAPRONI Italy

Cr & Cockade, Aut '65:215-227 (Bombers, with bibliography & photos)

632.	"Sauro" (Sorrel Horse)		
	Flight, Ag 17, '33:836	1	
633.	Ca 33		
	Flying, N '61:1	1	D
	Pop Avn, O '36:26		P
634.	Ca 36		
	Bomber with detachable wings		
	Avn Wk, S 24, '62:36-37	1	D*
635.	Ca 41 Triplane		
	Pop Avn, O '36:26		P
636.	Ca 60 "Capronissimo"		
	Air Pict, D '57:415	1	
	Pegasus, D '56:Supp:18		P
	RAF Fly Rev, S '63:41	1	

CAPRONI (cont.)

637. Ca 79
Acft Eng, My '29:88 1
Flight, Ja 10, '30:97-98 1

638. Ca 90
Acft Eng, Jl '32:182 1
Flight, Ja 9, '31:27-28 (-P.B.) 1
Pop Avn, Mr '31:16 (-P.B.) 1

639. Ca 101
Flight, Jl 15, '32:659-660 ("Nord Africa") 1 C
Fly Rev Intl, S '69:50 (Hungarian-) Pr*

640. Ca 113
Flight, Ap 1, '32:279 1

641. Ca 133 "Caprona" (She-goat) or "Vacca" (Cow)
Fly Rev Intl, D '69:53 (Austrian-) Pr*

642. Ca 135 "Asso" (Ace), "Tipo Peru" (Peruvian type) or "Tipo Spagna" (Spanish type)
Aeroplane, Mr 14, '41:308 (-bis) 1
Air Pict, Ja '62:8-11 1
Flight, Jl 25, '40:64a-c (-bis) 1
Fly Rev Intl, S '69:51 (-bis:Hungary) Pr*
RAF Fly Rev, F '59:22 (-bis) 1
Je '60:50 1

643. Ca 161
Held 1937 altitude record
Pop Avn, Ag '37:41 4

644. Ca 165
Can Avn, O '40:20 1
Fly Rev Intl, Je '69:100 1

645. Ca 309 "Ghibli" (Desert Wind)
RAF Fly Rev, S '62:39-40 1

646. Ca 310 "Libeccio" (Southwest Wind)
Fly Rev Intl, Ap '69:52 (Norwegian-) Pr*
RAF Fly Rev, S '62:39-40 1

CAPRONI (cont.)

647.	Ca 311		
	Flight, Ag 22, '40:144a-b	1	
	RAF Fly Rev, S '62:39-40	1	
648.	Ca 312		
	Experimental Ca 310		
	Flight, Ag 22, '40:144a-b	1	D
	RAF Fly Rev, S '62:39-40	1	
649.	Ca 313		
	Improved Ca 311		
	Air Prog, N '67:59		Pr
	Flight, Ap 8, '43:364b	1	
	RAF Fly Rev, S '62:39-40	1	
650.	Ca 314		
	Heavily armed Ca 313		
	RAF Fly Rev, S '62:39-40	1	
651.	Ca 331 "Raffica" (Squall)		
	Air Pict, Ag '60:286-287	1	
652.	Ca 355		
	RAF Fly Rev, S '61:42	1	
653.	Ca 380 "Corsaro" (Corsair)		
	RAF Fly Rev, Je '60:50	1	
654.	F-5		
	Avn Wk, D 8, '52:25-26	1	

CAPRONI-CAMPINI Italy

655.	C.C.-2 or N. 1		
	Aeroplane, D 26, '41:697	1	
	RAF Fly Rev, S '54:7-8	1	
	Ap '56:15	1	

CAPRONI REGGIANE Italy

See REGGIANE

CAPRONI-STIPA Italy

656.	Experimental Monoplane (1933)		
	Acft Eng, Ag '33:175-176	2	P
	Flight, O 20, '32:995	1	P
	Pop Avn, Ap '33:231	1	

CAPRONI-VIZZOLA Italy

657.	F. 4, F. 5, and F. 6		
	RAF Fly Rev, S '58:24+	1	

CASA (Construcciones Aeronauticas S.A.) Spain
Skyline, Vol. 14, No. 4, '56:4-9 (Work on NORTH AMERICAN F-86F)

658.	SF-5		
	Original design: NORTHROP N-156		
	Avn Wk, F 26, '68:42-47	3	
	Flight, My 30, '68:830-832	1	
659.	201 "Alcotan" (Falcon)		
	Aeroplane, Ja 13, '50:38-39	1	S
660.	202 "Halcón" (Falcon)		
	Avn Wk, Je 1, '53:24	1	C
661.	212 "Aviocar"		
	Flight, My 30, '68:832-833	1	
	Interavia, No. 7, '67:1101	1	
662.	2111 "Pedro"		
	Spanish-built HEINKEL 111		
	Flight, Ag 17, '56:247-248	4	
	N 27, '69:Inset		P*
	RAF Fly Rev, Je '59:24-26	1	

CASPAR Germany

663.	C. 32		
	Air Pict, S '65:326-327	1	

CASTAIBERT Argentina

664.	913-IV (1916)		
	Fly Rev Intl, O '68:52 (Uruguayan-)		Pr*

CASTEL-MOUBOUSSIN			France
665.	CM-10 Glider		
	Interavia, Ag '48:427-429	1	
CAUDRON			France
666.	B. (1912)		
	Aeronautics, F '46:75		C*
667.	C. 232		
	Flight, O 3, '30:1102	1	Pr
668.	C. 371		
	Aero Digest, S '39:66	1	
669.	C. 430		
	Aeroplane, My 2, '34:751	1	
670.	C. 440 and C. 445 "Goëland" (Seagull)		
	Air Pict, Ap '54:116	1	
	Flight, Je 14, '34:579	1	
	Ag 12, '43:176	1	
	RAF Fly Rev, Ja '62:45	1	
671.	C. 710, C. 720, C. 760, C. 770, and C. 780		
	Air Pict, Ja '55:7-8 (-760, -770, -780)	1	C
	RAF Fly Rev, N '56:11	1	
672.	C. 714 "Cyclone"		
	RAF Fly Rev, N '56:11	1	
673.	G. III		
	Air Prog, Ag/S '64:64-65	1	C
	Mr '68:58-59		Pr†
	Fly Rev Intl, D '68:48-49		P
674.	G. IV		
	Cr & Cockade, Sum '66:185-192 (Smithsonian's-)	1	Pr, S*
	Flying, My '61:71 (Fonck's-)	1	D
CAVALIER			USA
675.	"Mustang" Converted NORTH AMERICAN F-51D		
	Air Prog, Ag '68:18-23+	1	C

CAVALIER (cont.)

675.	"Mustang" (cont.)		
	Avn Wk, O 16, '67:67-73	1	C
	Flight, My 4, '67:695	4	
	Mr 7, '68:354-355	1	
	O 3, '68:542-544	4	
676.	"Mustang 2"		
	Flight, Mr 7, '68:354-355	1	
	Fly Rev Intl, Ap '68:175-177	1	C
677.	"Turbo Mustang 3"		
	Air Prog, My '69:33-35	1	C, P*
	Avn Wk, O 21, '68:69-73	1	C
	Flight, N 21, '68:833-836	1	C
	Fly Rev Intl, Ja '69:21-22	1	
678.	2000		
	Converted F-51 two-seater		
	Flying, Ap '61:38-39+	4	C
	F '69:54-58	4	C
	RAF Fly Rev, Jl '61:44	1	C

C.E. (Designed by Royal Aircraft Factory) GB

679.	1		
	Air Pict, F '62:50-52	1	

CENTRAL USA

680.	101 "Air Tractor"		
	Flight, N 5, '54:674-675	1	
	Skyways, Mr '54:22	1	
	O '55:49		D

CESSNA USA

Air Pict, O '60:350-353 (Pt 1) (Aircraft photos: 1911)
N '60:379-383+ (Pt 2) (Aircraft photos: 1911-1960)
Air Prog, Win '60:46-66 (Aircraft types)
Flying, N '54:21-23 (Aircraft genealogical chart)

681.	"Citation"		
	Earlier name: "Fanjet 500"		
	Can Avn, O '69:33	1	
	Flight, O 2, '69:521	1	
	Fly Rev Intl, D '69:28-29	1	
	Interavia, No. 11, '69:1786-1787	1	

CESSNA (cont.)

682.	"Crane"		
	RAF version of T-50		
	Flight, S 9, '43:283	1	
683.	"Fanjet 500"		
	Named "Citation" in September 1969		
	Avn Wk, O 21, '68:108-112	1	C
	Can Avn, N '68:14-15	1	
	Fly Rev Intl, D '68:31-32	1	
	Interavia, No. 12, '68:1557	1	
684.	A-37		
	Attack version of T-37		
	Air Force, Je '69:49-51	4	
	Air Prog, D '67:36-37+ (-A)	4	
	Avn Wk, F 19, '68:38-43 (-A)	4	
	Je 17, '68:40-41 (-B)	1	C
	Fly Rev Intl, S '68:44-45 (-B)	1	
685.	YAT-37D		
	Counterinsurgency (COIN) T-37		
	Avn Wk, Jl 1, '63:57-59	1	
	Jl 13, '64:36-37	1	C
	Jl 12, '65:66-81	4	
686.	AW Cabin Monoplane (1928)		
	Flying, Jl '64:25		D
687.	CH-1 "Skyhook"		
	Am Helicop, O '54:9	1	
	My '55:12	1	
	Jl '55:8-10+	1	S*
	N '55:11	2	C
	F '56:10-11	2	
	Mr '56:11	6 (Boundary Layer Control)	
	Jl '56:9+ (-A)	1	
	Mr '59:9+ (-C)	1	
	Avn Wk, Ag 8, '55:45	1	
	Ag 22, '55:54	1	C
	My 25, '59:80-84 (-C)	4	C
	Can Avn, N '61:34-35 (-C)	4	
	Flight, N 26, '54:777 (Proto-)	1	
	Flying, My '61:42-43+	4	C
	West Avn, N '54:21	1	
	Je '56:18 (-A)	1	

CESSNA (cont.)

688.	YH-41 "Seneca"			
	US Army Avn, D '57:33	1		
689.	L-19 "Bird Dog"			
	Boeing Mag, N '52:15 (X-/-B)	1		
	N '53:6-7 (X-/-B)	2		
	Can Avn, F '55:40 (-A)	1		C
	Flying, F '51:30-31+ (-A)	4		C
	US Army Avn, O '57:10-13	4		
	Je/Jl '58:17 (US Army's 1st-)	1		
	O '59:16-18	4		
	O '60:4-7	4		
	Mr '61:10-11	6	(Wing Stowage)	
	Ap '61:18-19	4		
	Ag '61:10-13	4		
690.	O-2			
	Civil designation: 337			
	Air Prog, O '69:23-26+	4		
	Avn Wk, S 18, '67:121 (-A&-B)	1		
	Ja 15, '68:38 (-A)	1		
	D 2, '68:20-21	1		
	Ap 14, '69:118 (-TT&-T)	1		
	Fly Rev Intl, Mr '69:23-24 (-TT)	1		
691.	T-37 "Tweety Bird" or "Tweet"			
	See also A-37 and YAT-37D			
	Air Force, Ja '58:62-64	1		
	Air Prog, Win '60:62-63			S*
	Ap/My '63:75 (X-)			C
	D '68:18-21+	4		
	Avn Wk, My 4, '53:23	1		S
	O 18, '54:15 (X-)	1		
	N 22, '54:46 (X-)	1		S
	Je 18, '56:119 (-A)	1		
	Ap 1, '57:48-60 (-A)	1		C
	Flight, O 29, '54:650 (X-)	1		
	Flying, Ap '60:54-55+ (-B)	4		
	Fly Rev Intl, D '63:32-36 (-B)	1		Pr, S*
	Skyways, N '54:37 (X-)	1		P
692.	T-41 "Mescalero"			
	US Army Avn, Ja '67:65 (-B)	1		P

CESSNA (cont.)

693.	T-50 British version: "Crane" USAF version: UC-78 US Navy version: JRC-1		
	Aviation, Ja '40:46-47	1	C, S
	Can Avn, Ja '40:48	1	
	Je '47:91	6 (Lube Chart)	S*
	Comm Avn, Ja '40:64	1	
	N '40:15-17+	1	C
694.	UC-78 "Bobcat" British version: "Crane" Civil designation: T-50 Military trainer designation: AT-17 US Navy version: JRC-1		
	Aero Digest, F '46:78-79	1	
	Flying, D '66:60-63	4	C
695.	185 "Skywagon" USAF designation: U-17		
	Flying, Je '61:34-35+ (-C)	4	C
	Interavia, No. 4, '69:400	1	
696.	195 USAF designation: LC-126		
	Skyways, Ja '50:16-17+	4	
697.	206 "Super Skywagon," "Turbo Skywagon" and "Super Skylane"		
	Can Avn, Ja '64:14-15	1	
	D '64:26-27	4	
	D '68:32-33+	4	
	Flight, Ag 4, '66:180-181	4	C
	Flying, D '64:63-64	1	
	Interavia, No. 4, '69:400	1	
698.	207 "Skywagon"		
	Air Prog, Je '69:23-25+ (Turbo-)	4	
	Flying, My '69:86-90	4	C
	Fly Rev Intl, Jl '69:28-29	1	
	Interavia, No. 4, '69:400	1	
699.	230 "Agwagon"		
	Avn Wk, Jl 4, '66:79	1	C
	Can Avn, Ag '66:12-13	1	

CESSNA (cont.)

700.	300 "Agwagon"		
	Avn Wk, Jl 4, '66:79	1	C
	Can Avn, Ag '66:12-13	1	
701.	336 "Skymaster"		
	Air Prog, Je/Jl '63:73-82+	1	C, Pr
	My '67:49-51	1	Pr†
	Flying, My '63:48-50+	4	C
702.	337 "Super Skymaster" and "Turbo Super Skymaster"		
	USAF designation: O-2		
	Air Prog, O '66:17-19+	4	C
	Flight, Ap 29, '65:666-667	4	C
	Ap 13, '67:545 (Turbo-)	1	
	Flying, Mr '65:26-27+	1	P*
	N '65:40-44	4	
	Jl '66:40-44	4	C, P*
	Jl '67:74-77 (Turbo-)	4	C
	O '69:36-41	4	
703.	401		
	Aeroplane, Jl 19, '67:20-21	4	
	Flight, F 23, '67:287-288+	4	C
704.	421		
	Aeroplane, F 14, '68:24-25	4	C
	Air Prog, Ap '69:22-23+	4	C
	Can Avn, S '69:24-25+	4	
	Interavia, No. 5, '69:624-625	4	C
705.	620		
	West Avn, S '54:14	1	
	S '56:34	1	

CHANCE-VOUGHT USA
See VOUGHT

CHASE USA

706.	YC-122 "Avitruc"		
	Development of XCG-18A Glider		
	Aero Digest, My '49:42-43+	1	
	Acft Eng, Mr '50:68-69+	1	
	Avn Wk, Ja 5, '48:23-26	1	
	Ja 3, '49:18-24	1	
	Flight, N 3, '49:593	1	

CHASE (cont.)

707. C-123 "Avitruc"
Later version: FAIRCHILD C-123
Aero Digest, S '49:74-76+ (X-) 1
Je '53:28-48 (-B) 1 S*
Air Force, D '49:44-45 (X-) 1
Acft Eng, Mr '50:68-69+ (X-) 1
Avn Opns, S '49:48-49+ (X-) 1 C*
Avn Wk, Ap 27, '53:15-16 (-B) 1
Skyways, Ja '50:22-23 (X-) 1

708. XCG-18A Glider
Developed into YC-122 "Avitruc"
Am Helicop, Jl '48:23 1
Skyways, S '48:34-35 1

CHILTON GB

709. Single-seat Monoplane (1938)
Air Pict, Ja '69:23-24 1
Flight, Ag 24, '39:183-184 (w. Train engine) 1
Pop Fly, Ap '38:21 1

CHINESE NATIONALIST AIR FORCE/PAZMANY Taiwan

710. PL-1
Fly Rev Intl, O '69:30-31 1

CHRISLEA GB

711. "Airguard"
Flight, S 15, '38:237-238 1 S

CIERVA GB
See also PITCAIRN-CIERVA, USA and AUTOGIRO, GB

712. W. 11 "Air Horse"
Aeronautics, My '52:28 P
Aeroplane, D 17, '48:749-750 1
Ap 8, '49:388-394 1 S*
Am Helicop, Mr '49:11-12+ 1 P
F '50:11-13+ (Pt 1) 1 S*
Mr '50:8-11+ (Pt 2) 1 C*, S*

CIERVA (cont.)

712. W. 11 (cont.)
Can Avn, Mr '50:27+ 4
Flight, Jl 4, '46:12-14 1 S
D 2, '48:665-668 1
Ja 20, '49:63-65 6(Landing Gear) S*
Ap 7, '49:402-403 (Pt 1) 1 S*
Ap 14, '49:427-431 (Pt 2) 1 C*, S*

713. W. 14 "Skeeter"
Later development: SAUNDERS ROE "Skeeter"
Am Helicop, Mr '49:11-12+ 1
Flight, O 21, '48:477-478 6 (Engine) C
O 27, '49:545 (-II) 1

CIVILIAN GB

714. "Coupe"
Air Pict, Jl '69:245+ 1
Flight, Ap 10, '31:308-309 1 C

CJC (C.J. Chu) Nationalist China

715. 3 Helicopter
Am Helicop, Ag '63:9+ 1

COANDA Rumania/France

716. Turbine Aeroplane (1910)
Flying, Jl '54:22+ 1

CODY GB

717. "Cathedral"
Aeroplane, D 12, '41:651 1
AAHSJ, Sum '60:138-145 1

718. "Flyer"
Flight, My 28, '54:700 1 P

COLUMBIA USA

719. XJL-1
Earlier version: GRUMMAN J2F-6 "Duck"
Fly Rev Intl, O '65:125 1
RAF Fly Rev, Ap '61:47 1

COMMAND-AIRE USA

720. Biplane (1928-1932)
Flying, Ap '57:46+ 1

COMMERCIAL USA

721. "Sunbeam" C-1
Aviation, F 1, '30:208 1

COMMERCIAL AEROPLANE WING GB

722. "Alula" or "Handalula" (1921)
Aeroplane, F 27, '42:251 1

COMMONWEALTH Australia

723. "Boomerang" CA-12, CA-13, CA-14 and CA-19
Developed from "Wirraway"
Fly Rev Intl, S '65:50 Pr*
F '68:108 1
RAF Fly Rev, N '57:7 5
Mr '61:38-39 (CA-12) 1 Pr*

724. "Ceres"
Air Pict, O '58:350 1

725. "Sabre" CA-27
Australian NORTH AMERICAN F-86
See also CANADAIR CL-13
Air Power, Ap '55:192, 196 P
Avn Wk, F 8, '54:32-38 1
Flight, O 9, '53:507-508 1
Ja 15, '54:65 1
Ja 29, '54:124 1
RAF Fly Rev, Ap '55:52-53 1 P*
My '60:16-17 1

726. "Wackett" CA-2 and CA-6
RAF Fly Rev, Jl/Ag '59:39 1
O '59:11 1
D '59:46 1

727. "Winjeel" CA-22
Flight, S 29, '49:448 1

COMMONWEALTH (cont.)

728. "Wirraway" (Challenger) CA-16
Fly Rev Intl, S '65:43 1
Pegasus, Ag '43:4 P
RAF Fly Rev, S '62:41 4
RAF Qtr, Je '44:144 P
Skyline, Vol. 3, No. 1, '42:36-37 1

729. "Woomera" CA-4 and CA-11
Air Pict, Jl '55:220 (-11) 1
O '56:367 1
Fly Rev Intl, Jl '65:59 1
RAF Fly Rev, N '56:15 (-11) 1
D '58:11 1

730. CA-15
Aviation, Jl '46:89 1
Can Avn, O '46:66 1
Flight, Ap 4, '46:331 1

731. CA-31
Flight, D 21, '67:1047 1

COMPER GB

732. "Kite"
Flight, My 10, '34:462-464 1 Pr, S
Jl 12, '34:706 6 (Gear Retract) S*

733. "Mouse"
Aeroplane, F 13, '42:193 1
Flight, S 28, '33:973-978 1 Pr, S
Pop Fly, N '33:423 1

734. "Streak"
Aeroplane, Ap 18, '34:693-696 1 S*
Flight, Ap 19, '34:377-381 1 Pr, S
My 3, '34:427 P
Jl 12, '34:706 6 (Gear Retract) S*

735. "Swift"
Aeroplane, D 23, '31:1429-1430 4
:1434-1435 3
O 3, '52:498-499 4
Air Pict, My '69:181 1

COMPER (cont.)

735. "Swift" (cont.)

Flight, Ja 17, '30:115-119	1	Pr, S
Je 3, '32:481	1	Pr
Pop Avn, S '31:12	1	
US Air Svcs, N '32:13-14 (Prince of Wales'-)	4	

CONROY USA

736. "Airlifter" CL-44-0

Conversion of CANADAIR CL-44

Avn Wk, D 15, '69:40-41	1	
Flight, D 18, '69:951	1	

737. "Stolifter"

Fly Rev Intl, Mr '69:24	1	

738. "Turbo-Three"

Conversion of DOUGLAS DC-3

Air Prog, N '69:27-29+	4	C
Fly Rev Intl, Ag '69:28	1	

CONSOLIDATED USA

See also CONVAIR and VULTEE

Aviation, Ap '43:122-125+ (20th Anniversary)

West Fly, O '40:12-14+

739. B-24 "Liberator"

Cargo version: C-87

Civil designation: CONSOLIDATED 39

Escort "Liberator": XB-41

Aero Digest, Ag '43:156-160	6 (Instruments)	C
N '43:270-271	6 (Anti-ice)	S*
N 15, '44:88-92	3	C, S*
D 1, '44:74-79	3	S*
Aeronautics, F '45:43-46	1	S
Aeroplane, N 19, '43:581-583 (-II)	1	S*
My 12, '44:530-534	1	S*
Air Cl, Ap '69:16-25	1	C, P
:24+	4	
:19 (456 Sq:"Lady Corinne")		D*
Air Pict, Ja '68:29 (-D:"Blue Streak")	5	
Air Prog, Ap/My '63:79 (-D)		C
Air Tech, N '42:29 (-D)		P
F '43:20-21	6 (Lndg Gear)	
Mr '44:19-25+	4	S*
O '44:50 (-J)		P

CONSOLIDATED (cont.)

739.	B-24 (cont.)			
	Acft Eng, Je '41:164-165	1		
	Aviation, Ja '42:66-69+	3		
	Jl '42:98-103+, 186-190	3		S*
	Ap '43:126-133	3		
	Jl '43:146-161+	3		S*
	F '44:146-148			S*
	F '45:198, 201			S*
	Jl '45:121-143	1		S*
	CAHS, Spg '65:16-18 (Canadian-)	4		
	Spg '68:27 (-III, -V, -VI)	1		
	Flight, Ag 22, '40:145+ (X-)	6	(Davis Airfoil)	
	Ag 27, '42:228b (-II)	1		
	N 1, '45:481	4		
	Flying, D '40:23	1		
	Jl '43:28-29	1		S*
	N '43:45 (-J)	1		
	Ag '44:61 (-J)			P*
	O '45:50-54+ (-J)	1		S
	N '66:11			D
	Fly Rev Intl, Ja '65:46 (-D, -J, -H, -L)			Pr*
	O '65:106-107 (-J)	6	(B-17G Nose Turret Conversion)	
	My '68:268 (-VI: 356 Sq)			Pr*
	N '68:43 (-J: Indian)			Pr*
	Roundel, My '63:33 (-V: Canadian)	1		P
	RAF Fly Rev, D '58:13-15 ("Lady Be Good"-)	4		
	My '61:30-31+ (-D)	4		
	Skyline, Vol. 4, No. 4, '43:6-8+ (North American built-)	3		
	Skyways, N '43:34-38+	1		S*
	Ap '44:23-24	6	(Anti-ice)	S*
	US Air Svcs, Je '43:30+	3		
	West Fly, Mr '43:34-35+	4		
	O '43:74-75+	3		
	:82-83+	3		
	D '43:52-53+	3		
740.	XB-41			
	B-24D escort conversion			
	RAF Fly Rev, Ap '56:11+	1		
741.	XB-46			
	Interavia, No. 2, '47:53	1		
	RAF Fly Rev, S '61:44	1		

CONSOLIDATED (cont.)

742.	C-87 "Liberator Express"		
	Aviation, Ap '43:126-133	3	
	Can Avn, Jl '44:87-94	3	
	Flying, Jl '43:28-29	1	S
743.	P2Y		
	Aero Digest, O '35:78 (X-/-1)		C
	Aviation, Mr '35:83-85 (-1)	1	
744.	XP3Y-1		
	Became PBY "Catalina"		
	US Air Svcs, Ag '35:32	4	
745.	XP4Y-1 "Corregidor"		
	Aeroplane, N 2, '39:550-551	1	
	Aviation, Je '39:38-39	1	
	Flying, Jl '43:80	1	
	Pop Avn, N '39:58	1	
	RAF Fly Rev, F '58:45	1	
	Mr '60:11	1	
746.	P-30		
	Flight, F 28, '35:230	1	
	Fly Rev Intl, Jl '65:57+	1	
	Pop Avn, F '37:22	4	
	RAF Fly Rev, Ap '60:11	1	
747.	XPB2Y-1		
	Aviation, Ja '38:30-31	1	
	Pop Avn, N '38:52-53+	1	P
	Mr '39:18-19+	4	
	RAF Fly Rev, F '58:42+	1	
	US Air Svcs, Ja '38:34	1	
	Je '39:18		P
	West Fly, Ja '38:30	1	
748.	PB2Y-3 "Coronado"		
	Aero Digest, S '43:182-189+	3	
	Jl 1, '44:82-85	1	S*
	Flight, Ja 14, '43:40a (-II)	1	
	Flying, Ja '42:109		P
	Jl '44:42-44+	1	
	RAF Fly Rev, F '58:42+	1	
	Skyways, Ag '44:50+ (Transport-)	1	

CONSOLIDATED (cont.)

749.	PB2Y-5 "Coronado"		
	RAF Fly Rev, F '58:42+	1	
750.	PB4Y "Privateer" or "Liberator"		
	Civil version: RY-3		
	Aeroplane, Mr 16, '45:310-311		P
	Flight, Ap 5, '45:364-365 (-2)	1	
	Flying, O '45:50+ (-1&-2)	1	
	Fly Rev Intl, Ja '65:46		Pr*
	Nav Avn News, Je 15, '44:25-35	4	C
	RAF Fly Rev, Mr '58:8+	1	
	Skyways, Ap '45:64-65+ (-2)	1	
	US Air Svcs, Mr '45:12 (-2)	1	P
	West Fly, Mr '45:46+ (-2)	1	
	O '45:72-73 (-2)	3	
751.	PBY "Catalina" or "Canso" (Canadian)		
	See also XP3Y-1		
	Civil designation: 28		
	Aero Digest, Jl '36:20 (-1)		P
	Air Cl, O '68:62-71 (-1 to -4)	1	P
	Pop Avn, Mr '40:28-30+	1	
	US Air Svcs, S '44:20-22+	4	
	West Fly, Je '43:39+	4	
752.	PBY-5 "Catalina" or "Canso" (Canadian)		
	Aero Digest, Ag '42:152-153		S*
	Ag 15, '44:90-95	1	C, S*
	S 1, '44:86-90	1	S*
	Aeroplane, Je 19, '42:696-697		S*
	Air Cl, O '68:62-71	1	C, P
	Air Pict, D '68:458	4	
	Air Tech, O '44:17-25+	1	S*
	Can Avn, S '42:37-38+ (-A)	1	
	:39+ (-A)	4	
	:40-51 (-A)	3	S*
	Mr '44:62-74+	3	
	Comm Avn, O '43:43-46 (Boeing Canada-)	3	S*
	Flight, O 1, '42:362b	1	
	Flying, Mr '41:40-41+	4	
	N '43:Supp:23+	6 (De-ice)	S*
	Mr '44:48		C
	O '44:52-53+	4	
	Ja '47:43-45+	4	
	Roundel, Mr '62:33	1	P
	US Air Svcs, Ja '40:23-24 (X-/-A)	1	

CONSOLIDATED (cont.)

753.	PBY-6A and PBN-1 "Nomad" (Prototype)		
	Air Cl, O '68:65, 71	1	
	Air Pict, D '68:458	4	
754.	PT-3 "Husky"		
	Air Prog, O/N '63:59-61	1	C, Pr
755.	RY-3 "Liberator Express"		
	US Navy version: PB4Y		
	Aviation, My '45:169	1	
	Comm Avn, Ap '45:59+	1	
	West Fly, My '45:98	1	
756.	16 "Commodore"		
	Aviation, Ja 11, '30:49-55	1	C
	US Air Svcs, Jl '30:28-35	4	P
757.	17 "Fleetster"		
	Aeroplane, Ag 27, '30:504-506	1	
	AAHSJ, Fall '62:206-207	1	
758.	21 A "Trainer"		
	Aero Digest, Jl '31:76-80	1	
	Aviation, Ag '31:488-489	1	
759.	28		
	US Navy designation: PBY		
	Aviation, Je '40:61		S*
	Flight, Ag 10, '39:132a-d	1	C
760.	37		
	USAF designation: CONVAIR B-36		
	Comm Avn, Ap '45:42-44	1	
	West Fly, Mr '45:72	1	
761.	39 "Liberator Liner"		
	Aero Digest, Ag 1, '44:88-90+	1	
	Bee-Hive, Jl/Ag '44:10		P
	Flight, Jl 13, '44:49	1	
	Fly Rev Intl, F '68:109	1	
	RAF Fly Rev, S '62:45	1	
	West Fly, O '44:72-73+	3	S
762.	111 Roadable Aircraft		
	Air Pict, Ag '58:286-287	1	
	Fly Rev Intl, Jl '68:413	1	

CONVAIR USA

See also GENERAL DYNAMICS/CONVAIR, CONSOLIDATED, and VULTEE

Flying, Je '54:23-25 (Aircraft genealogical chart)
Jl '60:36-40+ (Pt 1)
Ag '60:34-37+ (Pt 2)
S '60:36-39+ (Pt 3)
O '60:42-45+ (Pt 4)

Nav Avn News, S '57:12-15 (US Naval aircraft)

US Air Svcs, Jl '53:9-10 (30th Anniversary)

763.	"Flying Auto"		
	Skyways, S '48:24-25	1	
764.	B-32 "Terminator" and "Dominator"		
	Aero Digest, S 1, '45:149-151+	1	
	Air News, O '45:28-29+	1	P
	AAHSJ, Sum '68:80-92	1	P
	Aviation, S '45:168-169	1	
	Flying, O '45:50+	1	
	Fly Rev Intl, My '65:41	1	
	RAF Fly Rev, O '55:9	1	
	Skyways, O '45:20-21+	1	
	US Air Svcs, O '45:14	1	P
	West Fly, S '45:78+	1	
765.	XA-41		
	Air Pict, My '62:150-151	1	P
	Aviation, Je '46:95	1	
	US Air Svcs, My '46:11	1	P
766.	B-36 "Peacemaker"		
	Civil designation: CONSOLIDATED 37		
	Aero Digest, Jl '51:20-21+	2	
	Ja '54:28-36	1	S*
	Aeroplane, Ja 31, '47:132-135 (X-)	1	
	Mr 7, '52:270-271 (-F)	1	S
	Aerosp Hist, Win '66:161-163	1	
	Air Force, My '59:87-91	1	
	Air Pict, D '66:465 (R-)	1	
	Air Power, Spg '54:250 (FICON)		P
	Air Prog, '56/'57 Annual:48-49	1	C
	N/D '65:46-48	1	C
	Avn Age, Ag '52:33-37 (Pt 1)	1	C, S
	S '52:34-35 (Pt 2)	1	

CONVAIR (cont.)

766.	B-36 (cont.)		
	Avn Wk, O 18, '48:12-14	1	
	Ja 17, '49:34-37	3	
	Mr 14, '49:12-13	1	
	Ag 15, '49:13-16	1	
	Ag 22, '49:13-15	1	
	Ag 11, '52:21-28	1	
	D 1, '52:32	1	C*
	Ag 31, '53:12-13 (-D:FICON)	1	
	O 19, '53:28-42	6 (Lndg Gear)	
	Ag 1, '55:22-23 (GR-:FICON)	1	P
	Bee-Hive, Spg '50:22-23 (R-)	1	
	Sum '50:7-9	3	
	Ja '51:18-21	3	
	Can Avn, D '46:66	1	
	Flight, My 12, '49:568-570	2	
	Ja 25, '51:90-92 (-D)	4	
	F 8, '52:153-154 (-D)	4	S
	Jl 29, '55:159	1	C
	Flying, D '46:51-53 (X-)	1	
	Ag '47:35	1	
	Jl '49:16-17+ (-A to -D)	1	
	Ap '51:13-15+	4	S*
	My '51:36 (-D)		P*
	My '59:34-35+	1	
	Jl '61:38-39+	1	
	Interavia, No. 5, '51:248-249		P
	Pegasus, My '48:1		P
	Skyways, Ja '49:14-16+	1	
	US Air Svcs, Jl '47:18-19+ (X-)	1	P
	Ag '48:11	1	P
	N '50:22	6 (Hammocks)	
	West Avn, D '52:29 (-H)		C*
	West Fly, O '48:14+	1	
767.	B-58 "Hustler"		
	Later version: GENERAL DYNAMICS B-58		
	Air Force, Ag '61:48-52	4	
	Air Pict, F '57:54-55	1	Pr
	Ag '65:283	1	
	Air Prog, Ag/S '64:44-45	1	C
	Avn Age, Ja '57:34-39	6 (Weapons)	
	Avn Wk, Je 4, '56:50-55	6 (Engine)	
	D 17, '56:26-29	1	
	D 31, '56:30-31		P
	Ja 14, '57:50-59	6 (Inlet Control)	
	Mr 4, '57:76		D*

CONVAIR (cont.)

767.	B-58 (cont.)		
	Avn Wk, Jl 8, '57:133	6 (Nose Gear)	
	Jl 15, '57:28-29	1	
	S 15, '58:119-122	6 (Grnd Support)	
	D 15, '58:106-108	6 (Air Condition)	
	Ag 3, '59:76-85	6 (Bomb-Nav)	
	Ap 18, '60:64-67	1	
	My 23, '60:93 (T-/-A)	1	
	Jl 25, '60:33		C
	Bee-Hive, Spg '58:24-25	6 (Air Condition)	
	Can Avn, F '57:66-67		P
	Flight, S 21, '56:508-509	1	S*
	Ja 4, '57:16-17	1	
	Mr 5, '64:360-361	1	C
	Flying, S '57:30+	1	
	Interavia, No. 1, '57:33-36	1	P
	No. 7, '60:825-827	6(Bomb-Nav)C*, S*	
	RAF Fly Rev, S '61:33 (-A)	4	
	West Avn, N '56:22-23	1	
	D '56:18	1	
	Mr '58:17	6 (Photo)	
768.	YB-60		
	Jet-version B-36		
	Air Pict, S '68:348	1	
	Avn Wk, O 19, '53:28-42	6 (Lndg Gear)	
	Bee-Hive, Sum '52:6	2	
	RAF Fly Rev, N '57:7+	1	
769.	XC-99		
	Avn Wk, D 5, '49:14-15	1	
	Je 2, '52:18	1	
	Can Avn, F '47:84	1	
	Comm Avn, Ag '45:88	1	S*
	Interavia, No. 4, '46:76-77	3	S*
	Ja '48:20-26	1	C, P
	US Air Svcs, D '47:12-13	1	P
	D '50:9	4	
	Je '53:20	4	
	West Avn, Ag '52:16	1	
	West Fly, Ja '48:10+	1	C
770.	C-131 "Samaritan"		
	Avn Wk, Ap 5, '54:14-15 (-A)	1	
	Skyways, Je '55:8-10+ (Y-/-C)	1	S*

CONVAIR (cont.)

771.	XF2Y-1 and YF2Y-1 "Sea Dart"		
	Air Pict, Mr '53:71		P
	Avn Wk, Ag 3, '53:9, 15		P
	N 15, '54:20	2	
	Jl 11, '55:29	2	
	Ag 15, '55:32-33	1	P
	Interavia, No. 4, '53:211	1	
	No. 10, '53:591	1	
	Nav Avn News, F '53:12	1	
	RAF Fly Rev, D '62:45+	1	
	West Avn, Ja '53:9	1	
	Ag '53:30-31	1	
772.	XF-92A		
	Forerunner of F-102		
	Flying, S '49:24-25+	1	
	Interavia, Jl '49:428-429	1	P
	RAF Fly Rev, O '57:44	1	
	Skyways, Mr '50:14-15+	1	P
	US Air Svcs, Jl '52:6		P
773.	F-102 "Delta Dagger"		
	Developed from XF-92A		
	Aeroplane, Mr 12, '54:287 (Y-)		P
	Air Pict, My '54:137	1	
	Je '55:162 (Y-/-A)	1	
	Avn Age, Ja '57:30-31	6 (Weapons)	
	Avn Wk, O 26, '53:16	1	
	Ja 10, '55:9, 15 (Y-/-A)	1	P
	Mr 14, '55:72		P*
	Ap 18, '55:65 (Y-/-A)		P*
	Je 13, '55:51-56	3	
	Je 20, '55:19		D*
	Je 11, '56:50-64 (-A)	3	
	O 1, '56:30-31 (T-/-A)	1	
	Ag 19, '57:63-65 (-A)	1	
	Flight, N 14, '63:795-798	4	C
	RAF Fly Rev, Ap '54:15	1	
	D '57:34-37+ (-A)	1	Pr†, S*
	F '60:11 (T-/-A)	1	
	US Air Svcs, Je '55:16 (Y-/-A)		P
	D '56:21 (T-/-A)	6 (Ejection Seat)	
	West Avn, F '55:32 (-A)	1	

CONVAIR (cont.)

774.	F-106 "Delta Dart"			
	Avn Wk, D 16, '57:105+	6 (Navigation)		
	Mr 3, '58:71		P*	
	Ap 21, '58:33 (-B)	1		
	S 7, '59:126-133	4		
	F 5, '68:41-46 (-B)	1	C	
	Bee-Hive, Fall '57:34	1	P	
	Flying, Ag '61:46-47+ (-A&-B)	4	C	
	RAF Fly Rev, My '58:16-17 (-A)	1	Pr*	
	Sp/Aeron, N '58:72-77	2		
	Ap '60:83+	2		
	West Avn, S '59:14+ (-A)	4		
	My '60:22-27+	6 (Power Generating)		
775.	XFY-1 "Pogo Stick"			
	Air Power, Aut '54:4, 7-10	1	P	
	Am Helicop, Ap '54:9	1		
	Avn Wk, Mr 22, '54:14-18	1		
	Mr 29, '54:16-17	4		
	Je 21, '54:26-28	2		
	Mr 14, '55:73		P*	
	Flying, Jl '54:14-15+	4		
	Interavia, No. 5, '54:295-297	1		
	Nav Avn News, My '54:20	1		
	Ag '54:11	2		
	Ja '55:16-17	1		
	RAF Fly Rev, Je '62:43+	1		
	US Air Svcs, Jl '54:8	2		
	West Avn, D '54:16	4		
776.	L-13 "Sentinel"			
	Aeroplane, Mr 14, '47:253-255	1		
	Air Pict, Jl '51:199 (-A)	1	P	
	Ap '56:134 (-A conversion)	1		
	Aviation, Ja '47:66	1		
	Can Avn, Mr '47:35+	1		
	US Air Svcs, D '46:29	1		
	Je '47:25	6 (Engine)		
	N '47:23 (-B)	1		
	West Fly, D '46:50	1		

CONVAIR (cont.)

777.	XP5Y-1		
	Aero Digest, My '51:19+	1	
	Avn Wk, N 1, '48:13-14	1	
	Ag 7, '50:24-25	6 (Engine)	
	Ap 23, '51:21-29	6 (Engine)	
	Jl 27, '53:17	4	
	RAF Qtr, Ja '51:8		P
778.	XP-81		
	Aero Digest, Mr '46:80-81	1	S*
	Air News, Mr '46:33	1	
	Am Helicop, Mr '46:28-29	1	P
	Aviation, Mr '46:95	1	
	Flight, F 28, '46:216-217	1	S*
	US Air Svcs, F '46:31	1	
	Mr '46:39	1	P
	West Fly, Mr '46:40+	1	S*
779.	R3Y "Tradewind"		
	Avn Wk, Ja 25, '54:22-34	1	
	Je 21, '54:17-18 (-2)	1	
	Jl 16, '56:61-63	1	C
	Can Avn, Jl '56:43 (-2)	1	
	Flight, F 4, '55:139-141	6 (Engine)	S*
	Interavia, No. 2, '53:84-85 (-1)	3	
	No. 2, '54:91 (-1)	1	
	Nav Avn News, Ag '54:8 (-2)	1	
	F '58:38-39	6 (Refueling)	
	US Air Svcs, My '53:8 (-1)	1	
	West Avn, O '52:16-19	3	
780.	T-29 "Flying Classroom"		
	Avn Wk, D 13, '48:21	1	S*
	Flight, D 29, '49:820	1	S*
	Interavia, No. 5, '50:252-254	1	C
	US Air Svcs, O '52:10 (-B)	1	P
	West Avn, Mr '52:14 (-B)	3	
781.	TBY-1 and TBY-2 "Seawolf"		
	Earlier version: VOUGHT XTBU-1		
	RAF Fly Rev, O '57:11+	1	

CONVAIR (cont.)

782.	240 "Convair-Liner" and "Turbo-Liner"			
	Aeronautics, F '49:40-41	1		
	Aeroplane, S 10, '48:359-361	1		
	My 22, '53:669-671	4		C
	Acft Eng, Ja '49:9-13	1		S*
	Am Helicop, Ja '51:16-17	6	(Engine)	S*
	Avn Opns, Ap '50:20	6	(Anti-ice)	S*
	Avn Wk, S 29, '47:18-19	6	(Cowl)	S*
	Mr 8, '48:20-24	1		
	Je 28, '48:19-26	1		C
	N 21, '49:37-38	6	(Press)	
	Ja 8, '51:15-17	1		
	Can Avn, Je '48:24-25+	1		C
	Flight, F 24, '49:229	4		
	Ag 8, '52:157-159	6	(Engine)	
	F 13, '53:206-207	1		
	Fly Rev Intl, O '65:85-88	1		
	Interavia, No. 2, '47:18-24	1		C*
	No. 11, '50:588-590	1		
	Skyways, Ag '48:22-23	1		C
	Ap '51:30-31	1		S*
	US Air Svcs, F '46:21	1		
	Ap '47:14-15	1		P
	West Avn, S '54:12	1		
	West Fly, N '47:18-19+	3		
	My '51:24-25	1		
783.	340 "Convair-Liner"			
	Aero Digest, Ap '54:32-40	1		C
	Aeroplane, Ja 27, '56:117-121 (-Eland)	1		S*
	S 28, '56:485-488 (-Eland)	4		
	Air Prog, '53/'54 Annual:42-43+			S*
	AAHSJ, Fall '62:202			C
	Avn Age, S '55:18-25 (-Dart)	1		
	Avn Wk, Mr 24, '52:52-53	3		
	Ag 10, '53:13-15	1		
	S 20, '54:73-77	4		
	Ag 1, '55:79-81 (-Dart)	1		
	F 29, '60:71-77 (-Allison)	1		C
	Jl 18, '60:69-76 (-Allison)	4		
	Flight, F 13, '53:206-207	1		
	F 3, '56:136-137 (-Eland)	1		
	Je 1, '56:667-669 (-Eland)	4		C
	Jl 6, '56:34 (-Eland)	1		
	N 1, '57:708 (-Eland)	1		S*

CONVAIR (cont.)

783.	340 (cont.)			
	Flying, Ja '52:27+	1		
	Interavia, No. 5, '51:275	1		
	Skyways, Mr '53:10-12+	4		C
	West Avn, Ja '52:13	1		
	Je '52:16 (United-)	4		
	O '52:68 (Convertible-)	1		
784.	440 "Metropolitan"			
	See also CANADAIR 540 and CC-109			
	Aeroplane, My 5, '66:40-41 (-Dart)	1		
	Air BP, No. 4:27 (Sabena-)			Pr*
	No. 11:10 (Iberia-)			Pr*
	Air Pict, N '56:390-391	1		C*, Pr
	Avn Age, F '56:60-63	6	(Muffler)	
	Avn Wk, Ag 22, '55:17-18	1		
	Ag 29, '55:15	1		
	Flight, Jl 6, '56:35	1		
	S 9, '65:457 (-D)	1		S*
	Fly Rev Intl, O '65:85-88	1		
	Interavia, No. 10, '55:781	1		
785.	600 "Skylark"			
	Later version: 880			
	Aeroplane, O 3, '58:517-519	1		S*
	Avn Age, My '56:36-39	1		
	Avn Wk, Ap 9, '56:99	1		
	Je 18, '56:40	1		
	S 8, '58:50-57 (Pt 1)	1		
	S 15, '58:92-111 (Pt 2)	1		C
	Flight, N 20, '59:590-591			S*
	Fly Rev Intl, O '65:85-88	1		S
	Interavia, No. 11, '58:1195-1199	1		
	Sp/Aeron, S '59:127-128+	6	(Controls)	S*
786.	640			
	Interavia, No. 3, '67:373	6	(Engine)	
787.	880 "Golden Arrow"			
	Earlier version: 600 "Skylark"			
	Aeroplane, F 13, '59:189-194	1		S*
	Avn Wk, O 15, '56:57-62	1		
	Je 17, '57:82-91	1		
	F 17, '58:63-65	1		
	D 22, '58:34-35	1		
	O 17, '60:43 (-M)	1		
	My 29, '61:48-54	4		

CONVAIR (cont.)

787.	880 (cont.)			
	Flight, Jl 6, '56:68	1		
	N 1, '57:701-704	1		S*
	Jl 25, '58:138-140	1		S*
	Flying, Ja '59:47+	1		
	N '59:30-31	1		
	Fly Rev Intl, N '66:212-213	1		Pr
	Interavia, No. 11, '58:1195-1199 (-3&-25)	1		S*
	Sp/Aeron, D '58:30-31	6	(Controls)	S*
	S '59:127-128+	6	(Controls)	S*
	West Avn, F '59:16	4		
	Ap '60:76			P
	S '60:20-21+	6	(Sound-proofing)	
788.	990 "Coronado"			
	Aeroplane, F 22, '62:201			C, P
	Mr 1, '62:223-225	1		S*
	Air BP, No. 29:15 (-A:Swissair)			Pr*
	Avn Wk, D 5, '60:42-45	1		
	Mr 27, '61:38	1		
	F 26, '62:69-83	4		C
	O 29, '62:54-55 (-A)	1		
	Flying, S '61:24-25+	4		
	Interavia, No. 2, 61:176-177	1		
	No. 9, '62:1149-1156 (Swissair-)	4		C, S*
	West Avn, D '60:60	1		

COUZINET France

789.	"Aerodyne"			
	RAF Fly Rev, O '55:23-24	1		
790.	101			
	Flight, My 24, '34:522	1		

COX-KLEMIN USA

791.	XS-1			
	Fly Rev Intl, Ag '69:69	1		

CRAIGON Canada

792.	50-50			
	Comm Avn, D '39:19-21	1		Pr

CROUCH-BOLAS USA

793. "Dragonfly"
Aviation, Jl '34:230-231 1

CROWN USA

794. "Custombilt" B-3
Aviation, Ag '30:103 1

CSS (Centralne Studium Samolotów) Poland

795. 12
RAF Fly Rev, Mr '56:15 1

CTA (Centro Tecnico de Aeronautica) Brazil

796. "Bandeirante" (Explorer)
Avn Wk, Ja 13, '69:107 1

797. "Beija-Flôr" (Hummingbird)
Am Helicop, Ap '59:6-7 1

CUB Canada

798. J-5
See also PIPER, USA
Can Avn, My '41:23 1 C

CULVER USA

799. LCA and LFA
Skyways, O '44:28-29+ 1

CUNNINGHAM-HALL USA

800. PT-6F
West Fly, N '38:23 1

CUNLIFFE-OWEN GB

801. "Concordia"
Aeroplane, My 30, '47:563-567 1 S*
Air Pict, My '64:155 1
Flight, Mr 14, '46:261-262 1
Mr 29, '47:496a-500 1 S
Je 5, '47:526-527 1

CUNLIFFE-OWEN (cont.)			
801.	"Concordia" (cont.)		
	Interavia, No. 7, '46:75	1	
	RAF Fly Rev, Ja '59:42	1	
802.	"Flying Wing" OA-1		
	Aero Digest, Jl '39:109+	1	
	Aeroplane, My 17, '39:625-628	1	S*
	Flight, D 22, '38:572-575	1	C, S*
	Pop Fly, Jl '39:148		P
	RAF Fly Rev, Mr '61:11	1	
CURTISS			USA
	Comm Avn, D '39:11-14		
	Interavia, Jl '49:430-431		
	Nav Avn News, O '57:29-33 (US Naval aircraft)		
	Pop Avn, F '34:76-78+ (Early aircraft)		
	N '37:26-30+ (Patrol boats to 1937)		
	US Air Svcs, Je '31:36-38 (New Buffalo plant)		
803.	"Amphibion" (1934)		
	Aviation, Ag '34:269	1	
	West Fly, Ag '34:38	1	
804.	"Canada" C-1		
	CAHS, Win '67:93-95	1	P
805.	"Cleveland"		
	RAF designation for SBC-4		
	Acft Eng, Mr '42:79-80	1	P
	Flight, Ag 22, '46:192a-b	4	
806.	"Condor" Bomber B-2		
	Aero Digest, Ap '34:18		P
	Aeroplane, My 9, '34:783	1	
	Aviation, Mr '34:97	1	
	Flight, Ap 12, '34:365-366	1	
	Fly Rev Intl, My '67:579-582+	1	Pr*
	US Air Svcs, Mr '34:14	1	
	Ap '34:22	6 (Bombs)	
807.	"Condor" Transport		
	Aero Digest, F '33:42-43	1	
	My '33:35		C
	Ag '35:44		C*
	Aeroplane, Ja 17, '34:107-108	1	
	AAHSJ, Spg '67:3-27	1	P, Pr

CURTISS (cont.)

807. "Condor" (cont.)

Aviation, Je 7, '30:1128-1133 1 C
F '33:58-59 1
Ap '33:58-59 1
:132 3
Can Avn, D '33:10 (Byrd's-) 1
Flight, Je 15, '33:563-565 1 C
Fly Rev Intl, F '64:60+ 1
My '67:579-582+ 1 Pr*
Pop Avn, Jl '33:19 6 (Instruments)C*
Ap '34:216 (Byrd's-) 1
My '34:303 (Byrd's-) Pr
Ap '35:221-222 1
US Air Svcs, F '33:24 1
Mr '33:31 1
Ap '33:20-23 1
N '33:29 (Byrd's-) 1
F '35:34 (Cargo-) 1
Ap '35:19 (Cargo-) 1
West Fly, Mr '33:30 1
My '34:18 6 (Interior)

808. "Falcon Mailplane" or "Commercial Falcon"

Pop Avn, F '31:29 1

809. "Fledgling"

US Navy designation: N2C
Air Prog, Ja '67:48-51+ 4 C
Flying, Ja '67:52-55 4 C, P
Pop Avn, My '28:78 1

810. "Golden Flyer" and "Reims Racer"

AAHSJ, Fall '65:186-189 1 P
Flying, F '56:30-31+ 4

811. "Hawk"

RAF designation
US Army designation: BF2C
Fly Rev Intl, D '66:274-275 (-III&-IV) 1 Pr*
Pop Avn, My '34:288-290+ 1
Pop Fly, O '36:360 (-III) 1 P

812. "Hawk 75"

Export model
RAF designation: "Mohawk"
US Army designation: P-36 "Hawk"

CURTISS (cont.)

812. "Hawk 75" (cont.)

	Aviation, O '37:43	1	
	AAHSJ, Spg '66:13-30	4	Pr
	Flight, Ap 18, '40:350a-c ("French Hawk")	1	D, S
	Fly Rev Intl, N '66:218, 220		Pr*
	Ap '69:53 (-A-8:Norway)		Pr*
	RAF Fly Rev, Ap '55:9	1	
	Ja '61:18 (-A-3:Vichy AF)		Pr*
	:19-20+	1	

813. "June Bug"

	Hist Avn, Jl '69:16-19	1	

814. "Kingbird"

	Aeron (US), Je '30:457	1	P
	Flight, Ja 9, '31:29-30	1	

815. "Lark"

	CAHS, Win '65:98-101	1	Pr

816. "Mohawk"

RAF designation
US Army designation: P-36
See also "Hawk 75"

	Flight, F 27, '41:172a-c	1	S
	RAF Fly Rev, Ap '55:9	1	

817. "Oriole"

	Air Prog, Sum '61:49-51	1	Pr

818. "Pusher"

	Air Prog, O '69:27-30+ (Replica-)	4	
	Flight, D 2, '60:871-872	4	

819. "Red Wing"

	Air Pilot, N '44:34-35	1	

820. "Robin"

	Air Prog, O/N '63:49-51	1	C*, Pr
	Flying, Ap '58:54+	1	
	US Air Svcs, Ag '38:18 (Corrigan's-)	1	
	West Fly, Ag '38:12-14+ (Corrigan's-)	4	

CURTISS (cont.)

821.	"Tanager"			
	Aviation, F 8, '30:236-241	2		
	:242-248	1		
	:249-252	4		
	US Air Svcs, F '30:42-44+	1		P
	West Fly, F '30:57+	1		
822.	"Tripod Pusher"			
	Pop Avn, Ag '36:36			P
823.	A. 1			
	First US Navy aircraft			
	Air Prog, Win '61/'62:48-57	1		S*
	AAHSJ, Spg '64:67	1		
	Flying, F '62:54-55+	1		
	Nav Avn News, S '61:20-25 (Replica-)	1		
824.	A-8 "Shrike"			
	AAHSJ, Sum '65:129-139	1		Pr
	Can Avn, F '33:19 (Y-)	1		
	Pop Avn, F '35:93	1		Pr
825.	A-10 "Shrike"			
	A-8 variant			
	AAHSJ, Sum '65:129-139	1		Pr
826.	A-12 "Shrike"			
	Improved A-10			
	Air Prog, '56/'57 Annual:82-83			S
	'57/'58 Annual:92-93			S*
	AAHSJ, Sum '65:129-139	1		Pr
	Pop Avn, S '33:165	6	(Wing Slots)	
	F '36:112			P
	Pop Fly, D '33:485	1		D
	US Air Svcs, Jl '33:4	1		D
827.	XA-14 and Y1A-18			
	RAF Fly Rev, Ag '63:39+	1		
828.	A-25 "Helldiver"			
	US Navy designation: SB2C			
	Flight, Mr 18, '43:286	1		
	US Air Svcs, F '43:36	1		P
	West Fly, Mr '43:96	1		

CURTISS (cont.)

829.	AT-5		
	AAHSJ, Win '68:308		C*
830.	AT-9 "Fledgling" or "Jeep"		
	Flight, D 9, '43:642a	1	
	RAF Fly Rev, Ag '62:41	1	
831.	BF2C		
	Redesignated F-11C-3		
	British designation: "Hawk"		
	Fly Rev Intl, D '66:274-275 (-1)	1	Pr*
	US Air Svcs, N '35:13 (-1)		P
832.	BT-1 "Flying Lifeboat"		
	AAHSJ, Sum '65:102-104	1	Pr
833.	XBT2C-1		
	RAF Fly Rev, Mr '61:13	1	
834.	XBT-4		
	CURTISS O-1E variant		
	Aero Digest, Jl '30:117	1	
	Pop Avn, S '30:43	1	
	US Air Svcs, My '30:66	1	
835.	XBTC-1 and XBTC-2		
	RAF Fly Rev, Mr '61:13	1	
836.	C-46 "Commando"		
	Civil version: CW-20 "Commando"		
	Air Tech, O '42:31		P
	F '45:17-25+	1	S*
	Aviation, Ag '43:130-153	1	C, S*
	Avn Wk, My 2, '55:21	1	
	Ag 19, '57:71-75 (Super-)	1	
	Can Avn, S '42:52-53	1	
	Comm Avn, Je '42:34+	1	
	O '44:106	1	
	Ja '45:60-61	6 (Sound Proofing)	
	My '45:50	6 (Cockpit)	
	Flight, D 3, '42:608a	1	
	Skyways, Je '45:34-40+	1	S*
	F '53:18-20	6 (Jet Pod)	

CURTISS (cont.)

837.	C-76 "Caravan"		
	Aero Digest, Ja '43:186-188	1	
	Comm Avn, Ap '43:94-102	1	
	Flight, O 21, '43:451	1	
	RAF Fly Rev, N '60:13	1	
	Skyways, Mr '43:72-75	1	S
	West Fly, Ja '43:102	1	
838.	CW-20 "Commando"		
	USAAF designation: C-46 "Commando"		
	Aero Digest, Ag '38:48-49+	1	C
	D '38:53	2	
	N 15, '44:111-113+ (-E)	1	
	Aeroplane, D 19, '41:685	1	
	Air Pict, Jl '56:241 (-T)	1	
	Aviation, Ag '38:28-31+	1	C
	N '39:38	3	
	D '39:72-73	3	
	Mr '40:45-53	1	
	:54-57+	3	
	My '40:58	4	
	Mr '45:197-201 (-E)	1	C
	Can Avn, N '38:16	1	
	Mr '39:12+	1	C, S*
	D '44:185	1	
	Comm Avn, Ja '40:54+	1	
	N '44:80-84 (-E)	1	
	Flight, O 20, '38:338-340	1	C
	Ja 11, '40:28-30	1	
	D 18, '41:441-442	1	
	D 21, '44:662 (-E)	1	
	Flying, Je '45:59-62 (-E)	1	S
	US Air Svcs, D '38:16-17+	3	
	Mr '39:21	1	C
	Mr '40:33	1	
	My '40:20-21	4	P
	West Fly, N '44:80 (-E)	1	
839.	CW-21 "Demon" and "Interceptor"		
	Aero Digest, F '39:51-52	1	
	Aviation, F '39:84-85	1	
	F '41:127 (-B)		S*
	Can Avn, Mr '41:21-22 (-B)	1	S*
	Comm Avn, F '41:29-30+ (-B)	1	S*

CURTISS (cont.)

839.	CW-21 (cont.)		
	Fly Rev Intl, Ag '68:463	1	
	D '68:72	1	
	Pop Avn, Je '39:55	1	
	Pop Fly, Je '39:102-103	1	
	RAF Fly Rev, Ag '57:40	1	
	Ag '62:36 (-B:Netherland East Indies)		Pr*
	:37	1	
840.	CW-22 "Falcon"		
	Air Pict, F '67:61	1	
	Can Avn, F '40:64	1	
	Comm Avn, Ja '40:60+	1	
	US Air Svcs, F '40:20-21	1	P
841.	CW-23		
	Aviation, My '40:52-53+	1	S
	Can Avn, F '40:64	1	
	Comm Avn, Ja '40:58+	1	
	Pop Avn, Ap '40:13	1	C
842.	CW-24B		
	XP-55 prototype		
	Air Pict, Ja '58:21	1	P
843.	CW-32 "Sky Truck"		
	US Air Svcs, S '47:25	1	
	West Fly, Ja '47:46	1	S
844.	F-Boat (1913)		
	Can Avn, O '66:62	1	D*
845.	F6C "Hawk"		
	Aero Album, Win '69:31 (-1&-4)		P
	AAHSJ, Sum '67:156 (-4)		D
	Fall '69:145 (-1)		P
846.	F7C-1 "Sea Hawk"		
	AAHSJ, Ap/Je '56:42		P
	US Air Svcs, S '33:27		P
847.	F8C "Helldiver" and "Hawk"		
	See also SB2C		
	Aero Digest, D '30:60 (-5)	1	
	Ag '32:18		P
	Pop Avn, F '31:45 (-5:Command)	1	

CURTISS (cont.)

848.	F9C "Sparrowhawk"		
	AAHSJ, Ap/Je '58:70-109 (X-&-1&-2)	1	P, Pr, S*
	Flight, My 20, '32:447 (-2)	1	
	Pop Avn, Je '32:352 (-2)	1	
849.	F11C "Goshawk" or "Hawk"		
	Export version: "Hawk"		
	Aero Digest, N '33:50-54 (-2:Type 1)	1	C
	Je '35:46-47 (-2:Type 3)	1	
	Ag '35:43 (-2:Type 3)		C
	Air Cl, Mr/Ap '65:33 (-2)		Pr*
	AAHSJ, Spg '64:58-63	1	P
	:75 (-2)		D
	Fly Rev Intl, D '66:274-275 (-3)	1	
	Pop Avn, O '33:236 (-2)	1	
	F '34:95 (-2)	1	Pr
	RAF Fly Rev, Ja '60:42 (-1 to -3)	1	
850.	XF12C		
	RAF Fly Rev, Jl '61:57	1	
851.	XF13C		
	RAF Fly Rev, N '60:11+	1	
852.	XF14C		
	RAF Fly Rev, O '59:50	1	
853.	XF15C-1		
	RAF Fly Rev, Jl '62:45+	1	
854.	XF-87 "Blackhawk" or "Nighthawk"		
	Air Pict, D '69:476	1	
	Avn Wk, S 15, '47:11-12	1	
	Fly Rev Intl, S '68:111	1	
	RAF Fly Rev, O '56:11	1	
	US Air Svcs, O '47:13	1	P
	Mr '48:20	1	
855.	H-12 "Large America"		
	Nav Avn News, Jl '48:41		P
	RAF Qtr, Jl '37:Supp	1	D
856.	H-16		
	Avn Wk, Mr 27, '67:36-37	1	D*
	Nav Avn News, O 1, '44:20-21	4	

CURTISS (cont.)

857.	HA "Dunkirk Fighter"		
	AAHSJ, Win '61:259-260	1	Pr
858.	HS-1, HS-2, and HS-3		
	Aero Album, Fall '68:36 (-2L)		P
	Air Prog, Win '60:67-71	1	P, Pr
	Can Avn, Ap '57:157 (-2L)	4	
	CAHS, Fall '66:67-71 (-2L&-3L)	1	Pr
	Nav Avn News, O '48:33 (-2)		P
859.	JN-4 "Jenny" and "Canuck" (Canadian-built)		
	Air Pilot, N '44:14+	4	
	Mr '45:16-18	1	C
	Air Prog, '52 Annual:70-71 (-D)	1	S*
	AAHSJ, Win '61:Supp (-D)		P*
	Avn Wk, Ag 21, '61:52-53 (-D)	1	D*
	Can Avn, F '67:12-15	1	
	D '68:53 ("Canuck")		D*
	Flying, Mr '69:52-55	1	P*
	Hist Avn, Ja '69:12-14 (-D)	1	
	Pop Avn, F '38:35-36+	1	
	Roundel, My '62:23, 33 ("Canuck")	4	P
860.	N-9		
	Air Prog, F/Mr '66:6-15	1	Pr†
	Nav Avn News, Ap '66:10-11	1	
861.	XO-1G "Falcon"		
	Pop Avn, F '31:17	1	
	US Air Svcs, D '30:26		P
862.	O2C-1 "Helldiver"		
	Pop Avn, Jl '36:43		Pr
	US Air Svcs, Ag '31:46	1	P
863.	O-11 "Falcon"		
	Liberty-powered "Falcon"		
	Pop Avn, Mr '28:28		P
864.	YO-40 "Raven"		
	Flight, Ja 4, '34:20 (-A)	1	
865.	O-52 "Owl"		
	Flight, Ap 15, '43:392b	1	
	Fly Rev Intl, N '64:57	1	
	RAF Fly Rev, Ja '57:11+	1	

CURTISS (cont.)

866.	OC-2		
	Redesignated from F8C-3		
	AAHSJ, Jl/S '56:79		P
867.	P-1 "Hawk"		
	AAHSJ, Win '68:307 (Proto-)		C
	Spg '69:66		C*
	Pop Avn, N '35:296 (-A)	1	P
868.	P-6 "Hawk"		
	Air Prog, Ag '69:32-33 (-E:Replica)	1	P*
	AAHSJ, Fall '69:169-172 (X-/-A)	1	P
	Pop Avn, Ap '36:242 (-E:Skiis)	1	P
	RAF Fly Rev, F '61:41	1	
	Jl '62:34	1	
	:35 (-E:17 Pursuit Sq)		Pr*
	US Air Svcs, O '30:38 (Al Williams'-)		P
	Ja '32:42 (-E)	1	P
869.	XP-31 "Swift"		
	RAF Fly Rev, N '61:44	1	
870.	P-36 "Hawk"		
	Export model: "Hawk 75"		
	RAF name: "Mohawk"		
	Aero Digest, D '37:54-62 (-A)	1	
	Air Prog, Ap/My '36:72-73 (-A)		C
	Bee-Hive, Ag '37:5	1	
	Can Avn, My '41:18		C
	Comm Avn, Je '40:20 (-A)		C
	Flying, S '41:99 (-A)		P
	Fly Rev Intl, Mr '66:438 ('39 War Games)		P*
	N '66:218 (-A&-C)		Pr*
	Pop Avn, D '37:38	1	
	RAF Fly Rev, D '61:40-41 (-A:USAF Museum's)	4	
	US Air Svcs, O '36:18 (Y1-)	1	P
	Ag '37:16-17	1	P
871.	XP-37 and YP-37		
	RAF Fly Rev, O '57:11	1	
	Mr '58:50	1	
	US Air Svcs, Ja '38:23	1	P

CURTISS (cont.)

872.	P-40A to C "Tomahawk I & II"		
	Includes export Model 81A		
	Aeroplane, Je 27, '41:718		S*
	Air Pilot, N '43:24-26+ (Pt 1)	4	
	D '43:11-13 (Pt 2)	4	
	Air Prog, Jl '42:46-47		S*
	Ag/S '65:58-61+	1	D
	Acft Eng, Mr '41:60 (-A)		P
	AAHSJ, Win '67:254-261 (to -A)	1	
	Spg '69:26-29 (Pt 1)	1	
	Sum '69:130-133 (Pt 2)	1	
	Aviation, Ag '40:60-61	1	S
	F '41:127		S*
	S '42:156		S*
	D '42:96-101+	3	
	CAHS, Sum '66:42-44	1	
	Flight, F 27, '41:172a-c	1	S
	Mr 20, '41:224		S
	Flying, Mr '65:37	4	D
	Pop Avn, Ja '39:29		P
	RAF Fly Rev, Jl '58:6 (-C:AVG)	5	
	O '59:24 (-B:AVG)	1	
	:25 (-B:Olson's)		Pr*
	Mr '61:42	4	
	Jl '62:28-30+ (-C)	1	S*
	:31 (-IIA)		Pr
	D '62:47	4	
	Skyways, Ag '44:45+	1	
	US Air Svcs, Jl '42:37	4	
	West Fly, N '43:66-67	4	
873.	P-40D and P-40E "Kittyhawk I"		
	Aero Album, Fall '68:16-17 (RAAF-)	4	
	Air News, N '41:7 (-D)		D
	Air Prog, Mr '69:38-39 (-E:Truscott's)	4	Pr†
	Acft Eng, D '41:343, 345 (-D)	1	P
	Can Avn, Ag '41:31	1	
	CAHS, Spg '65:30 (-IA:RCAF 111 Sq)		Pr†
	Comm Avn, Ag '41:50 (-D)	1	P
	Flight, Je 18, '42:614a (-D)	1	
	Flying, My '65:72-76 (-E)	4	P*
	Fly Rev Intl, Mr '64:54 (-IA)		Pr*
	D '65:248 (-IA:New Zealand)		Pr*
	:250 (-E:"Flying Tigers")		Pr*
	RAF Fly Rev, O '59:24 (-E:AVG)	1	

CURTISS (cont.)

874.	P-40F and P-40L "Kittyhawk II" or "Warhawk"		
	Air Tech, O '42:34		P
	AAHSJ, Win '66:283-289	4	Pr
	Flight, My 14, '42:468, 474 (-F)	1	
	Fly Rev Intl, Mr '64:54 (-F)		Pr*
	D '65:248 (-F:3 Sq & Free French)		Pr*
	:250 (-F:33d Gp; 65 & 85 Sq; 325 Gp)		Pr*
875.	P-40K and P-40M "Kittyhawk III"		
	Fly Rev Intl, D '65:248 (-III:New Zealand; 112 Sq; 5 Sq)		Pr*
	:250 (-K:23 Fighter Gp)		Pr*
	My '68:272 (250 Sq-)		Pr*
876.	P-40N "Kittyhawk IV" or "Warhawk"		
	Air Prog, N '68:36-39+	4	C
	Fly Rev Intl, Mr '64:54		Pr*
	D '65:248 (-IV:Australia)		Pr*
	:250 (-N-25:398 Fighter Sq)		Pr*
877.	XP-46		
	RAF Fly Rev, N '57:9	1	
878.	XP-55 "Ascender"		
	Prototype: CW-24B		
	Air Force, F '45:37		P
	Air Pict, Ja '58:21	1	
	Flight, Ap 5, '45:360-361	1	
	Skyways, Ap '45:94	1	
879.	XP-60 and YP-60		
	Air Pict, Ap '61:106-109	1	P
	RAF Fly Rev, D '58:50	1	
880.	XP-62		
	Air Pict, Ja '60:24-25	1	
	RAF Fly Rev, D '58:50	1	
881.	XP-71		
	Fly Rev Intl, My '64:61-62	1	
882.	XP-87		
	See XF-87		

CURTISS (cont.)

883.	R2C Racer		
	AAHSJ, Spg '66:7-12	1	P
884.	SB2C "Helldiver"		
	See also F8C		
	US Army version: A-25		
	Aero Album, Win '68:3, 6, 8 (17 Navy Bomb Sq-)		P
	Air News, Mr '44:27		P
	Air Pict, My '54:149 (To -5)	1	
	Air Prog, Jl '42:21 (-1)		P*
	Can Avn, Ag '44:54-64+	3	
	Comm Avn, Ja '44:44-46 (-1)	1	
	Je '44:38-56 (-1)	1	
	Flight, My 8, '41:333-334 (X-/-1)	1	
	My 7, '42:454 (-1)	1	
	Mr 18, '43:284b (-1)	1	
	Flying, Ja '42:39 (-1)		P*
	Mr '44:30-31	1	
	N '44:60		P*
	S '45:51-54 (-4)	1	S
	Fly Rev Intl, S '66:90, 92 (-1, -3, -5)		Pr*
	Je '69:73 (-5:Greece)		Pr*
	Skyways, N '42:79		D
	O '44:34-39+	1	P*, S*
	My '45:66 (-4)	1	
	US Air Svcs, O '40:11 (X-/-1)		D
	S '42:12		P
	F '44:35-36	1	
	Mr '45:48 (-4)	1	
	West Fly, F '44:68+ (-1)	1	
885.	SBC-1, 2, and 3		
	Pop Avn, Mr '38:43 (-3)		P
	RAF Fly Rev, Jl '61:57+	1	
886.	SBC-4 "Helldiver"		
	RAF name: "Cleveland"		
	Acft Eng, Mr '42:79-80	1	P
	Flying, D '40:17		P
	RAF Fly Rev, Jl '61:57+	1	
887.	SC-1 "Sea Hawk"		
	Air Pict, Ja '66:27	5	
	Air Tech, Ag '45:31		P
	Comm Avn, Jl '45:62-64+	1	

CURTISS (cont.)

887.	SC-1 (cont.)		
	Nav Avn News, Je '48:8 (With wheels)		P
	RAF Fly Rev, S '61:44+	1	
	West Fly, Ag '45:94	1	
888.	SC-2 "Sea Hawk"		
	Avn Wk, S 8, '47:25	2	
	RAF Fly Rev, S '61:44+	1	
889.	SNC-1 "Falcon"		
	Aero Album, Fall '69:38-41	1	P, Pr†
	Aviation, Ag '41:105+	1	
	Can Avn, Ag '41:30+	1	
	Comm Avn, Ag '41:48+	1	
890.	SO3C "Seagull" or "Seamew"		
	AAHSJ, Sum '61:154		D
	Flight, Ja 21, '43:68b (-2)	1	
	US Air Svcs, Je '42:26 (-1)		P
891.	SOC "Seagull"		
	AAHSJ, Jl/S '58:167-174+	1	Pr
	Pop Avn, Ap '37:38 (-1)		Pr
	S '39:41 (-4)		P
	US Air Svcs, F '36:28 (-1)		P
892.	TS-1		
	AAHSJ, Jl/S '56:79		P
893.	VZ-7AP "Flying Jeep" or "Aerial Platform"		
	Avn Wk, N 16, '59:93-97	1	
	West Avn, N '59:30	1	
894.	X-19 VTOL Research		
	Civil designation: X-200		
	Avn Wk, Ag 6, '62:64-71	1	D
	Ag 5, '63:72-73	1	
	Flight, S 19, '63:509	1	
	Fly Rev Intl, N '64:23-24	1	
	US Army Avn, S '64:7	1	
895.	X-100		
	Avn Wk, Je 20, '60:277-283	1	
896.	X-200		
	USAF designation: X-19		

CURTISS (cont.)

896.	X-200 (cont.)		
	Avn Wk, Je 20, '60:277-283	1	
	Flying, N '62:44-46+	1	S*
897.	XFC-6 "Page Navy Racer"		
	Air Prog, S '68:45+	1	
898.	18-T "Curtiss Kirkham" Triplane or "Whistling Benny"		
	Air Prog, Je '68:54-57+	1	P, Pr†
	US Air Svcs, Mr '44:21	1	
899.	19		
	Aviation, N '35:27-29 (-L)	1	
	S '36:41-43 (-R)	1	
	Pop Avn, N '36:28+ (-R)	1	

CURTISS-BLEECKER USA

900.	Helicopter (1931)		
	Am Helicop, Ja '48:17+	1	P

CURTISS-REID Canada

901.	"Courier"		
	CAHS, Sum '69:54-56, 58, 64	1	Pr
	Flight, My 20, '32:442-443	1	
902.	"Rambler"		
	CAHS, Win '66:116 (-II)		Pr

CUSTER USA

903.	CCW-5 "Channel Wing"		
	Air Prog, O/N '64:30-31+	1	C
	Avn Wk, S 28, '59:113-115	1	
	Flight, Ja 4, '52:14	1	
	Flying, S '53:58	1	
	Ap '65:42-45	1	
	West Avn, F '52:17	1	
	D '52:18	1	

CW (Cronander and Waddington) GB

904.	"Cygnet Major"		
	Flight, F 3, '38:112-114	1	C

CW (cont.)

905.	"Cygnet Minor"		
	Flight, D 10, '36:628-630	1	S
	My 13, '37:481	1	
	O 28, '37:422-423	1	
906.	"Swan"		
	Flight, O 28, '37:421-422	1	

CZECHOSLOVAKIA Czech

907.	L-29 "Delfin" (Dolphin) [Maya]		
	Avn Wk, Ap 22, '63:89	1	
	Je 10, '63:174-175	1	C
	D 23, '63:62	1	
	Fly Rev Intl, Mr '65:17-19	1	S*
	Interavia, No. 7, '63:1080-1082	1	C
	No. 12, '64:1841-1846	1	C, S
	RAF Fly Rev, Ap '63:39	1	
908.	L-60 "Brigadyr"		
	Aeroplane, O 10, '58:562-563	4	C
	O 9, '59:325		P
	Flight, N 27, '59:646-647	4	
909.	L-410 "Turbolet"		
	Avn Wk, Je 2, '69:360	1	
	Jl 14, '69:29-32	1	C
	Interavia, No. 7, '67:1100	1	
	No. 8, '69:998-1000	1	C, S
910.	Z-37 "Čmelák" (Humble-Bee)		
	Air Prog, Ap '67:18	1	C
	Avn Wk, Ja 27, '64:105-115	1	
	Ag 16, '65:103	1	
	Fly Rev Intl, Je '69:38 (-2)	1	
	Interavia, No. 12, '64:1845-1846	1	

DAR (Darjauana Aeroplanna Rabotilnitza) Bulgaria

911.	10		
	Only Bulgarian combat design to enter service (1941)		
	RAF Fly Rev, Ja '62:45	1	

DARE USA

912. Variable Camber Monoplane (1921)
AAHSJ, Spg '63:54-56 1

DART GB

913. "Kitten"
Pop Fly, Ap '37:15 1

DASSAULT France

914. "Balzac"
Air Pict, S '62:287 1
Air Prog, Ja '64:82-83 1 S
Acft Eng, S '62:271 1
Avn Wk, Jl 30, '62:18-19 1
N 19, '62:64 1
Je 10, '63:100-102 1
Je 24, '63:68-69 1
Jl 29, '63:38-41 1
Boeing Mag, Ja '63:12-13 1
Flight, Ag 2, '62:174-175 1
N 15, '62:781-782 2
Jl 4, '63:11-12 1
Interavia, No. 1, '61:78-79 1
No. 8, '62:966-969 1
No. 2, '63:179 1
No. 9, '63:1370-1371 1
RAF Fly Rev, N '62:13-15+ 1 S*
F '63:15-16 1

915. "Etendard" (Flag or Banner)
Aeroplane, Ap 22, '60:479-481 (-IV)1
N 11, '60:637 (-IV) 1
Air Prog, Ja '65:37 (-IVM) Pr†
Avn Wk, O 15, '56:37 (-IV) 1
O 28, '57:50-60 (-4 & -6) 1
Ja 28, '63:106-107 (-4M) 4
RAF Fly Rev, Ap '57:8-9 (-IV) 1
Ag '60:19-21 (-IV) 1
:18, 20-21 (-IVM) Pr, S*
Ag '63:32-33 (-IVM) 1 Pr*

DASSAULT (cont.)

916.	"Fanjet Falcon"			
	Earlier version: "Mystere 20"			
	Aero Album, Spg '69:28-32	1		
	Can Avn, Ja '67:8-10	4		
	Flying, O '66:36-41	4		
	Fly Rev Intl, Ag '69:27	1		
	Interavia, No. 10, '64:1500-1501	1		
	No. 5, '67:767-769	4		
	No. 3, '68:290-292	4		C
917.	"Hirondelle" (Swallow)			
	Air Prog, Jl '69:51-55	1		S*
	Avn Wk, F 5, '68:71	1		
	Ap 8, '68:107-109	1		C
	Jl 22, '68:77	1		
	Ag 5, '68:89	1		
	Fly Rev Intl, Mr '69:21-22	1		
918.	"Mercure" (Mercury)			
	Avn Wk, N 11, '68:89	1		
	Je 30, '69:30-31	6	(Interior)	
	Flight, Ap 24, '69:656	1		
	Fly Rev Intl, Je '69:34-36	1		
	Interavia, No. 5, '69:623-624	1		
	No. 11, '69:1806-1807	1		S*
919.	"Milan"			
	Modified "Mirage 3R"			
	Interavia, No. 11, '69:1808-1810	1		S*
920.	"Minifalcon"			
	Fly Rev Intl, Ag '69:27-28	1		
	Interavia, No. 8, '69:994-995	1		D
921.	"Mirage"			
	See also "Milan"			
	Aeroplane, D 2, '60:736-741	1		S*
	F 3, '61:116-118 (-IV)	1		
	Ap 6, '61:367 (-IV)			P, S
	Aerosp Intl, Jl/Ag '69:26-27 (-F.1)	1		
	Air Force, Ag '69:19-20 (-F.1)	1		
	Air Pict, My '59:173-176 (-III)	1		Pr
	Je '63:176-177 (-IV)	1		Pr
	Mr '66:81-83 (-IV Spey)	1		

DASSAULT (cont.)
921. "Mirage" (cont.)

Air Prog,	Fall '60:12-13 (-III)		S*
	Ja '65:41 (-IIIR)		Pr†
	Jl '67:40-41+ (-I to -IV)	1	
	:42-43 (-III-RO)		S*
Acft Eng,	Ap '68:27-28 (-G)	1	
Avn Wk,	D 25, '61:21 (-IIIE & R)	1	
	Je 18, '62:74-75 (-3C&-4A)	1	
	S 23, '63:112-113 (-E)	1	
	D 19, '66:58-67 (-III)	4	
	Ja 16, '67:27 (-F1)	1	
	My 29, '67:261-263 (-F1)	1	
	Je 5, '67:36-37 (-3G)	1	
	Je 12, '67:96-97 (-3G)	1	
	Je 26, '67:92-93 (-F2&5)	1	
	D 18, '67:18 (-3G)	1	
	Ag 5, '68:30-31 (-G)	1	
Flight,	Ap 22, '60:558-562 (-3C)	1	C, S*
	Ja 4, '62:18-21 (-III&-IV)	1	S*
	F 14, '63:234-235 (-IV)	1	
	Je 20, '63:961-963 (-IIIB)	4	C
	S 5, '63:430-436 (-IIIC)	4	C
	Ja 2, '64:19 (-IIIR)	1	
Fly Rev Intl,	Ja '64:54-55 (-III)	1	Pr*
	Mr '64:48 (-IIIR)	1	Pr*
	My '64:13-16 (-III)	1	
	My '65:23-27 (-IVA)	1	Pr
	N '65:138-140+ (-III Series)	1	
	My '66:580 (-I)	1	
	N '68:26-28 (-III&-5)	1	Pr*
	F '69:45 (-IIIEP: Pakistan)		Pr*
	S '69:25-26 (-F1)	1	
Interavia,	No. 4, '60:463-470 (-IIIC)	1	
	No. 6, '63:784-785 (-III&-IV)	3	
	:786	1	
	No. 2, '64:193-194 (-IIIR)	1	S*
	No. 6, '64:804-806 (-IVA)	1	
	:807-809 (-IIIC)	4	
	:832 (-IIIBS)	4	
	No. 12, '65:1950-1952 (-IIIS)	4	
	:1952-1953 (-IIIS)	2	
	No. 8, '66:1188-1191 (-IIIS)	4	C*
	No. 5, '68:576-577 (-G)	1	

DASSAULT (cont.)

921. "Mirage" (cont.)
RAF Fly Rev, Ap '60:30-33 (-III) 1 Pr, S*
Jl '61:36-39 (-IV) 1 Pr, S*
My '63:18-19 (-IIIR) 1 P*
Sp/Aeron, D '60:57-59 (-III) 3 S

922. "Mystère"
Air Pict, Ja '52:36-37 1
Avn Wk, N 10, '52:21-37 (-2) 1
O 10, '55:26-32 (-4, -22, -24, -26) 1
Flight, Ja 7, '55:10-11 (-4) 1
Fly Rev Intl, N '68:43 (-IVA:India) Pr*
RAF Fly Rev, Ag '55:22-25 (-IVA) 1 Pr, S*

923. "Mystère 20" or "Falcon"
Later version: "Fanjet Falcon"
Avn Wk, Je 14, '65:260-267 1 C
Can Avn, My '64:24-25 1
Flight, Ja 30, '64:167-170 4 C
Flying, Jl '63:38-39+ 1
N '63:55+ 1 C
Fly Rev Intl, My '64:40-42 1
Interavia, No. 6, '63:782-783 1

924. "Ouragan" (Hurricane)
Aeroplane, My 14, '54:612-613 P
RAF Fly Rev, S '57:16 (Israeli-) Pr*
:17 1

925. "Spirale III"
Air Pict, Jl '62:206-207 1
Interavia, No. 7, '61:1006-1007 1
No. 7, '62:885-887 1

926. 316
RAF Fly Rev, My '63:43+ 1

DAVIS USA
Air Prog, Jl '65:16-27 (Pt 1) (Many aircraft photos)
Ag/S '65:44-48+ (Pt 2) (Many aircraft photos)

	DE HAVILLAND		GB
	Can Avn, O '60:13-20 (40th Anniversary)		
	Flight, S 26, '40:254-257 (Civil aircraft)		
	S 27, '45:342-348 (Aircraft list to D.H. 103 "Hornet")		
	S 23, '60:501-514 (Many aircraft photos)		
	Interavia, O '49:608-611		
	No. 9, '60:1080-1081		
	Pop Avn, My '38:38-42+		
927.	D.H. 1		
	Air Pict, My '60:162	1	
928.	D.H. 2		
	Air Cl, Mr/Ap '65:59 (Hawker's-)		Pr
	Ap '69:27		D*
	Air Pict, N '55:345	1	
	Avn Wk, Ag 5, '63:64-65	1	D*
	Pop Fly, Ap '34:28	1	D
	RAF Qtr, Jl '37:Supp	1	D
929.	D.H. 4		
	Aeroplane, My 21, '43:600	1	
	Aerosp Hist, Win '69:27 (-B)		P
	AAHSJ, Fall '61:179 (-B)		C*
	Win '62:243-253	4	P, Pr†
	Avn Wk, D 7, '59:96-97	1	D*
	Cr & Cockade, Spg '62:14-15	4	
	Flight, O 17, '52:506-510	1	
	Fly Rev Intl, My '68:268 (205 Sq)		Pr*
	Pop Avn, Ag '32:103+	1	
	Pop Fly, Ap '32:34		D
	S '32:364	1	
	RAF Qtr, Ja '39:Plate		D
930.	D.H. 5		
	Air Pict, S '56:317	1	
	Pop Fly, Mr '34:641	1	D
	Ja '38:584		P
	RAF Qtr, Ja '39:Plate		D
931.	D.H. 6 "Clutching Hand"		
	Air Pict, My '57:155	1	

DE HAVILLAND (cont.)

932.	D. H. 9 "Nine"		
	Flight, Ap 6, '56:385-388 (Pt 1)	1	
	Ap 13, '56:422-426 (Pt 2)	1	
	Fly Rev Intl, Je '69:72 (Greek-)		Pr*
	Pop Fly, O '33:372	1	D
	RAF Qtr, Ap '37:Supp	1	D
	Ap '38:194-200 ("De Voortrekker")	4	
	Ja '39:Plate		D
933.	D. H. 9A "Ninak" or "Nine-Ack"		
	Aeronautics, Je '55:53 (8 Sq)		P
	Air BP, No. 39:6-7		P
	Can Avn, N '65:1	1	D*
	Jl '69:54	1	D*
	Flight, My 25, '56:641-644 (Pt 1)	1	C
	Je 1, '56:677-680 (Pt 2)	1	
	Ag 1, '58:174		D
	Fly Rev Intl, My '68:269 (47 Sq)		Pr*
	Pop Fly, Jl '34:189	1	D
934.	D. H. 10 "Amiens"		
	Air Pict, Mr '61:70-73	1	
	RAF Qtr, Jl '37:Supp	1	D
935.	D. H. 17		
	Air Pict, Ja '68:29	1	D
936.	D. H. 18		
	Flight, Mr 1, '62:311	1	D
937.	D. H. 51		
	Aeroplane, Ag 26, '60:254-256	4	
938.	D. H. 53 "Hummingbird"		
	Air Pict, F '61:53-54	1	
	Ag '67:276	4	
	Pop Fly, My '35:76-77+	4	
939.	D. H. 60 "Gipsy Moth," "Moth," "Moth Trainer," and "Cirrus Moth"		
	Aeroplane, Ap 2, '30:565-572	1	
	:596-597	4	
	Je 24, '31:1208 (Trainer-)	1	
	My 30, '52:671-672	4	

DE HAVILLAND (cont.)

939.	D.H. 60 (cont.)		
	Air Pict, Ag '67:276	4	
	Acft Eng, Ap '29:44-47	3	S*
	Can Avn, Ag '69:10-11+ (CF-APA)	4	C
	Flight, Jl 3, '31:649-651 (Trainer-)	1	
940.	D.H. 75 "Hawk Moth" or "Moth Six"		
	Aeroplane, Mr 12, '30:444-448	1	S
	Can Avn, Mr '30:14-15	1	
	Flight, Mr 14, '30:291-293+	1	Pr, S
941.	D.H. 77 Interceptor		
	Air Pict, O '57:364-365	1	
	Flight, Je 20, '30:654-655	1	P
	Fly Rev Intl, Ja '67:332	1	
942.	D.H. 80 "Puss Moth"		
	Aeronautics, F '46:81 (-A)		C*
	Aeroplane, Ap 2, '30:574-582	1	S
	Jl 16, '30:195-196	4	
	D 17, '30:1400		C
	Je 13, '52:730-731	4	
	Acft Eng, S '32:226 ("Heart's Content")	1	P, S*
	D '32:302 ("Desert Cloud")	1	S*
	Can Avn, Jl '30:21-22	1	
	S '30:40	1	
	CAHS, Win '69:99-103 (Canadian-)	1	
	:108 ("Heart's Content")	4	
	Flight, F 7, '30:183-184 ("Moth III")	1	
	Ap 25, '30:453-458 ("Moth III")	1	C, Pr, P, S
	Ag 26, '32:795-798 ("Heart's Content")	4	
	D 15, '32:1192-1193 ("Desert Cloud")	1	S*
	D 22, '32:1207-1208 ("Desert Cloud")	4	
	Ag 22, '68:299 (-A)	1	P*
	Pop Avn, Je '31:35 (Prince of Wales'-)	1	
943.	D.H. 82A and D.H. 82C "Tiger Moth" Canadian-built version: 82C		
	Aeronautics, N '46:92-95	4	C*
	My '53:36-39	1	
	Mr '57:67-75	1	
	Aeroplane, N 11, '31:1131-1136	1	S*
	S 13, '33:470	4	

DE HAVILLAND (cont.)

943.	D.H. 82A and D.H. 82C (cont.)		
	Air Prog, S '69:39-41 (-A:Restored)	1	C, P*
	Can Avn, Ja '38:8	1	
	O '45:108-112+	3	S*
	CAHS, Win '69:111-115 (-C)	1	C, S*
	Comm Avn, Jl '39:25 (-C)	1	
	Flight, N 13, '31:1121-1124	1	C
	O 19, '33:1053-1054 (Fighter-)	1	C
	Ag 22, '52:205-207	4	
	F 27, '53:247-249	1	
	Flying, N '69:46-51	1	P*
	Fly Rev Intl, My '67:586 (Restored-)	1	P*
	Roundel, F '49:33	1	
	D '62:33 (-C)	1	P
944.	D.H. 83 "Fox Moth"		
	Aeroplane, Mr 16, '32:473-477	1	S
	:477-478	4	
	Can Avn, My '32:20-21	1	
	Jl '32:20	4	
	D '45:78+	1	
	CAHS, Fall '63:66-67 (Canadian-)	1	
	Flight, Mr 18, '32:231-234	1	C, Pr, S
	Mr 7, '46:241-242 (-C)	1	
	Flying, Jl '55:45	4	
	Pop Avn, Ag '32:85	1	
	Pop Fly, My '32:71	1	
945.	D.H. 84 "Dragon"		
	Aeroplane, D 21, '32:1210b-1216	1	S
	Ag 16, '33:306-308	4	
	Acft Eng, Ag '33:181-182 ("Seafarer")	1	P
	Can Avn, Ap '33:8-9+	1	
	Ja '35:23 (Seaplane-)	1	
	Flight, D 22, '32:1212-1216	1	Pr, S
	D 29, '32:1236	4	
	N 22, '62:810	4	
946.	D.H. 85 "Leopard Moth"		
	Aeroplane, N 1, '33:776-782	1	C, S*
	Ja 17, '34:104	4	
	F 6, '42:166	1	
	Acft Eng, O '33:229-231	1	
	Flight, N 30, '33:1191-1195	1	C, Pr, S
	O 17, '35:402		C
	Pop Fly, My '34:70-71	4	

DE HAVILLAND (cont.)

947.	D.H. 86 "Diana" or "Express Air Liner"		
	Aeroplane, F 21, '34:303-306	1	C*, S*
	Je 6, '34:900	1	
	Je 19, '35:715-716	4	
	Air BP, No. 46:15		C
	Acft Eng, S '34:241	1	
	Aviation, Je '34:196-197	1	
	Flight, F 22, '34:169-174	1	Pr, S
	F 14, '35:185		C
	O 17, '35:403		C*
	Ja 2, '36:17 (-A)	1	
	My 21, '36:536a	6 (Instrument)	C*
	N 25, '43:586 (-B)	1	
	US Air Svcs, Ap '34:21	1	
948.	D.H. 87 "Hornet Moth"		
	Aeroplane, Jl 24, '35:131	4	
	Can Avn, Jl '35:40	1	
	CAHS, Sum '66:58 (-A&-B)		Pr
	Flight, Jl 23, '34:706	1	
	Je 20, '35:670-673	1	S
	Jl 4, '35:Supp:a-d	4	
	Mr 12, '36:291-292	4	C
	Ja 30, '69:Inset	1	P*
	Pop Avn, My '36:340	1	
949.	D.H. 88 "Comet"		
	Aeroplane, S 19, '34:341-345	1	S*
	Flight, S 13, '34:938-939	1	
	S 20, '34:968-972	1	Pr, S
	O 25, '34:1119-1122	1	Pr, S
	Ap 7, '38:336-338	4	C*
	Fly Rev Intl, Ja '66:290-293	1	S*
	Pop Avn, Ja '35:15-16	4	
	Pop Fly, O '37:406	1	
	RAF Fly Rev, D '56:11+	1	
	US Air Svcs, N '34:21	1	
	West Fly, N '34:11	1	
950.	D.H. 89 "Dragon Rapide," "Dragon Six," and "Rapide"		
	Aeroplane, Ap 25, '34:711		P
	:719-721	1	C*, S*
	F 20, '35:223-224	4	
	Acft Eng, S '34:241	1	

DE HAVILLAND (cont.)

950.	D.H. 89 (cont.)		
	Can Avn, Ap '36:12+ (Skiis-)	1	
	Flight, Ap 26, '34:405-409	1	Pr, S
	My 2, '35:474-475 (Prince of Wales'-)	1	
	N 21, '35:526-527 (Military-)	1	
	D 12, '35:634-635 (Military-)	1	S*
	N 25, '43:587 (-A&-B)	1	
	N 20, '53:676-677 (Series 4-)	4	
951.	D.H. 90 "Dragonfly"		
	Aeroplane, Ja 22, '36:108-110	1	S*
	CAHS, Sum '69:46-49, 58	1	Pr
	Flight, Ja 23, '36:78-82	1	S*
	Mr 19, '36:311-312	4	
	Pop Fly, My '36:93	1	
952.	D.H. 91 "Albatross"		
	Aeroplane, N 16, '38:620-624	1	S*
	Air Pict, Ag '64:228-231 (Pt 1)	1	C
	S '64:292-294 (Pt 2)	1	
	Air Prog, F/Mr '66:53	1	Pr
	Flight, Mr 31, '38:331	6(Engine Cooling)	S
	S 29, '38:Supp:a-c	2	
	N 17, '38:450a-455	1	C, S*
	N 24, '38:470-471	4	C*
	Mr 11, '43:256a	1	
	Pop Avn, F '38:50	1	
	Pop Fly, Ja '39:496-499	1	C
	RAF Fly Rev, D '59:11+	1	
953.	D.H. 93 "Don"		
	Fly Rev Intl, Ag '68:463	1	
954.	D.H. 94 "Moth Minor"		
	Aeroplane, Je 28, '39:832-835	1	S*
	Can Avn, S '39:25		S*
	O/N '39:44B	1	
	Flight, Je 29, '39:656a-e	1	S*
	Ag 31, '39:202a-c	4	C*
	Je 10, '43:606a	1	
	Je 26, '69:Inset	1	P*
	Pop Avn, N '38:39	1	
	Pop Fly, D '38:464+	1	

DE HAVILLAND (cont.)

955.	D.H. 95 "Flamingo"			
	Aero Digest, Je '39:90+	1		
	My '40:48-49			S*
	Aeroplane, F 15, '39:214-218	1		S*
	Jl 19, '39:103	4		
	Mr 15, '40:354-359	2		
	Aviation, Ag '39:34-35	1		C
	Can Avn, My '38:9	1		
	D '38:14	1		
	F '39:20	4		
	Flight, S 22, '38:251-252	3		
	N 24, '38:474a	3		
	F 16, '39:151-159	1		S
	:168-169	4		C*
	Jl 20, '39:57 (Proto-)	4		
	S 14, '39:242h	6	(Radio)	
	Mr 14, '40:241-242	1		
	Pop Avn, Mr '39:59			P
	Ap '39:29	1		S
	Pop Fly, My '39:74-75+	1		
956.	D.H. 98 "Mosquito"			
	Aero Digest, Je '43:227-232+	1		S*
	Je 1, '44:102+	1		
	Aeronautics, My '45:32-34	6	(Armament)	
	Ag '45:43-46	1		S
	Aeroplane, My 7, '43:523, 532-539	1		C, S*
	Jl 2, '43:14-15 (-IV)			S*
	D 29, '44:733-734	4		
	Je 29, '45:746-749	4		C
	O 19, '45:449-450			D*
	Mr 1, '46:255-261 (-33 to -35)	1		D*
	Air BP, No. 25:12-15	4		C
	Air Cl, O '68:54-61+	1		P
	Air Pict, Mr '54:77-78 (Pt 1)	1		
	Ap '54:109-110 (Pt 2)	1		
	:118 (Correction)	1		
	My '54:152 (Correction)	1		
	Air Prog, O '43:54-55 (-IV)			S*
	Air Tech, S '44:34-38+	1		
	Acft Eng, Mr '44:64-67	1		
	My '47:163-174	3		S*

DE HAVILLAND (cont.)

956. D.H. 98 (cont.)

Aviation,	My '43:259-261	3	
	F '44:144		S*
	My '44:127-140 (Pt 1)	1	S*
	Je '44:123-137 (Pt 2)	1	C*, S*
	Ag '44:197, 199		S*
Can Avn,	My '43:47-52	1	
	:53-60+	6 (Wing)	
	Je '43:79-86+	1	C, S*
	F '44:54-55	3	
	S '44:57-59+	3	
	O '45:56-57+	2	C
CAHS,	Sum '64:35-38 (Canadian-)	1	
	:59 (-XX:Canadian)		S*
Comm Avn,	N '42:40-41	1	
	Ag '44:38-50	3	
	Jl '45:38-42+	3	
Flight,	D 17, '42:660a	1	
	Ja 14, '43:35-37	4	
	My 6, '43:467-472b	1	S*
	N 11, '43:524-525	1	
	Ap 20, '44:423-427	1	
	S 7, '44:257-262	4	
	N 23, '44:564 (-XVIII)	6 (Molins Gun)	
	Ja 4, '45:8-9 (-XVI)	1	
	F 22, '45:200-204 (Weather-)	4	
	F 28, '46:205-208+ (-33 to -36)	1	
	My 2, '46:442-444	4	C
	F 15, '51:186	1	
	Jl 4, '63:16	1	
	O 23, '69:26-32 (-T.3)	4	
Flying,	Ag '43:57 (-IV)		S*
	Mr '44:48		C
	Ag '44:49	1	
Fly Rev Intl,	Mr '66:432, 434 (-2, -6, -16, -18, -30, -35)		Pr*
	My '68:268 (-IV:139 Sq)		Pr*
	Ap '69:52-53 (-FB IV: Norway)		Pr*
Roundel,	Mr '64:33 (-30:RCAF)	1	P
RAF Fly Rev,	Ap '56:19-22	4	
RAF Qtr,	Je '46:144 (-RP&-16)		P
Skyways,	F '45:34-41	1	P*, S*

DE HAVILLAND (cont.)

957. D.H. 100 "Vampire"
Aeroplane, D 21, '45:716-721 (-I) 1 S*
Ja 4, '46:18-19 C*
Jl 11, '47:45-46 1
N 28, '47:702-705 4 C*
Air Pict, Jl '51:218 4
O '51:296-298 1
D '53:375 (-T. 11:Iraq) P
Can Avn, D '47:29+ (-III) 4
F '48:28-30+ (-III) 4
Flight, S 20, '45:305-308 1
D 20, '45:655-660 (-I) 1 C*, S*
N 27, '47:605-610 (-III) 4 C*
Ap 1, '48:347-348 (Ghost-) 4
Fly Rev Intl, My '68:272 (-F. 1:247 Sq) Pr*
N '68:43 (-NF. 54:India) Pr*
Mr '69:43 (-FB 6:Swiss) Pr*
Interavia, Je '48:340-343 1
Nav Avn News, S '48:41 (-F. 3) P
Je '55:20-21 (-T. 11) 4
RAF Fly Rev, Ap '53:37-38 (-T. 11)4
Ja '60:8 (-F. 1) 1
:9 (-F. 1:Dominican) Pr*

958. D.H. 100, Mk 20 "Sea Vampire"
Aeronautics, My '49:50-53 P
Avn Wk, F 16, '48:22-23 S*
Flight, D 13, '45:631-633 4
Ja 20, '49:72-73 1
Interavia, Je '48:340-343 1 S*

959. D.H. 103 "Hornet"
Aeroplane, Ja 25, '46:107-111 (-F.I) 1 C, S*
Air Pict, Je '54:179+ 1
Acft Eng, F '47:46-47 6 (Wing Spar)
Aviation, O '46:93 S*
D '46:115 S*
Can Avn, Ap '46:40-41 1 C
Flight, S 20, '45:305-308 1
Ja 24, '46:85-90 1 C, S*

960. D.H. 103, Mk 20 and Mk 21 "Sea Hornet"
Aeronautics, F '48:28, 31-33 P
Flight, Ja 24, '46:85-90 1 S*
Nav Avn News, S '48:41 P
N '48:1 P

DE HAVILLAND (cont.)

961.	D.H. 104 "Dove"		
	USA-built version: RILEY 400		
	Aeronautics, Ag '46:32-36	1	P
	F '47:38-39	4	C
	Aeroplane, O 12, '45:425-428	2	S*
	My 3, '46:521-522		D*
	My 31, '46:637-640	1	C
	My 28, '48:638-640	1	S*
	Je 25, '48:764-765	1	S*
	O 7, '60:504-506 (-8)	4	C
	O 11, '67:40	1	
	Air Pict, S '60:317-320	4	C*
	Acft Eng, N '46:396-397	3	
	D '47:393-401+	3	
	S '49:297-300	1	C*, S*
	Je '50:166-171	2	
	Can Avn, Ag '46:46-47+	1	C
	Comm Avn, Je '45:73	1	
	Ja '46:20-24	1	
	Flight, Ap 12, '45:399-400	1	
	O 11, '45:395-398	1	
	My 30, '46:546a-547	1	S
	:548-549	4	C
	Ag 14, '47:169	3	
	O 14, '60:615-616 (-8)	4	C
	F 3, '61:146-147 (-8)	1	C
	Flying, S '61:30-31+ (-Custom 800)	4	C
	Interavia, No. 3, '46:55-59	1	C, P
	Skyways, D '51:20-22+	4	
	D '52:10-13+	4	
962.	D.H. 106 "Comet"		
	Aero Digest, Ap '50:22-23+	1	
	Ag '50:88		C*
	Aeronautics, Je '50:44-72	1	D*, C*, P*, S*
	S '51:86-91	3	
	Je '52:48-72	1	
	D '58:74-83 (-4)	4	
	Aeroplane, Ag 5, '49:172-173		P
	O 28, '49:581-582	1	
	Ap 21, '50:469-470	1	C*
	O 26, '51:558		S*
	Ja 18, '52:74-78	6 (Radio)	S*
	My 2, '52:514-528	1	S*
	:539-546	4	

DE HAVILLAND (cont.)

962. D.H. 106 (cont.)

Aeroplane,	Jl 6, '56:12-17 (-4)	1		S*
	Jl 27, '56:125-128 (-4A)	2		
	S 6, '57:370-371 (-4B)			S*
	My 9, '58:657-661 (-4)	1		
	O 24, '58:610-624 (-4)	4		S*
	:625-631 (-4)	1		S*
Air BP,	No. 21:20 (-4C:MEA)			Pr*
	No. 26:17 (-4C:United Arab)			Pr*
	No. 34:17 (-4B:Olympic)			Pr*
Air Pict,	N '50:329 (Proto-)	1		
	Mr '51:79			C*
	S '64:284-287 (-4)	1		
Air Prog,	'56/'57 Annual:60-61			S*
Avn Age,	Ap '52:33-34	4		
	My '52:44-47	4		
	Je '52:34-35	3		
	D '52:32-33	6	(Engine Sync)	S*
Avn Opns,	Jl '50:17-18	1		C*
Avn Wk,	Ag 8, '49:12-13	1		
	O 24, '49:46-49	1		
	N 21, '49:16-17	1		
	Ap 3, '50:48-49	2		
	My 22, '50:48			C*
	Jl 17, '50:40-42	3		
	Mr 10, '52:30-37	4		
	Ag 25, '52:21-28	6	(Engine)	
	S 29, '52:20-29	1		
	O 6, '52:12-13 (-3)	1		
	Je 22, '53:18	4		
	Jl 20, '53:71-74 (RCAF-)	4		
	D 7, '53:13-14 (-2)	3		
	F 15, '54:60-63 (RCAF-)	4		
	N 1, '54:16-18	2		
	F 7, '55:28-42 (Pt 1)	2		
	F 14. '55:26-39 (Pt 2)	2		
	S 19, '55:140-141 (-3&-4)	2		S*
	Jl 30, '56:40-41 (-4&-4A)	1		
	Ag 20, '56:72-73 (-4A)	2		
	N 18, '57:91 (-4C)	1		
	Ja 13, '58:96-99	1		
	F 6, '59:41-43 (-4)	4		
	Jl 13, '59:52-61 (-4)	4		C

DE HAVILLAND (cont.)

962.	D. H. 106 (cont.)				
	Can Avn,	S '49:16-17+	1		
		Ja '50:24+	1		
		Ag '50:22			C*
		Ja '53:41-42+ (Pt 1)	1		
		F '53:38-40+ (Pt 2)	1		
		N '56:50-51 (-4A)	1		
		N '58:35 (-4)			D*
	CAHS,	Sum '64:51-52 (RCAF-)	1		Pr†
	Flight,	Mr 31, '49:361	1		
		Jl 28, '49:88-90	1		
		Ag 4, '49:120-121	2		
		O 27, '49:548-551	1		
		D 22, '49:783	2		
		D 29, '49:814-815	1		
		Mr 23, '50:364-366+	4		
		Ap 13, '50:452-453	2		C*
		Ja 18, '51:60-61	1		
		Ap 27, '51:500-503	3		
		S 7, '51:307-310	6	(Interior)	S*
		O 26, '51:534	1		
		F 1, '52:114-115	6	(Fueling)	
		Ap 25, '52:490-493, 503-505	4		
		:495-502, 510-511	1		C
		:506-509	2		
		O 17, '52:502 (-I to -III)	1		
		O 24, '52:525 (-III)	1		
		My 1, '53:535-539 (-I)	4		
		:550-555	1		S*
		:556-559	4		
		Jl 30, '54:132 (-Series 3)	1		
		My 18, '56:607-609 (-4)	1		
		Jl 6, '56:27-30 (-4)	1		S*
		Jl 27, '56:167-168 (-4&-4A)	1		
		My 17, '57:660-661 (RAF-)	4		
		My 24, '57:718-719 (-B)	1		
		N 1, '57:697-700 (-4&-4B)	1		S*
		N 15, '57:764 (-4C)	1		
		Ja 31, '58:159 (-4&-4B)	3		
		Jl 25, '58:121-123 (-4 to -4C)	1		S*
		S 5, '58:389 (-4)	1		
		O 17, '58:615-618 (-4)	4		
		O 24, '58:656-669 (-4)	1		C
		Ja 23, '59:123-125 (-4)	4		C
		S 4, '59:119 (-4)			P
		N 13, '59:548 (-4B)	1		Pr

DE HAVILLAND (cont.)

962.	D.H. 106 (cont.)			
	Flight, N 20, '59:598-599 (-4B)			S*
	D 11, '59:717-722 (-4B)	4		C
	Ja 15, '60:72-74 (-4)	4		
	F 12, '60:219-220	3		
	Mr 25, '65:465-466 (-4C)	1		
	Jl 31, '69:169-172	1		
	Flying, N '52:26-27	1		C, S*
	Interavia, No. 6, '52:309-312 (-I&-II)	1		C
	No. 10, '55:769 (-4)	1		S*
	No. 4, '59:373-375 (-4)	6	(Radio)	C, S*
	Roundel, My '59:2-6 (RCAF-)	4		
	O '63:33 (RCAF-)	1		P
	RAF Fly Rev, Jl/Ag '59:39 (RCAF-)	1		
	N '62:31 (-4C:RAF)	4		
	West Avn, F '56:33 (-3&-4)	1		
963.	D.H. 108			
	Aeronautics, Ag '48:34-37	1		P
	Aeroplane, F 13, '48:181	6	(Instrument)	S*
	Avn Wk, S 20, '48:13	1		
	Flight, Je 6, '46:561-562	1		
	O 3, '46:364-365	2		
	Ap 15, '48:404	4		
	S 16, '48:334	4		
	Fly Rev Intl, O '66:156	1		
	RAF Qtr, O '48:208			P
964.	D.H. 110 "Sea Vixen"			
	Aeroplane, Ap 4, '58:474-476	1		
	S 4, '59:116-117 (-FAW.1)			S*
	Avn Wk, S 15, '52:21-33	1		
	Flight, O 19, '51:509			P
	Ja 21, '55:75	1		
	Jl 13, '56:94-95	6	(Controls)	
	Ap 5, '57:433-436	1		Pr
	F 5, '60:179-186 (-FAW.1)	1		C, S*
	O 14, '60:611-613 (-FAW.1)			P
	Ja 24, '63:116 (-FAW.2)	1		
	Nav Avn News, N '62:39	1		P
	Roundel, F '52:49			P
	RAF Fly Rev, S '56:96-97	1		P*
	Ja '60:42 (-FAW.1)	1		
	S '60:40-43	1		Pr, S*

DE HAVILLAND (cont.)

965. D.H. 112 "Venom" and "Sea Venom"
"Sea Venom" designation: Mks 20 to 22
Aeroplane, Ap 20, '51:478-488 1 P
Flight, S 22, '49:376-377 1
N 23, '51:652 (-20) 1
Jl 23, '54:100-101 (-20) 4
F 8, '57:180 (-21) 6 (Ejection Seat)
Fly Rev Intl, Mr '69:43 (-FB 50:Swiss) Pr*
Interavia, No. 2, '68:216-217 2
Nav Avn News, F '54:1 (-3) P
RAF Fly Rev, S '58:46-49 (-22) 1 Pr, S*

966. D.H. 113 "Vampire Night Fighter"
Aeroplane, F 8, '52:156-160 (-X) 1 C, S*
Flight, S 22, '49:376-377 1
F 15, '52:183-186 1 C*, S*

967. D.H. 114 "Heron"
Canadian "stretched" version: SAUNDERS ST-27
Aeronautics, N '53:70-71 2
Aeroplane, F 17, '50:183-186 1
My 18, '51:601-607 1 C, P, S*
F 8, '52:165 2
D 26, '52:838-839 (-2) P
Air Pict, Mr '52:86 1
Avn Wk, Ja 8, '51:21-25 1
Ag 3, '53:28 (-II) 1
Flight, F 16, '50:233-235 1
Ag 3, '51:128-129 4 C
Ag 17, '51:193 1
F 1, '52:137 2
Ja 22, '54:97-102 (-2) 1 C*, S*
N 4, '55:720 (-2) 6 (Props)
Jl 6, '56:50 (-2) 1
Skyways, Ap '53:14-17+ 1 C, P, S*

968. D.H. 115 "Vampire Trainer"
Aeroplane, D 1, '50:485-490 1 C, S*
My 22, '53:660-661 (-11) P
Ja 6, '56:16-17 (-11) P
Flight, N 23, '50:462-467 1 C*, S*
My 8, '53:578-580 1 S*
My 22, '53:632-634 4
F 5, '54:161 1

DE HAVILLAND (cont.)

969. D.H. 121 "Trident"

Later development: HAWKER SIDDELEY "Trident"

Aeronautics, Ap '60:61			C*
Ja '61:58-61	4		C
Aeroplane, S 4, '59:109-111	1		
Ja 4, '62:12-15	1		
:16-18	6	(Flight Cntrl)	C
My 2, '63:46-49 (-1E&-1F)	1		
Acft Eng, Ja '62:1-6	1		
:7-19	6	(Engine & Flt Control)	C, S*
My '62:149-154	1		
D '62:377	2		
Avn Wk, Ag 31, '59:40-43	1		
Ja 25, '60:48-49	1		C
Je 3, '63:41 (-C, -E, -F)	1		
Can Avn, O '59:34-35	2		
F '62:36-37+	1		
Mr '63:31	6	(Auto Landing)	C
Flight, F 28, '58:267	1		
Jl 25, '58:120	1		
Ag 28, '59:91	1		
S 4, '59:102	1		
Ja 22, '60:102-104	1		
:120	6	(Auto Landing)	
Ag 30, '62:298-299	2		
Ja 3, '63:21-24	1		
My 2, '63:637-648 (-1C, -1E, -1F)	1		C, S*
Ap 2, '64:537-540	4		C
:541-543	3		
:544	6	(Flight Cntrl)	
:545-547	2		
Interavia, No. 3, '60:290	1		C
No. 2, '62:194-195	6	(Flight Cntrl)	C
No. 6, '63:821-825	1		S*
No. 11, '63:1605-1612 (-1C, E, F)	1		
:1612 (-1C, E, F)	2		
:1613-1615 (-1C, E, F)	6	(Engine)	S
:1616-1622 (-1C, E, F)	4		C*

DE HAVILLAND (cont.)

970. D.H. 125 "Jet Dragon"

Later development: HAWKER SIDDELEY 125 and "Dominie"

Air Pict, O '62:328	1	
Air Prog, Ap '65:41-43	1	C, S*
Acft Eng, F '63:32-36	1	C, S
:37-43	1	S*
Avn Wk, My 29, '61:74-75	1	
Jl 23, '62:68-71	1	
S 24, '62:112-113	1	
D 3, '62:110-112	1	
S 2, '63:52-55	4	C
Bris Sid Jnl, Aut '62:7-9	1	
Can Avn, N '62:24-25	1	
N '65:30-31	1	
Flight, My 25, '61:714-715	1	
Ja 11, '62:41-47	1	
Jl 26, '62:125-127	1	
Ag 23, '62:282-283	2	
D 6, '62:896-903	1	C, S*
Flying, N '62:56-57+	1	
O '64:52-55	4	C
F '66:84-90	4	C, P*
D '67:50-53 (-3AR)	4	
Interavia, No. 9, '62:1111-1112	1	
RAF Fly Rev, O '62:21-22+	1	
Jl '63:36-39+	1	Pr, S*

DE HAVILLAND AIRCRAFT PTY. Australia

971. D.H.A. 3 "Drover"

Avn Age, Ap '51:34	1	
Flight, Ap 29, '48:462	1	
RAF Fly Rev, My '60:9	1	

DE HAVILLAND CANADA Canada

Can Avn, Ap '53:47-61+
O '60:13-20
O '68:20-25
Comm Avn, O '39:10-12
Roundel, N '58:10-14

DE HAVILLAND CANADA (cont.)

972. AC-1 "Caribou"
US Army designation
See also DHC-4

Avn Wk, Ja 23, '61:56-69	4	C
US Army Avn, Mr '62:26-28 (-A)	1	C*

973. CV-7A "Buffalo"
US Army designation
See also DHC-5

Avn Wk, My 25, '64:84-89	1	
My 6, '68:48		P

974. DHC-1 "Chipmunk" and "Turbo-Chipmunk"
Acrobatic modification: HUNEAULT DHC-1

Aeronautics, Ja '48:52-53	4	C
F '52:30-32		S*
Aeroplane, Ag 22, '47:220-221	4	C
Mr 25, '49:335-337	4	C
D 2, '49:754-757 (-10)	1	C*, S*
Air Pict, My '60:154-156+	4	
Can Avn, Jl '46:42-43+	1	
Je '47:30-31	4	C
My '50:18-20 (Pt 1)	4	C*
Je '50:37+ (Pt 2)	3	S*
Mr '55:30-31	4	C
Mr '65:29+	1	P
F '68:8-10+ (Turbo-)	4	S*
CAHS, Fall '67:90		S*
Win '67:120		C
Flight, Je 20, '46:619	1	
Ag 21, '47:183-186	1	C*, S*
D 1, '49:710-712 (-10)	1	C
Jl 18, '58:69 (-23)	1	
O 26, '67:677-679 (Turbo-)	4	
Flying, N '68:56-60 (Turbo-)	4	P*
Interavia, No. 12, '47:40-42	1	C
RAF Qtr, Ja '49:xii		D*

975. DHC-2 "Beaver" and "Turbo-Beaver" (-3)

Aeroplane, Mr 3, '50:258-263	1	C*, S*
Mr 17, '50:316-319	4	C
Ag 14, '53:205-206	4	C
Ja 15, '54:70-71 (-2)	1	
S 1, '66:46-50 (-3)	4	

DE HAVILLAND CANADA (cont.)

975.	DHC-2 (cont.)		
	Air Pict, D '51:374	1	
	O '53:318 (-1&-2)	1	
	D '60:416-419	4	C
	Aviation, Je '47:63	1	
	Avn Age, D '54:144-149	4	
	Avn Wk, S 3, '51:18	1	
	Je 22, '64:80-87 (-3)	1	
	Ja 11, '65:84-89 (-3)	4	C
	Can Avn, Ap '47:24-25+	1	C
	S '47:22-23	1	
	Ap '48:28-29+	4	C
	My '48:64-66	3	
	Jl '50:19+ (Ag-)	1	
	O '54:33-35	3	
	Mr '64:20-23 (-3)	1	
	Mr '65:34-37 (-3)	4	
	Ja '66:12+ (Ag-)	4	
	Flight, Jl 3, '47:16	1	
	Ja 26, '50:92-93	4	C
	Ag 21, '53:219-221 (-2)	4	C
	F 19, '54:203-204	4	
	:204 (-2)	1	
	Ap 23, '54:507 (Ag-)	1	
	Je 6, '68:856a (RAF-)	1	P*
	Flying, O '49:34-35+	4	C, P
	Ap '55:58	1	
	RAF Fly Rev, Mr '58:8-9	1	P*
	Skyways, O '49:58	1	C
	Je '50:20-22+	4	C
	F '53:16-17+	4	
	US Army Avn, Jl '60:16-19	2	
	Ap '64:24-27	4	
976.	DHC-3 "Otter"		
	Aeroplane, Ja 1, '60:10-12	4	C*
	Air Pict, Je '53:187		P
	Air Prog, F/Mr '63:78-79		Pr†
	Ap '65:44-45	1	
	Army Avn D, My '55:28-32	1	C
	Avn Wk, F 18, '63:67-75	1	
	Flight, Ap 11, '52:428-429+	1	
	D 30, '55:973-974	4	C
	Jl 15, '60:105	2	
	Flying, Ja '56:67	4	
	F '59:62	1	

DE HAVILLAND CANADA (cont.)

976.	DHC-3 (cont.)		
	Roundel, Jl/Ag '60:20	1	
	US Army Avn, Mr '59:5-11	4	
977.	DHC-4 "Caribou"		
	US Army designation: AC-1		
	Aeronautics, Ja '60:36-37	1	
	Aeroplane, N 20, '59:514-517	4	C
	Avn Wk, Ap 1, '57:33-34	1	
	S 29, '58:60-73	1	
	O 19, '59:125-127	2	
	F 18, '63:67-75	1	
	Bee-Hive, Sum '61:29-33	1	
	Can Avn, Ap '57:152-153	1	
	Ag '58:26-27	3	
	O '58:24-31	1	C
	:32-34+	6	(Landing Gear)
	My '61:18-19	2	
	Je '66:32-33	4	
	Flight, Mr 15, '57:342	1	
	S 19, '58:482-483	1	
	N 20, '59:568	4	
	Ap 6, '61:436-438	1	
	Interavia, No. 2, '60:181-182	4	C
	No. 1, '62:76-78	4	C
	No. 9, '64:1318-1319	4	
	Roundel, Ja/F '61:33 (RCAF-)		P
	My '61:25-27 (RCAF-)	1	
	RAF Fly Rev, Ap '62:25-26	1	
	US Army Avn, N '59:1-4	1	C
978.	DHC-5 "Buffalo"		
	Canadian Defense Forces designation: CC-115		
	See also CV-7A		
	Avn Wk, Ja 13, '69:38-42	4	C, P
	Can Avn, Jl '64:16-19	1	
	Flight, My 14, '64:808-809	1	
	My 13, '65:747-748	1	
	Interavia, No. 10, '64:1536-1537	1	
	Roundel, Ja/F '65:3	1	D
	Sentinel, F '68:32-34 (CC-115)	2	
	Je '69:8-13	4	

DE HAVILLAND CANADA (cont.)

979.	DHC-6 "Twin Otter"		
	Aeroplane, S 1, '66:46-50	4	
	Avn Wk, Je 22, '64:80-87	1	
	Ja 4, '65:72-74	1	
	Ag 2, '65:105	1	
	F 14, '66:104-113	4	C
	Mr 7, '66:265		P*
	S 26, '66:79-80	1	
	Can Avn, O '64:47	1	
	My '65:30-31	1	
	My '66:14-16	4	
	My '67:34		P*
	Flight, S 24, '64:555-557	1	C, S*
	S 8, '66:442-445	4	C, S*
	Flying, O '66:64-65	1	P*
	Ag '68:40-42	1	
	Fly Rev Intl, Je '66:616-619	1	S*
	My '69:28-30 (-300)	1	
	Interavia, No. 11, '64:1663	1	S*
	No. 10, '65:1577	1	
	No. 8, '67:1266-1267	4	C
980.	DHC-7		
	Avn Wk, S 8, '69:40-43+	1	C
	Can Avn, N '68:23	1	
	Flight, O 24, '68:647-648	1	
	Je 19, '69:1007	1	
	Fly Rev Intl, D '68:32	1	
	Interavia, No. 1, '69:73	1	
	No. 9, '69:1549 (-Revised)	1	

DE LACKNER USA

981.	DH-4 "Heli-Vector"		
	Am Helicop, Ap '55:13+	1	

DE MARCAY France

982.	C-1		
	RAF Fly Rev, S '60:25	1	

DE SCHELDE Netherlands

983.	S. 20		
	RAF Fly Rev, Mr '59:45	1	

DE SCHELDE (cont.)

984.	S. 21		
	Aeroplane, Mr 1, '40:276-277	1	
	RAF Fly Rev, Ap '54:5	1	
	D '55:13	1	S
	My '59:11	1	

DESOUTTER GB

985.	Mk II "Sports Coupe"		
	Flight, S 12, '30:1011-1014, 1029	1	C, S, Pr

DEWOITINE France

986.	D. 27		
	Built in Switzerland		
	Fly Rev Intl, Mr '69:42 (Swiss-)		Pr*
	Pop Avn, Ap '31:42	4	
	RAF Fly Rev, N '59:46	1	
	RAF Qtr, Ap '31:284-285	1	
987.	D. 33		
	Flight, Ap 10, '31:314	1	
	Pop Avn, N '31:10	4	
988.	D. 332		
	Flight, Ag 10, '33:801	1	
989.	D. 333 "Antares"		
	Flight, Ja 3, '35:24	1	
990.	D. 338		
	Pop Fly, Ap '37:22		C
991.	D. 371		
	Pop Avn, My '36:318-319	1	P
	RAF Fly Rev, My '60:11 (Lithuanian-)	1	
992.	D. 500		
	Flight, Mr 15, '34:252	1	
993.	D. 510		
	See also D. 520		
	Pop Avn, My '36:318-319	1	P
	RAF Fly Rev, N '57:7	1	

DEWOITINE (cont.)

994.	D. 513 and D. 514		
	RAF Fly Rev, D '59:42+	1	
995.	D. 520		
	Refinement of D. 510		
	Air Prog, Je/Jl '63:46-47		Pr†
	Je '68:49 (Le Gloan's-)		Pr†
	Flight, Ap 11, '40:330e-g	1	D, S
	Fly Rev Intl, Ja '64:56-57		Pr*
	RAF Fly Rev, O '57:9+	1	
	S '58:31-32	4	
	Ja '59:6 (2-seat-)	1	
	Je '60:31-33	1	S*
	:30 (Vichy Groupe de Chasse)		Pr†
996.	D. 550		
	Fly Rev Intl, Ap '67:515	1	
997.	D. 551		
	Fly Rev Intl, Ap '67:515	1	
998.	D. 620		
	Acft Eng, My '36:136	1	
999.	D. 730 and D. 732		
	RAF Fly Rev, D '57:9	1	

DFS (Deutsches Forschungsinstitut für Segelflug) Germany

1000.	39 "Delta IV"		
	Fly Rev Intl, My '69:73	1	
1001.	230 Assault Glider		
	RAF Fly Rev, Ja '58:33-34	1	
1002.	346 Rocket Aircraft		
	Interavia, No. 5, '53:256-257	1	S*
	RAF Fly Rev, S '54:7	1	
1003.	582 Altitude Research Aircraft		
	Later versions: FFM 582 and DVL 582		
	Interavia, No. 5, '61:711	1	

DFW (Deutsche Flugzeug-Werke) Germany

1004.	C. V		
	Cr & Cockade, Sum '69:167 (Huppertz's-)	1	P
	:171 (Mann's-)	1	P
	Fall '69:281		P

DINFIA (Direccion Nacional de Fabricaciones e Investigaciones Aeronauticas) Argentina

1005.	I.A. 38 Tailless Transport		
	RAF Fly Rev, N '61:47	1	
1006.	I.A. 50 "Guarani II"		
	Avn Wk, D 30, '63:42-51	1	
	Mr 29, '65:77	3	
1007.	I.A. 58 "Delfin" (Dolphin) Counterinsurgency aircraft Military designation: A-X2		
	Avn Wk, O 27, '69:17	1	
	Fly Rev Intl, D '69:27-28	1	

DISTRIBUTER WING USA

1008.	DW-1		
	Avn Wk, S 14, '64:131-137	1	S*
	Jl 26, '65:91	1	

DOAK USA

1009.	VZ-4DA		
	Am Helicop, Ap '59:10	1	
	Avn Wk, Je 8, '59:34	1	
	Ag 17, '59:109-111	1	
	Nav Avn News, D '59:22	1	
	US Army Avn, Ag '59:38	2	

DOMAN USA

1010.	YH-31 Civil designation: LZ-5		
	Aeroplane, Jl 26, '57:128-129	1	
	Am Helicop, N '52:6-7+	1	S*
	My '56:4	4	
	Avn Age, Mr '53:32-33	1	S*
	Avn Wk, Mr 9, '53:25-26	1	S*
	My 11, '53:16	1	

DOMAN (cont.)

1011. LZ-1
Am Helicop, Ap '46:34-35 1

1012. LZ-2A "Pelican"
Am Helicop, Ap '49:11+ 1

1013. LZ-5
Military designation: YH-31
Am Helicop, N '52:6-7+ 1
My '54:10-11+ 1 S*
F '56:13 1
S '58:9 1
Avn Wk, Ja 7, '52:21-27 1
Mr 9, '53:25-26 1 S*
Can Avn, My '54:28-30+ 1 S*

DORNIER Germany and Switzerland

1014. "Libelle" (Dragonfly)
Flight, O 6, '32:930 1

1015. C. IV
Flight, My 24, '34:521 1

1016. D. I
Cr & Cockade, Win '68:391-395 2 P

1017. D-3800 to D-3803
Air Pict, Je '68:217 (-3802 & -3803) 1
Fly Rev Intl, Je '64:64 (-3800:Swiss) Pr*
Mr '69:43 (-3801:Swiss) Pr*
:43 (-3802A:Swiss) Pr*
Ap '69:72 (-3803) 1

1018. K
Flight, O 9, '31:1014 1
O 30, '31:1087-1089 3
Pop Avn, Ap '32:227 1

1019. R "Super Wal" (Super Whale)
Pop Avn, F '29:98-100 1 C

DORNIER (cont.)

1020.	Do. X		
	Aero Digest, D '30:66-67	1	
	Aeron (US), O '29:42-43	1	
	Aeroplane, N 12, '30:1075-1076	4	
	N 19, '30:1116-1126	4	P, S
	Ap 12, '33:650-654	1	
	D 18, '42:718	1	
	Air Cl, Ap '69:6-14+	1	P, Pr
	Acft Eng, D '29:339-341	1	
	Ag '32:193-196	1	
	Aviation, Ja 4, '30:4-11	3	
	Flight, F 21, '30:233-237	1	C
	N 14, '30:1246-1248	4	P
	My 27, '32:460	4	
	Jl 15, '32:667-669	4	
	Ap 13, '33:353-357 (Pt 1)	1	
	Ap 20, '33:374-378 (Pt 2)	1	
	Flying, Ap '62:44-45+	1	S*
	RAF Fly Rev, N '56:11+	1	
	US Air Svcs, F '30:22-25	4	C
	Ap '31:23-27	4	
	O '31:18-19	4	
	D '33:25-26	4	
1021.	Do. 11 or Do. F		
	Air Cl, Ap '69:49	1	
	Fly Rev Intl, Ja '66:317	1	
1022.	Do. 13		
	Fly Rev Intl, Ja '66:317	1	
1023.	Do. 17 "Flying Pencil" and "Eversharp" Later developments: Do. 215 and Do. 217		
	Aeroplane, D 22, '39:808-810	1	
	AF News Ltr, Ag/S '42:39 (-Z)		S*
	Aviation, Ag '42:196	6 (Guns)	S*
	O '42:154		S*
	Flight, Ap 14, '38:364a-366+	1	S*
	N 2, '39:350a-351	1	S
	RAF Fly Rev, N '54:37-38+	1	

DORNIER (cont.)

1024.	Do. 18		
	Aero Digest, O '36:58-69	1	
	Aeroplane, S 19, '41:317 (-K2)		S*
	Aviation, Mr '42:91 (-K2)		S*
	Flight, O 22, '36:421	1	
	Ap 7, '38:349	4	S*
	N 9, '39:372b-d	1	S*
1025.	Do. 19		
	RAF Fly Rev, Ap '55:10	1	
	My '61:48	1	
1026.	Do. 22		
	RAF Fly Rev, S '58:23	1	
	Jl '62:47	1	
	Ja '63:47	1	
1027.	Do. 23		
	Fly Rev Intl, Ja '66:317	1	
1028.	Do. 24		
	Acft Eng, Mr '38:83	1	
	Flight, Ap 27, '39:424c		C
	N 9, '39:372a-c	1	D
	My 13, '43:500b	1	
	RAF Fly Rev, My '61:25-26	4	
	Ag '61:14	4	
	Ja '63:37-38 (Spanish-)	4	
1029.	Do. 26		
	Aero Digest, Ag '39:138-141	1	C
	Aeroplane, S 7, '38:289-290	1	
	Acft Eng, S '39:357-363	3	C
	Aviation, F '41:122		S*
	Flight, S 8, '38:218	1	C
	Ja 28, '43:94b	1	
1030.	Do. 27		
	Aeroplane, My 11, '61:527-528	4	C
	Avn Wk, Mr 17, '58:85-89	1	
	Ag 3, '59:104-113	4	
	N 14, '60:135-139	1	
	Flight, Je 14, '57:796-797	4	
	Interavia, No. 7, '57:707-709	1	

DORNIER (cont.)

1031.	Do. 28 "Skyservant"		
	Aeroplane, My 11, '61:527-528	4	C
	Air Pict, Je '66:223-225 (-B-1)	4	S
	Air Prog, F '67:42-43 (-D)		S*
	Avn Wk, My 23, '60:102-105	1	
	N 14, '60:135-139	1	
	Ja 24, '66:103-106 (-D)	1	
	Je 19, '67:92	1	
	Flying, D '67:40-43 (-D)	4	C, P*
	Interavia, No. 4, '64:493-495 (-D)	1	
	No. 8, '65:1264 (-D)	1	
	No. 1, '68:47-49	1	C
1032.	Do. 29		
	Avn Wk, Je 1, '59:87	1	
1033.	Do. 31		
	Air BP, No. 45:30-33	1	P
	Air Force, F '68:83-87	1	
	Air Prog, S '67:24-25+	1	C
	Avn Wk, Mr 30, '64:30-32	1	
	Ja 3, '66:32-34 (-E)	1	
	F 27, '67:32	2	
	My 13, '68:78-79	1	
	Flight, My 28, '64:902-904	1	S
	My 9, '68:709-711	1	C, S*
	Interavia, No. 1, '64:59-62	1	
	No. 11, '65:1751	2	
	No. 1, '66:54-55	2	
	No. 6, '66:928-931	1	S*
	No. 12, '67:1852 (-E3)	4	
1034.	Do. 32		
	Avn Wk, Je 10, '63:160-167	1	
1035.	Do. 132		
	Avn Wk, Mr 11, '68:38-45	1	C, S*
	Interavia, No. 10, '67:1566-1567	1	
	No. 7, '69:910-911	1	S*
1036.	Do. 212		
	Fly Rev Intl, N '66:221	1	
	RAF Fly Rev, Ap '61:49	1	

DORNIER (cont.)

1037.	Do. 215		
	Development of Do. 17		
	Aeroplane, O 26, '39:511		D
	D 22, '39:808-810	1	S*
	Flight, N 2, '39:350a-351	1	S
	Fly Rev Intl, D '63:61	1	
1038.	Do. 217		
	Development of Do. 17 and Do. 215		
	Aeronautics, S '42:28-31 (-E)	1	S*
	Aeroplane, F 13, '42:183 (-E)	1	
	Je 26, '42:734-736 (-E)	1	S*
	Jl 17, '42:70-71 (-E)		S*
	Ag 21, '42:218		P
	Air Prog, Mr '43:56-57 (-E-2)		S*
	Air Tech, Ap '43:39-41+	1	
	Acft Eng, Ag '42:214-222 (-E-1)	1	S*
	:230-234	6 (Radio)	
	Aviation, N '42:154		S*
	Comm Avn, N '42:70-76+ (-E-1)	6 (Bombing)	S*
	D '42:42-44 (-E)	6 (De-ice)	S*
	Flight, Je 25, '42:635-639 (-E)	1	S
	Jl 2, '42:17 (-E)	6 (Dive Brakes)	S
	Ag 20, '42:205 (-E)	6 (Dive Brakes)	
	S 10, '42:282b (-E)	1	
	O 15, '42:412-414 (-E-2)	1	S*
	Flying, Ag '43:51 (-E)	6 (Dive Brakes)	
	Fly Rev Intl, Ap '67:512, 514 (-E-1, 2, 4, 5)		Pr*
	RAF Fly Rev, N '54:40	1	
	O '61:57	1	
1039.	Do. 335, Do. 435 and Do. 635 "Pfeil" (Arrow)		
	Do. 435: Proposed two-seat night fighter		
	Do. 635: Projected "twin" version		
	Air News, D '45:24	1	
	Air Prog, O '65:57 (-B-2)		Pr†
	Acft Eng, D '46:413-416	6 (Hydraulic)	S
	RAF Fly Rev, S '54:39	1	
	O '56:11+	1	
	D '60:30 (-A-1)		Pr
	:31-33	1	
	:32-33 (-A-6)		S*

DOUGLAS USA

See also MCDONNELL DOUGLAS
Aerosp Intl, S/O, '68:41-45 (Merger with MCDONNELL)
Can Avn, Jl '60:18-22 (Development of DC-1 to DC-8)
Flying, S '59:40-47+ (Development of DC- Series)
Interavia, No. 2, '54:92-96
Nav Avn News, My '56:14-17 (Naval aircraft, 1922 to 1956)
West Fly, Mr '30:60-62

1040.	"Cloudster"		
	Interavia, No. 1, '47:62-63	1	P
	US Air Svcs, Ja '47:15	1	
	West Fly, F '47:30	1	
1041.	"Dolphin"		
	Aero Digest, Mr '31:76-77	1	C
	S '35:44		C
	Aviation, Ap '31:248-249	1	
	Boeing Mag, Ja '69:12-13 ("Rover")	1	
	Flight, O 23, '31:1070-1072	1	
	F 15, '34:155-156	1	
	Pop Avn, My '31:42	1	
	US Air Svcs, Ja '31:30-31		P
	S '34:27 (-113)	1	
	D '35:31	4	
	West Avn, F '31:56	1	
	Ap '31:66	1	
1042.	A-1 "Skyraider"		
	See DOUGLAS AD		
1043.	A2D-1 "Skyshark"		
	Air Pict, Ag '50:243 (X-)		P
	Avn Wk, Je 12, '50:13-14 (X-)	1	
	O 23, '50:22-23	1	S
	Flight, N 16, '50:424-426 (X-)(Pt 1)	1	S
	N 23, '50:454-455 (X-)(Pt 2)	1	
	Nav Avn News, Jl '50:9	1	
	RAF Fly Rev, F '62:41	1	
	Skyways, S '50:23	1	P
	O '50:10-11+	1	
	US Air Svcs, Je '50:16		D
	N '50:6-8	1	P
	West Fly, Jl '50:10 (X-)	1	

DOUGLAS (cont.)

1044.	A3D "Skywarrior"		
	US Air Force designation: B-66 "Destroyer"		
	Air Pict, Mr '65:99	1	
	Avn Wk, D 13, '54:28-43	1	
	Mr 14, '55:75		P*
	Ag 13, '56:60-64	1	
	Ag 20, '56:66-70	1	S*
	Flight, F 18, '55:205-208 (-1)	1	S*
	F 15, '57:211-214 (-1)	1	S*
	Nav Avn News, Ja '53:26	4	
	My '57:9 (-1)	4	
	Ag '57:12-14	4	
	RAF Fly Rev, Jl '56:48-49 (-1)	1	
	Je '57:24-27 (-1)	1	Pr, S*
	West Avn, Ja '61:24-25+	4	
1045.	A-4 "Skyhawk"		
	Air Force, N '65:42-48	1	
	AF/Sp D Intl, Ja '66:11-18	1	
	Air Pict, Ag '54:231	1	
	Air Prog, S '68:40-44+	1	P*
	Avn Age, D '56:26-31	1	S
	Avn Wk, Je 14, '54:14-15 (-D)	1	
	Ja 24, '55:28-33 (-D)	1	
	Ag 26, '57:50-61 (-D-2)	1	S
	Ag 8, '66:31 (-F)	1	
	S 2, '68:23 (-C)	6 (Boron Flap)	
	Bee-Hive, Spg '67:22-25	4	
	Flight, Mr 7, '58:308-311 (-D-1&-2)	1	C, S*
	Ja 16, '64:100 (-E)	1	
	Flying, Ap '56:36-37 (-D)	1	S
	Fly Rev Intl, Mr '66:389-392	1	
	Interavia, No. 3, '55:172-178 (-D)	1	P, S*
	No. 5, '64:672-673	1	
	No. 4, '66:497-499 (-E&-F)	1	S*
	No. 8, '66:1207 (-F)		D
	Nav Avn News, D '56:27 (-D)	1	
	My '57:14 (-D)	2	
	O '60:13 (-D)	6 (Bomb Racks)	
	Ag '61:16 (-D-2)		D
	Mr '68:6-10 (-A to -F)	1	
	:11-13	4	
	Ap '68:11		D

DOUGLAS (cont.)

1045.	A-4 (cont.)		
	RAF Fly Rev, Ag '54:15	1	
	Mr '56:26-29 (-D-1)	1	P, Pr, S*
	Je '62:28-31+ (-D-2)	1	Pr, S*
	West Avn, Jl '54:14 (-D)	1	
1046.	A-20 "Havoc"		
	Export version: DB-7 "Boston"		
	Aero Album, Win '69:24-25 (-B)	4	P
	Aero Digest, My '43:180-181		S
	Air BP, No. 26:32		C
	AF News Ltr, Mr/Ap '42:17-18+	4	
	Air Tech, O '43:20-27+	1	C*, S*
	Aviation, Je '42:86-91	3	
	D '43:199		S*
	Ja '44:125-146	1	C, S*
	Comm Avn, Ap '44:112	6 (Armament)	
	Flying, Ag '66:23 (-G)		D
	Fly Rev Intl, Jl '68:408, 410 (-B)		Pr*
	Skyways, Ag '44:18-23+	1	S*
	West Fly, Mr '42:27+	4	
	O '43:76-77+	3	
1047.	A-24 "Dauntless"		
	US Navy designation: SBD-Series		
	Aeroplane, Jl 28, '44:98-101	1	S*
	Air Pict, O '64:325-328	1	
	AAHSJ, Spg '67:32-39	4	Pr
1048.	A-26 "Invader"		
	US Navy target-tug version: JD-1		
	Aero Digest, D 1, '44:72-73	1	
	Air Tech, Ja '45:46+	1	
	My '45:20-21+	3	
	Aviation, Ap '45:175-176+	1	
	My '45:160-161	6 (Flaps)	
	Comm Avn, Ja '45:67	1	
	Pegasus, F '45:5-7+	1	P
	Je '45:14-16	1	
	US Air Svcs, My '45:12-14	4	
	West Fly, N '44:82+	1	
1049.	AC-47 "Dragon Ship," "Spooky," "Puff the Magic Dragon"		
	Attack version of C-47		
	Avn Wk, Ap 17, '67:54-63	4	
	Flight, Ja 26, '67:124-128	4	

DOUGLAS (cont.)

1050.	AD "Skyraider" or "Able Dog"		
	Aero Digest, N '51:26-40 (-5)	1	S*
	Ap '54:50-56	1	
	Aeroplane, F 15, '52:187-191 (-4W)	1	P
	Air Pict, Ag '69:301	1	
	Aviation, Ap '47:78-80 (-1)	3	S*
	Avn Wk, N 19, '51:18 (-5)	1	
	Ag 24, '53:24-26 (-5)	1	
	Mr 11, '57:63 (-7)	1	
	Flight, O 22, '54:604-605 (-4W)	4	
	Flying, Ap '50:18-19+	1	
	F '52:30-31	1	
	Interavia, No. 2, '47:47-49	1	P
	Nav Avn News, O '46:17 (-1)	1	
	D '47:1-5 (-1)	1	P
	S '48:8 (-2)	1	
	D '49:20-21 (-1 to -4)	1	
	My '50:34 (-3W)	4	
	D '51:18 (-5)	1	
	Jl '57:18-19 (-1 to -7)	1	
	N '57:14-15 (-5W)	4	
	Ap '62:14-15	1	P
	S '67:10-11	4	
	Jl '68:18-27 (-1)	1	S
	RAF Fly Rev, D '62:18-20+	1	
	:18-19 (-4)		S*
	Skyways, Je '49:32-34+ (-1 to -4)	1	
	Ap '51:14-15+ (-1 to -4)	4	
	US Air Svcs, N '46:6 (-1)		P
	D '51:8 (-5)		P
1051.	YB-7 and XB-7		
	RAF Fly Rev, N '59:48	1	
1052.	B-18A "Bolo"		
	Canadian version: DB-280 "Digby I"		
	Air News, Je '41:18		C
	Air Prog, Ap/My '68:68		C
	Air Tech, N '42:31		P
	Pop Avn, D '37:26-27	1	P
	Ja '39:39	1	
	RAF Fly Rev, Ja '60:5+	1	

DOUGLAS (cont.)

1053.	XB-19 "Hemisphere Guardian"		
	Aero Digest, D '40:130-131	3	
	Air Cl, Mr/Ap '65:8-11+	1	C, P
	Air Force, Ag '49:32-33+	1	
	Air Pict, Ag '61:232-235	1	Pr
	Air Tech, O '42:30		P
	Aviation, O '40:58-59	1	
	Bee-Hive, Jl '41:2-4	1	
	Comm Avn, D '40:32-34	1	
	Jl '41:52	1	
	Flight, Ap 25, '40:391	1	
	O 3, '40:270a-b	1	
	Je 19, '41:416f-h	1	S*
	D 11, '41:420d-e	1	C
	Ja 28, '43:94a	1	
	Flying, Jl '42:50-52+	1	
	Fly Rev Intl, O '66:157	1	
	Pop Avn, Je '40:15	3	
	RAF Fly Rev, Jl '56:11+	1	
	US Air Svcs, Ap '40:36	1	
	Jl '41:25-26+	1	
	West Fly, D '40:18-21	3	
1054.	B-23 "Dragon"		
	RAF Fly Rev, Ap '61:51	1	
1055.	XB-42 "Mixmaster"		
	Aeroplane, My 3, '46:524	1	P
	Aviation, F '47:37-42	1	S*
	Fly Rev Intl, S '69:75-76	1	
	US Air Svcs, Ja '46:10	4	P
	West Fly, D '45:58+	1	
1056.	XB-43		
	Only two built; second aircraft named "Versatile II," "Frog," or "Grasshopper"		
	RAF Fly Rev, F '62:43	1	
	US Air Svcs, Ag '46:15	1	P
1057.	B-66 "Destroyer"		
	US Navy designation: A3D "Skywarrior"		
	Aeroplane, Mr 29, '57:442-443 (R-/-B)		P
	Air Pict, My '57:157 (R-)	1	
	Mr '65:99	1	

DOUGLAS (cont.)

1057. B-66 (cont.)

Air Prog, O '68:46-48+ (E-)	4	
Avn Wk, Ag 13, '56:60-64	1	
My 29, '61:61 (W-)	1	
Flight, Mr 29, '57:411-414 (R-/-B)	4	P
Ag 19, '60:262-263 (R-/-B)	4	

1058. BT-1

Air Cl, Ag '69:20-21, 24-25	1	C*
Pop Avn, F '39:30 (Tricycle-)	1	
RAF Fly Rev, S '62:45	1	
US Air Svcs, D '41:14		P

1059. BT-2

Air Cl, Ag '69:20-21, 24, 26+	1	P
US Air Svcs, Ja '38:10 (-B:Weather)		P

1060. BT2D-1 "Destroyer II"

Aviation, Ap '46:55-58 (Pt 1)	1	S*
My '46:77-79 (Pt 2)	1	S
RAF Fly Rev, D '60:11	1	

1061. BTD-1 "Destroyer"
Redesignation of SB2D "Destroyer"

RAF Fly Rev, D '60:11+	1	

1062. C-1

Flying, My '67:58		P

1063. C-9A "Nightingale"

Aerosp Intl, Ja/F '68:29-32	1	
Avn Wk, Ja 29, '68:36-37	1	

1064. C-47 "Skytrain," "Dakota," "Gooney Bird"
See also AC-47, XCG-17, DC-3, R4D and Y-123

Aero Digest, O 1, '45:73+	1	S*
Aeroplane, Jl 13, '45:48-50 (-I&-III)	1	
D 12, '52:776-780 (Dart-)	4	
Aerosp Hist, Aut '66:121-125	1	
Air Force, S '44:14-17+	4	
Air Prog, Ap/My '63:76 (-A)		C
Air Tech, S '44:15-22+	1	C*, S*
Fly Rev Intl, S '64:42		Pr*
RAF Fly Rev, Mr '59:41-42	5	
US Air Svcs, S '44:30 ("Old Miscellaneous")	4	
West Fly, O '44:108-109+	3	

DOUGLAS (cont.)

1065.	C-53 "Skytrooper"		
	Civil version: DC-4		
	US Navy versions: R4D and R5D		
	Flight, N 18, '43:552-553	1	
1066.	C-54 "Skymaster"		
	Civil version: DC-4		
	Aero Digest, O 15, '44:82-88	3	S*
	Air Tech, D '44:17-25	1	C, S*
	Bee-Hive, Spg '46:18-20	1	
	Flight, Jl 16, '42:75	1	
	Mr 4, '43:288b	1	
	N 29, '45:581-584 (-B: Churchill's)	1	S*
	Flying, Je '44:66-67	1	S*
	S '45:63		P*
	Skyways, Ja '45:34-40+	1	C, P*, S*
	US Air Svcs, Ja '46:13-14	3	
	West Fly, Jl '45:48-49+ (Roosevelt's-)	4	
	F '51:22 (-M)	1	
1067.	C-74 "Globemaster I"		
	Air Force, Je '48:18-19	1	
	Aviation, N '45:150-151	6 (Flaps)	
	Can Avn, N '45:36-37+	1	
	Interavia, No. 2, '46:56-57	6 (Flaps)	
	No. 7, '46:26-31	1	P
	West Fly, O '45:52-54+	1	
1068.	C-124 "Globemaster II"		
	Aero Digest, N '52:26-36 (-A)	1	S*
	Air Pict, O '69:389	1	
	Air Prog, Ap/My '63:73 (Y-)		C
	Avn Wk, Ja 22, '51:13-14 (Y-/-B)	1	
	N 17, '52:62-68	4	
	Bee-Hive, Spg '51:19-21	1	
	Interavia, No. 1-2, '50:57	1	
	No. 12, '51:679-682 (-A)	1	
	Skyways, F '50:28-29	1	
	US Air Svcs, D '49:12	1	P
	F '54:22 (Y-/-B)	1	
	West Avn, N '51:11	3	
	Mr '54:18 (Y-/-B)	1	
	West Fly, O '48:12+ (-A)	1	
	D '49:17-18	1	

DOUGLAS (cont.)

1069.	XC-132			
	Interavia, No. 4, '57:324	1		
	West Avn, S '56:29	1		
	N '56:26	1		
	Mr '57:16	1		
1070.	C-133 "Cargomaster"			
	Aeroplane, Je 21, '57:885-886	1		
	Avn Age, S '56:20-25 (-A)	1		S*
	Avn Wk, S 5, '55:28-39	2		
	F 6, '56:26-27 (-A)	1		
	F 18, '57:32-33 (-A)	6	(Cargo)	
	My 12, '58:70-83 (-A&-B)	1		C
	Ag 24, '59:70-82 (-A)	4		
	Flight, Ap 27, '56:506-508 (-A)	1		
	S 26, '58:519-523 (-A)	1		S*
	West Avn, Mr '56:32 (-A)	1		
1071.	XCG-17			
	DOUGLAS C-47 Glider			
	Air Pict, My '67:173	1		
	Comm Avn, Ag '45:60-64	1		
	West Fly, S '45:108	1		
1072.	CX-4			
	Projected six-engine transport			
	Interavia, No. 12, '63:1906-1907	1		D
1073.	D-558-1 "Skystreak"			
	Aeroplane, Ap 18, '47:396-398	1		
	Am Helicop, Ap '47:12+	1		S
	Aviation, My '47:54-56	1		
	Avn Wk, S 1, '47:10	4		
	S 8, '47:34-37	6	(Escape)	
	Flight, Ag 28, '47:225	4		
	Flying, D '47:32-34+	4		
	Interavia, No. 3, '47:26-27	1		
	Nav Avn News, Ja '47:24-26	1		S
	Ap '47:7	1		
	Ag '47:22-23	1		S*
	O '47:10	4		
	:11	6	(Escape)	S
	Pegasus, Je '52:10-11	1		S*
	Skyways, Ja '48:18-19+	4		
	US Air Svcs, Mr '47:12-13	1		S
	West Fly, Mr '47:18-19+	1		S

DOUGLAS (cont.)

1074. D-558-2 "Skyrocket"
Aeroplane, Jl 13, '51:46-47 1
Ap 25, '52:482-483 4 S
Avn Wk, N 17, '47:12-13 1
O 3, '49:10-11 1
Jl 16, '51:14-15 2
My 19, '52:25-35 2
N 30, '53:13-14 2
Flight, F 5, '48:147-149 1
Ja 5, '50:16-20 1
Flying, O '51:13-15+ 4
Je '53:12-14+ 4
F '54:16-17+ 4
Interavia, No. 3, '50:118-121 4
No. 10, '52:572-575 4
Nav Avn News, D '47:24 1
Ja '48:14 1
Je '49:8 1
D '51:34-35 2
Pegasus, Je '52:10-11 1
Skyways, Mr '48:20-21+ 1
Ag '50:10-11+ 1
US Air Svcs, N '47:27 1
West Avn, Ag '51:9-10 4
West Fly, Ja '50:12 1

1075. DB-7 "Boston"
USAF designation: A-20
Aeroplane, Ap 17, '42:448-449 (-III) 4
Air BP, No. 26:30-32 4 P
Air Prog, O '42:54-55 S*
AAHSJ, Spg '68:1-15, 80 (Free French-) 4 C, Pr
Flight, Je 6, '40:506a-b (-B3) 1
Mr 6, '41:190f-h (-II) 1 S
Ap 16, '42:372-373 (-III) 1
Jl 9, '42:40a (-III) 1
F 28, '46:214a-215 4
Flying, Jl '41:29 1
Ap '44:56-57 (-III) 1 S*
Fly Rev Intl, Ap '66:501 (-C) 1
My '68:269 (-III:144 Sq) Pr*
Jl '68:408, 410 (-III) Pr*
Pop Avn, Mr '40:35 1
Roundel, N '64:33 (-III:RCAF) 1 P

DOUGLAS (cont.)

1076. DB-8A
NORTHROP A-17 developed for export by DOUGLAS
Fly Rev Intl, Mr '65:41+ 1
RAF Fly Rev, Ja '57:15 1

1077. DB-280 "Digby I"
Canadian version of B-18
Roundel, Ap '65:33 (RCAF-) 1 P
RAF Fly Rev, Ja '60:5+ 1

1078. DC-1 "Airliner"
Aero Digest, O '33:45-47 1
Air Pict, S '69:349 1
Air Prog, Win '60:19-24 1 C
AAHSJ, Win '61:266 C
Aviation, O '33:331-332 1
Flight, Mr 1, '34:189-191 1
Pop Avn, D '33:377+ 1
Mr '34:161 6 (Flaps)
Ap '40:28-31+ 1
US Air Svcs, O '33:28-30 1
N '33:18 C
West Fly, S '33:12-15 1

1079. DC-2 "Airliner"
Aeroplane, N 15, '33:857-858 1
Air Pict, S '69:349 1
Comm Avn, O '40:70 (C-41) C
Flight, O 25, '34:1120-1121 1 S
N 1, '34:1159-1161 1
Pop Avn, F '34:86-88 1 Pr, S*
My '35:348-350+ 4
Mr '36:190+ (D.S.T.-) 1
D '36:24 1
Ap '40:28-31+ 1
RAF Fly Rev, Mr '62:31-32 ("Hanssin Jukka" bomber-) 1
US Air Svcs, O '35:32+ 4
F '36:22-23 (D.S.T.-) 1
West Fly, N '34:13+ (England-Australia) 4
Ja '36:11 (D.S.T.-) 1
Je '37:18 4

DOUGLAS (cont.)

1080. DC-3 "Dakota"

See also AC-47, C-47, R4D, and XCG-17

Aeroplane,	Ag 12, '36:214-215	1		
	Jl 12, '62:9-10	4		
Air Pict,	D '53:375 (Dart-)			P
	F '57:53 (Hi-Per-)	1		
Air Pilot,	Ag '44:6-8	1		C*
	S '44:14-16	6	(Fuel & Lube)	S*
	O '44:33-35	6	(Hydraulic)	S*
Air Prog,	My '69:55+ (Super-)	1		
Acft Eng,	D '50:356-360 (Super-)	1		
AAHSJ,	Win '61:266			C
	Spg '65:58-61	1		P, Pr
Aviation,	Ja '36:34	1		
	F '36:35-36	1		
	Ap '56:160-165	2		
Avn Opns,	Mr '49:17-20 (Super-)	1		
Avn Wk,	O 3, '49:38-41 (Mamba-)	1		
	O 17, '49:53 (Super-)	1		
	O 9, '50:50-53 (Super-)	4		
	D 25, '50:39-40 (Super-)	1		
	N 1, '54:70-76 (Hi-Per-)	1		
Can Avn,	Je '50:20-21+ (Mamba-)	2		
	Ag '59:36-38	1		
Flight,	Ap 8, '43:364a	1		
	S 22, '49:370-371+ (Mamba-)	4		C
	Ja 18, '51:67-70 (B.E.A.-)	1		C, S
	N 7, '52:578-579 (Dart-Twin-Wasp-)	1		
	Mr 23, '56:319-320	1		
	D 23, '60:981-985+	1		S*
Flying,	Je '42:44-46+ (TWA #370-)	4		
	Mr '44:48			C
	Ap '46:27-30+	1		
	O '56:47-49+	1		
Fly Rev Intl,	S '64:42			Pr*
Interavia,	N '49:673-676 (Super- & Mamba-)	1		C
Pop Avn,	F '37:42	1		
	Mr '37:30 ("Skylounge")	4		
	Mr '38:47	3		
	F '40:40-43+	3		
	Ap '40:28-31+	1		
Roundel,	Ap '57:30	1		

DOUGLAS (cont.)

1080. DC-3 (cont.)

Skyways, O '48:23 C
Ag '50:18-19+ (Mamba-) 1
D '50:10-12+ (Godfrey's-) 1 C*
O '52:28-31+ 3 C
N '52:14-15+ 6 (Palas Jet Boost) C
Je '53:24-25 (Godfrey's-) 1 C
O '53:24-25+ 2
Je '54:16-17 (Super-) 6 (Hydraulic) S*
D '54:14-15+ (Super-92-) 1
Ja '55:14-15 (Hi-Per-) 1
US Air Svcs, Jl '49:18-19 (Super-) 1 P
West Fly, N '37:34 ("Substratosphere Liner") 1
D '49:10-11 (Super-) 1

1081. DC-4

USAF designation: C-54
US Navy designations: R4D and R5D

Aero Digest, My '38:28-29+ 1 P
N '38:74 C*
Aeroplane, Jl 13, '38:53-60 4 C
D 7, '51:734-736 1
Aviation, My '37:37+ 1
Je '38:32-33 2
Jl '38:20+ 1
S '38:36-37, 53 2 C*
N '46:73-75 6 (Flight Cntrl)
Bee-Hive, Spg '46:18-20 1
Can Avn, F '38:8 1
S '38:23+ 6 (Radio)
O '38:8 2
N '44:72-73+ 1
Flight, Je 9, '38:568a 2
O 20, '38:339 C
Pop Avn, Je '37:16+ 1
D '37:35-36+ 1
Ag '38:27-28+ 1 P
D '38:17+ 2
F '39:24 C*
Jl '39:30-31+ 4
RAF Fly Rev, N '61:49 (Proto-) 1
US Air Svcs, Mr '37:23 1
Je '38:23-24 2
Jl '38:16 1

DOUGLAS (cont.)

1081.	DC-4 (cont.)			
	US Air Svcs, N '38:16	1		P
	Jl '39:20-21	4		
	West Fly, Mr '38:21	3		
	Jl '38:16-17+	1		
	O '38:24			C*
1082.	DC-5			
	Air Pict, Ap '55:123	1		
	Ag '66:299	4		
	Aviation, Ja '39:32-33	1		
	Can Avn, F '39:26	1		
	Comm Avn, Mr '39:39-40	1		
	Pop Avn, Ap '39:49	1		
	RAF Fly Rev, Ag '57:39+	1		
	F '58:45	1		
	US Air Svcs, My '39:22			P
	West Fly, Mr '39:21	1		
1083.	DC-6			
	USAF designation: C-118			
	Aero Digest, Ag '46:89+	6	(De-ice)	
	O '47:44-47+	1		S*
	Aeroplane, O 18, '46:451-452	1		
	Ag 8, '47:166-168	1		
	D 19, '52:814-816 (-B)	4		C
	Air Prog, '52: Annual: 86-87 (-B)			S*
	Aviation, Ap '47:49-51	2		
	Avn Opns, Je '50:27	6	(Engine)	S*
	Ja '51:22-23	3		S*
	Avn Wk, D 12, '49:28 (-A)	6	(Cargo)	
	Ap 7, '52:50-56 (-B)	6	(Instruments)	C
	My 12, '52:65-70 (-A)	4		
	Je 29, '53:66 (-C)	1		
	Can Avn, Mr '53:38-39+ (-B)	1		
	Comm Avn, O '45:53-58	1		
	Flight, O 31, '40:381	1		
	Ag 10, '56:222-224	1		
	Flying, D '47:24-25 (Truman's-)	1		
	Je '48:30+	1		
	Jl '61:96 (-A&-B)	1		
	Interavia, No. 1, '47:28-32	1		C
	No. 11, '47:44-48	2		
	O '48:555			C
	Skyways, Jl '51:16-17+	4		C

DOUGLAS (cont.)

1083. DC-6 (cont.)

US Air Svcs,	Je '46:30	1		
	D '46:4 (United-)	1		P
	Ap '47:22-24 (United-)	1		P
	Ag '53:20 (-C)	1		
West Avn,	O '53:26-27+	3		
West Fly,	N '45:54	6	(Interior)	
	O '47:16-17+	3		
	O '50:22+ (-A)	1		

1084. DC-7

Aero Digest,	My '53:26-42	1		S*
Aeroplane,	Je 5, '53:739	1		
	F 11, '55:178-181 (-C)	1		
	Mr 11, '55:298-300 (-C)	2		
	Mr 25, '55:388-390 (-C)	2		
AAHSJ,	Fall '62:202			C
Avn Age,	D '53:24-25	1		
	Ag '55:70-73	1		
	My '56:48-51 (-C)	2		
Avn Wk,	Je 16, '52:91-92	1		
	Ap 13, '53:66-70	6	(Air Cond)	S*
	My 25, '53:99-102	1		
	N 30, '53:15	4		
	D 14, '53:19			S*
	F 8, '54:62-74	4		
	S 20, '54:71-77	4		
	N 1, '54:11-15 (-C)	1		
	Ja 9, '56:99 (-C)	1		
	Je 4, '56:40-41 (-C)	1		
	O 15, '56:38-40	4		
Bee-Hive,	N '44:11	1		
Flight,	Jl 30, '54:143-147	1		C, S
	Jl 6, '56:51-54 (-C)	1		S*
Flying,	Je '44:27+	1		
	Ag '53:45	1		
	N '53:23-25	1		S*
Interavia,	No. 5, '58:402 (-C)	6	(Navigation)	C
	No. 4, '67:535-536 (-C)	6	(R Telemetry)	
Skyways,	Mr '52:55	1		
	F '54:9-11+	1		C
US Air Svcs,	N '44:58	1		
	F '52:25	1		
West Avn,	F '52:9	1		
	O '52:23-26	1		
	Jl '54:35 (-B)	1		
	O '54:65 (-C)	1		

DOUGLAS (cont.)

1085. DC-8

Aero Digest, N 1, '45:60-61	1		
Aeroplane, Ag 19, '55:277-280	2		
Je 22, '56:590-592	1		
Ja 17, '58:79-82	3		
N 25, '60:713-718	1		S*
Ja 25, '62:89 (-F)	2		
Mr 7, '63:9-10 (-F)	4		
Air BP, No. 36:8 (-40:Alitalia)			Pr*
AAHSJ, Fall '62:202			C
Avn Wk, Ag 11, '52:13	1		
Je 13, '55:13-14	1		
N 7, '55:107-109	1		
N 28, '55:29-35	1		
F 27, '56:103	1		
Je 4, '56:33	4		
Jl 9, '56:47			C
S 10, '56:38-39	1		
Ap 21, '58:102-123 (Pt 1)	1		
Ap 28, '58:60-69 (Pt 2)	1		
Je 9, '58:28-29	1		
D 8, '58:49	2		
Mr 30, '59:79-80	4		
Ap 4, '60:80-88	4		C
Ja 23, '61:42 (-50)	6	(Engine)	
Jl 10, '61:50-53 (-50)	4		
O 9, '61:39	6	(Hydraulic)	
Bee-Hive, Fall '59:2-5	1		
Can Avn, O '61:23-26+	4		C
Flight, Jl 6, '56:15-18	1		S*
N 1, '57:681-684	1		S*
Jl 25, '58:141-143	1		S*
Je 19, '59:830 (United-)	1		Pr
N 20, '59:594-595			S*
N 25, '60:851	1		
Flying, Je '59:24-26+	2		
Fly Rev Intl, Mr '69:38-41+ (-10 to -50)	1		P*, Pr
Interavia, No. 3, '56:184	1		
No. 11, '56:895 (Mock-up)	1		
No. 9, '57:931-932	3		
No. 3, '58:222	3		
No. 5, '58:403	6	(Antennas)	
Sp/Aeron, Ap '59:55-57	3		
My '59:85-86+	6	(Seals)	
Je '60:165-167	6	(HF Antenna)	

DOUGLAS (cont.)

1085. DC-8 (cont.)

US Air Svcs, Jl '55:11	1		
West Avn, Jl '55:21	1		
O '55:10-11+	2		
Mr '56:18-19	1		
My '56:6-11	2		
Jl '60:34-35+	6	(Lockbolt Fastener)	

1086. DC-8, 60 Series

Aerosp Intl, S/O '67:26-27 (-61, -62, & -Super 63)	1		
Air BP, No. 19:16 (Japan Air Lines)			Pr*
No. 45:10 (-61:Eastern)			Pr*
AF/Sp D Intl, O '66:43	1		
Avn Age, F '56:18-23	1		
Avn Wk, Ap 12, '65:54-67	1		
Ja 31, '66:32-33	1		
F 7, '66:36-37 (-61)	1		
Jl 4, '66:31-32 (-62)	1		
Mr 13, '67:30 (-63)	2		
D 4, '67:56-62	4		
S 9, '68:44-50 (-61 to -63)	3		
Flight, My 2, '68:660 (-63)	1		P*
Fly Rev Intl, My '68:273-275 (-61 to -63)	1		P*, Pr
Interavia, No. 3, '66:282A (-61)	1		
No. 12, '66:1891-1895 (-61 to -63)	4		

1087. DC-8F "Jet Trader"

Avn Wk, Ap 17, '61:42	1		
Interavia, No. 12, '62:1616-1618	1		

1088. DC-9

Aeroplane, Ja 21, '65:16+	1		
Ap 7, '66:4-10	4		C
Air Pict, My '69:166-167+	4		
Air Prog, Ap '65:34-36	1		C
My/Je '65:2			P
Avn Wk, Ag 11, '47:18	1		
S 3, '56:26-27	2		
Je 29, '59:30-31	1		
Ap 15, '63:40-42	1		
My 27, '63:42-43	1		
N 18, '63:39			C
My 4, '64:44-45	1		

DOUGLAS (cont.)

1088.	DC-9 (cont.)				
	Avn Wk,	S 14, '64:94-95	3		
		Ja 18, '65:30-33	1		C
		Mr 8, '65:29 (-B)	1		
		N 1, '65:37-50	4		
		S 12, '66:42-47	4		
		Ja 1, '68:28-29	3		
		N 4, '68:38-44	3		
		My 19, '69:70-72+	6	(Rocket Assist)	
	Bee-Hive,	Ja '66:2-7	1		
	Can Avn,	Ja '64:6-8	1		C
		Je '64:26-27	1		
		F '65:24-25	1		
		Mr '65:22+	1		C
		Je '66:20-21	1		C
	Flight,	Ja 28, '65:129-130	1		
		Mr 25, '65:442	2		
		Mr 3, '66:341-348	1		C, S*
	Fly Rev Intl,	O '65:96-99	1		S*
		O '67:895-896	1		Pr
	Interavia,	No. 9, '47:27-29	1		
		No. 5, '63:564	1		
		No. 10, '63:1506-1509	1		C, S*
		No. 7, '64:1014-1015	3		
		No. 3, '65:361-363	1		C
		No. 6, '65:723-725	4		
		:726-738	1		C, S*
		:739-742	2		
		No. 3, '67:377-379	4		C
	West Avn,	Jl '59:24	1		
1089.	DC-10				
	Aeroplane,	F 28, '68:14-15 (-A&-B)	1		
		My 22, '68:14-15	1		C
		Jl 3, '68:26-27	1		
	Aerosp Intl,	Jl/Ag '68:54-55	1		D
	AF/Sp D Intl,	O '66:43	1		
	Avn Wk,	F 14, '66:37-39	1		
		F 21, '66:40-41	1		
		My 16, '66:47-52	1		
		Je 26, '67:26-29 (-3&-4)	1		
		N 13, '67:36-39 (-A&-B)	1		
		F 26, '68:26-29	4		
		My 13, '68:36-39	1		C
		N 18, '68:38-39	6	(Food Service)	
		D 2, '68:87-103	6	(Auto Landing)	C

DOUGLAS (cont.)

1089. DC-10 (cont.)

Avn Wk, Mr 31, '69:78-81	6	(Cabin Air)	C
Je 16, '69:64-68	6	(Engine)	S*
O 20, '69:55-57	3		
N 24, '69:50	3		
D 15, '69:25	3		
Bee-Hive, Win '69:1 (-20)			D
Flight, Jl 13, '67:42-43	1		
F 29, '68:290	1		
My 23, '68:777	1		C
N 7, '68:731 (-20)	1		
Mr 13, '69:407-416	1		C*, Pr, S*
Fly Rev Intl, Ag '68:427	1		
Mr '69:26-27	1		
Ag '69:21	1		
N '69:28-29	1		
Interavia, No. 4, '68:503	1		D
No. 6, '69:721-724	3		C, S
:725	2		
Sp/Aeron, Ag '68:42-54	1		C, S*

1090. DF Flying Boat

Aviation, Ja '37:37-39	1	
Pop Avn, D '36:27	4	
West Fly, O '36:21	1	

1091. DWC "World Cruiser"

Aeroplane, Ag 27, '43:254	4	
Air Pict, My '56:160-161	1	Pr
Ag '56:289-290	5	
Ja '57:32	5	
Ag '57:283-284	5	
US Air Svcs, O '43:9-10	4	
O '44:18-20+	4	

1092. EA-1E "Guppy"

Nav Avn News, Ap '67:16	4	

1093. F3D "Skyknight"

Aero Digest, Ap '52:24-31	1	S*
Aeroplane, Ag 24, '51:219 (-1)		P
Avn Wk, O 11, '48:12-13	1	
Mr 6, '50:23-24	1	
F 9, '53:13-14	4	
Flying, Jl '51:22-23	1	
Interavia, No. 11, '50:576-578+	4	

DOUGLAS (cont.)

1093.	F3D (cont.)			
	Nav Avn News,	N '48:11 (X-)	1	
		My '50:24-25	1	
		My '51:20-21 (-1)	1	P
		Jl '51:12 (-1)	4	
	RAF Fly Rev,	D '61:59	1	
	Skyways,	S '50:12-13+	4	
	US Air Svcs,	Ja '49:10 (X-)	1	
1094.	F4D "Skyray"			
	Avn Wk,	F 5, '51:16 (X-)	1	
		O 26, '53:18 (X-)	1	P
		N 30, '53:18	2	
		Je 6, '55:40-49	1	
	Bee-Hive,	Ja '55:3-6	1	
	Flight,	F 25, '55:251	1	
	Flying,	Ja '54:58	4	
	Interavia,	No. 8, '54:534		P
	Nav Avn News,	N '53:12	4	
		F '54:11	2	
		Jl '58:21 (-1)	4	
	RAF Fly Rev,	O '59:30-33+ (-1)	1	Pr, S*
	US Air Svcs,	F '51:12-13 (X-)	1	
		D '53:20	4	
1095.	F5D-1 "Skylancer"			
	RAF Fly Rev,	Mr '60:11	1	
1096.	JD-1			
	US Navy designation			
	USAF designation: A-26			
	Nav Avn News,	Je '46:14-15	1	C
		Ag '46:27	5	
1097.	M-2 Mailplane			
	AAHSJ,	Win '61:262		C
		Fall '68:197		C
	CAHS,	Spg '66:30 (MO-2B)		Pr
1098.	O-2			
	CAHS,	Spg '66:30 (-B)		Pr
1099.	O-25			
	Pop Avn,	Ja '31:28 (-B)		C*

DOUGLAS (cont.)

1100.	O-31		
	AAHSJ, Fall '63:205-209 (Pt 1)	1	P
	Win '63:276-282 (Pt 2)	1	P, Pr
1101.	YO-35 and XO-35		
	Pop Avn, Je '33:372	1	
1102.	O-38		
	Air Force, F '69:56-60 (AF Museum's-)	4	
	US Air Svcs, N '32:26		P
1103.	O-43		
	AAHSJ, Fall '63:205, 207-208 (Pt 1)	1	P
	Win '63:276, 280-283 (Pt 2)	1	Pr
1104.	O-46		
	AAHSJ, Fall '63:205, 209 (Pt 1)	1	P
	Win '63:276, 281-282 (Pt 2)	1	Pr
	Pop Avn, My '37:42 (-A)		Pr
1105.	YOA-5		
	US Air Svcs, Ag '36:16-18+ ("Big Duck")	4	
1106.	OD-1		
	AAHSJ, Win '68:304		P
1107.	R4D-8 "Skytrain"		
	US Navy designation		
	USAF designations: C-47, C-53 and C-117		
	Air Prog, My '69:55+	1	
	Nav Avn News, Ag '54:2		C
1108.	R5D "Skymaster"		
	US Navy designation		
	USAF designation: C-54		
	Civil designation: DC-4		
	Nav Avn News, Ag '50:18-19 ("Navy Delta")	1	
	Ag '53:28 (Navy's oldest-)	3	C
1109.	SB2D "Destroyer"		
	Later designation: BTD-1		
	RAF Fly Rev, D '60:11+	1	

DOUGLAS (cont.)

1110.	SBD "Dauntless"		
	US Navy designation		
	USAF designation: A-24		
	Air Cl, Ag '69:20-37	1	P*, S*
	:26 (-1)		C*
	Air Pict, O '64:325-328	1	
	Air Prog, Jl '42:23		P*
	Air Tech, Ap '44:17-23+	1	S*
	AAHSJ, Spg '67:32-39 (-5)	4	Pr
	Flight, N 26, '42:578a (-1&-3)	1	
	Flying, F '43:148		P
	Ag '43:127		S*
	My '44:62-63	1	S*
	Ja '45:30-31+	1	
	Mr '66:25	1	D
	Fly Rev Intl, Ja '67:330 (-1, -3, -5)		Pr*
	Nav Avn News, F 15, '44:16		P
	S 15, '44:11	4	
	S '69:15		P
	US Air Svcs, Ag '44:46	4	
1111.	XT-30		
	Not built		
	Flight, D 22, '49:799-800	1	
1112.	TB2D "Devastator II" or "Skypirate"		
	RAF Fly Rev, D '60:11+	1	
1113.	TBD-1 "Devastator"		
	Air Cl, O '68:28-43	1	C, D*, Pr*
	Air Prog, Ag '67:18 (Torpedo 8 Sq-)		Pr†
	Pop Avn, F '38:17 (X-)	1	
	Ap '39:18	1	
	RAF Fly Rev, Ag '58:6+	1	
1114.	X-3 "Flying Stiletto"		
	Aeroplane, N 27, '53:707	1	
	Air Pict, F '54:37	1	
	Air Prog, Ap/My '63:70		C*
	Avn Age, F '54:27-29	6 (Cooling)	S*
	Avn Wk, N 23, '53:13-15	1	
	Ja 4, '54:56-59	6 (Cooling)	
	N 29, '54:35	1	
	Flight, F 26, '54:233	6 (Cooling)	S
	West Avn, D '53:14	1	
	O '54:56	6 (Wing)	

DOUGLAS (cont.)

1115.	Y-123 "Duck" or "Bloomer Girl" C-47 with floats		
	Aero Digest, O 1, '44:74-76	6 (Floats)	S*
	Aerosp Hist, Aut '68:65	1	P
	Spg '69:27	1	
1116.	2086		
	Flight, Ap 26, '62:671-673	1	
	Flying, S '62:33+	1	
	Interavia, No. 5, '62:551	1	

DUNNE GB and USA

Flight, My 27, '43:556-558

1117.	Swept-wing Biplanes		
	Can Avn, Ag '65:1	1	D*
	CAHS, Fall '66:62-66	1	
	Pop Avn, My '36:334	1	P
	Roundel, O '59:17-21	1	P
	Ja/F '63:26-27	1	

DW (Dudley Watt) GB

1118.	2		
	Flight, My 23, '30:555-558	1	S

DWL (Doswiadczalne Warsztaty Lotnicze) Poland

1119.	RWD-11 and RWD-13		
	Air Pict, D '58:440-441	1	
1120.	RWD-25		
	Air Pict, O '63:329	1	

DYLE ET BACALAN France

See BORDELAISE

EDGAR PERCIVAL GB

1121.	E.P. 9 "Prospector"		
	Aeroplane, D 30, '55:1011-1015	1	S*
	F 3, '56:146-147		P
	Ja 4, '57:15-18	4	C
	Air Pict, My '59:186-187	4	

EDGAR PERCIVAL (cont.)

1121. E.P. 9 (cont.)

Flight,	D 16, '55:890-891	1	
	D 30, '55:969-972+	1	S*
	S 7, '56:427	1	
	O 26, '56:666-667	4	
	Ap 5, '57:440-443+	3	
	My 13, '60:654-655	4	

EDO USA

1122. XOSE-1

Pegasus, O '46:6-9	1	P
RAF Fly Rev, My '61:48+	1	

EFW (Eidgenössischen Flugzeugwerke) Switzerland
See FEDERAL AIRCRAFT FACTORY

EKW (Eidgenössisische Kriegstechnische Werkstätten) Switzerland

1123. C-35

Fly Rev Intl, D '67:1068	1	
Mr '69:43		Pr*

ELLEHAMMER Denmark

1124. Rotary Wing (1912)
Am Helicop, My '48:9+ 1

1125. "Standard" (1909)
Interavia, No. 4, '50:192-193 (Replica-) 1 P

ENGINEERING DIVISION, U. S. AIR SERVICE USA

1126. GA-1 or GA-X
See BOEING GA-1

1127. PW-1 or VCP-2
See also BARLING
AAHSJ, Fall '61:179 C

ENGLISH ELECTRIC GB

1128. "Ayr" M. 3
Air Pict, Ja '66:26 1

ENGLISH ELECTRIC (cont.)

1129. "Canberra"

US licensee: MARTIN B-57

Aero Digest,	Ap '51:17-18+	1	
Aeronautics,	F '51:44-45		P
	Jl '53:58-59		P
	My '56:52-53		P
Aeroplane,	Ag 26, '49:262-263		P
	D 2, '49:748-749	1	
	F 10, '50:153-162	1	S*
	Mr 30, '51:383		P*
	Ag 31, '51:246-247	1	
	S 7, '51:290		P
	Mr 27, '53:362	4	
Air Power,	Jl '56:246 (-P. R. 7)		P
Am Helicop,	My '52:13+	1	
Avn Wk,	F 13, '50:21-25	1	S*
	O 30, '50:32-33 (-2)	3	
	Ag 27, '51:44 (-2)	1	S
	Ap 27, '53:20-23	4	
Flight,	Ag 25, '49:216-217	1	
	D 1, '49:700-701	1	
	D 15, '49:765-772	1	S*
	Mr 16, '50:343	3	
	Mr 2, '51:246-248	4	
	Ja 18, '52:58-59+	4	
	:71-74	1	
	N 12, '52:645 (-T4)	1	
	Ap 3, '53:418 (Olympus-)	1	
	:430-433	4	
	My 15, '53:594-595 (Olympus-)	4	
	Je 5, '53:726-731	1	
	Ag 28, '53:257-258	4	
	Jl 23, '54:110-111 (-P. R. 7)	2	P
	Ag 6, '54:167 (-B. 8)	1	
	Je 10, '55:794		P
	F 17, '56:180-181	1	
	Ja 11, '57:37-41 (-P. R. 7& B(I). 8)	4	
	D 27, '57:975-976 (-B(I). 8)	4	
	Mr 14, '58:332-339	1	S*
	D 12, '58:905-906 (-U. 10)	4	
	My 8, '69:757-761+	1	
Fly Rev Intl,	My '68:268 (-II:10 Sq)		Pr*
Interavia,	No. 1-2, '50:71-74 (-B. I)	1	S*
Nav Avn News,	Je '51:20-21	1	P

ENGLISH ELECTRIC (cont.)

1129. "Canberra" (cont.)

RAF Fly Rev, My '53:19-21 4
S '53:16-17 (London-N.Z. race) 4
N '53:15-16 (London-N.Z. race) 4
Jl '55:52-53 1 P*
Ja '58:4 1
:5 (-B.8:Venezuela) P*
N '58:23-25 (VHF controlled-) 4
N '59:19-21 1
:18 (-P.R.9) Pr
:20-21 (-P.R.7) S*
Skyways, Ag '51:10-12+ 1
US Air Svcs, O '51:18-19 4

1130. "Kingston" P. 5

Air Pict, Ag '62:232-236 1
Ja '66:26 1

1131. "Lightning" P. 1

Name introduced O '58
Later version: BAC "Lightning"

Aeroplane, Jl 26, '57:101-103 1
Ja 10, '58:49-52 3 S
My 2, '58:622-625 2
My 18, '61:538-539 (-F.1) P
My 25, '61:584-585 (-F.1) P
Jl 13, '61:42-50 (-F.1) 1 S*
Air Force, N '59:42-43 1
Air Pict, D '54:361-363+ 1 Pr
Ag '57:270-271 (-B) 1 Pr
D '59:441-443 (-F.1) 1 Pr
S '64:276-277 4 Pr
Air Power, Sum '57:264 P
Air Prog, Ja '65:44 (-F.3) Pr
Avn Wk, S 30, '57:50-59 1
D 23, '57:68-71 2
Ag 14, '61:65 (-2 to -4) 3
Flight, Ag 13, '54:202-203 2
S 16, '55:474-475 1
D 2, '55:837+ 1
Mr 9, '56:258-259 1 P
Ap 26, '57:541-545 (-A&-B) 1
Jl 19, '57:81-85 2

ENGLISH ELECTRIC (cont.)

1131. "Lightning" (cont.)

Flight,	Ja 10, '58:47-51	6 (Wing)	S*
	O 31, '58:693-695	1	P
	Ja 23, '59:131-134	1	
	F 24, '61:230-231 (-F.1)	4	
	Jl 13, '61:44-51 (-F.1)	1	S*
	Ag 3, '61:153-156 (-F.1)		P
	N 16, '61:765		P
	S 7, '67:413	1	
	Ap 17, '69:639-640	1	
Fly Rev Intl,	Ag '64:35-39	1	Pr
Interavia,	No. 3, '56:192-193	1	
	No. 6, '58:559 (-B)		P
	No. 9, '59:1081-1085	1	
	No. 9, '61:1210-1211	3	
	No. 9, '68:1074-1075	1	
RAF Fly Rev,	Ag '55:18	1	
	D '56:33-34	4	
	S '57:38-39	1	
	N '58:30+	1	
	My '61:42-43+	4	P*
	N '61:32-35 (-1A)	1	Pr, S*

1132. "Lightning Trainer" P. 11

Aeroplane,	Je 19, '59:696-697	1	P
Flight,	S 4, '59:121		P
	S 18, '59:257		P
	My 31, '62:875-877	4	
	Ap 18, '63:570-572 (-T.5)	4	
	Jl 9, '64:75-76 (-T.4)	4	
Fly Rev Intl,	Ag '64:39	1	
Interavia,	No. 8, '62:1013-1015 (-T.4)	2	

ENSTROM USA

1133. F-28A

Flying,	S '68:68-70+	1	
	S '69:46-49	4	C

ETRICH Germany

1134. "Taube"

Air Prog,	Ag '66:83		D
Flight,	Ja 17, '30:127	1	

EUROPEAN GB/France/Germany

1135.	A-300 "Airbus"			
	Aeroplane, Ap 26, '67:24-25	1		
	Jl 5, '67:16-19	1		
	N 15, '67:27	1		
	My 8, '68:22	1		
	S 18, '68:15	1		
	Air Pict, F '69:40 (-B)	1		
	Acft Eng, Mr '69:33-35 (-B)	1		
	Avn Wk, D 16, '68:29 (-B)	1		
	N 3, '69:33-34 (-B)	1		
	Flight, F 16, '67:229-230	1		
	Ap 27, '67:655	1		
	My 9, '68:697-698	1		
	O 31, '68:684	6	(Flight Deck)	C
	D 19, '68:1009-1010, 1013-1016 (-B)	1		
	F 27, '69:319-321 (-B)	1		
	Fly Rev Intl, Ag '68:431	1		
	F '69:28 (-B)	1		
	Interavia, No. 3, '67:328-329	1		
	No. 12, '67:1871-1873	1		
	:1874-1875	6	(Engine)	S*
	No. 11, '68:1429-1431	1		S

EWR SÜD (Entwicklungsring Süd) Germany

1136.	VJ-101C			
	Air Prog, Ag/S '64:53	1		C
	Avn Wk, My 27, '63:70-75	1		
	Jl 1, '63:69	1		
	N 18, '63:59	1		
	Flight, My 23, '63:746-748	1		C
	O 24, '63:699	2		
	Fly Rev Intl, F '64:31-33	4		
	Interavia, No. 6, '63:910-911	1		
	No. 7, '63:1076-1078	1		C
	No. 1, '64:53-58	6	(Engine)	S*
	No. 5, '67:779 (& -D)	1		
	RAF Fly Rev, Ag '63:13-15	1		C

FABRE France

1137.	"Canard"			
	World's first seaplane (1910)			
	Flying, Je '62:35+	1		

FABRE (cont.)

1138. "Hydro-Aeroplane" (1911)
Air BP, No. 35:30-33 1 P

FABRICA MILITAR DE AVIONES Argentina
See MILITARY AIRCRAFT FACTORY

FAHLIN USA

1139. "Plymocoupe"
See NB "Plymocoupe"

FAIRCHILD USA
Interavia, N '49:651-653 (Managerial fight: 1949)
Pegasus, N '43:8-12+ (Pt 1 - Beginnings)
D '43:5-7+ (Pt 2 - Building FC-1)
Ja '44:14-16 (Pt 3 - FC-1 and FC-2)
F '44:13-16 (Pt 4 - Air mail and aerial photography)
Mr '44:13-16 (Pt 5 - 1929 and 1930)
My '44:13-16 (Pt 6 - 1931 to 1944)
S '44:1-16 (1916 to 1944)
D '52:1-19 (Cargo aircraft: 1926 to 1952)

1140. "Argus"
Civil model: 24
British name for UC-61
Aeroplane, Je 27, '47:681 S*
Flight, Je 3, '43:580a (-I) 1
Pegasus, Mr '43:18 P

1141. XA-942 Amphibian
Civil model: FAIRCHILD 91
Aero Digest, My '35:34-36 1
Acft Eng, Jl '35:175-177 1
Aviation, My '35:183-185 1 S
Flight, My 9, '35:503 1
Pop Avn, Ap '36:268 1

1142. XAT-13 "Yankee-Doodle"
Aviation, S '42:225 1
Pegasus, Ja '43:14-15 1
RAF Fly Rev, Ag '57:35 1
US Air Svcs, S '42:31 1

FAIRCHILD (cont.)

1143.	XAT-14 "Yankee-Doodle"			
	Can Avn, Ja '43:104	1		
	Comm Avn, D '42:102	1		
	Pegasus, Ja '43:14-15	1		P
	RAF Fly Rev, Ag '57:35	1		
	US Air Svcs, Ja '43:40	1		P
	West Fly, Ja '43:104	1		
1144.	AT-21 "Gunner"			
	Air Prog, Jl '43:20-25	3		P*
	Air Tech, Jl '44:17-23	1		C*, S*
	Flying, D '43:160	3		
	Fly Rev Intl, Jl '67:748-749	1		
	Pegasus, Je '44:3			P
	RAF Fly Rev, Ag '57:35	1		
1145.	XC-31			
	Aero Digest, D '34:44-45	1		C
	Aviation, D '34:411-412	1		
	Flight, Mr 21, '35:311	1		C
	Pegasus, O '43:6-8	1		C
1146.	C-82 "Packet" or "Flying Boxcar"			
	Later development: C-119			
	Air Tech, Jl '45:15-22+	1		
	Am Helicop, My '46:35	1		
	Aviation, Ag '45:115-139 (Pt 1)	1		S*
	S '45:115-127 (Pt 2)	1		C*, S*
	Avn Wk, D 8, '47:24, 26	1		
	Jl 5, '48:21-22	6	(Landing Gear)	
	Comm Avn, N '44:41+	1		
	Ap '45:58+ (-Ambulance)	1		
	Je '45:58+	1		
	Interavia, No. 6, '46:29-36	1		S
	Pegasus, D '44:1-5	1		
	:6-9	4		
	Mr '45:14-15	6	(Litters)	
	Ap '45:1-3+	1		S
	My '45:1-3+	2		
	Je '45:6-7	4		
	Ag '45:4-7+	6	(Anti-ice)	S*
	F '46:1-4	2		
	Mr '46:12-13	3		
	Ap '46:10-12	6	(Mail)	S
	N '46:8-9	4		
	Mr '49:1-5	4		
	D '57:12-15 ("Thing")	1		

FAIRCHILD (cont.)

1146.	C-82 (cont.)			
	Skyways, N '45:34-40+	1		S*
	Mr '49:36-37	4		
	O '50:20-21+	4		
	US Air Svcs, Ja '53:18-22	4		
	O '54:10	6	(J-44 Boost)	
	West Fly, O '44:140	3		
	D '44:134	1		
1147.	C-119 "Skyvan," "Packet," or "Flying Boxcar" Development of C-82 See also STEWARD-DAVIS "Stolmaster"			
	Aero Digest, Ag '52:24-47 (-H)	1		S*
	Avn Age, D '52:34-35 (-H)	1		S*
	Avn Wk, D 8, '47:24-26 (-A&-B)	1		
	N 27, '50:20-21	3		
	F 5, '51:15-16 (-C)	1		S*
	Jl 30, '51:17-18 (-H)	1		
	N 12, '62:127-131	4		
	S 23, '68:40-50 (-A)	1		
	Can Avn, F '49:27	1		
	Nav Avn News, Ag '48:33 (-A)			P
	Pegasus, F '48:Back Cover			P
	Ap '49:1-6+, 8-9, 17	3		P, S
	Ag '49:Back Cover			P
	Mr '50:1-4	1		
	Jl '52:Back Cover (-H)			P
	O '55:1-6	1		
	Skyways, Mr '49:36-37	4		
	US Air Svcs, D '49:18	2		
	Jl '52:9 (-H)			P
	D '52:16-17	4		
	Ja '53:18-22	4		
	S '53:11			P
1148.	XC-120 "Pack Plane"			
	Air Pict, N '50:333	1		P
	Ap '51:109			P
	Avn Wk, Je 26, '50:12-14	1		
	S 11, '50:16-17	1		
	N 13, '50:22-31	1		
	Flight, S 21, '50:332-334	1		
	Flying, Ap '51:24-25+	1		
	Interavia, No. 11, '50:560-563	1		P
	Pegasus, Ap '49:15	6	(Photo Pod)	
	Jl '50:1-5	1		

FAIRCHILD (cont.)

1148. XC-120 (cont.)
Skyways, O '50:19 1
West Fly, Ag '50:17 1
O '50:24 1

1149. C-123 "Provider" and "Avitruc"
Earlier version: CHASE C-123
Later Development: STROUKOFF YC-134
Avn Wk, Ag 8, '55:18 (Y-/-E) 1
Ap 9, '56:28-31 4
Ag 13, '62:84-89 (-H) 1
Pegasus, Jl '56:6-11 4
US Air Svcs, Ap '55:21 (-B:"Flying Lighter") 1
West Avn, O '54:60 (-B) 1
My '55:32 (-B:"Flying Lighter")1

1150. F-27 "Friendship"
Original design: FOKKER F-27
See also FAIRCHILD HILLER FH 227
Avn Wk, S 15, '58:72-91 4 C
Mr 2, '59:28-30 4
Jl 10, '61:37 (-F&-G) 1
S 19, '66:49-50 1
Flying, O '58:56-57+ 4
Fly Rev Intl, F '67:385-386+ 1 Pr

1151. FC-1
Pegasus, D '43:5-7+ 1

1152. FC-2
Can Avn, Mr '65:1 D*
Flying, S '64:17 D
Roundel, Jl/Ag '62:24 (-W2:Canada)1

1153. M-62
Air Prog, My '69:36-37+ 1
Comm Avn, N '39:33-34 1

1154. M-186
Flight, D 4, '53:738 1
West Avn, Ja '54:16 1

1155. M-232
Avn Wk, D 19, '55:28-32 1
Flight, Ja 6, '56:5 1

FAIRCHILD (cont.)

1156.	PT-19 "Cornell" Open Cockpit		
	Civil designation: FAIRCHILD 62		
	Aero Digest, O '40:130-137	3	
	Air Prog, O/N '63:52-55		C*, Pr†
	Air Tech, My '44:19-24+	1	S*
	Aviation, My '40:54+	1	S
	Can Avn, Mr '43:45-51	3	
	Comm Avn, Ag '43:46-56 (Fleet-)	1	
	:64-68 (Fleet-/-A)	6 (Landing Gear)	
	Pegasus, Mr '43:1-3	1	
1157.	PT-23 "Cornell" Open Cockpit		
	Civil designation: FAIRCHILD 62		
	Air Prog, O/N '63:52-55		C*, Pr†
	Air Tech, My '44:19-24+	1	S*
	Can Avn, Mr '43:45-51	3	
1158.	PT-26 "Cornell" Canopy		
	Civil designation: FAIRCHILD 62		
	Aero Digest, Ag '43:214-215		S*
	Aeroplane, Mr 9, '45:284-285	4	
	Air Prog, O/N '63:52-55		C*, Pr†
	Air Tech, My '44:19-24+	1	S*
	Aviation, Je '42:209	1	C
	Can Avn, Mr '42:22	1	C
	Mr '43:45-51	3	
	Comm Ávn, Ap '43:112	6 (Wing)	
	Ag '43:46-56 (Fleet-)	1	
	:64-68 (Fleet-/-A)	6 (Landing Gear)	
	Flight, Ap 22, '43:413-416	1	
	My 27, '43:554a	1	
	Fly Rev Intl, Ap '69:53 (Norway-)		Pr*
1159.	R4Q "Workhorse"		
	US Marine Corps designation		
	USAF designation: C-82 and C-119		
	Nav Avn News, D '54:11	4	
1160.	T-31		
	US Navy designation: XNQ-1		
	Avn Age, D '50:22-23	1	S*
	Avn Wk, Ap 4, '49:15	1	C
	D 11, '50:18	1	
	Flight, D 22, '49:800	1	
	Flying, N '49:24-25+	6 (Cockpit)	C
	US Air Svcs, Ag '50:26	1	

FAIRCHILD (cont.)

1161.	UC-61 "Forwarder"		
	British name: "Argus"		
	Civil model: FAIRCHILD 24		
	Air Prog, Ag '66:41		Pr
	Air Tech, Ap '45:15-19+	1	C*, S*
1162.	XNQ-1 Trainer		
	USAF designation: T-31		
	Flight, N 6, '47:513-514	1	C
	Pegasus, F '47:1-4+	6 (Cockpit)	C, P
	My '47:Back Cover		P
	Jl '47:13-15	2	
	US Air Svcs, N '46:22	1	
	West Fly, D '46:54	1	
1163.	24		
	British name: "Argus"		
	Military version: UC-61 "Forwarder"		
	Air Pilot, N '44:33	4	C*
	Air Prog, Ag '66:40-43	1	C*, Pr
	Flying, N '45:51-54	1	S
	Ag '67:44-49 (-R)	4	
	Pegasus, S '45:1		P
	D '45:5-9	4	C
	Ja '46:Back Cover		P
	F '46:8-12+	4	
	My '47:5-7 (w. floats)	4	P
	Je '47:6-8	6(Stall Warning)	C
	O '47:1-4	4	C, P
	My '48:6-9	4	
	Skyways, Jl '46:23-27+	1	S*
	West Fly, Mr '35:19-21+	3	S*
	F '39:21	1	
1164.	45		
	Aviation, S '35:28-30	1	
	Flight, O 17, '35:415	1	
	Pop Avn, Ag '36:27+	1	
	West Fly, S '35:23-24	1	
	Jl '36:28+	1	C
1165.	46		
	Pegasus, My '44:13		P

FAIRCHILD (cont.)

1166. 62 "Cornell"
Military designations: PT-19, PT-23 and PT-26

Aero Album, Win '69:34-37	1	P, Pr†
Air Prog, O/N '63:52-55	1	C*, Pr†
Aviation, Jl '39:55	1	
Can Avn, Ja '40:54	1	
Ap '41:35-36	1	
:69	2	
Mr '42:22	1	C
:23-25+	4	C
Ap '42:90+	3	
Comm Avn, Ap '41:34+	1	

1167. 71B

Can Avn, S '30:21	1	

FAIRCHILD, LTD. Canada

1168. "Bolingbroke"
Canadian-built BRISTOL "Blenheim"

Can Avn, D '39:40	1	
F '41:15-17+, 37-43+	3	
Comm Avn, Jl '41:56-58	6 (Prop)	S*
Jl '42:44-48 (-IV)	1	
Roundel, O '62:33	1	P
Jl/Ag '64:8-12	4	
RAF Fly Rev, Ag '63:39	1	

1169. "Husky" F-11

Can Avn, Jl '46:77	1	
Ag '46:40-41+	1	C
Fly Rev Intl, Ja '68:53	1	

1170. 22

Can Avn, D '32:17, 19	4	

1171. 45-80 "Sekani"

Can Avn, N '37:7	1	

1172. 71

Aero Digest, Ag '35:46	1	
Aeroplane, My 2, '34:749-750 (-C&-CM)	1	
Aviation, Jl '35:36-40 (Super-)	1	
Can Avn, My '34:15+	1	
D '34:2, 13	1	P

FAIRCHILD, LTD. (cont.)

1173. 82
Can Avn, Ag '35:8-9 (-A) 1

FAIRCHILD HILLER USA

1174. F 228
Development of FOKKER F. 28 "Fellowship"
Aeroplane, F 9, '66:34 1
Je 26, '68:34 1
Avn Wk, F 6, '67:37-38 1
F 12, '68:31 3
Flight, F 9, '67:194-195 1 S*
Mr 9, '67:347 2
Je 27, '68:942 1

1175. FH 227
Long-fuselaged FAIRCHILD F-27
Avn Wk, D 13, '65:42 1
F 7, '66:41 1
F 28, '66:43 1
Ap 25, '66:38 (-B) 1
Ag 1, '66:40-41 1
Flight, F 17, '66:249 1
Fly Rev Intl, Ap '66:477 1
Interavia, No. 7, '66:1036-1037 (-B) 1

1176. FH 1100
Aeroplane, Ag 5, '65:32-33 4 C
Avn Wk, Ag 2, '65:80-82 1 S*
Ag 30, '65:80-87 4
Flight, F 25, '65:298 1
Ag 5, '65:221-223 4 C
Flying, N '65:52-56 1 S
F '67:39 1
N '67:48-52 4 C
Interavia, No. 4, '65:503-504 1

FAIRCHILD HILLER/PILATUS USA/Switzerland

1177. "Heli-Porter" and "Armed-Porter"
Air Prog, Ag '68:44-45+ (Heli-) 4 C
Avn Wk, S 25, '67:67-74 (Armed-) 1 C

FAIRCHILD HILLER/PILATUS (cont.)

1178.	"Turbo-Porter"		
	See also PILATUS PC-6 "Turbo-Porter"		
	Avn Wk, O 5, '64:84-89	4	C
	Flight, Mr 21, '68:429-430	1	
	Flying, D '65:30-35	4	
FAIREY			GB
	Comm Avn, D '44:153-155+		
	Flight, F 20, '41:151-154		
	Jl 22, '55:119-136 (40th Anniversary)		
1179.	"Albacore"		
	Aeronautics, S '40:45	1	
	Air Cl, Mr/Ap '65:50-53+ (Pathfinder-)	4	
	Acft Eng, Ap '41:103, 106	1	P
	Flight, F 4, '43:120b	1	
	Roundel, Ja/F '65:33 (RCAF-)	1	P
	RAF Fly Rev, O '57:47	1	
1180.	"Barracuda"		
	Aeronautics, Je '45:44		P*
	Aeroplane, Ap 14, '44:414-415		P
	Comm Avn, S '44:64-66	1	
	Flight, Ap 20, '44:416-419	1	
	Ag 10, '44:148-150	3	
	D 21, '44:660-661	1	S*
	Flying, Jl '44:46	1	
	RAF Fly Rev, N '58:48	1	
1181.	"Battle"		
	Aeroplane, Mr 25, '36:381-384	1	P
	Ag 18, '37:197-199	1	S*
	O 19, '39:483		D
	Air BP, No. 35:4-7	4	
	Air Pict, Ag '60:288-292 (Pt 1)	1	Pr†
	O '60:362-364 (Pt 2)	1	
	Flight, Jl 8, '37:49		P
	Ag 19, '37:188a-192	1	S*
	Jl 18, '46:60-61	4	
	Fly Rev Intl, Ap '65:49-51	4	Pr*
	Je '65:43-44	1	
	My '68:268 (218 Sq-)		Pr*
	Pop Avn, O '37:23		P
	Pop Fly, Ja '38:569-570	1	
	Roundel, Ap '63:33 (Canadian-)	4	P
	RAF Fly Rev, Ap '58:58+	1	
	Ag '62:16	4	

FAIREY (cont.)

1182.	"Campania"		
	Air Pict, My '63:142-145	1	C
1183.	"Delta One" F.D. 1		
	Aeroplane, Mr 23, '51:343		P
	Air Pict, My '51:144		P
	O '53:316	1	P
	Air Power, Spg '54:258		P
	Flight, Mr 23, '51:348-349	1	
1184.	"Delta Two" F.D. 2		
	Ogee wing variant: BRISTOL 221		
	Aeroplane, Ap 6, '56:216-217		P
	F 22, '57:275-276	1	
	Air Pict, D '54:361-362+	1	
	Ag '56:258-260	1	
	Air Power, Ap '56:204-205	4	P
	Flight, Ap 6, '56:383	1	
	F 22, '57:235	1	
	Mr 1, '57:266-267	2	
	Ap 25, '58:571-575	1	S*
	RAF Fly Rev, F '56:17-18	4	
	S '59:40-43	1	Pr, S*
1185.	"Fantôme" (Phantom)		
	Belgian-designed		
	Aeroplane, Je 26, '35:732-733	1	C*, S*
	Pop Avn, N '35:286	1	
	RAF Fly Rev, O '56:17	1	
1186.	"Ferret"		
	Air Pict, F '56:58-59	1	Pr†
1187.	"Firefly"		
	Aeronautics, D '44:34-35		P
	F '47:44 (-T.1)		P*
	Aeroplane, My 4, '45:517-521	1	S*
	My 11, '45:550-553	1	C, S*
	N 9, '45:533 (-IV)	1	
	S 12, '47:365 (-IV)		P*
	N 20, '53:684-689 (-I to -VII)	1	C*, S*
	Mr 18, '55:356		P
	Air Pict, F '66:67	4	
	Air Tech, My '45:38-40+	1	S*
	Can Avn, Ja '45:54-55	1	

FAIREY (cont.)

1187.	"Firefly" (cont.)		
	Flight, Jl 18, '30:800-801 (-IIM)	1	P
	N 9, '44:496-497	1	
	D 7, '44:609-612	1	S*
	Ap 19, '45:423-427	1	S*
	N 8, '45:493-494 (-IV)	1	S
	S 5, '46:235-236 (-T.1)	1	
	O 17, '46:421 (-IV)	2	
	D 26, '46:704-706 (-IV)	1	
	Ja 23, '47:90a-91	1	
	Ap 6, '50:432-435 (-V)	4	
	Fly Rev Intl, S '66:86-89 (-II)	1	Pr*
	Interavia, Mr '49:152-154 (-T.1)	4	C, S*
	RAF Qtr, Je '46:Plates (-I&-IV)		P
1188.	"Flycatcher"		
	RAF Qtr, Ap '38:Supp	1	D
1189.	"Fox"		
	Aeroplane, Mr 7, '34:400 (Belgian-)		P
	Flight, F 6, '31:109		P
	Ap 8, '55:441	1	
	RAF Fly Rev, Je '63:22 (-VIC, 5 Esc. Belgium)		Pr*
	:23-24+ (Belgian-)	1	
1190.	"Fulmar"		
	Aeroplane, N 29, '40:608-609	1	P
	Aviation, Jl '41:58	1	
	Flight, Ap 3, '41:254f-g		P
	RAF Fly Rev, Jl '56:11	1	
1191.	"Gannet"		
	Prototype: FAIREY 17		
	Aeronautics, Je '53:28-29		P
	O '53:59	1	
	Je '54:56-59	1	P
	S '55:66-69		P
	F '56:34-35		P
	Aeroplane, Ja 11, '52:45	1	
	Jl 4, '52:13		P
	Jl 13, '56:48-53	1	
	D 6, '57:851-854 (-5)	4	
	S 26, '58:494-495 (-AEW.3)		P
	Air Pict, D '55:377 (-T.2)		P
	Avn Wk, Ap 7, '52:25-28	1	

FAIREY (cont.)

1191. "Gannet" (cont.)
Flight, Ja 11, '52:44 1
Jl 4, '52:14-15 (Proto-) 1 P
My 6, '55:574-575 (-A.S. 1) 4
D 6, '57:861-862 (-T.5) 4
Mr 11, '60:343-344 (-AEW. 3) 4
RAF Fly Rev, F '56:48-49 1 P*
My '59:26-29+ 1 Pr
:28-29 (-A.S. 4) S

1192. "Gyrodyne"
Aeroplane, Mr 18, '49:302-303 4
Am Helicop, Je '48:13-14+ 1
Flight, Jl 15, '48:78-79 (Pt 1) 1
Jl 22, '48:106-107 (Pt 2) 1
Ap 21, '49:453-459 1 C*, S*
Interavia, Ag '49:489-492 1 S*

1193. "Hamble Baby"
Flight, N 29, '57:845-848 1

1194. "Hendon"
Air Pict, Jl '56:244-245 1 Pr
Flight, D 1, '32:1145 1 P
Fly Rev Intl, My '68:268 (-II:38 Sq) Pr*

1195. "Jet Gyrodyne"
Aeronautics, Ja '56:42-43 P
Flight, O 5, '56:570-571 1

1196. "Long Range Monoplane"
Aeroplane, O 28, '31:1012-1016 1
F 15, '33:268-270 1 S*
D 4, '42:662 1
Acft Eng, Mr '33:57-58 1 S*
Flight, Jl 10, '31:657 P
O 30, '31:1082-1085 1
F 19, '32:147 D
F 9, '33:117 4
F 16, '33:142-148 (World Distance Record) 4 S*
My 11, '33:438-439 (Across Africa) 4
Pop Fly, Ap '33:Insert D*
RAF Qtr, Ap '33:133 P

FAIREY (cont.)

1197.	"Pintail"		
	Aeroplane, Ja 8, '43:54	1	
1198.	"Primer"		
	Aeroplane, F 6, '48:158-159	4	C
	N 26, '48:670-672	3	
	My 19, '50:856		P*
	Flight, F 5, '48:139-140	1	S*
	F 12, '48:167-168	4	C*
	N 25, '48:619-621	3	C*, S
1199.	"Rotodyne"		
	Aeronautics, F '58:52-53		P
	Ag '58:24-31	1	S
	Aeroplane, Ag 17, '56:208-208c	1	
	N 15, '57:700-703+	1	S*
	:726-727		S*
	D 13, '57:886-888	1	
	N 14, '58:727-728	1	
	Air Pict, F '59:34-35+	1	
	Am Helicop, D '56:6-7+	1	
	Je '58:6-8+	1	C
	Avn Wk, D 2, '57:54-63	1	
	Je 23, '58:55-57	1	
	O 6, '58:42-43	1	
	Can Avn, N '58:26-27	1	C
	Flight, Jl 31, '53:131 (Proto-)	1	
	Ag 17, '56:244-245	1	
	Ag 9, '57:191-197	1	C*, S*
	N 15, '57:762-763	2	C
	D 20, '57:953-958	1	C, S
	:959-960	6 (Engine)	
	Je 6, '58:761-762	1	
	N 14, '58:758-759	2	
	Ja 16, '59:102-103	1	
	Ag 9, '62:200-202	1	S*
	:203-204	4	
	Interavia, No. 6, '58:571-574	1	C, S*
	RAF Fly Rev, Ag '58:32-33	4	C
	Sp/Aeron, F '59:55-57	1	S*
1200.	"Seafox"		
	Aeroplane, D 8, '37:706-708	1	S*
	Air Pict, Je '67:213	1	
	Flight, D 9, '37:570-574	1	S

FAIREY (cont.)

1201.	"Seal"		
	Flight, Mr 16, '33:242-243	1	
	RAF Qtr, O '33:353		P
1202.	"Spearfish"		
	Aeronautics, Ap '46:30-31	1	P
	Aeroplane, F 15, '46:190-193 (-I)	1	S*
	Flight, F 14, '46:166a-169	1	S*
1203.	"Swordfish" or "Stringbag"		
	See also BLACKBURN "Blackfish"		
	Air BP, No. 42:18		C
	Air Prog, Jl '66:84		P
	Flight, F 4, '43:120a	1	
	Fly Rev Intl, Mr '66:438		P
	Je '67:688 (-I to -III)		Pr*
	Pop Avn, Je '39:35		P
	RAF Fly Rev, Jl '58:11+ (-I to -IV)	1	
	Je '62:21	4	
	Sentinel, Jl/Ag '66:7		D
1204.	"Ultra-Light" Helicopter		
	Flight, O 7, '55:588-589+	1	S*
1205.	F.C. 1 Airliner		
	Flight, Ja 19, '39:69	1	
	Jl 13, '39:32d	1	
1206.	G. 4/31		
	Air Pict, My '65:156-157	1	Pr
1207.	N. 4 "Atalanta"		
	Air Pict, Jl '57:247-248	1	
1208.	P. 4/34		
	Pop Avn, Ap '38:33		P
1209.	III D		
	Air Pict, F '65:63	1	
1210.	III F		
	RAF Fly Rev, O '56:17 (Mk IV M)	1	
	D '58:8 (Mk III B)	1	

FAIREY (cont.)

1211.	17 Anti-Submarine Aircraft Successor: "Gannet"		
	Aeronautics, Ap '51:32-33		P
	Aeroplane, Ag 4, '50:134-135		P
	Avn Wk, Mr 5, '51:28-30	1	
	Flight, Ag 3, '50:152	1	
	Ap 6, '51:399-403	1	
	RAF Qtr, Ap '51:120		P

FARMAN France

1212.	F. 50		
	Cr & Cockade, Aut '66:272-273	1	
1213.	F. 220		
	Pop Avn, O '33:238	1	C
1214.	F. 222 and F. 223		
	Air Pict, D '66:465	1	
	Fly Rev Intl, D '65:251 (223.4)	1	
1215.	F. 231		
	Flight, Ja 23, '31:69	1	
1216.	F. 250		
	Flight, My 6, '32:401-402	1	
1217.	F. 300		
	Acft Eng, Ap '30:82	1	
1218.	F. 301		
	Flight, Mr 27, '31:265	1	
1219.	F. 390		
	Flight, O 26, '33:1079	1	
1220.	F. 1000 "Stratosphere"		
	Pop Avn, O '32:225	1	
1221.	F. 1001		
	Flight, S 12, '35:296	4	P

FARMAN-RENAULT France

1222.	Racing Monoplane		
	Pop Avn, N '33:317	1	

FARMAN, HENRI France

1223. II (1909)
Pegasus, Ag '55:8-9 S*

FARMAN, MAURICE France

1224. "Longhorn" M.F. 7
Avn Wk, Ja 25, '65:58-59 1 D*
RAF Qtr, Ja '39:89-92, Plate 4 D

1225. "Shorthorn" or "Rumpitee" M.F. 11
AAHSJ, Sum '64:143 (Tallman's-) 1
Flying, F '68:38-43 (Tallman's-) 1
RAF Qtr, Jl '37:Supp 1 D

F.E. (Designed by Royal Aircraft Factory) GB

1226. 2
Air Pict, Ja '56:25 (-B) 1
Avn Wk, N 28, '60:100-101 (-B) 1 D*
Mr 9, '64:46-47 (-D) 1 D*
Cr & Cockade, Fall '69:253, 264 (-D) P
Flight, D 12, '52:724-728 1
Flying, S '55:36-37+ 4
Pop Fly, F '33:637 1 D
O '33:361-362+ (-D) 4
Mr '37:638-640 (-B) 4 P
RAF Qtr, Ap '37:Supp (-D) 1 D
Jl '39:300-305 (-D) 1

1227. 4
Air Pict, S '59:335 1

1228. 8
Air Pict, My '58:170 1
RAF Qtr, Jl '39:306-307, Plate 1 D

1229. 9 and 12
Air Pict, Jl '61:202-206+ 1

FEDERAL AIRCRAFT FACTORY Switzerland

1230. C-3600 to C-3605
Air Prog, Ja '69:44-45 1
Fly Rev Intl, S '65:43-44 1
F '69:52-55 1 Pr*

FEDERAL AIRCRAFT FACTORY (cont.)

1231. N-20 "Arbalèté" (Crossbow)
Air Pict, Jl '64:203-204 1
Fly Rev Intl, Ja '65:50-51 1

1232. N-20-01 "Aiguillon" (Sting)
Fly Rev Intl, Ja '65:50-51 1

1233. N-20-20 "Harpon" (Harpoon)
Fly Rev Intl, Ja '65:50-51 1

FELIXSTOWE GB

1234. "Fury" and "Porte" F. 1 to F. 5
Flight, D 2, '55:842-846 (Pt 1) ("Porte") 1 S
D 16, '55:895-898 (Pt 2) (F. 3) 1 Pr
D 23, '55:929-932 (Pt 3) 1
Ja 20, '56:87-89 1
Lock Hor, 7th issue, Ja '69:5 (F. 2A) D*
Roundel, Je '55:38-43 4

FFA (Flug- & Fahrzeugwerke AG) Switzerland

1235. P. 16
Later development: AFA P-16 and AFU AJ-7
Avn Wk, Jl 4, '55:35-37 1
Interavia, No. 6, '55:442-443 1
RAF Fly Rev, Ja '60:31-33 1
:33 (-3) S*

FFM (Flugwissenschaftliche Forschungsanstalt e. V. München) Germany

1236. 582
Earlier designation: DFS 582
Later version: DVL 582
Interavia, No. 4, '64:484 1

FFVS (Flygförvaltningens Flygverkstad) Sweden

1237. J 22
Fly Rev Intl, N '64:48 (-A:F9) Pr*
:49-51 1 P*
RAF Fly Rev, D '56:11 1

FGP (Fluttechnische Fertigungsemeinschaft Prag) Czech

1238. FG-227
Flying scale model of BLOHM & VOSS Bv-238
RAF Fly Rev, F '58:11+ 1

FIAT Italy
Air Pict, My '58:188-190 (Pt 1)
Jl '58:250-252 (Pt 2)
Ag '58:282-285+ (Pt 3)
Flight, Ja 25, '57:111-114
Interavia, No. 6, '59:706-708

1239. B.R. 20 "Cicogna" (Stork)
Flight, Ag 1, '40:90-92 1
Ag 5, '43:144 1
RAF Fly Rev, Ap '62:37-38 1

1240. C.R. 20
Fly Rev Intl, Ap '64:61 1
S '69:50 (Hungarian-) Pr*

1241. C.R. 30
Flight, Mr 22, '34:279-280 1
Fly Rev Intl, D '69:53 (Austrian-) Pr*

1242. C.R. 32
Refined C.R. 30
Can Avn, O '40:25 1
Fly Rev Intl, Mr '65:49-51+ 1 Pr*
S '69:51 (Hungarian-) Pr*
RAF Fly Rev, My '57:16-19 (Morato's-)4 Pr*

1243. C.R. 42 "Falco" (Falcon)
Often mistakenly named "Freccia" (Arrow)
Acft Eng, Ag '44:242-243 6 (Fuel Tank)
Can Avn, O '40:23-24 1
RAF Fly Rev, N '56:9 1
S '60:65-66 1

1244. G. 12
Later version: G. 212
Flight, Ag 5, '43:145 1

1245. G. 18
Flight, My 16, '35:Supp:a 1
Je 9, '38:579-580 (-V) 4

FIAT (cont.)

1246.	G. 46		
	Aeronautics, F '51:65-66	1	
	Interavia, Mr '49:164-165	1	C
1247.	G. 49		
	Aeroplane, O 24, '52:578-579 (-1)	1	
	Avn Age, Ja '55:32-33	1	C, S
	Flight, O 24, '52:537	1	
1248.	G. 50 "Freccia" (Arrow)		
	Acft Eng, S '38:285-286	1	
	Can Avn, O '40:24	1	
	RAF Fly Rev, D '58:6+	1	
	S '59:62	4	
1249.	G. 55 "Centauro" (Centaur)		
	RAF Fly Rev, D '58:6+	1	
1250.	G. 56		
	RAF Fly Rev, D '58:8	1	
1251.	G. 80		
	Aeronautics, Mr '52:75	1	
	Aeroplane, D 21, '51:782	1	
	D 28, '51:814		C
	Acft Eng, Mr '52:72-75	1	C
	Avn Age, S '52:30-31	1	C
	Avn Wk, Jl 31, '50:28	1	
	Ja 14, '52:21-22	1	
	Mr 17, '52:28	3	
	Interavia, No. 12, '51:678	1	
	No. 5, '52:275-279	2	
1252.	G. 82		
	Interavia, No. 9, '54:613-615	1	C, S
	RAF Fly Rev, O '55:30-33	1	Pr, S*
1253.	G. 91		
	Aeroplane, My 8, '59:543-547	1	C, S*
	Air Force, D '59:42-43	1	
	Air Pict, N '61:360-361	1	
	Air Prog, Jl '65:28-31+ (-T3)	4	C, S*
	N/D '65:40 (-Y)	1	
	Je '69:44-47+ (-Y)	1	C, S*

FIAT (cont.)

1253. G. 91 (cont.)

Avn Wk, Je 10, '57:50-60 1
Je 2, '58:54-57 1 S
Je 29, '59:56-61 4
Ja 11, '60:90-91 3
Jl 19, '65:62-66 (-Y) 1
N 11, '68:98 (-Y) 6 (JATO)

Flight, F 7, '58:175 (-A) 1
My 23, '58:696 1
Ap 3, '59:461 1
Jl 19, '62:94 (-T3) 1 C
Ja 5, '67:36 (-Y) 1
S 7, '67:414 (-Y) 2
Je 20, '68:931-932 (-Y) 1

Fly Rev Intl, Ag '64:56 (-PAN:"Frecce Tricolori") Pr*
N '65:180-182 (-Y&-YT) 1 Pr*
Ap '66:498 (-R&-T) Pr*
Ag '68:422-424 (-Y) 1 S

Interavia, No. 1, '59:51-53 1 S*
No. 9, '59:1116-1117 4 C
No. 2, '62:225 (-T4) 6 (Electronics) S*
No. 6, '63:826-827 (-T) 1 C*
No. 12, '65:1934-1935 (-Y) 1
No. 8, '68:963-964 (-Y) 1
No. 9, '69:1525-1527 (-Y) 6 (Weapons) P

RAF Fly Rev, Mr '59:30-33+ 1 C, Pr, S*
F '61:32-33 4
Ag '61:42-43 (-R) 1 Pr*

1254. G. 95

Interavia, No. 8, '60:981 1
No. 6, '62:680 (-/1 to -/6) 1

RAF Fly Rev, S '62:32-33 (-/1 to -/6) 1 S*

1255. G. 212

Development of G. 12

Interavia, Ap '48:205-208 1 S

1256. R. 700

Aero Album, Win '69:26-30 1 P

Air Pict, N '65:407 1

1257. R.S. 14

Flight, Ag 19, '43:203 1

FIAT (cont.)

1258.	7002		
	Am Helicop, N '57:6-7+	1	
	Avn Wk, Ja 25, '60:112	1	
	Flight, Ja 24, '58:110	1	

FIESELER Germany

1259.	Fi 98		
	RAF Fly Rev, Ap '60:9	1	
1260.	Fi 103 "Reichenberg" FZG 76 Piloted V-1 bomb		
	Fly Rev Intl, O '67:940	1	
	RAF Fly Rev, My '53:11-12	4	
	My '55:25-27+	1	
	D '57:50+	1	
1261.	Fi 156 "Storch" (Stork)		
	Acft Eng, F '38:38+	1	
	Flight, O 26, '39:334c-d	1	
	F 11, '43:148b	1	
	Fly Rev Intl, Ag '68:458 (-C&-D)		Pr*
	RAF Fly Rev, Ja '60:37-38	1	
	West Fly, Mr '40:17	1	
1262.	Fi 158		
	Fly Rev Intl, Jl '69:65	1	
	RAF Fly Rev, Ap '60:9	1	
1263.	Fi 167		
	RAF Fly Rev, Ag '58:42	1	
	O '58:6	1	
264.	Fi 333		
	RAF Fly Rev, O '58:6	1	

ILPER USA

265.	"Beta 200"		
	Air Prog, S '66:84		P
	Flight, Je 30, '66:1108	1	C

FIRESTONE USA

1266. XR-9B
Am Helicop, My '46:13+ 1
Ag '46:29+ 2
Aviation, My '46:99 1
N '46:89 S*
D '46:113 C*

FISHER USA

1267. XP-75 "Eagle"
Air Pict, Jl '59:258-259+ 1
RAF Fly Rev, D '56:13+ 1

FIZIR Yugoslavia

1268. A.F. 2
Flight, My 1, '31:376-377 1

FLANDERS GB

1269. B. 2
Air Pict, S '61:269-273 1

FLEET Canada

1270. 2 Biplane (1930)
AAHSJ, Spg '68:55-59 1 Pr
Can Avn, Ap '30:28 1
West Fly, O '35:23-24 3

1271. 16 "Finch"
Comm Avn, Jl '39:26 1
Roundel, Jl/Ag '63:33 1 I
RAF Fly Rev, Je '60:9+ 1

1272. 50
Aviation, My '38:34-35 1
My '41:60-61 1
Can Avn, N '37:14-15 (-J&-K) 1 S
Spg '69:3-15 1 C, Pr, S
Fly Rev Intl, Ap '65:44 (-K) 1

FLEET (cont.)

1273.	60 "Fort"			
	Aero Digest, Je '40:65-66+	1		
	Aviation, My '41:60	1		
	Can Avn, Ap '40:25	1		
	My '40:24-25+	1		P
	S '41:22-23	1		
	:40-42+	3		
	Comm Avn, Mr '40:46	1		
	Ap '40:19-22	1		
	Ag '41:57-60	3		
	Roundel, My '65:33 (RCAF-)	1		P
	RAF Fly Rev, Je '60:9	1		

FLEETWINGS USA

1274.	"Seabird"			
	Air Prog, Mr '68:48-49 (Restored-)	1		C
	Aviation, O '36:38-41	1		S*
	Comm Avn, F '39:37	1		
	Flight, O 22, '36:424	1		
	Pop Avn, D '36:30+	1		
1275.	BT-12			
	Aero Digest, D '42:186-187			S
	Aviation, F '41:128 (X-)			S
	D '42:108-112	3		S
	My '43:205	6	(Engine)	P
	O '43:119-132+	1		S
	Comm Avn, F '41:60 (X-)	1		
	Mr '41:72 (X-)	1		
1276.	33			
	Aviation, Ap '42:112-115			S*

FLETCHER USA

1277.	BT-22			
	Avn Age, N '51:40-41	1		S
1278.	FBT-2			
	Aero Digest, F '41:165-166	1		
	Comm Avn, Ap '41:44-45+	1		

FLETCHER (cont.)

1279.	FD-25 "Defender"		
	Air Pict, N '52:329	1	P
	D '52:365	1	
	Air Prog, Ja '65:4-11+ (-B)	1	C, S*
	Avn Wk, Jl 30, '51:16	1	
	RAF Qtr, Ja '53:10-11	1	
	Skyways, F '52:20-21+	4	
1280.	FL-23		
	Flying, Mr '51:27+	1	
	Skyways, Jl '51:50	1	

FLETCHER — New Zealand

1281.	FU-24 "Utility"		
	Avn Wk, Ag 23, '54:21	1	
	Flying, My '62:66-67+	4	C
1282.	FU-1060 and FU-1160		
	Fly Rev Intl, O '68:25-26	1	

FLETTNER — Germany

1283.	FL-265		
	Fly Rev Intl, F '65:49-51	1	Pr*
	RAF Fly Rev, S '56:75	1	
1284.	FL-282 "Kolibri"		
	Am Helicop, Je '47:18-19+	1	S*
	S '49:10-12+	2	
	Aviation, Je '46:60-64	1	
	Fly Rev Intl, F '65:49-51	1	Pr*
	RAF Fly Rev, S '56:75-76	1	

FMA (Fabrica Militar de Aviones) — Argentina
See DINFIA

FOCKE-ACHGELIS — Germany
Am Helicop, Ja '47:14-16+

1285.	Fa 223 "Drache" (Dragon)		
	Am Helicop, Ja '47:14-16+	1	C
	RAF Fly Rev, Jl '55:12	1	
	Ap '59:37-38	1	
	Jl/Ag '59:39	1	

FOCKE-ACHGELIS (cont.)

1286.	Fa 269		
	RAF Fly Rev, D '57:52	1	
1287.	Fa 330 "Strandläufer" (Sandpiper)		
	Am Helicop, F '47:10-11+	1	

FOCKE-WULF Germany

1288.	"Flitzer"		
	RAF Fly Rev, D '60:9+	1	
1289.	A. 20 and A. 28 "Habicht" (Hawk)		
	Acft Eng, Mr '30:55-56	1	
	Pop Avn, Mr '31:36	1	
1290.	A. 38 "Möwe" (Gull)		
	Flight, Ag 21, '31:843-844	1	C
1291.	A. 47		
	Acft Eng, Ja '33:17-18	1	
	Flight, D 29, '32:1237	1	C
1292.	F 19a "Ente" (Duck)		
	Aero Digest, Ja '31:72-74	1	
	Aeroplane, N 18, '31:1182	1	
	Air Pict, Jl '64:223	1	
	Acft Eng, F '31:39-40	1	
	Flight, Ja 2, '31:4-7	1	
	N 13, '31:1137	1	
	Pop Avn, Ja '31:19	1	
	Jl '32:19-20	1	C
1293.	FW 56 "Stösser" (Falcon Hawk)		
	RAF Fly Rev, Ja '58:9	1	
	F '63:45+	1	
1294.	FW 57		
	RAF Fly Rev, Ja '62:47	1	
1295.	FW 58 "Weihe" (Kite)		
	Aeronautics, D '39:58	1	S*
	Fly Rev Intl, D '69:53 (-B:Austria)		Pr*
	RAF Fly Rev, O '61:67 (Fixed gear-)	1	
	D '61:57	1	

FOCKE-WULF (cont.)

1296.	FW 61		
	Aeroplane, My 25, '38:639-643	1	
	Pop Fly, My '39:72-73+	1	
	RAF Fly Rev, D '62:43+	1	
1297.	FW 159		
	RAF Fly Rev, F '56:9	1	
	O '57:11	1	
1298.	FW 186 Gyroplane		
	RAF Fly Rev, D '60:9	1	
1299.	FW 187 "Falke" (Falcon)		
	Acft Eng, Ap '41:105-106	1	
	Flight, O 24, '40:359	1	D
	RAF Fly Rev, My '57:43-44	1	
1300.	FW 189 "Uhu" (Owl)		
	Flight, O 2, '41:221	1	
	Ap 15, '43:392a	1	
	RAF Fly Rev, O '56:53-54	1	
1301.	FW 190		
	Later development: Ta 152		
	Aero Album, Win '68:22-23 (-G)		C, P
	Spg '69:44-45 (2-seater-)	1	
	Aero Digest, Ap 1, '44:110-111		S
	Aeronautics, Je '43:38-41 (-A3)	1	C
	O '44:34-35	2	
	Aeroplane, Ag 14, '42:186-187 (-A3)	1	
	Ag 21, '42:216-217 (-A3)		C
	Ag 28, '42:248-249, 262 (-A3)		S*
	Ja 29, '43:122-124 (-A3)	1	C
	Air Pict, D '57:439-440 (-A5:Japan)	1	
	Ag '59:293-296+ (-A)	1	Pr
	Ja '66:27 (Torpedo-)	5	
	Air Prog, F '43:54-55 (-A3)		S*
	Ag/S '65:42-43 (-A3)	1	S*
	Air Tech, N '42:16-17+	1	S*
	Acft Eng, Mr '43:76-81	1	C
	Ag '44:243-244	6 (Drop Tank)	
	Aviation, F '43:151 (-A3)		S*
	O '44:127-152	1	C*, S*
	Mr '45:156-157	6(Gear Retract)	S*

FOCKE-WULF (cont.)

1301. FW 190 (cont.)

Can Avn,	O '42:38+ (-A3)	1	
	Mr '43:54-55 (-A3)	1	S*
Flight,	My 14, '42:472-473	1	
	Ag 13, '42:182-185	1	
	Ag 27, '42:224-225	1	S*
	Ja 28, '43:99-100 (-A3)	1	C
Fly Rev Intl,	D '64:46 (-A3, -A6, -F8, -G)		Pr*
	D '65:244 (-A:Bär's & Priller's)		Pr*
	:246 (Wurmheller & Mayer's-)		Pr*
RAF Fly Rev,	Je '57:16-20+ (Pt 1)	1	Pr*
	Jl '57:27-31 (Pt 2)	1	
	Ag '60:11 (-D)	5	
Skyways,	D '42:14-17+	1	
	Je '43:46		P
	Jl '44:24-28+	1	S*

1302. FW 191

RAF Fly Rev,	Ja '59:39-40	1	

1303. FW 198

Aircraft erroneously attributed to FOCKE-WULF
See DE SCHELDE S. 21

1304. FW 200 "Kondor" (Condor)

Aero Digest,	S '38:77-78	1	C
Aeroplane,	N 30, '45:630-631 (Himmler's-)	1	
Flight,	Je 30, '38:639-640	4	
	Ag 18, '38:148 (Berlin-N.Y. flight)	4	
	Ag 27, '42:228a	1	
Pop Fly,	Je '39:129-130	4	C
RAF Fly Rev,	Ag '56:35-36	1	
	O '60:11+ (Hitler's-)	1	
	My '61:30-31+ (-C-3)	4	

1305. FW 200 "Kurier" (Courier)

Military version of "Kondor"

Aeroplane,	Mr 14, '41:310 (-K)	1	
Air News,	F '43:15 (-C)		D
Air Prog,	Je '43:15 (-C)		D
Air Tech,	N '44:44-45+ (-K2)	1	

FOCKE-WULF (cont.)

1306. S 24 "Kiebitz" (Lapwing)
Acft Eng, Je '29:128 1

1307. Ta 152
Developed from FW 190
RAF Fly Rev, Jl '57:26, 28-29 (-C) Pr†, S*
:27-31 1
F '56:13 1

1308. Ta 154 "Moskito" (Mosquito)
RAF Fly Rev, F '56:13 1
Ap '57:41-42 1

1309. Ta 183
Air Pict, S '63:306 1
RAF Fly Rev, My '63:45 1

1310. Ta 283
Reached design stage only
RAF Fly Rev, My '56:13 1

1311. Ta 400
RAF Fly Rev, Ag '61:50 1

1312. 102
Acft Eng, Ja '33:17 (L-&W-) 1

1313. 260
Avn Wk, Je 4, '62:103-104 1

FOKKER Germany, Netherlands, and USA
Air Pict, Ja '61:2-7 (1920-1940 fighters)
Air Prog, Spg '59:48-63 (WW I types; photo story)
AAHSJ, Sum '66:79-92 (Pt 1)(Amer-built Fokkers)
Win '66:253-265 (Pt 2)
Fall '67:174-188 (Pt 3)
Avn Wk, Ag 8, '49:21-24
Cr & Cockade, Sum '66:175-179 (E. series)
Flight, O 28, '48:511-515+ (Types in 1948)
D 30, '60:1014-1016 (Airliners:F.II to F.27)
Interavia, No. 10, '69:1626-1630 (50th Anniversary)
Pegasus, S '52:9-11
RAF Fly Rev, My '59:30-34 (V. 1 to V. 45, with photos)
O '60:9 (In Spanish Civil War)
N '60:28-30+ (E. I to E. IV)

FOKKER (cont.)

1314.	"Spin" (Spider)		
	Air Prog, Mr '67:29-31 (-1 to -3)	1	Pr
1315.	"Super-Universal"		
	AAHSJ, Fall '67:232 (Seaplane-)		D
	CAHS, Sum '69:42-45	1	Pr
	Flying, My '64:20		D
1316.	XB-8 and YB-8		
	See also XO-27		
	Pop Avn, F '31:22	1	
1317.	C. I		
	Fly Rev Intl, S '69:50 (Hungarian-)		Pr*
1318.	C-2		
	US Army version of F. VII/3m		
	Air Prog, D '67:52-55+	1	
	Ja '68:49-50+ ("Bird of Paradise")	4	Pr†
	:49+, 51 ("America")	4	Pr†
	Pop Avn, S '27:5-8 ("America")	4	C, S*
	Mr '29:35-40+ ("Question Mark")	4	
1319.	C. V		
	Air Prog, F '68:50-51		Pr†
	Fly Rev Intl, S '68:62 (Danish-)		Pr*
	Mr '69:43 (-E:Swiss)		Pr*
	Ap '69:52 (-D:Norway)		Pr*
1320.	C. X		
	Fly Rev Intl, D '68:50 (Finnish-)		Pr*
	Pop Avn, Mr '36:182	1	
	RAF Fly Rev, Ap '62:45+	1	
1321.	D. III		
	Air Prog, O '68:25 (Boelke's-)		Pr†
1322.	D. V		
	Cr & Cockade, Aut '64:213-221	1	P, Pr
	Spg '67:64-65	1	P

FOKKER (cont.)

1323.	D. VII		
	Aeroplane, Ag 14, '42:188	1	
	Air Cl, O '68:14-22	1	P, Pr
	:23 (Replica-)		P*
	Ap '69:52	1	D
	Air Prog, D '69:39-42+ (Replica-)	3	P*
	AAHSJ, Win '61:Supp		P*
	Avn Wk, Mr 13, '61:224-225	1	D*
	Je 1, '64:36-37		D*
	Cr & Cockade, Aut '61:290		Pr*
	Win '61:295-315	1	Pr, S
	Sum '64:117-144 (Smithsonian Inst's-)	1	P
	Win '66:409 (Noltenius'-)		Pr†
	Fall '69:280 (Wüsthoff's-)	1	P
	:281 (Blumenthal's-)	1	P
	Win '69:374-378	1	P
	Flying, D '53:22+	1	
	O '64:63-66	4	C, P
	Mr '67:102-107	1	S
	Hist Avn, S '69:28-29	1	
	Pop Fly, Ja '32:579	1	D
	Roundel, N '53:32-34	1	
	RAF Fly Rev, Ja '57:16-20	1	Pr*
1324.	D. VIII		
	Pop Fly, Ag '34:258	1	D
1325.	D. 17		
	Air Prog, Je '68:34-35	1	S
	Flight, Mr 29, '34:310-311	1	S
	Pop Avn, Ag '32:80	1	S
	RAF Fly Rev, O '62:28-29	1	Pr*
	N '62:39	4	
1326.	D. 21		
	Flight, Ja 9, '36:43	1	
	Fly Rev Intl, N '63:38 (Denmark, Finland, Java, Netherlands-)		Pr*
	:43	4	
	D '68:50-51 (Finnish-)		Pr*
	RAF Fly Rev, Mr '58:16-20+	1	Pr*
	S '62:28-31	1	Pr†, S*

FOKKER (cont.)

1327.	D. 23		
	Aviation, F '41:132		S
	Flight, Mr 14, '40:234i		S*
	Fly Rev Intl, Je '68:356	1	
	RAF Fly Rev, S '55:21	1	S
	S '61:40	1	
1328.	Dr. I Triplane		
	Air Cl, Ap '69:33	1	D
	Avn Wk, Ap 13, '59:171		D*
	My 11, '59:155+	1	D*
	Ja 7, '63:36		D*
	Cr & Cockade, Aut '62:196-203 (Serial Nos.)	4	
	Spg '64:1-29	1	P, Pr, S
	Sum '69:184 (Stapenhorst's-)	1	P
	Win '69:374 (-141/17)	1	P
	Flight, My 2, '68:661	1	P*
	Flying, Je '60:73 (Voss'-)	1	D
	Hist Avn, D '69:4-7	1	Pr
	Pop Fly, Ag '32:283	1	D
	RAF Fly Rev, Ap '59:41 (Stapenhorst's-)	5	Pr
	My '59:31 (Richthofen's-)		Pr*
1329.	E. I		
	Flying, Ja '59:67	1	D
	Hist Avn, My '69:18-19	1	
	RAF Fly Rev, N '60:29-30+	1	
1330.	E. III		
	Air Cl, Ap '69:26		D*
	Avn Wk, N 28, '60:100-101	1	D*
	Cr & Cockade, Win '62:329		D, Pr
	:384 (Replica-)		P
	Flying, My '64:36-39+	4	C, P
	RAF Fly Rev, N '60:28-30+	1	Pr*
1331.	E. IV		
	Air Pict, O '67:373	5	
1332.	F. VII/3m		
	British-licensed: AVRO 618 "Ten"		
	US Army designation: FOKKER C-2		

FOKKER (cont.)

1332. F. VII/3m (cont.)

Aeroplane, D 16, '31:1375	1		
F 19, '43:226 ("Southern Cross")	1		
Air BP, No. 4:27 (Sabena-)			Pr*
Air Pict, Ap '58:136-138	1		
O '58:362+ ("Southern Cross")	1		
Ap '66:142-143 ("Southern Cross")	1		
N '66:416	1		
D '67:52-55+	1		
Air Prog, Ag/S '64:76-77 ("Southern Cross")			Pr†
AAHSJ, Fall '59:194-206 ("Southern Cross")	1		
Flight, My 9, '30:499-500 ("Spider")	4		
Jl 4, '30:757-761 ("Southern Cross")	1		C
Ja 16, '31:49	6	(Cockpit)	C
Je 19, '31:550	6	(Wing)	
Jl 17, '31:693-696	6	(Wing)	
D 18, '31:1234	6	(Wing)	
Ag 9, '45:141-143 ("Southern Cross")	1		
Flying, Je '53:34-35+ ("Southern Cross")	4		
Fly Rev Intl, D '63:59 (G-EBYI)	1		
Interavia, No. 5, '60:589-591 ("Josephine Ford")	4		
US Air Svcs, Ag '30:24-27 ("Southern Cross")	4		

1333. F. IX

Flight, My 30, '30:580-582	1	

1334. F. XI "Universal"

Flying, Je '63:1		D

1335. F. XII

Aeronautics, F '46:81		C
Aeroplane, F 11, '31:244-246	1	
Acft Eng, Ap '31:97-98	1	
Flight, F 13, '31:132-134	1	

1336. F. XIV

Bee-Hive, Mr '30:13	1	P

FOKKER (cont.)

1337.	F. XVIII		
	Flight, Jl 29, '32:711-712	1	C
1338.	F. XX		
	Aero Digest, Ja '34:46-47	1	
	Flight, Je 22, '33:611	1	
	O 5, '33:993-995	1	C
	US Air Svcs, F '33:35-36	1	
1339.	F. 22		
	Bee-Hive, Mr '34:6	1	
	Jl '35:8	1	
	Flight, My 2, '35:482		C
	Pop Avn, Ap '34:232	1	
	S '35:153 (Swedish-)	1	
1340.	F. 27 "Friendship" See also FAIRCHILD F-27 and FAIRCHILD-HILLER F-27		
	Aeronautics, D '55:38-41	3	
	Ap '57:26-39	2	S*
	F '59:34-35		P
	Aeroplane, Ap 11, '58:509-515	1	S*
	Ag 8, '58:203-206	4	C
	Air BP, No. 9:7 (Trans-Australia-)		Pr*
	Air Pict, D '55:369-371	1	
	My '59:177-179	1	S*
	D '63:407 (Trans-Australia-)	5	
	Acft Eng, S '55:313-315	3	
	Avn Age, Ap '53:30-31	1	S*
	F '56:34-43	1	
	O '56:220-223	4	C*
	Je '57:218-221	1	
	Avn Wk, D 27, '54:75-78	1	S*
	Ja 9, '56:33-34	1	
	F 6, '61:45-47	1	
	Can Avn, Mr '57:26-27	1	
	:28-29	4	C
	Flight, D 25, '53:830	1	
	Ap 23, '54:526 (Proto-)	1	S*
	Je 3, '55:762-767+	1	C*, S*
	Jl 6, '56:47-49	1	S*
	S 28, '56:560	2	
	D 28, '56:987-990	4	C
	N 1, '57:712-714	1	S*
	Jl 25, '58:104-106	1	S*
	Jl 15, '60:94-95	1	

FOKKER (cont.)

1340.	F. 27 (cont.)			
	Fly Rev Intl, D '64:22-25	1		S*
	F '67:385-386+	1		Pr
	Interavia, No. 10, '55:775-777	1		S*
	No. 4, '57:310	4		
	No. 9, '68:1140-1142	1		
	RAF Fly Rev, O '62:24-27 (-M)	1		Pr, S*
	US Air Svcs, F '56:21	1		
1341.	F. 28 "Fellowship"			
	Later development: FAIRCHILD-HILLER F 228			
	Aeroplane, N 17, '66:9-18	1		S
	N 15, '67:21	2		
	Air Pict, Ja '66:24-25	1		S
	Air Prog, S '69:29-31	1		S*
	Acft Eng, Je '67:9-12, 17-36	1		S*
	:15-17	2		
	:37-40	6	(Interior)	C*, S
	:41-45	6	(Engine)	S*
	Jl '67:17-23	3		
	:24-34	1		S*
	Avn Wk, Je 14, '65:246-251	1		C
	Ag 22, '66:28-29	3		
	My 29, '67:224-229	1		
	Ja 22, '68:44-46+	4		C
	Flight, Ag 27, '64:319-322	1		S*
	Je 15, '67:975, 977-984	1		C*, S*
	Ap 18, '68:583-588	4		C
	Flying, Ja '63:25+	1		
	Fly Rev Intl, O '68:34-39	1		S
	Interavia, No. 6, '62:679	1		
	No. 8, '63:1188-1190	1		
	No. 10, '64:1542	1		S*
	No. 11, '64:1706	3		
	No. 6, '67:934-936	1		
	No. 11, '67:1799	2		
	No. 3, '68:319-323	4		C
1342.	F. 32			
	Aeroplane, Jl 31, '42:141	1		
	Acft Eng, N '29:313	1		P
	AAHSJ, Sum '66:110-117, 151	1		C, P, Pr†
	Win '68:263-264			P
	Bee-Hive, Ja '31:8-9	6	(Engine Cowl)	
	Flight, Ap 24, '31:363-365	1		C

FOKKER (cont.)

1343.	F. 36		
	Flight, Jl 26, '34:763	1	
1344.	G. I "Le Faucheur" (The Grim Reaper)		
	Air Pict, My '58:157-158 (-A&-B)	1	
	Aviation, Mr '37:33-34	1	
	Jl '41:61	1	
	Fly Rev Intl, Ap '67:507-511	1	Pr*
	Pop Avn, Je '37:19-20+	1	
	RAF Fly Rev, Mr '55:11-12	1	
	Je '55:9	1	
	D '59:9+	1	C
1345.	K. 1		
	Fly Rev Intl, Ja '69:68	1	
1346.	M. 7		
	Avn Wk, O 9, '61:86		D*
1347.	XO-27		
	See also XB-8 and YB-8		
	US Air Svcs, O '30:42	1	P
1348.	S. 11 "Instructor"		
	Aeroplane, F 18, '49:180-181	4	
	Avn Age, Ap '51:26-27	1	C, S
	Can Avn, Ap '50:19	1	
	Flight, F 24, '49:218-221	4	C*, S
1349.	S. 13		
	Air Pict, S '50:273		P
	Avn Wk, O 2, '50:28-29	1	S
	Flight, My 25, '51:608-609+	4	
1350.	S. 14 "Mach Trainer"		
	Aeroplane, Jl 20, '51:69-71	1	S*
	Ap 29, '55:568-571	4	C
	Air Pict, S '51:285	1	
	Air Prog, '53/'54 Annual:90-91+		S*
	Avn Age, Ap '51:27		S*
	Avn Wk, Jl 30, '51:29-33	1	C*, S
	Flight, N 24, '49:687-688	1	S*
	Ja 15, '54:68-72	4	C*
	Interavia, No. 9, '51:510-511	1	C*, S*
	RAF Fly Rev, Je '55:30-33	1	Pr, S*
	West Avn, F '56:34	1	

FOKKER (cont.)

1351. T-2
US Army designation of F. IV
Flying, F '57:54+ (Macready & Kelly's-) 4
Hist Avn, F '69:10-15 (Macready & Kelly's-) 1 C*, Pr†
Pop Avn, O '27:19-21+ (Macready & Kelly's-) 4

1352. T. V
Aero Album, Spg '69:3-10 1 P, Pr†, S
Air Prog, Je '68:34-35 1 S
Acft Eng, Mr '38:83-84 1
RAF Fly Rev, Jl '63:55 1

1353. T. VIII
RAF Fly Rev, Ag '57:40 (-L&-W) 1

FOKKER/REPUBLIC Netherlands/USA

1354. D. 24
Interavia, No. 5, '62:619-620 1

FOLLAND GB
Air Pict, Ag '54:244

1355. "Gnat" Fo. 141
Later version: HAWKER SIDDELEY "Gnat"
Aeroplane, Jl 29, '55:176-177 P
N 22, '57:771-778 1 S*
Air Pict, Jl '58:239-241 1 Pr
Air Power, Ja '55:90-99 1
Avn Age, Jl '53:26-27 1
Avn Wk, N 8, '54:28-36 3
O 17, '55:34-45 1
Je 2, '58:61-62 1
Ag 10, '59:101-103 1
Bris Sid Jnl, Aut '62:18-19 4
Flight, Ap 3, '53:425-426 1
O 14, '55:607 1 S*
N 22, '57:799-808 1 C*, S*
:809-812 6 (Engine)
Flying, Je '55:23-25 1 S
Fly Rev Intl, N '68:43 (Hindustan-built-) Pr*
Interavia, No. 9, '58:918 (-1&-2) 1

FOLLAND (cont.)

1355.	"Gnat" (cont.)			
	RAF Fly Rev,	Ap '54:9-10	1	
		D '57:4	1	
		:5 (Finnish-)		Pr*
		D '58:31-34	1	
		:30, 32-33 (-I)		Pr, S*
		Mr '59:6 (Indian & Yugoslavian-)	5	
1356.	"Gnat Trainer" Fo. 144			
	Later version: HAWKER SIDDELEY "Gnat Trainer"			
	Aeroplane,	S 4, '59:120-122	1	S*
	Bris Sid Jnl,	Win '59-'60:30-33	2	S
	Flight,	N 27, '59:640-645	1	C*, S*
		S 15, '66:472-474+	4	
	Fly Rev Intl,	D '63:27-30	1	S*
	Nav Avn News,	D '62:22	1	P
	RAF Fly Rev,	N '60:20-21	1	P*
		O '61:36-39	1	Pr, S*
1357.	"Midge" Fo. 139			
	Aeronautics,	Ja '55:28-30	4	C, P*
	Aeroplane,	Ag 20, '54:240-241	1	P
		Ag 27, '54:269-270	1	C
	Air Power,	Ja '55:90-99	1	P
	Flight,	Ag 20, '54:228-229	1	
	Flying,	Je '55:23-25	1	
	RAF Fly Rev,	O '54:19-20+	4	
1358.	E. 28/40			
	Flight,	Ja 3, '46:14-15	1	
1359.	43/37 Test Bed			
	RAF Fly Rev,	N '57:44	1	

FORD USA

1360.	2-AT "Air Transport"			
	Hist Avn,	S '69:14-15	1	
1361.	3-AT "Air Pullman"			
	Hist Avn,	S '69:16-17	1	

FORD (cont.)

1362.	4-AT "Tri-Motor"		
	Flight, N 14, '30:1233-1237 (-E)	1	C
	Hist Avn, S '69:11-23	1	
	Pop Avn, S '27:19-22 ("Stanolind")	1	
	US Air Svcs, My '30:48	1	
1363.	5-AT "Tri-Motor" or "Tin Goose"		
	Air BP, No. 44:25-27 (Island Airways-)	4	P*
	Air Pict, Mr '57:104 ("WZ"-)	1	
	Air Prog, Je/Jl '64:59-67+	4	P
	Ag '68:40	1	
	Acft Eng, O '29:277	1	
	AAHSJ, Win '62:278-282 (-B)	4	
	Avn Wk, Mr 3, '69:44-45 (Byrd's-)	4	D*
	Bee-Hive, Mr '30:10-11 (-C-5)	1	
	Fall '55:24	1	
	Flight, N 14, '30:1233-1237 (-C)	1	C
	F 20, '31:165	6	(Mail Compartment)
	O 16, '31:1046		P
	Flying, Ja '56:43	1	
	S '63:46-47+	4	C
	N '63:76	1	D
	Hist Avn, S '69:11-23	1	C
	Pop Avn, F '39:20-23+	1	
1364.	7-AT		
	US Air Svcs, Ja '30:29	1	
1365.	8-A		
	Flight, Ag 7, '31:795-796	1	
1366.	9-A and 11-A		
	US Air Svcs, Je '30:58	1	
1367.	14-A		
	Aero Digest, Ap '32:64-68	1	
	Aviation, Ap '32:180-184	1	

FOUGA France

1368.	"Cyclone" CM. 8-R13		
	Aeroplane, Ap 21, '50:463-467	4	C
	Flying, S '49:13-15+	1	
	Skyways, Ap '50:28-29+	1	

FOUGA (cont.)

1369. "Gemeaux" (Gemini) CM. 88-R
Avn Wk, Ap 28, '52:35 (-4) 1

1370. "Magister" (Schoolmaster) CM. 170 and CM. 175
Air Force version: CM. 170
Naval version: CM. 175
Later version: POTEZ-HEINKEL CM. 191
Aeroplane, Jl 2, '54:14-19 4 C
Ja 16, '59:70-71 4
Air Pict, My '54:131-132 (-170R) 4
Avn Age, D '57:34-35 1 S
Avn Wk, Ag 25, '52:28-29 (-170R) 1
O 8, '56:33-34 (-170) 4
Ap 13, '59:106-107 (-170R) 3
Can Avn, O '56:56-57+ (-170) 4
Flight, Ap 2, '54:401-406 (-170) 4 C, S
Je 14, '57:797-798 (-175M) 4
Interavia, No. 1, '55:48-50 (-170) 1 C*, S
No. 2, '58:144-145 (-170) 4
RAF Fly Rev, Ag '57:26-29 (-170R) 1 Pr, S*

FOUND Canada

1371. "Centennial 100"
Flying, Ag '68:42-43+ 4

1372. FBA-2
Air Prog, Ap '67:14-15 1
Avn Wk, Je 15, '64:98-102 (-C) 4
Can Avn, O '60:27 (-A) 1
F '61:52-53 (-B) 1
Ag '64:16-17+ (-C) 4

FRIEDRICHSHAFEN Germany

1373. D. I
Cr & Cockade, Sum '67:206 P

1374. FF. 33
Air Prog, Ja '66:48-49 (-E&-J) Pr†
Fly Rev Intl, Je '64:46-50 ("Wölfchen") 4

1375. G. III
Cr & Cockade, Sum '63:168 Pr

FRIESLEY USA

1376. "Falcon"
AAHSJ, Sum '65:127-128 1 P

FRYE USA

1377. F. 1 "Safari"
Avn Wk, S 26, '55:108-109 1
Ap 16, '56:39-41 1
Je 18, '56:37 1
Can Avn, N '55:107 1
Flight, O 7, '55:600 1
Ap 13, '56:434 1
Jl 6, '56:64-65 1 S*

FUJI Japan

1378. T1F1-2
Air Pict, My '59:184-185 (-2) 1
RAF Fly Rev, Mr '60:34-37 (-2) 1 Pr, S*
Ag '63:18 (-1&-2) 1

FUJI KOKUKI Japan

1379. "Ohka" (Cherry Blossom) [Baka] (Fool)
See YOKOSUKA MXY7

FULTON USA

1380. "Airphibian" FA-2
Can Avn, N '50:14-15+ 4
Flight, Ag 5, '48:160-161 1
Flying, Ag '50:24-25+ 4
Skyways, F '47:31+ 1
Ag '50:14-17+ 4

FUNK USA
Pop Avn, Ja '40:42-43

G & A USA
See FIRESTONE

GALEAO Brazil

1381. PB-61
Am Helicop, My '52:6-7+ 1

GALEAO (cont.)

1382. PB-63
Am Helicop, S '53:12+ 1

GATES USA
See LEAR JET

GEE-BEE (Granville Bros.) USA
Air Pict, Mr '56:74-75 (Pt 1)(Aircraft types produced)
Ap '56:116-119 (Pt 2)(Aircraft types produced)
Je '56:216 (Addendum)(Aircraft types produced)
Air Prog, Fall '59:57-62
Pop Avn, D '39:26-29+

1383. C-8 "Eightster"
Aero Digest, Jl '33:38-39+ 1

1384. R-1 "Super Sportster"
Aeroplane, Ap 12, '33:645-648 (Doolittle's #11-) 1
O 1, '43:398 1
Air Prog, Fall '59:57-62 1 P
Acft Eng, N '32:285-286 1
AAHSJ, Win '62:274-277+ 1 D
US Air Svcs, O '32:12 (Doolittle's #11-) 4 P

1385. Z
Aero Digest, O '31:70-72 1
Aeroplane, O 1, '43:398 1
Air Prog, Fall '59:57-62 1 D, P

GENERAL AIRBORNE USA

1386. XCG-16
Fly Rev Intl, F '65:43 (-A) 1
West Fly, Ja '45:44+ 1

GENERAL AIRCRAFT GB
See also MONOSPAR

1387. "Cygnet" G.A.L. 42
Aeroplane, Jl 12, '39:69-72 4 S*
Air Pict, Mr '65:99 (-II) 1
Flight, Jl 13, '39:42-45 1 C*, S

GENERAL AIRCRAFT (cont.)

1388.	"Hamilcar"		
	Aeroplane, D 15, '44:667-673	1	C*, S*
	N 16, '45:573-574 (-10)	1	
	Flight, D 14, '44:634-638+	1	S*
	N 15, '45:532-534 (-10)	1	S
	RAF Fly Rev, S '58:20	1	
1389.	"Hotspur"		
	Flight, O 7, '43:392	1	
	Fly Rev Intl, My '66:581	1	
1390.	"Monospar"		
	Later development: See MONOSPAR		
	Aeroplane, Ap 20, '32:691-710 (ST. 4)	1	S
	Acft Eng, My '32:111 (ST. 4)	1	
	Jl '32:177-179 (ST. 4)	3	
	Flight, Jl 10, '31:667-670 (ST. 3)	1	S
	O 9, '31:1013-1014 (ST. 3)	1	
	Ap 22, '32:341-348 (ST. 4)	3	Pr, S
	Je 17, '32:525		P
1391.	"Owlet"		
	Aeronautics, Ja '41:46-49	1	S*
	Aeroplane, N 29, '40:603-606	1	S*
	Aviation, My '41:64-65	1	S
	Flight, N 28, '40:446a-g	1	C
	Je 10, '43:606b	1	
	RAF Fly Rev, D '60:9	1	
1392.	"Universal Freighter" G.A.L. 60		
	Aeroplane, Ap 29, '49:482-483	1	
	Air Pict, Ja '51:15	1	
	Flight, F 5, '48:150-153	1	C, S*
1393.	G.A.L. 38		
	RAF Fly Rev, My '59:9	1	

GENERAL AIRCRAFT CORPORATION USA

1394.	GAC-100		
	Avn Wk, Je 3, '68:102-105	1	
	Can Avn, Ja '69:27	1	
	Fly Rev Intl, S '69:29-30	1	

GENERAL AVIATION USA

1395. F-15 and FLB-51
Bee-Hive, Jl '32:5-6 1
Flight, Ag 24, '33:844-845 1
US Air Svcs, Je '32:35 1

1396. G.A. 43
Aero Digest, F '33:44-46 1
My '34:36 1
Je '34:36 1
Air Pict, N '63:364-366 1
Bee-Hive, My '34:6-7 (-J) 1
Flight, Mr 2, '33:203 1
Pop Avn, Jl '33:22 1

1397. OJ-2
Predecessor: BERLINER-JOYCE XOJ-1
Flight, My 10, '34:468 1

GENERAL DYNAMICS USA

1398. B-58 "Hustler"
See CONVAIR B-58

1399. F-111 and TFX
Aerosp Intl, Ja/F '68:6-14 (-A) 4 C*
S/O '68:24 1
Air Force, F '66:37-38 (FB-) 1
D '67:45-55 (-A) 4 S*
Ag '69:62-65 (-D) 6 (Avionics)
S '69:40-41 4
D '69:73 (FB-) 1
AF/Sp D Intl, Mr '66:26-27 (FB-) 1
Air Prog, My/Je '65:48-49 (-A) Pr
Ja '66:4-6 1
Acft Eng, O '65:305-308 1
Avn Wk, Je 1, '64:21-22 (-A&-B) 1
Ag 17, '64:77-85 6 (Navigation)
O 19, '64:26-29 1
O 26, '64:30-33 1 C, S*
N 2, '64:40-41 1
D 28, '64:18-19 2
Mr 15, '65:67 P*
My 17, '65:35 (-B) 1
Ag 16, '65:40-41 1
S 27, '65:93-97 1
O 4, '65:73-76 3

GENERAL DYNAMICS (cont.)

1399. F-111 and TFX (cont.)

Avn Wk,	Ap 18, '66:62-63 (-A)	2	
	My 30, '66:66-75 (-A)	3	
	Je 6, '66:93-103 (-A)	6 (Avionics)	
	D 12, '66:76-83 (-A&-B)	6 (Flight Cntrl)	
	Ja 23, '67:84-89 (-A)	6 (Radar)	
	F 27, '67:69-72 (-B)	6 (Instruments)	C
	O 23, '67:29 (-B)	1	
	Ja 1, '68:17-18 (-A&-B)	2	
	Mr 18, '68:67 (-A)		P*
	Mr 25, '68:16-17 (-A)	1	
	My 27, '68:80-92	1	
	Jl 15, '68:80 (-D)	6 (Avionics)	
	F 3, '69:42-44+ (-A)	4	
Bee-Hive,	Ja '65:34		P
	Sum '65:38		P
Can Avn,	My '65:38-39	1	
Flight,	F 8, '62:207-208	1	S
	Je 18, '64:1024 (-A&-B)	1	
	O 29, '64:764-765 (-A)	1	C
	D 24, '64:1093-1095 (-A)	1	S*
	Ja 7, '65:13 (-A)	2	
	Ap 29, '65:663-664	1	
	O 28, '65:727-729	2	
	F 24, '66:301-307	1	S*
	Ag 25, '66:312-313	1	
	N 23, '67:877	3	
	Ja 18, '68:106-107	1	
	F 15, '68:244-245	1	
	Mr 7, '68:353-354 (-C)	1	
Flying,	My '69:S-1 to S-11+	1	
	:S-12 to S-16+	4	
Fly Rev Intl,	N '64:42-43	1	
	F '65:21-23+ (-A&-B)	1	Pr
	F '66:327-330+ (-A&-B)	1	C, S*, Pr
	My '69:34-37+	1	Pr*
Interavia,	No. 4, '63:419	1	
	No. 12, '64:1776-1779 (-A&-B)	1	
	No. 2, '65:264-265 (-A)	1	
	No. 7, '65:1126-1127 (-B)	1	
	No. 1, '66:52 (-A&-B)	2	
	No. 2, '66:166-169	1	C
Nav Avn News,	Ag '64:10-11	1	
	Jl '65:11 (-B)	1	
RAFs Qtr,	Spg '69:65-68 (-C:RAF)	1	
Skyline,	Vol. 27, No. 4, '69:42-47	6 (Afterburner)	

GENERAL DYNAMICS (cont.)

1399. F-111 and TFX (cont.)

Sp/Aeron, Je '63:72-83	1	
Ag '65:46-54 (-A&-B)	1	

GENERAL DYNAMICS/CONVAIR USA

1400. 48 "Charger"
Counterinsurgency (COIN) aircraft

Air Prog, Ap '65:37	1	
Avn Wk, Je 1, '64:74-81	1	
S 28, '64:22-23	3	
D 7, '64:23	2	
Ap 5, '65:33	1	S
S 6, '65:98	1	
S 20, '65:95	1	S*
Flight, Ja 14, '65:54-55	1	S
Fly Rev Intl, Je '65:31-35	1	S*
Interavia, No. 1, '65:103	1	S

1401. 660

Aeroplane, Ja 31, '68:27	1	
Avn Wk, Ja 22, '68:28-30	1	
Flight, D 7, '67:932	1	
Interavia, No. 1, '68:78-79	1	S*

GIL (Glowny Instytut Lotnictwa) Poland

1402. Helicopter (1950)

Am Helicop, F '50:19+	1	

GLATFELTER USA

1403. XRG-65

Air Prog, Ag '68:51+	1	C

GLOSTER GB

Flight, My 27, '55:713-732 (40th Anniversary)
Interavia, S '49:548-550
RAF Fly Rev, Ap '58:58 (Floatplanes: I to VI)

1404. "Gamecock"

Air Pict, Je '57:198-199 (-I:23 Sq)	5	Pr
:199 (-I)	1	
Flight, My 23, '58:703		D
Fly Rev Intl, My '68:265 (-I:23 Sq)		Pr*
D '68:51 (-II:Finland)		Pr*
RAF Fly Rev, Jl '58:42 (-I&-II)	1	

GLOSTER (cont.)

1405.	"Gauntlet"		
	Aeroplane, Mr 14, '34:439		P
	Flight, D 14, '33:1259-1260	1	
	Ap 26, '34:399		P
	Je 7, '34:558 (F. 7/30)	1	
	D 27, '45:676-677	4	
	Fly Rev Intl, My '68:265 (-II:3 Sq)		Pr*
	S '68:63 (Danish-)		Pr*
	Pop Fly, S '35:325		D
	O '35:396	1	P
	RAF Qtr, Ja '34:1		P
1406.	"Gladiator"		
	Aeronautics, Mr '61:57		P
	Air Prog, '39/'40 Annual:56		P
	Ja '65:38 (-I)		Pr†
	Bris Sid Jnl, Win '60/'61:56	1	P
	CAHS, Win '67:96-99	4	
	Flight, N 21, '35:527	1	
	Je 25, '36:668-669		S*
	D 27, '45:676-677	4	
	Fly Rev Intl, N '63:36 (-I:72 Sq)		Pr*
	:37+	1	
	Jl '65:68 (-I&-II)		Pr*
	My '68:265 (-II:25 Sq)		Pr*
	D '68:51 (-II:Finland)		Pr*
	Ap '69:52 (-II:Norway)		Pr*
	Pop Avn, D '37:31 (Belgian-)		P
	Pop Fly, Jl '38:176		P
	RAF Fly Rev, O '55:37-38+	4	
	Ja '58:16 (-I:Belgium, 1st Sq)		Pr*
	:17-20+	1	P
	Ag '61:26	4	
1407.	"Gnatsnapper"		
	Flight, My 9, '30:515		P
1408.	"Grebe"		
	Air Pict, S '65:310-313	1	Pr
	RAF Qtr, Ap '38:Supp (-II)	1	D
1409.	"Javelin"		
	Developed from G.A. 5		
	Aeroplane, D 2, '55:858-863	1	
	Ap 13, '56:250-251		P

GLOSTER (cont.)

1409. "Javelin" (cont.)

Aeroplane,	N 14, '58:720-726	1	S*
	N 27, '59:534-536	4	
Air Pict,	Ja '67:20-21	1	
Acft Eng,	Mr '55:68-72	2	
Avn Wk,	O 20, '52:21-30	1	
	My 23, '60:94-95 (-7)	1	
Flight,	Jl 13, '56:88-93 (-FAW. 1)	4	
	Je 22, '61:867-870 (-FAW. 5)		P
Fly Rev Intl,	My '68:272 (-FAW. 9:64 Sq)		Pr*
Nav Avn News,	My '57:24-25		P
RAF Fly Rev,	Ja '55:13-14	4	
	S '58:16-17 (-7&-8)	1	Pr*
	Ap '59:26-29 (-FAW. 8)	1	Pr, S*
	Ja '61:30-31 (-FAW. 9)	1	P*

1410. "Meteor"

Variant: ARMSTRONG WHITWORTH "Meteor"

Aeronautics,	My '46:34-39 (-IV)	1	P
	O '47:44-45 (-IV)	1	P*
Aeroplane,	N 16, '45:566-567 (-IV)		S*
	F 15, '46:200-204 (-IV)	1	C*, S*
	Jl 5, '46:14-16 (-IV)	4	
	Mr 26, '48:365-367 (Trent-)	4	
	My 21, '48:604 (-VII)	1	
	N 11, '49:644-648 (-VII)	4	
	My 11, '51:565-568 (-8)	4	C
	Je 8, '51:685-691 (-8)	1	S*
Air Pict,	O '50:301 (-8)		P
	Ja '51:6 (To NF-11)	1	
	Je '51:165, 175 (-8)	1	P
	Je '64:187 (-III)	4	
Air Power,	Spg '54:320 (-8:Prone Pilot)		P
Aviation,	Ja '46:69-72	1	
Avn Wk,	O 16, '50:27-28 (-PV)	1	
Can Avn,	N '45:90-92	4	
	Ap '48:30+ (Trent-)	4	
Comm Avn,	My '45:45	1	
	N '45:43-44	4	
Flight,	Jl 19, '45:70-73	1	
	Ag 2, '45:116	6 (Lndg Gear)	
	O 25, '45:444-446 (-IV)	1	S
	Ja 16, '46:74 (-IV:Short Span)	1	S*
	F 14, '46:157-159 (-IV)	1	
	Je 13, '46:599	2	
	Ap 10, '47:321 (-IV)	1	

GLOSTER (cont.)

1410. "Meteor" (cont.)

Flight,	F 12, '48:192 (-VII)		S*
	Mr 18, '48:299-301 (Trent-)	4	
	Ap 1, '48:349-351 (-VII)	1	C
	My 20, '48:541-543 (-VII)	4	
	Je 30, '49:763-764 (-VII)	4	
	O 6, '49:465-469 (-VIII)	1	S*
	Jl 13, '50:85 (-9&-10)	1	
	D 14, '50:556-557 (-VIII)		P
	Mr 30, '51:358-359 (-VIII)	1	C*, S
	Ag 31, '51:254 (-VIII)	1	P
	Jl 2, '54:15-17	1	
	Mr 30, '56:345-348 (-8: Prone Pilot)	4	C
Flying,	S '46:48		P
Fly Rev Intl,	Ag '64:56 (-8:Brazil)		Pr*
	My '68:272 (-VIII:500 Sq)		Pr*
	S '68:62 (-F8:Denmark)		Pr*
Interavia,	No. 2, '51:76-78 (-T.7)	4	
Nav Avn News,	Ag '51:20-21 (-8)	1	P
RAF Fly Rev,	S '55:37-38 (-8: Prone Pilot)	1	
	N '59:16-17+ (-In WW II) (Pt 1)	4	
	D '59:27-29+ (-In Korean War) (Pt 2)	4	

1411. A.S. 31 Survey Biplane

Flight,	Ja 24, '30:143	1	
	Ja 31, '30:161	4	

1412. C. 16/28

Flight,	Je 17, '32:531-532	1	
	Jl 29, '32:xviii		P

1413. CXP-1001

Projected jet fighter for China, 1946

RAF Fly Rev,	My '62:39+	1	

1414. E. 1/44 "Ace"

Flight,	Ap 22, '48:429-432	1	C*, S*
	Jl 1, '48:19	1	

GLOSTER (cont.)

1415. E. 28/39
First British jet to fly; predecessor of "Meteor"
Flight, O 27, '49:553-558 1
RAF Fly Rev, My '53:19-20 4

1416. F. 5/34
RAF Fly Rev, F '62:43+ 1

1417. F. 7/30
See GLOSTER "Gauntlet"

1418. F. 9/37
Aeroplane, My 19, '44:557 1
RAF Fly Rev, Ap '58:61 1

1419. G.A. 5
Developed into "Javelin" FAW. 1
Avn Wk, D 10, '51:15 1
Flight, D 7, '51:712-713 1
Nav Avn News, Mr '53:41 P
RAF Fly Rev, Ja '55:13-14 4
RAF Qtr, Ap '52:176 P

1420. S.S. 19
B version was prototype of "Gauntlet"
Flight, F 27, '31:176 1

GLUHAREFF USA

1421. Strap-on Helicopter
Am Helicop, Ja '58:9+ 1

GOODYEAR USA

1422. F2G-1
Developed from VOUGHT F4U
Air News, Mr '46:58-64+ 1 C*, S*
Aviation, Jl '46:66-67 6 (Lube) S*
Comm Avn, O '45:92 1
RAF Fly Rev, D '55:11 1

1423. FG-1
RAF Fly Rev, D '55:11 1
US Air Svcs, Mr '43:32 1

GOODYEAR (cont.)

1424.	GA-1		
	AAHSJ, Spg '60:91-108	1	P
1425.	GA-2 "Duck"		
	AAHSJ, Spg '60:91-108	1	P
	Avn Wk, Jl 7, '47:53	1	
	Can Avn, N '47:26-27	1	
	Flight, Mr 18, '48:307-308	1	
	Flying, S '47:54-55+	4	
	Skyways, N '46:32-35+	1	C
	O '47:43	1	
1426.	GA-22 "Drake"		
	AAHSJ, Spg '60:91-108	1	P
	Avn Wk, N 6, '50:43	1	
1427.	GA-400R "Gizmo"		
	Am Helicop, N '57:8+	1	
	Avn Wk, O 7, '57:155 (-3J)	1	

GOTHA Germany

1428.	"Ursinus" UWD		
	Cr & Cockade, Win '60:54-55		Pr†
	RAF Fly Rev, S '60:21	1	
1429.	G-I to G-X		
	Air Prog, F/Mr '63:18-23 (-I to -X)	1	
	:21 (-V)		C
	Airpwr Hist, O '61:194-206 (-IV)	4	
	Avn Wk, F 29, '60:36 (-IV)		D*
	Cr & Cockade, Win '63:354-359 (-VI)	1	
	Pop Fly, Mr '34:620-623 (-IV)(Pt 1)	4	
	Ap '34:30+ (-IV)(Pt 2)	4	
1430.	Go. 147		
	RAF Fly Rev, N '58:48	1	
1431.	Go. 149		
	Fly Rev Intl, S '68:111	1	
1432.	Go. 229		
	Air Pict, Ap '59:131-133	1	
	:132-133 (-A-O)	5	Pr

GOTHA (cont.)

1433. Go. 242
Flight, O 8, '42:390b 1
RAF Fly Rev, Mr '58:41-42 1

1434. Go. 244
Flight, S 16, '43:310b 1
RAF Fly Rev, Mr '58:41-42 1

1435. WD 44
RAF Fly Rev, My '58:42+ 1

GOTHA-KALKERT Germany

1436. Ka 430
Prototypes: GOTHA Go. 242 and Go. 244
RAF Fly Rev, Mr '58:41-42 1

GOURDOU-LESEURRE France

1437. GL-832
RAF Fly Rev, Ap '62:41 1

GRAHAME-WHITE GB

1438. "Charabanc"
Aeroplane, Ja 2, '42:26 1

GREAT LAKES USA
Air Prog, N '66:10-18+ (Many photos)

GRUMMAN USA
AAHSJ, Win '69:267-271 (Attack aircraft since 1935)
Nav Avn News, My '48:1-5 (US Naval fighters since FF-1)
Jl '56:14-17 (US Naval aircraft: 1929-1956)

1439. "Ag-Cat" G. 164
Avn Wk, N 25, '57:32-33 1
D 2, '57:95 1 C
My 12, '58:113-117 1
F 21, '66:73 1
Flying, Mr '63:24-25+ 4
Interavia, No. 10, '68:1250 1

GRUMMAN (cont.)

1440. "Albatross" G. 64

Aero Digest, N '53:28-40	1	S*
Avn Wk, Jl 17, '50:22-29+	1	
Flying, N '56:41+	4	
Fly Rev Intl, O '68:26 (Turbo-)	1	
Ap '69:53 (Norwegian-)		Pr*
RAF Fly Rev, N '62:24-27 (SA-16B)	1	Pr, S*
Sentinel, S '67:24-27 (RCAF-)	4	

1441. "Goose" G. 21
USAAF designation: OA-9 and OA-13
US Navy designation: JRF
Later development: MCKINNON "Goose"

Aero Digest, Jl '37:64-66	1	
Aviation, Jl '37:36-37	1	
Flight, My 13, '43:500a	1	
West Fly, S '37:44	1	

1442. "Gulfhawk" G. 22

Aviation, F '37:44	1	
Flight, Jl 7, '38:22-23	1	
Hist Avn, Ap '69:16-17 (-II)	1	
Pop Avn, Ap '37:21-22	1	
Ag '38:45	1	C

1443. "Mallard" G. 73

Aeroplane, N 28, '47:706-707	1	
Aviation, Ja '47:82-83 (Pt 1)	1	
F '47:62-63, 75 (Pt 2)	1	S*
My '47:83, 85		S*
Flight, Ja 2, '47:17	1	C
Interavia, No. 8/9, '46:34-35	1	C
Skyways, N '48:20-22+	4	S
US Air Svcs, N '46:18	1	

1444. "Gosling" and "Widgeon" G. 44
US Navy designation: J4F "Widgeon"

Aeroplane, D 8, '44:642-643 (-I)		P
Avn Age, Ap '53:132-133	1	
Flight, O 28, '43:478	1	
RAF Fly Rev, F '61:9	1	
West Avn, N '52:18	2	

GRUMMAN (cont.)

1445.	A2F-1 "Intruder"		
	Later designation: A-6		
	Aeroplane, Jl 8, '60:36-37	1	
	N 11, '60:666		P
	Air Pict, F '61:56-57		P
	Avn Wk, My 9, '60:94	1	
	O 24, '60:66-67	1	
	N 5, '62:87-93	6 (Electronics)	
	Bee-Hive, Sum '60:33	1	P
	Flying, Jl '60:50	1	
	Interavia, No. 10, '60:1234-1235	1	
	Nav Avn News, Je '60:28	1	
	RAF Fly Rev, Ap '62:34-35	1	P*
1446.	A-6 "Intruder"		
	Earlier designation: A2F		
	Air Prog, My '68:22-25+	4	
	AAHSJ, Sum '69:112-114 (-A)	1	
	Avn Wk, Ag 17, '64:46-52 (-A)	6 (Avionics)	
	D 27, '65:18-20 (-A)	4	
	Ap 24, '67:67-74 (-A)	4	
	N 4, '68:101-109	1	
	Bee-Hive, Spg '65:2-7 (-A)	4	
	Flight, Je 20, '68:936 (E-/-B)	1	
	Fly Rev Intl, Jl '68:367-369+ (-A)	1	P*, Pr, S*
	Nav Avn News, Je '67:6-11 (-A)	4	
	Ap '68:7		D
	Je '68:6-9 (-A)	4	
1447.	AF-2 "Guardian"		
	Aero Digest, My '51:26-28+	1	
	Nav Avn News, Ap '51:12-13 (-S&-W)	1	
	O '51:20-21 (-S&-W)	1	P
1448.	AO-1 "Mohawk"		
	Later designation: OV-1 "Mohawk"		
	Aeroplane, Ap 15, '60:467-468	1	C
	Avn Wk, O 28, '57:26	1	
	Jl 27, '59:59-65 (Y-)	1	
	Flight, Jl 6, '61:10-12 (-AF)	4	C
	Interavia, No. 11, '60:1402-1405	1	C
	No. 7, '61:990-992	1	P
	RAF Fly Rev, F '60:33	1	
	Sp/Aeron, F '60:65-66+ (Y-)	1	
	D '61:195		D*

GRUMMAN (cont.)

1448.	AO-1 (cont.)			
	US Army Avn,	S '60:1-5 (-AF)	1	C
		O '61:13-17 (-AF)	6 (Ejection Seat)	
		F '62:12-14 (-AF to -CF)	6 (Rescue)	C, S*
		Ap '62:24-27	6 (Ejection Seat)	
		Jl '62:41-43 (-A)	2	
		Je '63:20-23	4	
1449.	C-2A "Greyhound"			
	Aeroplane, Je 21, '67:24		1	
	Air Prog, Ap '65:38-40		1	C*
	Avn Wk, D 7, '64:20-22		1	
1450.	E-1B "Tracer"			
	Earlier designation: WF-2			
	Nav Avn News, Ag '65:6-8		1	
1451.	E-2A "Hawkeye"			
	Earlier designation: W2F-1			
	Air Prog, Mr '69:83			P
	Avn Wk, Mr 4, '63:68-77		4	
	Ag 1, '66:56-63		4	
	Nav Avn News, N '63:6-9		1	
	Jl '66:14-15		4	
1452.	F2F-1			
	Aero Digest, F '36:56		1	
	AAHSJ, Sum '69:97-98		1	
1453.	F3F			
	Air Pict, Mr '58:90-91 (-1:7th Ftr Sq)		5	Pr
	Air Prog, S '68:35 (-2)			P*
	AAHSJ, Spg '67:49 (-1)			C
	Aviation, D '40:79			S*
	Pop Avn, Ja '37:22 (-1)		1	Pr
	RAF Fly Rev, F '60:42		1	
	O '61:50		1	
	:51 (-3:5th Ftr Sq)			Pr*
1454.	F4F "Wildcat" or "Martlet" (RAF)			
	Air Pict, D '61:380-381		1	Pr
	Air Prog, Jl '42:19 (-3)			P*
	AAHSJ, Win '61:223-245		1	Pr
	Spg '62:46-47+		1	
	Aviation, Ap '44:223			D*

GRUMMAN (cont.)

1454.	F4F (cont.)		
	Avn Age, Ap '50:32-33	1	
	Flight, Je 25, '42:Supp:b (-3)	1	
	Flying, Je '65:61	4	D
1455.	XF5F-1 "Skyrocket"		
	USAF designation: XP-50		
	Air Tech, N '42:27		P
	Can Avn, My '40:50	1	
	RAF Fly Rev, Je '56:9	1	
1456.	F6F "Hellcat"		
	Aero Digest, N 15, '45:40+	4	
	AAHSJ, Sum '69:122-124 (-3&-5)		Pr†
	Fall '69:185 (-3:Royal Navy)		Pr†
	Win '69:248-249 (-5)		Pr†
	Aviation, Ap '46:105, 107		S*
	My '46:107		S*
	Je '46:97		S*
	Ag '46:97		S*
	Flight, N 18, '43:556	1	
	Flying, D '43:53	1	
	Fly Rev Intl, Ap '66:496 (-3, -5, Mks I & II)		Pr*
	O '68:53 (-5:Uruguay)		Pr*
	Nav Avn News, F 1, '45:9		P
	S '62:27 (Last US Navy-)	1	
	RAF Fly Rev, D '61:36-37 (-5)		S*
	:36-38	1	
	:38 (-3)		Pr
	Skyways, D '43:18-21+	1	
1457.	F7F "Tigercat"		
	Aviation, Jl '45:170	1	
	Comm Avn, Jl '45:76+	1	
	Nav Avn News, Ap '48:33 (-3N)		P
	RAF Fly Rev, Jl '58:6+	1	
	Skyways, Ag '45:65	1	
	US Air Svcs, Jl '45:32-33	1	P
1458.	F8F "Bearcat"		
	Air Prog, Ja '69:34-36+ (-2: Greenamyer's)	4	
	Flying, F '64:46-47+ (-2)	4	
	Fly Rev Intl, Ja '66:310-313	1	Pr*
	Nav Avn News, Je '49:9	2	
	:29	6 (Wing Tip Jettison)	

GRUMMAN (cont.)

1458. F8F (cont.)

Nav Avn News, D '69:20-23 (-2: Greenamyer's) 4
RAF Fly Rev, D '58:8 1
:9 (-1D:Royal Thai AF) Pr*
US Air Svcs, N '48:12-14 ("Gulfhawk 4") 1 P

1459. F9F "Panther"

See also F9J
Aero Digest, Jl '48:28-30 (-2) 1 C
D '50:24-27+ 1
Avn Opns, Ap '50:32-33 4
Avn Wk, D 15, '47:12-13 (X-/-2) 1
Nav Avn News, Ag '48:33 P
N '51:21 (-5) P
S '52:35 (-5P) 1
Skyways, My '48:28-30+ (-2) 1
US Air Svcs, D '47:29 (X-/-2) 1

1460. F9F-6 "Cougar"

Nav Avn News, F '60:38 (-K:"Ol '23") 1

1461. F9F-8 "Cougar"

Aero Digest, S '54:33-40 1 S*
Air Pict, S '56:310-311 1 Pr
Air Prog, '56/'57 Annual:86-88 Pr, S*
Avn Wk, Jl 9, '56:37 (-P) 1
Ja 28, '57:66-75 (-T) 1
Nav Avn News, Ap '54:27 1
RAF Fly Rev, Ja '56:52-53 1 P*
F '57:26-29 1 Pr†, S*
West Avn, My '54:28 1
Je '56:14 (-P) 1

1462. F9F-9 "Tiger"

Avn Wk, Ag 16, '54:381 1
Ag 23, '54:26-30 1
Ag 30, '54:9 1
My 16, '55:15 1 P
US Air Svcs, S '54:17 1
West Avn, S '54:13 1

1463. F9J "Cougar"

Swept-wing F9F
Nav Avn News, My '67:16 (T-) 1

GRUMMAN (cont.)

1464. XF10F-1 "Jaguar"

Avn Wk, My 20, '63:105 1
Interavia, No. 5, '62:617-619 1
RAF Fly Rev, Je '55:10 1
F '56:15 1
Ag '60:11 1

1465. F11F "Tiger" and "Super Tiger"

Aero Album, Spg '68:12-14 (-1) 2 P
Air Pict, Ap '57:126-127 (-1F) 1 Pr
Avn Wk, Ja 2, '56:30-31 1
O 29, '56:28-29 1 P
N 5, '56:30-31 2
D 24, '56:26-27 (-1F) 1
Interavia, No. 5, '58:439-441 (-1F)1
No. 3, '59:282-283 (-1F)3
RAF Fly Rev, Ag '56:40-41 (-1) 1 P*

1466. F-14 VFX "Tomcat"

Aerosp Intl, S/O '68:22-24 1
Air Prog, Jl '69:58-59 (-A) 1
Avn Wk, S 2, '68:16-17 1
D 16, '68:15 1 D
Ja 20, '69:24 1 P
F 24, '69:18-21 1 P
Mr 17, '69:46-47+ 3
Je 30, '69:17 1
Jl 28, '69:89-95+ 6 (Armament)
Bee-Hive, Fall '69:33 P
Flight, Mr 6, '69:370-371 1
Fly Rev Intl, My '69:25-26 1
Interavia, No. 4, '69:384-385 1
Nav Avn News, Ap '69:20-21 (-A) 1
Sp/Aeron, F '67:74-85 1
N '68:50-58 1

1467. FF-1

Canadian licensee: CANADIAN CAR & FOUNDRY "Goblin I"

Air Prog, O '66:14-16 1 C, Pr
Fly Rev Intl, O '66:158 P*
F '67:398 1 P*
Nav Avn News, Je '54:16 1
Ag '67:19 1
RAF Fly Rev, F '60:11+ 1

GRUMMAN (cont.)

1468. G. 32
Air Prog, Ap '65:33 (-A) 1 C
S '68:32-34+ 1 C

1469. J2F "Duck"
Flying, Ja '42:112 P
RAF Fly Rev, Ag '61:50 1

1470. J4F "Widgeon"
Civil version: "Widgeon" and "Gosling" G. 44
Aero Digest, S 15, '44:98-102 (-2) 1 S*
Air Tech, My '45:13-19+ 1 C, S*
Pegasus, Ap '43:10 C
S '45:13-15+ 1
Nav Avn News, Jl '48:25 ("Petulant Porpoise") 2
RAF Fly Rev, F '61:9 1

1471. JF-1 and JF-2 "Ugly Duck"
Aviation, D '34:391-393 1 S
Flight, Mr 28, '35:340 1 S
Pop Avn, Ja '40:41 (-2) P

1472. OA-14 "Widgeon"
See J4F and "Widgeon" and "Gosling" G. 44

1473. OV-1 "Mohawk"
Earlier designation: AO-1
Air Pict, Mr '69:41-45 (-B) 1 P*, S*
Avn Wk, F 28, '66:70-81 4
My 13, '68:90 (-D) 6 (Radar)
Fly Rev Intl, F '67:341-346 (-A to -C) 1 Pr, S*
Interavia, No. 3, '66:338-339 (-A to -C) 6 (Surveillance)
RAF Fly Rev, Je '63:34-35 (-A to -C) 1 P*
US Army Avn, Ja '64:8-11 4
Je '65:23-24 2
O '66:50 (-C) 1
Ap '67:15-16 (-B&-C) 1
Ja '68:62 (-B) 6(Voice Warning) S*
O '68:18-23 4

1474. XP-50 "Skyrocket"
US Navy designation: XF5F-1
RAF Fly Rev, Je '56:9 1
Je '60:53 1

GRUMMAN (cont.)

1475. S2 "Tracker"
Canadian version: CS2F1
Trainer version: TF-1 "Trader"

Aeroplane, Ap 5, '57:471-474 (C-/-F1)	1	P
Avn Wk, Mr 29, '54:9 (X-/-F1)		P
Mr 4, '57:75 (C-/-F1)		D*
Jl 1, '57:46-49	1	
Can Avn, F '55:24-25 (C-/-F1)	1	
:26-31 (C-/-F1)	3	
Ag '56:30-32 (C-/-F1)	1	
S '57:42-43+ (C-/-F1)	4	
S '59:18-19 (C-/-F2)	1	
Interavia, No. 3, '63:323 (-F3)	1	
Nav Avn News, My '54:34-35 (-F)	1	
S '55:1-5 (-F)	4	
Jl '58:26-27 (-F)	4	
Ja '67:8-12 (-E&-F)	4	
Pegasus, Je '55:1-7 (-F)	4	
RAF Fly Rev, Ag '57:12-13 (-F1)	1	
Mr '62:22-24 (-F1 to -F3)	1	
US Air Svcs, D '55:14 (-F2)	1	P

1476. SA-16 "Albatross"
See "Albatross" G. 64

1477. TBF "Avenger"

Air Cl, F '69:37-52	1	P*, Pr
:53	4	
Air Tech, N '44:19-25	1	C*, S*
Nav Avn News, Jl 15, '44:40-47	4	
Je '54:12-13	4	
Skyways, Jl '44:18-19+	4	

1478. TBM-1 "Avenger"

Aviation, Jl '43:168-173+	3	
Fly Rev Intl, Je '66:642, 644 (-C)		Pr*
Nav Avn News, O 1, '44:28		P
Je '54:12-13	4	

479. TBM-3 "Avenger"

Aero Digest, Mr 15, '45:82-89	3	S*
Ap 1, '45:88-93+	3	S*
Air Cl, F '69:51		S*
Fly Rev Intl, Je '66:642, 644		Pr*
Nav Avn News, N '45:20	6 (Rescue Gear)	
Je '54:12-13	4	

GRUMMAN (cont.)

1480. TC-4C "Gulfstream"
Avn Wk, My 1, '67:84-91 1

1481. TF-1 "Trader"
Trainer version of S2
Avn Age, Ap '56:44-47 1
Avn Wk, Jl 1, '57:46-49 1

1482. U-16 "Albatross"
See "Albatross" G. 64

1483. W2F-1 "Hawkeye"
Later designation: E-2A
Air Pict, F '61:58 P
Avn Wk, N 7, '60:117-119 1
D 5, '60:85 1
Nav Avn News, My '62:11 1

1484. WF-2 "Tracer"
Later designation: E-1B
Avn Age, D '57:62-65 1
Avn Wk, Mr 4, '57:27 1
Jl 1, '57:46-49 1
Ap 14, '58:76 1
Ja 11, '60:85-89 1

GWINN USA

1485. "Aircar"
Aviation, O '37:36-37 1 C
Pop Avn, N '37:16 1 C
Pop Fly, D '37:515 P
US Air Svcs, D '37:16-17+ 1
West Fly, N '37:31 1

GYRODYNE USA

1486. G.C.A. 2
Am Helicop, S '50:6-8+ 1
Avn Age, My '52:38-39 1 S
Avn Wk, Ap 28, '52:25-32 (-C) 1 S

1487. G.C.A. 7 "Helidyne"
Am Helicop, D '49:8+ 1
Mr '50:6-7+ (-A&-B) 1
Can Avn, D '50:60 1

GYRODYNE (cont.)

1488.	XRON-1 "Rotorcycle"		
	Am Helicop, My '56:8	1	
	Avn Age, N '56:32-33	1	C*
	Avn Wk, Ag 20, '56:64-65	1	
	West Avn, D '58:18	1	

GYROPLANE USA
See GWINN

HAFFNER GB
Fly Rev Intl, D '63:44-46

1489.	"Gyroplane" A.R. III		
	Flight, F 8, '37:160-162 (-II)	1	C, S*
1490.	"Gyroplane" A.R. IV		
	Aeroplane, S 7, '38:290	1	

H.A.L. (Hindustan Aeronautics, Ltd.) India
See HINDUSTAN

HALBERSTADT Germany

1491.	CL. II		
	Cr & Cockade, Sum '61:130-149	1	Pr, S
1492.	CL. IV		
	Pop Fly, My '34:83	1	D
1493.	D. III		
	Avn Wk, D 9, '63:62-63	1	D*
	Cr & Cockade, Sum '69:164-165 (Scheiner's-)		C, P

HALL USA
Fly Rev Intl, Ja '69:58-62

1494.	PH-2		
	Pop Avn, Je '38:11		P
	N '39:35		P
	RAF Fly Rev, Ag '62:41	1	
1495.	XP2H-1		
	Aviation, O '34:317-319	1	

HAMBURG Germany
See BLOHM & VOSS

HAMILTON METALPLANE USA
Boeing Mag, Ap '68:7

HAMMOND USA

1496. Y
Later design: STEARMAN-HAMMOND Y

Aeroplane,	N 17, '37:583	4	
Aviation,	My '36:31-33	1	
Pop Avn,	S '36:44+ (-1)	1	C
	O '36:14-16+	4	
	F '37:42	3	
	O '37:32+	4	
West Fly,	N '34:18-19	1	
	My '36:32	1	

HANDLEY PAGE GB
Air Pict, Je '59:201-207 (50th Anniversary)
Acft Eng, My '68:25-31
Flight, Je 22, '39:630h-637
Je 23, '49:733-739 (Aircraft photos and descriptions)
O 30, '59:461-462
Je 19, '69:1009-1026 (50th Anniversary - Many aircraft photos)
Interavia, No. 1-2, '50:68-70
RAF Fly Rev, Je '59:Supp:1-12 (List of models with profiles)

1497. "Gugnunc"

Flight,	Mr 7, '30:267-270	1	P, Pr

1498. "Halifax"
Civil version: "Halton"

Aero Digest,	Je '42:128 (-II)		S*
Aeronautics,	Ag '44:77 (-III)		D*
	D '44:43-46 (-III)	1	S
Aeroplane,	S 19, '41:306-307	1	
	N 7, '41:506-507	1	C
	Ap 24, '42:466-471 (-II)	1	S*
	My 29, '42:613	1	
	Je 18, '43:692 (-II)	1	

HANDLEY PAGE (cont.)

1498.	"Halifax" (cont.)		
	Air Pict, D '54:367-369 (Pt 1)	1	
	Ja '55:11-12 (Pt 2)	1	
	:12-13	5	
	F '55:48-50 (Pt 3)	1	
	Air Tech, Ap '43:30-33	3	
	Aviation, Ag '42:211+	1	S*
	S '43:232 (-II)	1	
	Can Avn, N '42:44-46 (-II)	1	S*
	Flight, S 18, '41:173-175	1	
	N 6, '41:322-324	4	
	Ap 23, '42:398-405	1	C, S*
	Ag 6, '42:146-147, 148a	1	S
	Je 17, '43:640 (-Ia)	1	
	Jl 22, '43:92-93 (-II, 1. A)	1	
	Ag 26, '43:233 (-II, 1. A)	1	
	F 17, '44:168-170 (-III)	1	
	N 15, '45:530-531 (-A-IX)	1	C
	Flying, My '43:36-37+	3	
	Roundel, D '54:16-19	1	
	S '63:33 (Canadian-)	1	P
	RAF Qtr, Je '46:145		P
	West Fly, D '42:45-46	1	C
1499.	"Halton"		
	Bomber version: "Halifax"		
	Air Pict, F '55:50	1	
	Flight, My 31, '45:577-580	1	
	O 2, '47:385-388	1	S
1500.	"Hampden"		
	Aero Digest, O '42:195-196	1	C, S
	Aeroplane, My 3, '39:569-574	1	S*
	Air BP, No. 23:20-23	4	C
	Air News, Mr '43:32		D
	Air Pict, N '58:378-381	1	
	Can Avn, Jl '39:28+	1	
	S '40:18-20+	3	
	N '40:64		C
	Comm Avn, Jl '39:30	1	
	Flight, Je 30, '38:647	1	
	D 29, '38:591-593	2	
	My 4, '39:444-445	1	S*
	:452a-454	3	C, S
	My 16, '40:450-452b+	4	
	Ja 1, '42:7		C

HANDLEY PAGE (cont.)

1500.	"Hampden" (cont.)			
	Flight, Je 11, '42:588b	1		
	D 13, '45:634-635	4		C
	Fly Rev Intl, My '68:269 (-I:44 Sq)			Pr*
	Pop Avn, Jl '39:19	1		
	Pop Fly, Jl '39:164-165	1		
	Roundel, Mr '63:33 (Canadian-)	1		P
	RAF Fly Rev, My '58:44	1		
1501.	"Harrow"			
	Aeroplane, Ag 4, '37:138-142	3		S*
	Flight, N 12, '36:497	1		
	My 20, '37:494-497	3		S
	Ag 26, '37:208			C
1502.	"Hastings"			
	Aeronautics, S '51:82-83			P
	Can Avn, O '47:73	1		
	Flight, Je 6, '46:566-567	1		
	Je 27, '46:653-655	6	(Hydraulic)	S
	S 25, '47:359-363 (C.I)	1		S
	F 8, '52:156-157 (C.III)	1		C
	RAF Fly Rev, Jl '61:35 (202 Sq)	4		
1503.	"Herald"			
	Aero Digest, N '55:46-50	1		S*
	Ag '56:44-57	1		S*
	Aeroplane, S 3, '54:306-307	1		S*
	My 6, '55:596-599	1		S*
	S 9, '55:428-429 (-R3)			S*
	Jl 27, '56:119-124 (-R3)	1		P
	Mr 21, '58:388-392 (Dart-)	1		S*
	Mr 27, '59:380-382 (Dart-)	4		C
	N 21, '63:20-23 (Dart-)	1		
	Jl 22, '65:15-18	1		
	S 18, '68:28 (-200)	1		
	Air Pict, Je '58:199-200	1		
	S '61:275-282	4		C, S*
	Avn Age, Mr '56:24-25	1		S*
	Avn Wk, N 1, '54:55	1		S*
	Can Avn, Ja '62:12-21+ (Dart-)	1		S*
	Flight, S 10, '54:413	1		S*
	Ag 19, '55:260-265 (Proto-)	1		S*
	S 23, '55:518-519 (Proto-)	1		C
	Jl 6, '56:59-61	1		S*
	Jl 27, '56:149-152, 154-155	1		C
	:153, 157-158	2		
	:156	6	(Engine & Prop	

HANDLEY PAGE (cont.)

1503.	"Herald" (cont.)			
	Flight, My 17, '57:652 (Dart-)	1		
	N 1, '57:709-711	1		S*
	My 2, '58:608-614 (Dart-)	1		C, S*
	Jl 25, '58:132-134 (Dart-)	1		S*
	Mr 27, '59:430-431 (Dart-)	4		C
	My 18, '61:662-665 (-200)	2		
	D 28, '61:976-977 (Military-)	1		
	Fly Rev Intl, Je '65:37-39	1		S*
	Interavia, No. 10, '55:779-780	1		S
1504.	"Hereford"			
	Flight, N 2, '39:343 (-I)	1		D
	D 13, '45:634-635	4		
	Pop Fly, Ag '34:248-249	4		D
	RAF Fly Rev, My '58:44	1		
1505.	"Hermes"			
	Aeronautics, Ja '46:53, 55	1		
	O '48:30-37	1		C*
	Aeroplane, Ap 5, '46:401-402			D*
	My 24, '46:606-613			C, S*
	O 1, '48:447-452 (-IV)	1		C, S*
	Ja 13, '50:40-41 (-IV to -VI)	1		
	O 26, '51:556			S*
	D 31, '64:18-19	1		
	Can Avn, Ja '45:73-74	1		
	Comm Avn, D '44:164-165	1		
	Jl '45:82-84	1		
	Flight, O 19, '44:420a-421	1		S
	My 31, '45:577-580 (Cargo-)	1		
	Ja 10, '46:38a-40	1		S*
	Je 27, '46:653-655	6	(Hydraulic)	S
	Ag 19, '48:214-215 (-IV)	6	(Interior)	
	S 30, '48:403-408+ (-IV)	1		C*, S*
	S 1, '49:260-261 (-V)	1		
	Mr 2, '50:285-289+ (-IV)	4		C
	N 30, '50:504 (-V)	6	(Flaps)	
	Je 15, '51:721 (-V)	4		
	N 9, '51:588 (-VB)	1		
	Interavia, No. 4, '46:20-24 (-I to -III)	1		
	US Air Svcs, Ja '45:32	1		
	West Fly, D '44:60	1		

HANDLEY PAGE (cont.)

1506.	"Heyford"		
	Aeronautics, Je '55:51		P
	Aeroplane, Mr 14, '34:425-427		P
	Flight, Jl 6, '33:666-670	1	Pr, S
	F 15, '34:160	1	
	Je 25, '36:672-673	1	S*
	Fly Rev Intl, My '68:269 (-I:99 Sq)		Pr*
	Pop Avn, Ap '34:214-215	1	
	Pop Fly, My '33:73	1	P
	RAF Qtr, Ap '35:Supp	1	
1507.	"Jetstream"		
	Aeroplane, My 10, '67:20-31	1	C
	S 18, '68:34	1	
	Acft Eng, S '66:38+	1	
	S '67:10-28	1	S
	:33-37+	6 (Engine)	S
	:38-41	1	S*
	Avn Wk, Jl 17, '67:113	1	
	Je 2, '69:348-349	3	
	Can Avn, Ag '67:14-16	1	
	Flight, S 22, '66:512-513+	1	
	My 11, '67:737-746	1	C*, S*
	Jl 13, '67:52-53	1	C
	Ja 4, '68:34 (-3M)	1	
	Ag 29, '68:332-344	1	S*
	Je 12, '69:Inner cover		P*
	Je 19, '69:1017 (Proto-)		P
	O 23, '69:632-637 (-I)	4	C
	D 25, '69:992		C
	Flying, Jl '67:68-70	1	
	F '69:50-53	1	
	Fly Rev Intl, Ap '66:477	1	
	S '67:829-832	1	S
	My '69:30-31 (-3M)	1	
	Interavia, No. 11, '66:1159-1160	1	
	No. 3, '68:315-318	1	
1508.	"Marathon"		
	Earlier version: MILES "Marathon"		
	Aeronautics, Ap '53:34-37 (-I)	1	P
	Acft Eng, Ja '51:15-18 (-I) (Pt 1)	1	S*
	F '51:51-53 (-I) (Pt 2)	1	
	Flight, My 19, '49:591-595 (-I)	1	C, S
	Jl 21, '49:62-63 (Prototype II-)	1	
	Ag 4, '49:143-144 (-II)	1	
	My 9, '52:548-549	1	

HANDLEY PAGE (cont.)

1509.	"Victor"			
	Aeronautics, S '53:88-89			P
	Aeroplane, Ja 2, '53:6-7	1		
	S 5, '58:378-380	3		S*
	D 12, '58:848-850 (-B.1)	1		S*
	Mr 27, '59:372-378	1		S*
	Ap 22, '60:475 (-B.2)			P
	Air Prog, Spg '62:12-13			S*
	Avn Age, F '53:21-22	1		
	F '54:24-26	1		
	Avn Wk, N 9, '53:28-48	1		
	S 29, '58:73-77	4		
	Mr 16, '59:53-58	3		S*
	Can Avn, F '53:26-27	6	(Wing)	
	Flight, My 24, '57:719 (Civil-)	1		
	S 5, '58:388	1		S*
	S 19, '58:493-496	4		
	Mr 20, '59:372-373	1		
	Mr 27, '59:432-433 (-B.2)			P
	S 4, '59:126 (-B.1)			P
	Ap 8, '60:471-472	4		
	F 15, '62:247-250	4		
	Fly Rev Intl, My '68:268 (-II)			Pr*
	Interavia, No. 2, '53:102-103	1		
	RAF Fly Rev, N '53:11-12	2		S*
	Je '59:20-23+ (-B.1)	1		Pr, S*
	Jl '62:22-23	1		
1510.	V/1500			
	Aeroplane, Ap 24, '42:479	1		
	Air BP, No. 24:2-5 ("Atlantic")	1		
	Air Pict, Ap '62:108-114	1		C
	Can Avn, N '57:64-65 ("Atlantic")	4		
1511.	W. 8b			
	Air Pict, My '67:170-172 (Pt 1)	4		
	Je '67:210-212 (Pt 2)	4		
1512.	W. 10			
	Air Power, Sum '54:382 (Refueling-)			P
1513.	O/100			
	Avn Wk, Ap 8, '63:36-37	1		
	Flight, F 27, '53:254-259	1		

HANDLEY PAGE (cont.)

1514.	O/400		
	Avn Wk, Ap 8, '63:36-37	1	D*
	Cr & Cockade, Spg '63:180-188	4	
	Flight, F 27, '53:254-259	1	
	Ja 13, '61:53-54	1	
	Fly Rev Intl, My '68:268 (207 Sq)		Pr*
	Pop Fly, S '33:322	1	D
1515.	H.P. 21		
	Three built; one for US Navy as SD 24		
	Fly Rev Intl, O '68:76	1	
1516.	H.P. 42 "Hannibal"-Class Airliner		
	Aero Digest, Ja '31:70-71	1	
	Aeroplane, N 19, '30:1137-1140 ("Hannibal")	1	P
	Flight, N 21, '30:1370-1371	1	
	N 28, '30:1381-1385	1	S
	Jl 17, '31:700-702 ("Hannibal")	4	C
1517.	H.P. 46		
	Air Pict, Ag '67:285	1	D
	D '67:451	1	
1518.	H.P. 47		
	Air Pict, Jl '62:219	1	
	Flight, Ap 18, '35:408, 411, 417-418	1	P
1519.	H.P. 100		
	Fly Rev Intl, Ja '68:52	1	
1520.	H.P. 111		
	Cargo aircraft based on "Victor" bomber		
	Aeroplane, Ag 22, '58:252-253+	1	S*
	Sp/Aeron, D '58:32-37	1	S*
1521.	H.P. 113		
	Interavia, No. 8, '61:1105	1	S*
1522.	H.P. 115		
	Avn Wk, Ag 28, '61:34	1	
	S 18, '61:78-79	1	
	Flight, Ag 24, '61:248		P

HANNOVER Germany

1523. CL. II
Cr & Cockade, Sum '62:124-137 1 C, P, S
Sum '69:182-183 1 P
Fall '69:274 (Nock's-) 1 P

1524. CL. IIIa
Pop Fly, Ja '34:543 1 D

HANRIOT France

1525. D. I
Aeronautics, F '46:78 C*

1526. H. 110
Flight, F 22, '34:179 1
Pop Avn, S '33:161+ 1

1527. H. 230, H. 231, and H. 232
RAF Fly Rev, Ag '61:46 1

1528. H. 510, H. 511, and H. 530
RAF Fly Rev, O '57:47 1

1529. NC-600
RAF Fly Rev, O '60:46 1

HANRIOT-DUPONT France

1530. HD-1
Air Cl, D '69:56-63 1 C, P, Pr†
Air Pict, Ja '57:21 1
Air Prog, Ja '66:45 Pr†
O '68:40-45+ (Restoration-) 1 C, P*
N '68:57+ (Restoration-) 4
AAHSJ, Win '61:Supp P*
Avn Wk, Mr 5, '62:54-55 1 D*
Cr & Cockade, Aut '64:Back cover
(-Belgium: 9th Escadrille) D
Flying, F '60:79 (Coppens'-) 1 D
Fly Rev Intl, D '66:278 P*
Mr '69:42 (Swiss-) Pr*
Pop Avn, Ap '35:233 (Nungesser's-) 1
RAF Fly Rev, D '60:16-17 (-Belgium:
1st Escadrille) 5 Pr*
:16-18 1

HANRIOT-DUPONT (cont.)

1531. HD-2
Float version of HD-1
Air Prog, Ja '66:45 Pr†

HANSA-BRANDENBURG Germany

1532. W. 19
Cr & Cockade, Sum '60:8-9 Pr†

HARLOW USA

1533. PC-5

Aviation, S '39:34-35	1	C
S '42:231 (-A)	1	C
Comm Avn, Ag '39:30	1	C
N '39:52	1	C
US Air Svcs, Ag '39:29	1	

1534. PJC-1

Aviation, Mr '39:58-59	1	
Pop Avn, Ap '38:24	1	

HAWKER GB
Flight, N 30, '51:677-696 (Many aircraft photos)
Interavia, Je '49:334-337 (Includes Sopwith aircraft)
RAF Fly Rev, N '60:25-27+

1535. "Audax"
"Hart" variant

Aeroplane, Je 8, '32:1035		P
Je 22, '32:1143		P
N 9, '32:877		P
Flight, Ap 29, '32:x		P
S 23, '32:viii		P

1536. "Dankok" (Danecock)
Development of "Woodcock" by Royal Danish Naval Dockyard
Fly Rev Intl, S '68:62 (Danish-) Pr*

1537. "Dantorp"
Export version of "Horsley"
Air Prog, F/Mr '63:80-81 1 Pr†

HAWKER (cont.)

1538.	"Demon"		
	Aeroplane, Ap 12, '33:630		P
	My 12, '37:567		P
	Can Avn, Je '33:12		P
	Fly Rev Intl, My '68:265 (-604 Sq)		Pr*
	Interavia, Je '49:335		P
	Pop Fly, S '33:321		D
	Jl '38:177		P
1539.	"Fury"		
	Aeroplane, N 16, '32:923		P
	Ag 9, '33:248 (High speed-)		P
	Mr 21, '34:485 (Super-)		P
	F 10, '37:156-159		P
	Ap 21, '37:463		P
	F 1, '46:137-144	1	P
	O 24, '47:558 (-Trainer)	1	S
	Air Cl, O '69:38-55+	1	D*, P, P*, Pr
	:43 (Republican Spain & Persian-)		Pr*
	Air Pict, N '57:390-391+ (-II)	1	Pr
	Flight, Je 19, '31:547		P
	My 6, '32:xviii		P
	My 4, '33:417		P
	F 11, '37:Supp:b-c		P
	O 10, '46:392-394	1	
	Fly Rev Intl, My '68:265 (-I:1 Sq)		Pr*
	F '69:45 (-FB60: Pakistan)		Pr*
	Interavia, Je '49:335		P
	Pop Fly, F '39:554-555+	4	
	RAF Fly Rev, Mr '56:13+	1	
	F '58:11 (Sabre-)	1	
	O '58:8 (-I&-II)	1	
	:9 (-I:1 Sq)		Pr*
	Jl '61:25-27 (Republican Spain-)	1	Pr*
	N '61:29	4	
1540.	"Hardy"		
	Air Pict, My '56:162	1	
	Flight, Je 28, '34:642-643	1	S*
1541.	"Harrier"		
	RAF Qtr, Ja '30:96		P

HAWKER (cont.)

1542.	"Hart"		
	See also "Audax"		
	Aeronautics, My '54:49		P
	Aeroplane, N 11, '31:1119		P
	Je 29, '32:1205		P
	Ja 24, '34:127 (-33 Sq)		P
	Air Prog, Jl '69:38-41 (Restored-)	4	P*
	Acft Eng, N '34:285		S*
	Flight, F 28, '30:245-246	1	P
	N 24, '32:xxii		P
	D 13, '34:1325	1	
	O 19, '61:639	1	P
	F 22, '68:257	1	P*
	O 23, '69:37 (-57 Sq)		P*
	Fly Rev Intl, My '64:54-55	1	P*
	My '68:269 (-33 Sq)		Pr*
	Pop Fly, Ja '34:544-545		S*
	RAF Fly Rev, Ap '62:22 (Finnish-)		Pr*
	S '62:18	4	
1543.	"Hector"		
	Pop Fly, Ag '37:281		P
	RAF Qtr, Jl '37:xxxi		P
1544.	"Henley"		
	Aeroplane, Mr 1, '39:274-277	1	S*
	Flight, Jl 8, '43:38	1	
	Jl 18, '46:60-61	4	
	RAF Fly Rev, F '58:42	1	
	Jl '62:21	4	
1545.	"Hind"		
	Aeroplane, Je 16, '37:725		P
	Flight, F 3, '38:21 (Portuguese-)		P
	Pop Avn, My '38:47		P
1546.	"Horsley"		
	See also "Dantorp"		
	Aeroplane, D 23, '31:1415 (-100 Sq)		P
	N 16, '32:922		P
	F 5, '43:167	1	
	Flight, Je 27, '30:702		P
1547.	"Hotspur"		
	RAF Fly Rev, F '58:42	1	

HAWKER (cont.)

1548.	"Hunter"		
	Aeronautics, F '54:32-35		P
	D '54:31		P
	S '55:80-81		P*
	Jl '57:36-37		P
	:50-51 (-T. 7)	4	
	Aeroplane, Ja 21, '55:78-79 (-I)		P
	My 17, '57:707-710 (-T. 7)	4	C*
	Ja 3, '58:15-18 (-T. 7)	1	C
	My 2, '58:619-621	1	S*
	Air Prog, F/Mr '63:76-77 (-T. 7&-T. 8)		Pr†
	Avn Wk, Je 29, '53:18	1	
	Flight, F 25, '55:237-243	1	
	N 25, '55:809 (-T. 7)	1	
	My 17, '57:662-664 (-T. 7)	4	C
	My 9, '58:633-641 (-F. 6)	1	C, S*
	F 13, '59:228	4	
	Je 17, '60:833-836 (-T. 7)		P
	Fly Rev Intl, Ap '64:11-19	1	Pr*, S*
	My '68:272 (-F. VI:56 Sq)		Pr*
	S '68:63 (-F. 51:Denmark)		Pr*
	N '68:43 (-F. 56:India)		Pr*
	Mr '69:42-43 (-58:Swiss)		Pr*
	Interavia, No. 9, '54:596-599	3	
	No. 9, '58:923-924	1	
	RAF Fly Rev, N '54:15-16	2	
	My '55:52-53	1	P*
	F '57:16-19 (-T. 7)	4	Pr*
	S '57:48 (-F. 6)		Pr
	:49-51+	1	S*
	N '58:11 (-F. 56:India)	5	
	S '60:56 (-FGA. 9)	1	
	:57 (-FGA. 9:8 Sq)		Pr*
1549.	"Hurricane"		
	See also HILLSON F.H. 40 Mk 1		
	Aero Album, Fall '69:42-43 (-I: Ortmans')	4	Pr†
	Aeronautics, F '41:26-31	1	
	N '43:36-37		P
	Aeroplane, My 11, '38:573-576	1	S*
	S 6, '40:254-265 (-I)	1	S*
	O 17, '41:419-421 (-II)	1	
	S 17, '43:328-329 (-II)		P
	Mr 3, '44:15		P
	Ag 18, '44:186	1	

HAWKER (cont.)

1549.	"Hurricane" (cont.)		
	Air BP, No. 21:9-12	4	C
	Air News, F '43:35		D
	Air Pict, S '52:267-270	1	
	Acft Eng, F '39:61-71 (-I)	3	S*
	N '41:307, 318 (-IIB&-IIC)	1	P
	N '46:391-394 (Pt 1)	3	S*
	D '46:427-428 (Pt 2)	3	S*
	Aviation, F '41:137		S
	O '41:93	1	S
	CAHS, Sum '67:35-40 (RCAF-)	1	D
	:34 (-12A)		Pr†
	Comm Avn, Jl '39:32	1	
	Flight, Mr 10, '38:222-223	4	
	My 12, '38:467-473	1	S*
	F 29, '40:194c		D
	O 31, '40:369-370h	1	S*
	:371-373	4	
	Ja 1, '42:8a		D
	Jl 23, '42:92a (-IIB)	1	
	Ap 1, '43:336b (-Tropic IIC)	1	
	Ap 29, '43:442-444	4	
	Je 24, '43:655 (-IID)	1	
	Ag 24, '44:208-209	4	
	S 26, '46:346-347	4	
	S 11, '69:Inset (-I)	1	P*
	Flying, Mr '43:28-29+	1	
	Fly Rev Intl, S '65 (Vol. 21):56, 58		Pr*
	My '68:272 (-I:73 Sq)		Pr*
	:302		P*
	N '68:42 (-IIB:India)		Pr*
	D '68:51 (-I:Finland)		Pr*
	Pop Avn, Ag '39:13 (Yugoslavian-)		P
	Roundel, S '62:33 (-I:Canada)	1	P
	RAF Fly Rev, N '54:27-29+	4	
	N '55:35-36+ ("Hurricats"-)	4	
	Ap '58:29-31+	1	
	:28, 30-31 (-I)		Pr, S*
	D '60:50 (Belgian-)	1	
	Ag '63:31+ (Science Museum's, London-)	4	P
1550.	"Nimrod"		
	Aeroplane, My 25, '32:932		P
	S 14, '32:515		P
	Ap 26, '33:716		P
	Ap 25, '34:709, 712 (Danish-)		P

HAWKER (cont.)

1550.	"Nimrod" (cont.)		
	Aeroplane, D 26, '34:762		P
	F 24, '37:222		P
	Air Cl, O '69:43 (-800 Sq)		Pr*
	Air Pict, Mr '65:86-89	1	Pr
	Flight, My 20, '32:viii		P
	Fly Rev Intl, S '68:63 (Danish-)		Pr*
	Pop Fly, D '32:500-501	1	D
1551.	"Osprey"		
	Aeroplane, Ja 6, '32:13		P
	Ag 3, '32:235		P
	Ag 17, '32:329		P
	Pop Fly, Ap '33:17	1	D
1552.	"Sea Fury"		
	Aeronautics, My '48:26-29 (-X)		P
	My '49:38-41 (-XX)		P
	Aeroplane, F 1, '46:137-144 (-X)	1	S*
	F 15, '46:195		P
	Ja 28, '49:105-107 (-XI)	4	C
	F 4, '49:130-132 (-XX)	1	
	Acft Eng, S '48:252-266	3	
	CAHS, Fall '67:71-73 (Canada-)	1	
	Flight, Ja 31, '46:111-116 (-X)	1	S*
	O 10, '46:392-394	1	
	Je 3, '48:595-596 (-XX)	1	
	Fly Rev Intl, D '63:48 (-II:Canada Navy)		Pr*
	:49-50	1	P*
	Interavia, No. 5, '46:37-40 (-X)	1	
	RAF Fly Rev, Je '62:45+	1	
1553.	"Sea Hawk"		
	Development of N 7/46 and P. 1040		
	Later development: ARMSTRONG WHITWORTH "Sea Hawk"		
	Air Power, Aut '57:60-61 (-4:Germany)	4	P
	Avn Wk, N 15, '48:20	1	
	Flight, N 10, '49:613-618	1	S*
	F 19, '54:209-215	4	
	RAF Fly Rev, O '53:18-20+ (-806 Sq)	4	
	Ja '57:30-33	1	Pr, S*
	Jl '58:16-17 (German-)	1	Pr*

HAWKER (cont.)

1554.	"Tempest"		
	Aeronautics, Ap '45:34-37		P
	O '45:34-37 (-II)		P
	N '45:75 (-V)		C*
	Aeroplane, O 20, '44:450-451		P
	D 1, '44:620		P
	Mr 2, '45:254-257	1	
	Mr 9, '45:279-283	3	C, S*
	S 7, '45:272-275 (-II)	1	P
	O 12, '45:418 (-II)		P
	Jl 11, '47:44		D*
	Air News, Jl '45:31+ (-V)	1	S*
	Acft Eng, D '46:432-433	3	S*
	Aviation, Jl '45:191, 193 (-V)		S*
	Flight, O 19, '44:419	1	
	O 26, '44:448-449	1	
	N 16, '44:538		D
	Mr 1, '45:229-234 (-V)	1	S*
	Ag 23, '45:196-197 (-II)	1	
	O 7, '48:421-423 (-VI)	4	
	Fly Rev Intl, My '68:272 (-II:33 Sq)		Pr*
	N '68:42 (-II:India)		Pr*
	F '69:45 (-II:Pakistan)		Pr*
	RAF Fly Rev, Ap '57:49 (-II)	1	
	F '58:9 (-I)	1	
	RAF Qtr, Mr '47:Plate		P
1555.	"Tornado"		
	Flight, Ap 13, '44:391	1	P
	RAF Fly Rev, Je '55:12	1	
1556.	"Typhoon"		
	Aero Digest, Mr 1, '44:106-108		S*
	Aeronautics, Mr '45:45	1	P*
	Aeroplane, Ap 30, '43:513-515	1	
	My 14, '43:558	1	
	Jl 9, '43:44-45	4	
	Ja 21, '44:64-65	3	
	F 11, '44:153-161	1	S*
	D 15, '44:667		P
	S 14, '45:303-304		D*
	Air Prog, S '43:8-9	1	P
	Acft Eng, D '46:428-433	3	S*
	Aviation, Ag '41:108	1	
	My '44:201, 203		S*

HAWKER (cont.)

1556.	"Typhoon" (cont.)		
	Can Avn, Je '44:78-79	1	C
	Flight, Ap 29, '43:440-441 (-I)	1	
	My 13, '43:492-494	1	
	Jl 8, '43:33-35	4	
	Ja 20, '44:61-68 (-IA&-IB)	1	C, S*
	Ap 13, '44:388-390	4	S
	Ap 19, '45:412		P
	Fly Rev Intl, My '68:272 (-IB:198 Sq)		Pr*
	Roundel, N '63:33 (RCAF-)	1	P
	RAF Fly Rev, Mr '57:15-19+ (-609 Sq) (Pt 1)	4	
	Ap '57:35-38 (-609 Sq) (Pt 2)	4	
1557.	"Woodcock"		
	See also "Dankok"		
	RAF Qtr, Ap '38:Supp	1	D
1558.	N 7/46		
	Later development: HAWKER P. 1040		
	Nav Avn News, F '49:33		P
1559.	P. 1040		
	Development of HAWKER N 7/46		
	Aeronautics, S '48:80-83		P
	Ag '49:45		P*
1560.	P. 1052		
	Aeronautics, My '50:45		P*
	Aeroplane, Jl 8, '49:43		P
	Avn Wk, Je 11, '51:34	1	S
	Flight, F 9, '50:189-194	1	S
1561.	P. 1067		
	Unfinished project; three prototypes		
	Flight, Ag 3, '51:145	1	
	O 5, '51:454-455	1	P
	Roundel, Ja '52:49		P
1562.	P. 1081		
	Development of P. 1052 for Australia		
	Flight, Ag 3, '50:148-149	1	P
1563.	P. 1083		
	Development of "Hunter"; one prototype		
	RAF Fly Rev, Ja '58:36-37	1	D

HAWKER (cont.)

1564.	P. 1121			
	Aeroplane, S 11, '59:180-184	1		C, S*
	RAF Fly Rev, N '58:33	1		
1565.	P. 1127			
	See also later HAWKER SIDDELEY P. 1127 and P. 1154			
	Aeroplane, D 9, '60:766-767	1		
	F 15, '62:165+	2		
	Air Pict, Ja '63:4-5	4		C
	Acft Eng, Ja '61:23	1		
	D '62:346-349	1		
	Avn Wk, D 5, '60:33-34	1		
	S 17, '62:54-58	1		P
	O 22, '62:78	1		
	Bris Sid Jnl, Sum '61:100-103	6	(Engine)	P
	Spg '63:58-61	1		C, S
	Sum '63:98-99	2		
	Flight, D 2, '60:864-865	1		S*
	Mr 23, '61:360			P
	O 11, '62:595-597+	1		C, S*
	Ap 25, '63:602-603	1		
	Jl 4, '63:11-12	1		
	Interavia, No. 1, '61:77 (Proto-)	1		
	No. 5, '62:596-597	1		
	No. 2, '63:175	4		C, P
	:176-179	1		
	RAF Fly Rev, Ja '61:14	1		
	My '61:17-19	1		
	Ja '62:20-21+	4		
	N '62:32-34	1		S*
1566.	P.V. 4			
	Flight, Ag 27, '36:237	1		

HAWKER SIDDELEY GB

1567.	"Andover"			
	Commercial version: 748			
	Earlier designation: AVRO 748			
	Flight, Ag 31, '67:333-338	1		C*, S*
	:339-342	4		S*
	Fly Rev Intl, S '65:13-15+ (-I)	1		S*

HAWKER SIDDELEY (cont.)

1568. "Argosy"
Earlier designation: ARMSTRONG WHITWORTH "Argosy"

Acft Eng, Ag '65:232-239	1		
:240-246	2		
:247-257, 260-263	1		C, S*
:258-259	3		
Flight, Ja 9, '64:51	1		
Ja 28, '65:141-148 (-220 Series)	1		

1569. "Buccaneer"
Earlier version: BLACKBURN "Buccaneer"

Aerosp Intl, S/O '68:31-36	1		
Air Pict, O '65:357-358	4		
Acft Eng, S '65:278-279	6	(RATO)	S
Bris Sid Jnl, Sum '63:86-90	1		
Flight, D 23, '65:1082-1085 (-S. 2)	4		
N 3, '66:754-756	6	(Boundary Layer Control)	
Jl 25, '68:127-133	1		C*, S*
Fly Rev Intl, S '65:21-25+	1		Pr
Interavia, No. 9, '68:1076-1078 (-S. 2)	1		C*, S*

1570. "Dominie"
RAF version of 125

Air Pict, Je '65:199-200	1		
Avn Wk, Ap 19, '65:93	1		C
Flight, F 4, '65:178a-b (-T. 1)	1		P
My 13, '65:737-744 (-T. 1)	1		C, S*, Pr†
Ap 21, '66:683-684	4		
Fly Rev Intl, D '66:229-231	4		C
Interavia, No. 4, '65:526 (-T. 1)	1		

1571. "Gnat Trainer"
Earlier designation: FOLLAND "Gnat Trainer"

Avn Wk, D 23, '63:58-60	1		C

1572. "Nimrod"
RAF designation of 801

Acft Eng, Je '68:16+	6	(Sonobuoy)	
S '69:8-12 (-MR. 1)	1		
:13-14+ (-MR. 1)	6	(Equipment)	
Avn Wk, S 16, '68:30-32	1		

HAWKER SIDDELEY (cont.)

1572. "Nimrod" (cont.)

Flight, Ap 21, '66:683-684 1
Je 1, '67:876 2
:920 1 S*
O 5, '67:584 2
Je 13, '68:887-893 (-MR.1) 1 C, S*
Fly Rev Intl, O '67:891-893 1 P*, Pr
N '68:25-26 (-MR.1) 1
Interavia, No. 1, '69:54-58 1

1573. "Trident"

Earlier version: DE HAVILLAND D.H. 121 "Trident"

Aeroplane, Mr 19, '64:Supp:3-5 1
:6-13, 22-23 4
:14-16 6 (Auto Landing)
:24-35 6 (Generator)
O 29, '64:25 (-1E) 1
N 4, '64:4-6+ (-1C) 4
Ap 5, '67:22-23 (-3B) 1
Mr 13, '68:4-7 (-1C) 4
Ap 3, '68:24 (-2E) 1
Jl 24, '68:14-15 (-3) 1
S 18, '68:20 (-2E) 1
:22 (-3B) 1
Air BP, No. 40:8 (-1E: Pakistan Intl) Pr
Air Pict, Mr '67:85-88 6 (Auto Landing) C
Acft Eng, Je '64:157-159, 166-171, 176-177,
184-198 1 C*, S
:160-165 6 (Auto Landing) C
:172-175, 178-181 6 (Flight Cntrls) S
:182-183 6 (Engine)
:194-195 3
Jl '66:32-33 6 (Interior)
Ap '69:6-9+ (-2E&-3B) 1
:10-13 (-2E&-3B) 2
:14-24 (-2E&-3B) 6 (Systems)
:27-28 (-2E&-3B) 6 (Engine)
:29-31 (-2E&-3B) 6 (Auto Landing)
Avn Wk, S 7, '64:48-49 (-1E) 3
S 14, '64:83-87 (-1C&-1E) 4
Ap 10, '67:47-48 (-3B) 1
Flight, Je 20, '63:989 (-1F) 1
F 11, '65:203-205 1
:205-209 (-1E) 1 C
S 2, '65:364-365 (-2E) 1
Je 9, '66:978-982 6 (Auto Landing)
Ap 6, '67:502-503 (-3) 1

HAWKER SIDDELEY (cont.)

1573.	"Trident" (cont.)			
	Flight, Ap 17, '69:620-622	6	(Auto Landing)	
	O 30, '69:670-672	6	(Auto Lndg)	C*
	D 18, '69:952-957	1		S*
	Fly Rev Intl, Je '64:35-44	1		
	F '69:29-30	1		
	S '69:46-49+ (-2)	1		C, P*
	Interavia, No. 5, '67:582 (-3B)	1		
	No. 5, '68:604-605 (-3B)	1		
1574.	P. 1127 and P. 1154 "Harrier" and "Kestrel"			
	See also earlier HAWKER P. 1127			
	Aerosp Intl, Mr '67:18-23	1		
	S/O '69:6-8+	4		
	Air Force, Ja '67:61-66	1		
	O '69:50-53	4		
	Air Pict, N '65:376-378	4		
	Air Prog, Jl '66:6-7+	4		C
	Acft Eng, D '69:5-16, 19-27, 36-39	1		S*
	:16-19	4		
	:27-30	6	(Air Intake)	S*
	:30, 35-36	6	(Wing)	
	Avn Wk, S 7, '64:72-75	1		
	Je 12, '67:102-103	1		
	Ja 29, '68:52-55	1		C
	Je 24, '68:204-207	4		
	O 14, '68:38-39	4		
	D 23, '68:48-49	3		
	:55	6	(Nav/Attack)	C
	Ap 7, '69:54-55 (-T-2)	3		
	My 5, '69:26 (-T-2)	2		P
	My 12, '69:28-30	4		
	Bris Sid Jnl, Vol. 7, No. 1, '66:11-16	2		
	Flight, O 14, '65:654-657	4		
	N 2, '67:725-733	1		C*, S*
	Ja 11, '68:68-70	1		C
	D 19, '68:1039	4		
	Mr 20, '69:444-445	3		
	Flying, Ap '63:28-29+	1		C
	Fly Rev Intl, F '64:26-27	1		S*
	Jl '64:15-20	1		Pr
	Je '65:17	1		
	D '65:224-228	4		Pr
	S '68:42-43 (-E&-II)	1		
	S '69:36-41+	1		P*, Pr*

HAWKER SIDDELEY (cont.)

1574.	P. 1127 and P. 1154 (cont.)			
	Interavia, No. 11, '66:1794-1795	1		
	No. 3, '68:336-338	1		
	No. 5, '69:569-572 (-50)	1		C*, S*
	:597	6	(Nav/Attack)	
	Nav Avn News, Jl '69:26-27 (-V)	1		
	RAFs Qtr, Aut '67:184-192	1		
	US Army Avn, S '64:9	1		S*
1575.	125			
	RAF designation: "Dominie"			
	Earlier designation: DE HAVILLAND 125			
	Aeroplane, S 18, '68:33 (-400)	1		
	Air Pict, Je '65:199-200	1		
	Avn Wk, Ap 4, '66:97 (-3)	1		C
	Ag 28, '67:111 (-3AR)	4		
	Bris Sid Jnl, Vol. 6, No. 1, '65:4-7	1		
	Can Avn, My '64:22-23+	4		C
	Ja '65:18-21+	4		C
	Mr '68:19-20 (-3AR)	4		S*
	Flight, Jl 30, '64:179-182	4		C
	My 13, '65:737-744	1		C, S*, Pr†
	D 23, '65:1089-1090	1		
	Ap 7, '66:560-563	4		C
	Mr 2, '67:321-328	4		
	D 25, '69:992			C
	Fly Rev Intl, O '69:28-30 (-400A&-400B)	1		
	Interavia, No. 10, '65:1540-1542	4		C
	Sp/Aeron, O '69:73-75	6	(Lift-Dump)	
1576.	681			
	Can Avn, S '63:36	1		
	Flight, Mr 14, '63:361	1		
	Interavia, No. 4, '63:419	1		
1577.	748			
	RAF version: "Andover"			
	Earlier designation: AVRO 748			
	Aeroplane, D 26, '63:15-17 (-2)	1		
	S 18, '68:30 (-2A)	1		
	Air Pict, O '68:372-376	1		
	Acft Eng, Mr '65:69-82	1		S*
	:82-85	2		
	:86-87, 90-91	6	(Engine)	S*

HAWKER SIDDELEY (cont.)

1577. 748 (cont.)

Acft Eng, Mr '65:93-99, 104-106	6 (Various)	S*
:99-100	6 (Cockpit)	C*
:101-103+	3	
:88-89		S*
My '67:45-48	1	S*
Can Avn, Mr '68:14-15+	4	C
Flight, O 9, '69:562-563		P

HAYDEN USA

1578. "Bushmaster"

Air Prog, Mr '67:50-51 (-2000)	1	C
West Avn, D '56:22	1	

HEINKEL Germany

See also POTEZ-HEINKEL

1579. H.D. 44

Acft Eng, Ja '30:20	1	

1580. He 8

Fly Rev Intl, S '68:62-63 (Greenland-)		Pr*

1581. He 46

Fly Rev Intl, S '69:50 (-eUn:Hungary)		Pr*

1582. He 51

Fly Rev Intl, O '68:54-57	1	P, Pr*
RAF Fly Rev, N '61:26 (Richthofen JG-)		Pr*
:27	1	
D '61:61	1	

1583. He 57

Aero Digest, Je '31:82-84	1	
Bee-Hive, Je '31:17	1	
Flight, My 29, '31:472-474	1	
Pop Avn, Mr '33:159-160+	1	

1584. He 59

RAF Fly Rev, Jl '61:51+	1	

1585. He 60

RAF Fly Rev, Jl '62:37-38	1	

HEINKEL (cont.)

1586.	He 64		
	Acft Eng, S '32:227	1	
	Flight, S 9, '32:848-850	1	
	My 4, '33:420	1	
	My 11, '33:454-455	2	
	Pop Avn, Mr '33:161 (-D)	1	
1587.	He 70 "Blitz" (Lightning)		
	Later development: He 170 and He 270		
	Aeroplane, Mr 8, '33:421-424	1	
	Je 20, '34:963	4	
	Aviation, Ap '33:126-127	1	
	Flight, Mr 9, '33:223-226	1	C
	Fly Rev Intl, O '65:116-119	1	Pr*
	Pop Avn, Je '33:367-368	1	
	RAF Fly Rev, F '57:13+	1	
1588.	He 100		
	-D model also incorrectly called HEINKEL He 113		
	Air Prog, D '66:41-43+	1	
	Fly Rev Intl, My '65:43	1	
	Jl '67:742 (-D&-V)		Pr*
	RAF Fly Rev, Mr '55:39 (-D)	1	
	Ag '56:13+ (-D)	1	
	My '61:51 (-D)	1	
	F '63:34 (-D-1)		Pr*
	:35-36+ (-D&-V)	1	
1589.	He 111		
	Spanish-built version: CASA 2111		
	Z model named "Zwilling" (Twin)		
	Aeroplane, O 26, '39:511 (-K)		D
	N 2, '39:542-544	1	S*
	O 11, '40:410-411	1	
	Air Pict, My '66:181 (Czech-)	5	
	Acft Eng, D '42:342-357	6 (Radio)	
	Ag '44:241-242	6 (Fuel Tank)	
	Aviation, Ag '42:197	6 (Gun)	S*
	O '42:157		S*
	Flight, Ja 16, '36:63	1	
	S 21, '39:256a-d	1	S
	N 21, '40:428a-c (-K)	1	S*
	Jl 2, '42:12a (-K)	1	
	N 27, '69:Inset	1	
	Flying, Mr '41:17+ (-K)	1	

HEINKEL (cont.)

1589.	He 111 (cont.)		
	Fly Rev Intl, Je '64:69+ (-Z)	1	
	Ag '64:56 (-H:Czech)		Pr*
	D '67:1064, 1066 (-H&-P)		Pr*
	RAF Fly Rev, Mr '56:9-10 (-Z)	1	
1590.	He 112		
	Aeroplane, N 16, '39:600-601	1	S*
	N 30, '39:672	1	
	Air News, Jl '44:19-21	1	
	Air Pict, Je '60:197-200+	1	Pr
	Can Avn, Ag '40:17-19	6 (Gun)	S*
	Flight, O 5, '39:286c-287	1	S, D
	Fly Rev Intl, My '65:43 (-U)	1	
	F '67:387-391	1	Pr*
	RAF Fly Rev, Ag '55:12	1	
	My '61:51 (-B)	1	
1591.	He 113		
	Correct designation: He 100D		
	Air News, Jl '44:19-21	1	S*
	Flight, My 23, '40:470-471	1	D
1592.	He 114		
	RAF Fly Rev, Mr '61:53-54	1	
1593.	He 115		
	Aeroplane, N 2, '39:541		D
	O 25, '40:455	1	
	Flight, Ag 19, '43:202	1	
	RAF Fly Rev, O '60:39-40	1	
1594.	He 116		
	Acft Eng, Ja '38:17	1	
	Flight, S 2, '37:232	1	
	Fly Rev Intl, D '63:59	1	
	RAF Fly Rev, N '56:15	1	
1595.	He 118		
	Fly Rev Intl, O '63:60+	1	
	RAF Fly Rev, Je '56:13	1	
1596.	He 119		
	RAF Fly Rev, Ap '55:13	1	
	Ja '61:23-25+	1	

HEINKEL (cont.)

1597. He 162 "Volksjäger" (People's Fighter) and "Salamander"

Air News, D '45:22	1		
Air Pict, My '58:166-167	1	Pr	
Fly Rev Intl, Je '67:684-687 (Pt 1)	1	Pr*	
Jl '67:747 (Pt 2)	1		
Interavia, No. 7, '52:372-375	1		
RAF Fly Rev, Ap '53:16-17	1		
N '55:39-40	1		
US Air Svcs, N '45:26		P	

1598. He 170

Hungarian export version of He 70

See also He 270

Fly Rev Intl, O '65:116-119	1	Pr*

1599. He 176

RAF Fly Rev, O '54:18	1	

1600. He 177 "Greif" (Griffon)

Aeroplane, My 11, '45:546-547	1	
My 18, '45:571-572	1	
Air Pict, O '59:369-373 (Pt 1)	1	Pr
N '59:390-393 (Pt 2)	4	
D '59:432-434 (Pt 3)	1	
Flight, Mr 26, '42:285	1	
D 31, '42:716b	1	
My 10, '45:498-500	1	C
RAF Fly Rev, My '56:43-44+	1	

1601. He 178

World's first turbojet aircraft

Air Pict, Mr '60:86-87	1	Pr
RAF Fly Rev, N '56:13	1	

1602. He 211

Avn Wk, Ap 30, '62:43-45	1	
Interavia, No. 4, '62:393-394 (-A&-B)	1	

1603. He 219 "Uhu" (Owl)

Air Prog, Ja '65:15-16+	1	Pr†
RAF Fly Rev, F '57:43-44	1	
F '62:28-30+	1	S*
:31 (-A-5/R2)		Pr

HEINKEL (cont.)

1604.	He 270			
	Only a prototype built			
	Fly Rev Intl, O '65:116-119	1		
1605.	He 280			
	RAF Fly Rev, N '55:9	1		
	O '58:23-25+ (-V1 to -V3)	2		
1606.	He 343			
	RAF Fly Rev, Mr '59:53	1		
1607.	P. 1077 "Julia"			
	Experimental rocket aircraft			
	RAF Fly Rev, O '57:9	1		
1608.	P. 1078			
	RAF Fly Rev, D '55:12-13	1		

HELICOPTER CLUB OF FRANCE — France

1609.	"Helicopta 1"			
	Am Helicop, Ap '48:11+	1		S*

HELIO — USA

		4		
1610.	"Courier" H-295			
	Flying, Ag '69:44-49	4		C, P
1611.	"Courier" H-391			
	Aeroplane, Mr 11, '60:310-312	4		C
	Air Prog, D '68:44-45+	4		C
	Avn Wk, Jl 12, '54:21-22	1		
	D 16, '68:99	6	(Flaps)	C
	Flying, S '55:28-29+	4		
	F '61:39+	4		
	Skyways, S '55:10-12+	4		
1612.	"Courier II"			
	Air Prog, Ja '69:56-57	4		
1613.	"Helio Twin" or "Twin Courier" U-5/H-500			
	H-500 is civil designation			
	Air Prog, Ag/S '64:68-69	1		C
	Avn Wk, Mr 2, '64:87	1		
	Ag 31, '64:74-77	4		

HELIO (cont.)

1614.	"Helioplane"		
	Flying, Ap '50:14-15+ (Proto-)	1	
	Pegasus, Ag '49:6-9	1	
	Skyways, Jl '49:31	1	
	Ag '49:14-15+	1	
1615.	"Stallion" or "Courier" U-10/HST-550		
	Air Prog, Ag/S '65:84		P
	F '67:28-30+	4	C
	Je '68:51+	1	
	Avn Wk, Je 1, '64:92-98	1	C
	Ag 31, '64:74-77 (-X)	4	
	Je 7, '65:69 (-B)	4	
	Flying, Jl '66:102-108	1	P*
1616.	"Strato Courier" H-392		
	Avn Wk, Mr 11, '57:101-104	1	
1617.	H-6		
	Avn Wk, N 23, '53:37	1	
1618.	H-634		
	Avn Wk, Ap 1, '68:70-73	1	
1619.	H-1320		
	Avn Wk, Je 22, '64:20-21	1	

HELWAN — Egypt

1620.	HA-300		
	First jet fighter built in Egypt		
	Fly Rev Intl, Je '64:7+	1	
	Ap '67:491-492	1	

HENDY — GB

1621.	302		
	Flight, Ag 22, '30:940-942	1	S

HENSCHEL — Germany

1622.	Hs 123		
	Flight, S 28, '39:270c-d	1	D
	RAF Fly Rev, My '57:11+	1	
	Ap '60:41-42	1	

HENSCHEL (cont.)

1623.	Hs 124		
	RAF Fly Rev, Ap '59:9	1	
1624.	Hs 126		
	Aeroplane, N 30, '39:666-668	1	S*
	Can Avn, Ap '40:24+	3	
	Flight, O 26, '39:334a-c	1	D
	Fly Rev Intl, Je '69:73 (-A:Greece)		Pr*
	RAF Fly Rev, Jl '61:48	4	
1625.	Hs 128		
	Fly Rev Intl, N '67:1005	1	
1626.	Hs 129		
	Air Pict, F '60:49-51	1	Pr
	Flight, Mr 25, '43:310b	1	
	RAF Fly Rev, Ag '57:21-22	1	
1627.	Hs 130		
	RAF Fly Rev, D '57:9+	1	
1628.	Hs 132		
	Air Pict, Ja '64:25	1	
	Fly Rev Intl, Jl '65:57	1	
	RAF Fly Rev, N '55:9	1	

HERRICK USA

1629.	"Vertaplane"		
	Aviation, S '37:53+	1	

HESTON GB

1630.	"Phoenix"		
	Aeroplane, O 23, '35:501-504	1	S*
	Flight, O 24, '35:Supp:a-d	1	C, S
	Pop Fly, F '37:559	1	P
1631.	A.O.P. (Air Observation Post)		
	Aeroplane, Ja 7, '49:13-15	1	
	Flight, Mr 11, '48:284	1	
	RAF Fly Rev, Ja '58:41	1	

HFB (Hamburger Flugzeugbau GmbH) Germany

1632.	314		
	Interavia, No. 4, '60:484-485	1	
1633.	320 "Hansa Jet"		
	Air Prog, Ja '69:37-40	1	C, S*
	Acft Eng, Ag '64:248-251	1	
	O '65:316-318	6 (Gear & Hydr)	S
	Avn Wk, My 29, '67:351-356	4	
	Flight, Ag 6, '64:213-214+	4	C
	Ap 8, '65:539	1	C
	Flying, Ja '65:74-77	1	C
	Jl '66:100-101	1	P*
	O '67:46-51	4	
	Interavia, No. 5, '62:552	1	
	No. 6, '63:832-834	1	
	No. 7, '64:960-961	6 (Wing)	
	:962-964	1	C*
	No. 10, '65:1537-1539	2	
	No. 6, '67:884-887	4	C
1634.	330 "Hansa Fan Jet"		
	Interavia, No. 11, '69:1781	1	

HILLER USA

1635.	E-4 "Station Wagon"		
	Flying, N '60:32-34+	4	
1636.	H-23 "Raven"		
	Commercial version: 360 and UH-12		
	Aeroplane, N 30, '51:700 (-A&-B)	1	C
	Am Helicop, Ja '56:10 (-B)	4	
	Avn Wk, D 3, '51:41 (-A&-B)	1	C
	Flight, N 30, '51:676 (-B)	1	C
	Skyways, Mr '52:13 (-B)	1	C
	West Avn, D '51:18 (-A&-B)	1	
1637.	XH-44 or Hx-44 "Hiller-Copter"		
	Flying, D '44:57+	1	
	Skyways, Ag '45:46-47+	1	
1638.	HJ-1		
	Civil version: HJ-2 "Hornet"		
	Am Helicop, Je '55:6-8+	2	

HILLER (cont.)

1639.	HJ-2 "Hornet"		
	Military version: HJ-1		
	Am Helicop, F '51:8+	1	
	Ap '51:10-11+	1	
1640.	L-4		
	Flying, Ap '63:34-35+	4	
1641.	OH-5A		
	Avn Wk, S 9, '63:87	1	
	Ja 20, '64:64-71	1	
	Flight, My 7, '64:773-775	1	
	US Army Avn, Ag '64:8-9+	1	
1642.	UH-12		
	See also 360 "Hornet"		
	Military designation: H-23 "Raven"		
	Aeroplane, Ap 13, '56:252-255 (-B)	4	C
	S 23, '60:453-455 (-E)	4	C
	Am Helicop, Ap '57:6-7+ (-C)	6 (Supercharger)	
	Avn Wk, O 19, '59:98-103 (-E)	4	
	Je 7, '65:78-85 (-L4&-SL-4)	4	C
	Flight, O 14, '60:616-617 (-E)	4	
1643.	X-18 "Propelloplane"		
	Air Pict, F '59:68	1	
	Am Helicop, D '57:8+	2	
	D '58:6+	1	
	Avn Wk, D 5, '58:34-36	1	
	D 28, '59:22	2	
1644.	360 "Hornet"		
	See also UH-12		
	Military designation: H-23		
	Aeroplane, F 24, '50:211-216	4	C, S
	Am Helicop, Ja '48:6-7	1	
	Ap '50:11+	1	
	Avn Wk, D 15, '47:25-30	1	
	N 1, '48:22-24	2	
	My 8, '50:35	1	
	Can Avn, Ag '49:16-17+	4	C
	Flight, N 10, '49:604-605+	4	
	F 23, '51:206-207	1	
	O 9, '53:508-509	4	
	Flying, My '49:34-35+	1	
	Interavia, O '48:546-548	1	

HILLER (cont.)

1644. 360 (cont.)
Skyways, F '50:22-24 4
West Fly, Mr '51:28 1

1645. 1099
Interavia, No. 2, '62:212-213 1

HILLSON GB

1646. "Bi-Mono"
Aeroplane, Jl 7, '44:14-15 1
RAF Fly Rev, Ja '57:15 1

1647. F.H. 40 Mk 1 Slip-wing HAWKER "Hurricane"
RAF Fly Rev, Ja '59:42 1

HINDUSTAN India
RAF Fly Rev, F '58:11
West Fly, F '45:46-47+

1648. HF-24 "Marut" (Wind Spirit)
First India-designed jet fighter
Avn Wk, Jl 17, '61:89 1
Fly Rev Intl, D '64:35-37 1 Pr

1649. HJT-16 "Kiran"
Avn Wk, O 5, '64:25 1
Flight, Mr 11, '65:371 1

1650. HT-2
Aeroplane, Mr 21, '58:398-399 4 C
Flight, Ag 31, '51:250 1
D 19, '58:941-943 4

HISPANO Spain

1651. HA-100 "Triana"
Avn Wk, Ag 27, '56:85 (-E&-F) 1
RAF Fly Rev, F '58:4-5 (-E&-F) 1 P*

1652. HA-200 "Saeta" (Arrow)
Air Prog, Jl '68:38-39 (-E:Super) 1 C
Avn Wk, S 10, '56:132 1
Interavia, No. 8, '58:836-837 (-R1) 1 S
RAF Fly Rev, Ja '59:26-29 (-R1) 1 Pr, S*

HISPANO (cont.)

1653. HA-1109, HA-1110, HA-1112
Original design: MESSERSCHMITT Me-109
Air Prog, O '69:44-47+ (-1112-M1L) 4 C
Flight, D 18, '69:Inset (-1109) P*
Fly Rev Intl, Jl '68:412 1
RAF Fly Rev, Mr '56:35-36+ 1
Jl '58:27-29 1
:26, 28-29 (-1112-M1L) Pr, S*

HODKINSON USA

1654. Transport (1930)
Aviation, F 1, '30:208-209 1

HOFFMAN USA

1655. "Flying Wing"
Pop Avn, Mr '35:163+ 1 Pr

HOLLANDAIR Netherlands

1656. HA-001
Air Pict, O '57:362 1
Flight, S 6, '57:415 1

HÖNNINGSTAD Norway

1657. 5A
Flight, Ja 17, '46:53-54 1

HOPPI-COPTER USA

1658. 102
Am Helicop, N '47:18-19+ 1
My '48:15 4
S '56:10 1
Flight, O 14, '48:466-467 1

HOPPI-COPTERS, LTD. GB

1659. Mark II
Am Helicop, O '50:6-7 1

HORTON Germany

1660. Ho. IX
Developed into GOTHA 229
Air Pict, Ap '59:131-133 1
RAF Fly Rev, F '55:9 1
F '57:15 1
S '58:73-74 1

1661. HW-X-2652
Avn Wk, F 1, '54:17-18 1
West Avn, Ag '52:26 1

HOWARD USA

1662. DGA-6 "Mister Mulligan"
Air Prog, My/Je '65:41+ 1
Aviation, O '35:15-16 1
Flight, Mr 28, '35:339 1
Pop Avn, Mr '36:187 3 P

1663. DGA-8
Can Avn, Je '37:48 1
Pop Avn, Je '36:425-426+ 1 C
Mr '37:22+ 1

1664. DGA-15
Aviation, Jl '40:54-55 S*
US Air Svcs, Ap '40:30+ (-P) 1

1665. DGA-125
Aviation, D '40:68 1
Comm Avn, N '40:52+ 1
Flight, F 6, '41:118-120 1

1666. DGA-160
Aviation, S '41:91 1

HUGHES USA

1667. Racing Monoplane (1936)
Pop Avn, Jl '36:18 1
Ag '37:45-46+ 4

1668. H-4 or HK-1 "Hercules" or "Spruce Goose"
Aero Digest, S 1, '45:97-98 1
Avn Wk, N 10, '47:11-12 2
Ap 2, '51:17-18 1

HUGHES (cont.)

1668.	H-4 or HK-1 (cont.)		
	Can Avn, O '45:58-60+	1	
	Comm Avn, Ja '46:41-44+	3	
	Flight, Jl 14, '49:55	1	
	Flying, Ap '46:51-53+	1	
	Fly Rev Intl, S '69:75	1	
	Interavia, No. 5, '46:26-33	3	P
	No. 12, '47:48-52	2	P
	Nav Avn News, F '47:21	1	
	RAF Fly Rev, Je '57:40+	1	
	West Fly, S '45:44-45+	3	C
1669.	OH-6A "Cayuse"		
	Civil version: 500		
	Avn Wk, F 3, '64:72-81	1	
	S 28, '64:69	6 (Blade Folding)	
	Je 28, '65:62-70	4	
	S 12, '66:97-101	1	
	My 20, '68:26-28	4	
	Flight, F 24, '66:316-318	1	S
	D 15, '66:1046	1	
	Fly Rev Intl, Ja '66:261-263	1	
	US Army Avn, Ag '64:10-11+	1	S*
	D '66:48	4	
	My '67:31	1	
	Jl '67:5	1	
1670.	XF-11 or XR-11		
	Avn Wk, O 13, '47:23-27	1	
	Fly Rev Intl, S '69:75 (XR-/-A)	1	
	RAF Fly Rev, F '57:15	1	
	West Fly, My '47:48	1	
1671.	XH-17 "Flying Crane"		
	Am Helicop, Ja '56:8+	2	P
	Avn Wk, Ap 3, '50:15	1	
	RAF Fly Rev, Ap '60:50	1	
	West Avn, Je '52:30	1	
	D '52:30	1	
	Ja '56:14	1	
1672.	XV-9A		
	Avn Wk, Je 22, '64:60-65	1	S*
	F 1, '65:40-41	1	C
	Interavia, No. 4, '65:501-502	1	S*

HUGHES (cont.)

1673.	269 "Tiny Tim"		
	Avn Wk, Ja 15, '62:107-113 (-A)	1	
	Can Avn, F '64:24 (-A)	1	
	Flying, Mr '61:32-33 (-A)	4	C
	Mr '63:38-39+ (-A)	4	
	West Avn, F '57:20	1	
1674.	300		
	Can Avn, F '64:24-25	1	
	Flight, Ja 30, '64:165-166+	4	C
	Flying, Je '64:36-37+	4	
1675.	500		
	Army designation: OH-6A "Cayuse"		
	Flight, D 29, '66:1096+	4	C
	Flying, Mr '66:62-65	1	
	F '67:32-33+	4	
	Ja '69:80	1	
	Fly Rev Intl, Ja '66:261-263	1	
	Interavia, No. 9, '67:1410	1	C

HUNEAULT Canada

1676.	DHC-1 "Chipmunk"		
	Acrobatic version of DE HAVILLAND CANADA DHC-1		
	Fly Rev Intl, D '69:31	1	

HUNTING GB

See HUNTING PERCIVAL

HUNTING PERCIVAL GB

See also PERCIVAL

1677.	"Jet Provost" P. 84		
	Later version: BAC "Jet Provost"		
	Aeronautics, My '57:30-31	2	C*
	My '59:60-61	4	
	Jl '60:24-25 (-T. 3)		P
	Aeroplane, S 24, '54:480-481 (-T. 1)		P
	Ap 15, '55:475-479 (-I)	1	S*
	F 22, '57:266-269 (-2)	1	C, S*
	Je 21, '57:877-879	4	C
	Air Pict, Je '55:176-177 (-T. 1)		S*
	Air Prog, F/Mr '63:74-75 (-I to -III)		Pr†
	Avn Wk, Ap 18, '60:140-141 (-3)	4	C

HUNTING PERCIVAL (cont.)

1677.	"Jet Provost" (cont.)		
	Bris Sid Jnl, Sum '62:104-107	1	C, P
	Can Avn, N '55:40+ (-I&-II)	4	C
	N '56:41+ (-II)	4	C
	N '58:28-29+ (-III)	4	
	Flight, Ag 6, '54:181 (-I)	1	
	My 6, '55:582-587 (-T.1)	1	C, S*
	Mr 8, '57:300-302+ (-I to -III)	4	C
	My 31, '57:746-748+ (-III)	4	C
	Je 13, '58:795-799 (-T.3)	1	C*, S*
	Ap 1, '60:443-446 (-T.3)	4	
	Ag 12, '60:219-222 (-T.3)	4	P
	Flying, Ag '56:38-39+	4	C*
	Interavia, No. 9, '57:883-884 (-T.3)	1	C
	RAF Fly Rev, S '55:46-49 (-T.1)	1	Pr, S*
	My '57:8-9 (-T.1 to -T.3)	1	P*
	Je '60:15-17 (-T.1 to -T.3)	4	P*
	Mr '63:28-29 (-T.4)		P*, S*
	:30+	1	
	:31 (-T.51)		Pr
1678.	"Pembroke" P. 66		
	Flight, F 26, '54:234 (-C. I)	1	
	Mr 26, '54:365-371 (-C. I)	1	C*, S*
	RAF Fly Rev, F '57:4	1	
	:5 (-C. 51: Belgium)		P*
1679.	"President"		
	Mark V version of PERCIVAL "Prince"		
	Aeroplane, O 18, '57:589-591	1	C
	Flight, Jl 6, '56:70	1	
1680.	H 107		
	Interavia, No. 10, '59:1255-1256	2	
1681.	H 126		
	Jet flap research aircraft		
	Acft Eng, Je '63:166-167	1	
	Avn Wk, S 3, '62:21	1	
	Flight, Ap 4, '63:454-455	1	S
	S 12, '63:474	1	S*
	Interavia, No. 7, '63:1073	1	

HUNTING PERCIVAL (cont.)

1682.	P. 74		
	Aeroplane, Ag 5, '55:203-206	1	
	My 11, '56:381		P
	Flight, D 23, '55:938-942	1	C, S*
	RAF Fly Rev, Ag '58:11	1	
1683.	P. 105		
	Aeroplane, Ag 5, '55:204-206	1	S*
	Am Helicop, S '55:11, 13	1	S*

HUREL-DUBOIS France

1684.	H.D. 10		
	Flying, S '50:25+	1	
1685.	H.D. 31		
	Aeronautics, My '54:70-75	1	
	Aeroplane, Ap 2, '54:404-405	1	C
	Avn Wk, S 8, '52:21-29	1	
	Flight, N 28, '52:676-677	1	
	Je 18, '54:793		C
1686.	H.D. 32		
	Aeroplane, Je 18, '54:793-794	2	
	Ap 15, '55:496-497	1	C
	Ap 6, '56:226-228	2	
	Avn Wk, Ja 4, '54:46-47	1	
	Flight, N 28, '52:676-677	1	
	Je 18, '54:793-799	1	C, S*
1687.	H.D. 34		
	Flight, S 20, '57:479-480	4	
1688.	H.D. 321		
	Flight, Jl 6, '56:62	1	
1689.	HDM-105 "Caravan"		
	Aeroplane, S 6, '57:390		S*

IAI Israel

See ISRAEL AIRCRAFT INDUSTRIES

IAR (Societate Anonima Industria Aeronautica Romana) Rumania

Became part of URMV-3

IAR (cont.)

1690. 37, 38, and 39
Fly Rev Intl, N '69:79 1

1691. 80 and 81
RAF Fly Rev, S '59:15 (-80) 1
N '62:16 (-81C) Pr*
:17 (-80&-81) 1

ICAR (Intreprindere Pentru Contructii Aeronautice Romane) Rumania

1692. Commercial Monoplane (1935)
Flight, Ap 11, '35:394 1

IKARUS Yugoslavia

1693. B-5 "Pionir"
Fly Rev Intl, S '65 (Vol. 21):60 1

1694. IK. 3
See also IKARUS S-49
RAF Fly Rev, F '57:9 1
Je '57:40 1

1695. IK. 451
Prone pilot aircraft
Fly Rev Intl, S '65 (Vol. 21):60-61 1
RAF Fly Rev, N '60:9 1

1696. IK. 452
Fly Rev Intl, S '65 (Vol. 21):60-61 1

1697. S-49
Development of IK. 3
Fly Rev Intl, My '68:298-299 1
RAF Fly Rev, F '57:9 1

IL (Aircraft Institute) Poland

1698. TS-8 "Bies" (Devil)
Aeronautics, S '57:81 1

ILYUSHIN Russia
Fly Rev Intl, D '66:268-270 (Pt 1)
Ja '67:315-316 (Pt 2)

ILYUSHIN (cont.)

1699. IL-2 "Stormovik" [Bark]
Successor: ILYUSHIN IL-10
Flight, Mr 18, '43:276-277 1
My 20, '43:526a (-C) 1
Fly Rev Intl, S '65:51-54 (Pt 1) 1 Pr*
S '65 (Vol. 21):51-54 (Pt 2) 1
Interavia, N '49:643 P

1700. IL-4 or DB-3F
RAF Fly Rev, Ap '63:43+ 1

1701. IL-10 "Ruszko" [Beast]
Predecessor: ILYUSHIN IL-2
Czech version: AVIA B. 33
Fly Rev Intl, S '65 (Vol. 21):51-54 1 Pr*
S '69:51 (Hungarian-) Pr*
RAF Fly Rev, F '59:42 1

1702. IL-12 [Coach]
Aeroplane, S 23, '49:444-445 1
Acft Eng, N '49:363 1
Avn Age, D '53:80-81 1
Avn Wk, Ja 23, '50:23-25 1
Flight, Je 24, '48:696-697 1
S 22, '49:395-396 4 C
Flying, D '50:18-19+ 1
S '51:16-17+ 4
Interavia, No. 8-9, '50:444-445 1
RAF Fly Rev, Ja '57:48 1

1703. IL-14 [Crate]
Czech version: AVIA 14
Aeroplane, D 6, '57:841, 844-847 1
Avn Wk, Ja 28, '57:41 (-M) 1 S
Ap 22, '57:64-71 2
Flight, Mr 4, '55:285 1
RAF Fly Rev, Ja '57:48-49 1

1704. IL-18 "Moskva" (Moscow) [Coot]
"Moskva" is name for 1957 prototype only
Aeroplane, My 16, '58:677-678 1
Air BP, No. 18:10 (Malev-) Pr*
No. 35:9 (Tarom-) Pr*
No. 39:8 (Tabso-) Pr*
Air Force, N '60:46-48 1

ILYUSHIN (cont.)

1704.	IL-18 (cont.)			
	Air Prog, Ag/S '65:62-63	1		C
	N/D '65:17 (-D)	1		C
	Avn Wk, O 7, '57:43-45	1		
	Je 15, '59:46-47	1		C*
	Je 25, '62:55 (-I)	1		
	My 3, '65:36-37	3		
	Flight, Ap 25, '58:560	6	(Cockpit)	C
	Jl 25, '58:110-111	1		S*
	S 4, '59:187	1		
	O 16, '59:405	1		Pr
	Jl 1, '60:13-17	1		C, S*
	Ja 16, '64:85	6	(Engine)	
	N 12, '64:823	4		
	Fly Rev Intl, My '66:545	1		
	Interavia, No. 6, '59:737	1		
	No. 6, '65:959	3		
	RAF Fly Rev, S '57:23+	1		
	D '57:38+	1		
1705.	IL-28 [Beagle]			
	Earlier code name: [Butcher]			
	Avn Wk, Mr 30, '53:30-33	1		
	Ja 11, '54:14 (-2)	1		
	Flight, S 7, '56:463	1		P
	Fly Rev Intl, Jl '64:42-45	1		Pr
	Nav Avn News, Ap '57:35	1		
1706.	IL-62 [Classic]			
	Aeroplane, S 15, '66:4-7	1		S*
	Ag 7, '68:25	4		
	Ag 28, '68:9	4		
	Air Pict, F '64:44-45	1		Pr
	Air Prog, N/D '65:16-17	1		C
	Avn Wk, Ag 17, '64:32-33	1		C
	Ag 24, '64:40-41	2		C
	Jl 19, '65:31	1		
	O 17, '66:54-57	1		C, S*
	Jl 22, '68:28-30	4		
	Fly Rev Intl, My '66:546	1		
1707.	IL-40 and IL-140 [Blowlamp]			
	Air Force, Ja '58:68-69	1		
	Avn Wk, O 14, '57:90-91	1		
	Ap 6, '59:28-29	1		

IMAM (S. A. Industrie Meccanichee Aeronautiche Meridionali) Italy

1708. Ro. 41
Pop Avn, S '39:52 1

1709. Ro. 43 and Ro. 44
RAF Fly Rev, Jl/Ag '59:9+ 1

1710. Ro. 51
RAF Fly Rev, Ja '61:9+ 1

1711. Ro. 57
RAF Fly Rev, Ja '58:38 1

INDONESIA Indonesia

1712. NU-200 "Sikumbang" (Bee)
Flight, N 5, '54:683 1

INSTITUTO AEROTECNICO Argentina

1713. "Calquin" (Royal Eagle) I. AÉ 24
RAF Fly Rev, Mr '59:6 1
O '61:65 1

1714. "Ňamcú" I. AÉ 30
RAF Fly Rev, Mr '58:6 1

1715. "Pulqui" (Arrow)
Avn Wk, Ag 21, '50:21 (-I) 1
Fly Rev Intl, O '64:76-77 (-I&-II) 1

INTA (Instituto Nacional de Tecnica Aeronautica) Spain

1716. HM-1
Fly Rev Intl, Ag '69:69 1

INTERSTATE USA

1717. L-6 "Grasshopper"
Aviation, Ap '43:185 S
Flight, D 30, '43:723 1

IRBITIS Latvia
Air Pict, Ag '57:254-257 (With table of acft types)

IRELAND USA

1718. "Privateer"
Flight, Ap 10, '31:310-311 1

ISRAEL AIRCRAFT INDUSTRIES Israel
Fly Rev Intl, Ja '68:15-16+

1719. "Arava"
Aeroplane, Mr 9, '67:27-30 1
Air Prog, F '68:52+ 1
Avn Wk, Jl 8, '68:73-76 1
Fly Rev Intl, O '68:30 1

ISRAEL AIRCRAFT INDUSTRIES/SOCATA Israel/France

1720. "Sherpa"
Development of "Arava"
Interavia, No. 9, '69:1548 1

IVL (Industria Valtion Lentokonetehdas) Finland
See STATE AIRCRAFT FACTORY

JACKAROO GB

1721. "Paragon"
Flight, F 27, '59:288 1

1722. "Thruxton Jackaroo"
Flight, Jl 12, '57:44 4
O 3, '58:540 1

JACOBS USA

1723. 104
Am Helicop, Je '52:6-8+ 1
West Avn, Jl '52:18 1

JACUZZI USA

1724. Monoplane (1921)
AAHSJ, Sum '68:114-115 1

JENSEN USA

1725. 21
Am Helicop, S '48:10-11 1 P
Avn Wk, Ag 2, '48:25-26 1
West Fly, D '48:18 1

JL (John Larsen) USA

1726. 12 Monoplane (1921)
Aerosp Hist, Spg '68:63 1

JONA Italy

1727. J. 6
Flight, O 10, '35:378-379 1 S

JOV (Jovanovich) USA

1728. 3
Enlarged into MCCULLOCH MC-4
Am Helicop, Mr '48:20+ 1 P
Avn Wk, Ja 19, '48:21 1

JOVAIR USA

1729. 4E and 4ES "Sedan"
Development of MCCULLOCH MC-4
Air Prog, Ag/S, '64:70-71 1 C, S

JUNKERS Germany, Russia, and Sweden

1730. A-50 "Junior"
Flight, Ja 24, '30:141-142 1

1731. EF-61
RAF Fly Rev, N '61:49 1

1732. EF-150
Avn Age, O '52:33-35 1
N '52:41 1 S*

1733. F. 13
Aeroplane, Jl 30, '43:141 1

1734. G. 38
Air Pict, S '66:346-347 1
Acft Eng, F '30:40-41 1
Aviation, Ja 18, '30:113-117 1
Flight, Je 19, '31:561-563 1
Jl 1, '32:607 (-D2500) 1
Ag 20, '36:210 1
Pop Avn, F '33:88-89 (-D2500) 1
RAF Fly Rev, Je '56:13 1

JUNKERS (cont.)

1735.	J. 1 "Blechesel" (Sheet Metal)		
	Aeroplane, My 28, '43:630	1	
1736.	Ju. 52 "Iron Annie"		
	Air Pict, Mr '65:99 (Argentinian-)	4	
	Acft Eng, N '31:291-292	1	
	Bee-Hive, Je '35:6-7 (-/3m)	1	
	Can Avn, Ja '32:18+	1	
	CAHS, Spg '63:19 (-/1m:"Old Farm")	1	
	Flight, F 27, '31:186-187	1	
	O 12, '39:300e	1	
	O 8, '42:388 (-/3m)	1	
	My 8, '69:768-769 (-/3m)	1	
	Fly Rev Intl, F '65:52 (-/3m)		Pr*
	Pop Avn, Ap '33:220 (Prince of Bibesco's-)	1	S*
	RAF Fly Rev, S '57:77-78 (-/3m)	1	
	Ja '63:49 (-/1m)	1	
1737.	Ju. 60		
	Can Avn, D '33:7	1	
1738.	Ju. 86		
	Aeroplane, O 19, '39:478-479 (-K)	1	S*
	Flight, Je 18, '36:657	1	
	O 12, '39:300a-d (-K)	1	S
	Fly Rev Intl, S '69:50-51 (-K-2:Hungary)		Pr*
	RAF Fly Rev, N '55:13	1	
1739.	Ju. 87 "Stuka"		
	Aeronautics, Mr '43:43-44, 61	1	S*
	Aeroplane, F 23, '40:248-252	1	S*
	Air News, S '41:13		D
	Air Prog, Spg '59:39-44	1	P
	Acft Eng, Je '42:164-167 (-B2)	6 (Radio)	
	Flight, S 28, '39:270a-d	1	S
	O 17, '40:322-324	1	S
	N 26, '42:578b (-B)	1	
	Fly Rev Intl, F '64:52 (-B-2:3/SG 1)		Pr*
	:53-54 (Pt 1)	1	
	Mr '64:49-53 (Pt 2)	1	S*
	:52 (-D-5:Hungary)		Pr*
	Mr '68:160, 162 (-B, -D, -G, -R)		Pr*

JUNKERS (cont.)

1739.	Ju. 87 (cont.)			
	RAF Fly Rev,	D '54:31	1	
		My '59:37	4	
		O '59:53	4	
		D '60:43-44 (-B)	5	Pr
		N '61:39-41+	4	
1740.	Ju. 88			
	Aero Digest,	My '42:255+ (-A-6)	1	
	Aeronautics,	O 11, '40:408-409 (-A-1)	1	C
		Mr '43:34-41	1	
	Aeroplane,	Mr 29, '39:391	4	
		F 6, '42:154 (-A-6)	1	
	Air Force,	Ja '44:12-16	4	
	Air Pict,	Ja '56:4-9 (Pt 1)	1	P, Pr†, S
		F '56:43-46 (Pt 2)	1	Pr
		Mr '56:79-82+ (Pt 3)	1	Pr
		Ap '56:113-115 (Pt 4)	1	Pr
		N '58:389-391+ (-C)	1	Pr
		D '69:476 (-A-4)	5	
	Air Prog,	Ap/My '63:72 (-D)		C
	Air Tech,	Jl '44:45-47+	1	
	Acft Eng,	Mr '42:64-68 (-A-6)	1	S*
		:68-70 (-A-2)	2	
		Ag '42:230-234	6 (Radio)	
		D '42:342-357	6 (Radio)	
	AAHSJ,	Win '67:288-295 (-A)	4	Pr
	Aviation,	Ag '42:196	6 (Gun)	S*
		D '42:179 (-A-1)		S*
		Je '43:187, 191		S*
		S '43:189		S*
		N '43:177, 179		S*
		F '45:200		S*
		Mr '45:191		S*
	Comm Avn,	Ap '42:28-34 (-A-1, -5&-6)	1	S*
		Je '42:28-32 (-A-6)	2	
	Flight,	O 12, '39:300a-c	1	S
		N 23, '39:416	1	
		F 29, '40:193	1	
		Ap 11, '40:325		D
		O 3, '40:270e-271	1	S*
		D 5, '40:468b-f	1	S*
		F 12, '42:138-141 (-A-6)	1	
		Jl 2, '42:12b	1	
	Flying,	Mr '41:66-68	1	
	Fly Rev Intl,	N '65:184-185 (German & Finnish-)	1	Pr*

JUNKERS (cont.)

1740.	Ju. 88 (cont.)		
	RAF Fly Rev, S '54:9	1	
	Ja '56:15 (-B)	1	
	Je '63:20-21+	4	
1741.	Ju. 88 "Mistel" or "Father and Son"		
	Carried MESSERSCHMITT Bf 109F as warhead		
	Fly Rev Intl, N '65:189	1	
1742.	Ju. 89		
	Flight, O 12, '39:300d-e	1	
	RAF Fly Rev, O '55:11	1	
1743.	Ju. 90		
	Developed into Ju. 290		
	Aeroplane, S 29, '37:391-392	1	
	Can Avn, Ja '38:22	1	
	Flight, S 23, '37:323	1	
	Ja 14, '43:40b	1	
	Flying, My '43:80	1	
	Fly Rev Intl, O '67:941 (Lufthansa's-)	1	
	RAF Fly Rev, F '63:43	1	
1744.	Ju. 188		
	Development of Ju. 88		
	Air Pict, Mr '56:79-80, 86	1	P
	RAF Fly Rev, Ja '56:15 (-S-1/U)	1	
1745.	Ju. 252		
	Development of Ju. 52		
	RAF Fly Rev, Jl '58:37-38	1	
1746.	Ju. 287		
	Flying, Ap '49:24-25+	1	P
	Fly Rev Intl, Ja '65:43-45	1	Pr*
	Interavia, No. 7, '50:380-382	2	
	RAF Fly Rev, D '55:49-50	1	
1747.	Ju. 288		
	Air Pict, Mr '56:81+	1	Pr
	RAF Fly Rev, Jl '55:10	1	
	Jl '60:45-46	1	D
1748.	Ju. 290		
	Development of Ju. 90		
	Air News, Ja '46:69	1	
	Flight, O 14, '43:420b	1	

JUNKERS (cont.)

1748. Ju. 290 (cont.)
Fly Rev Intl, Je '66:640-641+ 1
RAF Fly Rev, My '55:45-46 1

1749. Ju. 322 "Mammut" (Mammoth)
Troop glider
RAF Fly Rev, Ap '62:47 1

1750. Ju. 352 "Herkules" (Hercules)
Development of Ju. 252
RAF Fly Rev, Jl '58:37-38 1

1751. Ju. 388 "Störtebeker"
Development of Ju. 88 and Ju. 188
Air Pict, Ap '56:113-115 1 Pr
RAF Fly Rev, Mr '55:15 1
O '55:35-36 1

1752. Ju. 390
Development of Ju. 290
RAF Fly Rev, N '56:13 1

1753. Ju. 488
Air Pict, Ap '56:114-115 1 Pr

1754. K. 43
Reconnaissance bomber
Development of W. 34
Fly Rev Intl, D '68:72-73 1

1755. K. 47
Aero Digest, My '30:150 1
Flight, Mr 29, '34:311-312 1

1756. W. 33 and W. 34
Aeroplane, S 24, '43:355 1
Pop Avn, Je '28:13-16 (-L:"Bremen") 4
Jl '28:70-71 (-L:"Bremen") 1

JURCA France

1757. M.J. 7 and M.J. 77 "Gnatsum"
2/3 and 3/4-scale NORTH AMERICAN "Mustang"
Fly Rev Intl, D '69:29 1

KAISER-FLEETWINGS USA

1758. XBTK
Fly Rev Intl, N '66:217+ 1
RAF Fly Rev, S '61:40 1

KALININ Russia

1759. BICh-14
Fly Rev Intl, O '64:76 1
Ja '67:333 1
RAF Fly Rev, Ag '56:9 1

1760. K-7
Fly Rev Intl, N '66:217 1

1761. K-12 or BS-2 "Zhar-Ptitsa" (Firebird)
Fly Rev Intl, Ja '67:333 1

KAMAN USA
Nav Avn News, Ag '58:32 (US Naval aircraft)

1762. H-43 and HH-43 "Huskie"
Air Force designation
Civil designation: K. 600
US Navy designation: HOK-1
See also K. 1125
Avn Wk, Je 30, '58:36 D*
D 22, '58:71-75 (-B) 1 S*
F 23, '59:82-91 (-A) 4
My 18, '59:133 (-B) D*
Mr 7, '60:245 (-B) P*
Mr 5, '62:46-56 (-B) 4 C
My 3, '65:71-75 (-F) 4
Flying, Jl '61:30-31+ (-B) 4 C

1763. HOK-1
US Navy designation
Air Force designation: KAMAN H-43
Civil designation: K. 600
Am Helicop, D '56:26 2
Avn Wk, My 11, '53:15 1

1764. HTK-1
Civil designation: K. 240
Air Pict, Je '53:167 P
Am Helicop, Je '54:9 1
Avn Age, N '52:42-43 1 C, S*
Avn Wk, D 8, '52:20-22 3

KAMAN (cont.)

1765. HU2K
Later designation: UH-2
Avn Wk, Ja 20, '58:79 1
Jl 21, '58:36 D*
O 26, '59:92-103 (-1) 1 C

1766. K. 16
Air Prog, F '67:34-35 (-B) 1 P

1767. K. 17
Avn Wk, Ap 14, '58:73-75 1

1768. K. 18
Avn Wk, Ja 18, '60:65 1 S*

1769. K. 190
Am Helicop, S '48:6-7+ 1 C, S*
Avn Wk, Ja 31, '49:20-24 1 S*
Can Avn, Ja '50:40+ 1
Flight, Ag 26, '48:229-230 1 S*
US Air Svcs, Ag '48:16 1
Je '49:23 6 (Crop Dusting)

1770. K. 225
Development of K. 190
Avn Wk, Ag 18, '52:35-39 (Turbine-) 1

1771. K. 600
Civil designation
US Navy designation: HOK-1
Air Force designation: H-43 and HH-43
See also K. 1125
Am Helicop, Jl '56:7+ 1
My '57:10-11+ (-3&-4) 2
Avn Wk, Ap 15, '57:74-77 (-3&-4) 1
Ap 26, '57:537 (-4) 6 (Engine)
Flight, Jl 27, '56:173 1

1772. K. 700
Fly Rev Intl, My '69:31 1

1773. K. 800
Fly Rev Intl, My '69:31 1

KAMAN (cont.)

1774.	K. 1125 "Huskie III"		
	Aeroplane, O 11, '62:29	1	
	Avn Wk, O 8, '62:102-109	1	C, S*
	Boeing Mag, N '62:12-13	1	
1775.	UH-2 "Seasprite"		
	Earlier designation: HU2K		
	Air Prog, D '66:38-39 (-C)	1	S*
	Avn Wk, Ap 8, '63:80-88 (-A)	4	C
	F 8, '65:74-75	2	
	F 26, '68:50	4	
	F 17, '69:17 (-C)	1	P
	Fly Rev Intl, F '65:5		D*
	Ja '69:25-26 (-C)	1	
	Nav Avn News, Jl '67:41 (-A)		P
	D '67:14 (-C)	1	

KAMOV Russia

1776.	Ka-15 [Hen]		
	Avn Wk, Je 9, '58:51-54	6 (Rotor)	S*
1777.	Ka-20 [Harp]		
	Civil counterpart: Ka-25		
	Air Pict, N '62:352-353	1	
1778.	Ka-22 "Vintokryl" (Screw-Wing) [Hoop]		
	Air Pict, Ag '62:240-241	1	Pr
1779.	Ka-25 [Hormone]		
	Military counterpart: Ka-20		
	Fly Rev Intl, O '69:27-28	1	
1780.	Ka-26 [Hoodlum]		
	Avn Wk, Ap 20, '64:109	1	
	Ja 24, '66:109	1	
	F 6, '67:103	1	
	Mr 31, '69:90-91	1	P
	Fly Rev Intl, My '66:546	1	
	Ap '67:461-462	1	P*

KAWANISHI Japan

RAF Fly Rev, Ap '56:49-50 (WW II fighters)

1781.	E15K1 "Shiun" (Violet Cloud) [Norm]		
	RAF Fly Rev, Mr '56:13	1	

KAWANISHI (cont.)

1782.	H6K1-5 Type 97 [Mavis]		
	Flight, Ag 26, '43:232	1	
	RAF Fly Rev, Ag '60:16-17	4	
1783.	H8K1-4 Type 2 [Emily]		
	Air Pict, Ag '69:280-284	1	C
	Air Prog, Ja '67:46-47+	1	
	Nav Avn News, O '46:24	3	
	Ag '47:24	1	P
1784.	K. 12 "Sakura"		
	Air Pict, F '63:46-47	1	
1785.	N1K1-J "Shiden" (Violet Lightning) [George]		
	RAF Fly Rev, N '56:15	1	

KAWASAKI Japan

1786.	KDA-4 Type 88		
	Flight, Jl 6, '33:679	1	
1787.	KDA-5 Type 92		
	Flight, Ja 5, '33:13	1	
	Fly Rev Intl, N '69:79	1	
	RAF Fly Rev, Ja '61:40	1	
1788.	Ki. 32 Type 97 [Mary]		
	Flight, Jl 15, '43:71	1	
	RAF Fly Rev, Jl '63:51	1	
1789.	Ki. 45 "Toryu" (Dragon Killer) [Nick]		
	See also Ki. 102		
	RAF Fly Rev, N '59:41-42	1	
	S '62:22 (-KAIc:53 Ftr Sentai)		Pr*
	:23+	1	
1790.	Ki. 48 Type 99 [Lily]		
	RAF Fly Rev, D '60:39+	1	
1791.	Ki. 61 "Hien" (Swallow) [Tony]		
	RAF Fly Rev, S '55:65-66	1	
	Ja '62:42 (-2A:224 Ftr Sq)		Pr*
	:43+	1	
	Je '63:28-30	1	
	:28-29 (-IIa)		S*
	:31 (-IB)		Pr

KAWASAKI (cont.)

1792. Ki. 64 [Rob] or [Bob]
RAF Fly Rev, S '55:65-66 1

1793. Ki. 78
RAF Fly Rev, S '55:66 1
D '59:46 (-KEN III) 1

1794. Ki. 100 Type 5 [Tony 2]
Air Prog, O '65:55 (-1b) Pr†
Fly Rev Intl, D '63:57 1
RAF Fly Rev, N '58:41-42 1

1795. Ki. 102 [Randy]
Development of Ki. 45
RAF Fly Rev, Je '62:39-40 1

1796. Ki. 119
Air Pict, F '60:46-47+ 1

1797. P2V-Kai
Later designation: P-2J
Original design: LOCKHEED P2V
Fly Rev Intl, S '66:64 1

1798. T-33
See also LOCKHEED T-33
Avn Wk, F 20, '56:57-68 3

KAY GB

1799. "Gyroplane"
Aeroplane, D 26, '34:772-775 1 S*
Flight, D 27, '34:1377-1378 1 S*

KELLETT USA

1800. XG-1 and YG-1
Civil designation: KD-1A
RAF Fly Rev, Je '60:53 1

1801. XH-10 and XR-10
See also KH-2A
Am Helicop, Ag '47:26 (XR-) 1
Je '48:8 (XR-) 1
F '49:14+ (XH-) 1
S '49:13+ (XH-) 2

KELLETT (cont.)

1801.	XH-10 and XR-10 (cont.)		
	Avn Wk, Jl 14, '47:11-12 (XR-)	1	
	Interavia, Jl '49:425-427	1	P
1802.	K-2		
	Aviation, Jl '31:408-415	1	
	Pop Avn, Ag '31:13	1	
	D '31:109	1	
1803.	KD-1A		
	Military designation: XG-1 and YG-1		
	Aviation, D '35:30-32	1	S
	West Fly, D '34:16-17	1	P
1804.	KD-1B "Cropmaster"		
	Am Helicop, D '58:8	1	
	Mr '59:10+	1	
	Avn Wk, My 18, '59:144-145	1	
1805.	KH-2A		
	Civil designation of XH-10		
	Am Helicop, Mr '50:12	1	
1806.	KH-15 "Stable Mabel"		
	Am Helicop, S '55:8	1	
	West Avn, S '55:24	1	
1807.	XO-60 and YO-60		
	Development of XR-2 and YG-1		
	RAF Fly Rev, Je '60:53	1	
1808.	XR-8 "Synchropter"		
	Aero Digest, Jl 1, '45:88-90+	1	
	Am Helicop, Ja '46:36	1	
	Can Avn, Jl '45:106	1	
	Comm Avn, Jl '45:80	1	
	Flight, Jl 12, '45:35	1	
	West Fly, S '45:98	1	

KEYSTONE USA

1809.	"Commuter"		
	Aviation, Jl 5, '30:12-16	1	
1810.	PK-1		
	Aero Digest, O '31:64	1	

KFF (Kungl. Flygförvaltningens Flygverkstad)			Sweden
	See SWEDEN		
KIEBITZ			Germany
	See MERCKLE		
KINNER			USA
1811.	"Invader"		
	Pop Avn, Ap '36:269	1	
1812.	"Security Airster"		
	West Fly, Ag '33:17	1	
KLEMM			GB
1813.	"Swallow"		
	Aeroplane, F 21, '34:302	4	
	F 28, '34:348	4	
	:351-354	1	S
	Flight, Mr 1, '34:195-197	1	
	Pop Fly, Ap '34:18-19	4	
KOKUSAI			Japan
1814.	Ku. 7 "Manazuru" (Crane)		
	RAF Fly Rev, S '57:27+	1	
KONDOR			Germany
1815.	"Taube"		
	Cr & Cockade, Aut '67:314		P
KOOLHOVEN			Netherlands
1816.	F.K. 40		
	Flight, F 14, '30:203-204	1	C
1817.	F.K. 48		
	Aeroplane, Je 13, '34:937	1	
	Flight, Ja 17, '35:77-78	1	
1818.	F.K. 49		
	Flight, N 28, '35:553	1	

KOOLHOVEN (cont.)

1819. F.K. 50
B model was never built
Flight, O 3, '35:366 1
Ag 27, '36:222-223 (-B) 1
RAF Fly Rev, Ja '61:9 1

1820. F.K. 52
Aeroplane, Ag 3, '38:147-148 1
Fly Rev Intl, Ap '69:72 1

1821. F.K. 55
Pop Avn, Jl '37:27 1
RAF Fly Rev, F '59:11 1

1822. F.K. 56
Flight, Mr 10, '38:243 1

1823. F.K. 58
Flight, O 6, '38:298 1
RAF Fly Rev, Jl '57:39 1
O '57:9 1

KOVO Czech
See LET

KRAUSS Germany

1824. Light Autogyro (1969)
Fly Rev Intl, Ap '69:30-31 1

KYUSHU Japan

1825. J7W1 "Shinden" (Magnificent Lightning)
RAF Fly Rev, Jl '56:31-33 1 S*

1826. K10W1 Type 2 [Oak]
Air Pict, Mr '69:101 1

1827. K11W1 "Shiragiku" (Chrysanthemum)
RAF Fly Rev, Mr '62:47 1

LACAB (Les Ateliers de Constructions Aéronautiques Belges) Belgium

1828. G.R. 8
Pop Avn, Ag '37:29 1 I

LAGG Russia
See LAVOCHKIN

LAIRD USA
Air Prog, My '67:26-29+ (Pt 1)
Je '67:45-47+ (Pt 2)
Jl '67:37+ (Pt 3)
Ag '67:52-54 (Pt 4)
S '67:44-45 (Pt 5)

1829. "Solution" LC-DW 300
Air Prog, Ag '67:52 (1930 Thompson Trophy-) Pr†
S '67:44 (Dale Jackson's-) Pr†

1830. "Whippoorwill"
Pop Avn, Ap '28:25 P

1831. L-RT
Air Cl, Ag '69:70 (Turner's-) 1
Pop Avn, N '37:14-15 (Turner's-) 1 P

1832. 400
Pop Avn, N '33:291-292+ 4

LAISTER-KAUFFMAN USA

1833. XCG-10A "Trojan Horse"
Comm Avn, Ja '45:62-63 1
Fly Rev Intl, Mr '64:61 1
West Fly, Ja '45:44+ 1

LAKE USA

1834. "Skimmer"
Nav Avn News, S '64:24 1

LAMBERT USA
Pop Avn, O '36:27+ (Operations in 1936)

LANDGRAF USA

1835. H-2
Am Helicop, D '45:19 1
O '46:14-15+ 1
Aviation, O '44:186-189+ 1
N '44:205 S*
F '45:193 S*
Avn Wk, Mr 29, '48:24-25 1
West Fly, Ag '45:48-49 1 C

LANGLEY USA

1836. "Aerodrome"
Pop Avn, S '38:51-52+ 1
US Air Svcs, N '42:12-15 1

1837. "Aeronaut"
Airpwr Hist, Jl '59:173-188 1 D

LANGLEY AVIATION USA

1838. Monoplane (1941)
All plywood construction
Comm Avn, N '41:128+ 1
Flight, O 30, '41:296a 1
Ja 8, '42:28h 3
Flying, D '41:23+ 1

LANIER USA

1839. "Paraplane"
Flying, S '49:22-23+ 1
O '50:25+ 1

1840. X-L-1 and X-L-2
Pop Avn, Ja '31:21 1

LATÉCOÈRE France

1841. 28
Flight, My 16, '30:549-550 (-3) 4

1842. 298
Fly Rev Intl, Ag '68:460-462 1 Pr*
RAF Fly Rev, Ag '58:8+ 1

1843. 299
Air Pict, S '56:331+ 1

1844. 300
Flight, Ja 1, '32:11-13 1

1845. 521
Only one built, "Lt. de Vaisseau Paris"
Acft Eng, My '35:109-112 1
Flight, Ja 24, '35:93-94 1 C
Pegasus, Ja '45:7 I

LATÉCOÈRE (cont.)

1845. 521 (cont.)
Pop Avn, Mr '35:148 1
My '35:284 1
N '38:19 P
RAF Fly Rev, S '62:47 1

1846. 582
Acft Eng, My '36:134-135 1 S*

1847. 631
Aeronautics, Ja '48:46-49 1 P
Aeroplane, Ja 26, '45:114-115 1
Jl 25, '47:107-108 1
Air Pict, F '54:53 1
Flight, Jl 24, '47:87-88 (-04) 1
RAF Fly Rev, Mr '60:9 1

LAVOCHKIN Russia

Fly Rev Intl, Ap '68:220-222+ (Pt 1)(Last airplanes, w. photos)
My '68:289-291 (Pt 2) (Last airplanes, w. photos)
Je '68:349-351 (Pt 3) (Last airplanes, w. photos)

1848. La-5
RAF Fly Rev, Je '57:13 1

1849. La-7 [Fin]
Aero Album, Spg '68:22-27 1 C*, P, Pr†
Fly Rev Intl, Ag '66:774 (Kozedub's-) Pr*
RAF Fly Rev, Je '58:8-9 1
:9 (Kozedub's-) Pr*

1850. La-9 [Fritz]
Flight, F 17, '49:193 1 D

851. La-11 [Fang]
Flight, Jl 14, '49:48-49 1
Fly Rev Intl, Ag '66:774 Pr*

852. La-15 [Fantail]
Often incorrectly referred to as La-17
RAF Fly Rev, Jl '54:13 1

LAVOCHKIN (cont.)

1853.	La-17		
	Avn Age, My '51:21-22	1	Pr, S*
	Skyways, S '51:10-11+	1	
1854.	La-156		
	Fly Rev Intl, Ap '68:222	1	P
1855.	La-160		
	Later development: La-15		
	RAF Fly Rev, Ag '62:41	1	
1856.	LaGG-3		
	Aeroplane, Ag 13, '43:175	1	
	Acft Eng, O '43:289-292	1	S*
	Flight, Ag 12, '43:178-180	1	S*
	S 23, '43:341	1	
	RAF Fly Rev, D '56:13	1	
	S '59:47-48 (Japanese captured-)	1	

LEARJET USA

1857.	"Twinjet Helicopter"		
	Avn Wk, S 15, '69:34-35	1	C
	S 22, '69:80-84+	1	C
	Flight, S 25, '69:488-489	1	
	Fly Rev Intl, N '69:27-28	1	
	Interavia, No. 11, '69:1784-1785	1	
1858.	23		
	Flying, Je '62:40-41+	1	Pr, S
	Ap '64:34-37+	4	
	Fly Rev Intl, Ap '65:27-30	1	S*
	Interavia, No. 5, '61:676-677	1	
	No. 8, '64:1057	1	
	No. 10, '65:1534-1536	4	C
1859.	24		
	Air Prog, O '66:41-43	1	P
	O '67:48-51+	4	S
	Fly Rev Intl, D '65:229	1	
	Ap '69:27-28 (-B)	1	
	D '69:29-30	1	
	Interavia, No. 11, '69:1782-1783 (-C&-D)	1	

LEARJET (cont.)

1860.	25		
	Air Prog, F '69:50-51 (Stretched-)	1	S
	Avn Wk, N 4, '68:52-65	1	C
	Flight, Je 19, '69:1004	1	
	Flying, Ja '69:82	1	
	F '69:32-37	4	
	Fly Rev Intl, D '69:29-30	1	
	Interavia, No. 12, '68:1547-1549	4	C, S
	No. 11, '69:1782-1783 (-C)	1	
1861.	28 "Lear Liner"		
	Avn Wk, Je 29, '64:29-30	1	
	Flight, Je 25, '64:1049-1050	1	S*
	Interavia, No. 8, '64:1057	1	
1862.	40 "Lear Liner"		
	Aeroplane, O 13, '66:25-26	1	
	Avn Wk, My 30, '66:60-65	1	
	Flight, O 21, '65:681	1	
	My 26, '66:867	1	
	Interavia, No. 7, '66:1038	1	

LEDUC — France

1863.	0.10		
	Flying, Mr '50:22-23+	1	
	Interavia, No. 6, '52:341	1	
	No. 7, '53:400-403	1	C, P
	Skyways, Ag '52:16-17+	1	
1864.	0.21		
	Air Pict, Je '52:165	1	D
	Air Power, Aut '55:18		P
	Interavia, No. 7, '53:400-403	1	P
	RAF Fly Rev, Mr '54:15-16	1	
	My '55:30-33	1	Pr, S*
	Skyways, Ag '52:16-17+	1	D
1865.	0.22		
	Avn Wk, Ja 21, '57:32-34	1	
	N 4, '57:50-63	1	
	RAF Fly Rev, Ap '55:21	1	D
	S '63:39	1	
1866.	RL-21		
	Air Prog, Mr '68:42-43	1	S*

LEE-RICHARDS GB

1867. Annular-Wing
Aeronautics, S '57:174-177 1

LEPERE USA

1868. Packard-Lepere Biplane (1921)
AAHSJ, Spg '69:10 C

LET (Letecké Zavody Narodni Podnik) Czech

1869. "Aero 45"
Aeroplane, Ag 12, '49:193-194 4

LETOV Czech

1870. S 32
Flight, Ja 8, '32:36-37 1

1871. S 50
RAF Fly Rev, Ja '63:47 1

1872. S 231
Flight, Ja 24, '35:97 1
RAF Qtr, Ja '35:20-21 6 (Guns) P

1873. S 328
Fly Rev Intl, N '66:214-216 1 Pr*

LEVASSEUR France

1874. P. L. 14
Pop Avn, My '33:293 1

LFG (Luftfahrzeug Gesellschaft) Germany
See ROLAND

LINCOLN USA

1875. "Special"
Pop Avn, Je '37:21-22 1

1876. "Sport"
Air Prog, Ag '67:50-51 1 S*

LING-TEMCO-VOUGHT USA

1877. A-7 "Corsair II"
Aerosp Intl, Mr/Ap '68:21-29 (-D) 1
Air Force, F '68:32-39 (-D) 1
Ap '68:72 (-A) P*
AF/Sp D Intl, Mr '66:34-35 (-A) 1
Air Pict, Mr '66:96 (-A) 1
Air Prog, Ja '68:38-39+ (-A) 1
Ag '68:41-43+ (-D) 1 P*
Avn Wk, My 3, '65:99 (-A) D*
N 8, '65:30-33 (-A) 1
N 22, '65:62-63 (-A) 1
Mr 7, '66:67 (-A) P*
Jl 18, '66:74-76 (-A) 1
Ag 28, '67:123 (-A&-D) 1 S
S 18, '67:79-91 (-A) 3
F 19, '68:24 (-B) 2
Jl 1, '68:20-21 (-D) 6 (Armament)
N 11, '68:84-87 4
Mid-D, '68:139 P*
Ja 6, '69:93-95+ (-D&-E) 6 (Nav/Weapon Delivery)
Ag 18, '69:75 (-D) 6 (Brakes)
Flight, Je 18, '64:1025 (-A) 1
N 25, '65:943 (-A) 1
S 21, '67:495-497 1
Fly Rev Intl, Ja '66:265-266 1 P*
Ja '68:7-10 (-D) 1
Interavia, No. 9, '64:1370-1371 (-A) 1 S*
No. 10, '65:1576 (-A) 3
No. 12, '65:1956 (-A) 1
No. 1, '66:89-91 (-A) 1 S*
No. 6, '69:740-743 1 S*
Nav Avn News, S '64:6-9 (-A) 1
Ap '65:13 (-A) D
D '65:20-21 (-A) 1
F '67:14-15 (-A) 2
Jl '67:12 (-A) 4
My '68:6-9 4
Jl '68:7-9 (-A) 3
Mr '69:31 (-E) 6 (Attack)

1878. F-8 "Crusader"
Air Prog, Ap '69:40-41+ 1
:42-43 (-H) P*
Avn Wk, F 26, '68:23 (-J) 2
F 17, '69:22-23 (-A) 6 (Transonic Wing)

LING-TEMCO-VOUGHT (cont.)

1878. F-8 (cont.)

Nav Avn News, Mr '65:10-11 (-A)	4		
Je '66:35 (-C)	6 (Direct Lift Control)		

1879. XC-142A

Earlier version: VOUGHT-HILLER-RYAN XC-142

Air Prog, Ap '65:27-29	1	C
Avn Wk, Je 22, '64:16-19	1	
F 8, '65:18-19	1	
F 15, '65:27-28	4	
My 30, '66:28-29	2	
Bee-Hive, Sum '64:1		P
Fly Rev Intl, N '64:24-26	1	S*
Interavia, No. 1, '66:52-53	2	
Nav Avn News, Mr '65:24-25	2	P
Jl '66:20-21	2	
US Army Avn, S '64:9-11	1	

LIORE-ET-OLIVIER France

1880. CL 10

Flight, D 29, '32:1233	4	S

1881. H-27

Acft Eng, My '34:121-122	1

1882. H-242

Acft Eng, My '34:121-122	1

1883. H-246

Acft Eng, Ap '38:120	1

1884. LeO-30

Flight, F 15, '34:155	1

1885. LeO-45, LeO-451, LeO-453, and LeO-455

Air Pict, Je '54:181	1	
Flight, My 16, '40:452c-453	1	D
Fly Rev Intl, Jl '65:65-67 (Pt 1)	1	Pr*
Ag '65:39 (Pt 2)	1	

1886. LeO-208

Acft Eng, O '35:257-258	1
Flight, Ap 11, '35:394	1

LIPPISCH Germany

1887. Tailless Monoplanes
Aeroplane, N 18, '31:1180 1
Acft Eng, N '31:274 1
Flight, O 9, '31:1008-1009 1
Pop Avn, Ja '32:32+ 1
RAF Fly Rev, Ap '56:13 1

1888. DM-3
Air Pict, O '68:385 1

LOCKHEED USA
Flying, Mr '60:24-27+ (Pt 1)
Ap '60:44-48+ (Pt 2)
My '60:42-46+ (Pt 3)
Interavia, No. 6, '57:582-583 (25th Anniversary)
Nav Avn News, Jl '57:12-15 (US Naval aircraft)
Pop Avn, O '37:16-18+
West Fly, F '34:18-20

1889. "Air Express"
Air Prog, Jl '66:24-26+ 1
AAHSJ, Spg '64:28-39 1 P
:38 ("Gilmore Lion") Pr†
Pop Avn, Ap '29:15-16 ("Texaco 5")4

1890. "Alcor"
Aeroplane, Je 20, '34:957-959 1
Flight, Jl 5, '34:693 1

1891. "Altair"
Aeroplane, O 16, '35:462 1 S*
Pop Avn, Ja '35:10 ("Lady Southern Cross") 4
Pop Fly, Ja '37:508 1
West Fly, D '34:22 ("Lady Southern Cross") 4

1892. "Constellation" L-049 to L-749
See also "Super Constellation"
Military versions: C-69, C-121, PO-1W
Aero Digest, Je '51:26-29+ 1
Air Cl, Je '69:26-43 1 P*
Air Pict, Ag '54:246 ("Old 1961") 1
Air Pilot, Jl '44:30-31, Cover 1 C*
Air Prog, Je '69:35-37+ (-749) 4 C
Acft Eng, N '50:340-341 1

LOCKHEED (cont.)

1892.	"Constellation" (cont.)			
	Can Avn, Mr '43:56	1		
	Mr '46:32-34+	1		
	Ap '46:76-78+	6	(Cntrl Boost)	C
	Comm Avn, F '43:110-118	1		
	Ag '45:44-50	3		
	Flight, F 25, '43:198-201	1		
	O 14, '43:420a	1		
	Jl 11, '46:29-32 (Pt 1)	4		
	Jl 18, '46:55-58 (Pt 2)	4		C
	Je 26, '47:600	1		
	Flying, Jl '44:67-68+	1		
	My '45:67-70	1		C, S
	Interavia, No. 1, '46:38-44	1		C
	O '48:554			C
	Skyways, Mr '43:34-35	1		
	O '44:20-21+	1		
	US Air Svcs, F '43:16-17+	1		
	Mr '47:22	2		
	West Avn, Ag '54:12 ("Old 1961")	1		
	West Fly, F '43:40+	1		
	O '45:50-51+	6	(Engine)	S
	Mr '48:15+	6	(Lndg Gear)	S*
1893.	"Constitution"			
	Aeroplane, Je 20, '47:652-653	1		
	Air Prog, F '67:31-33+	1		C
	Aviation, O '46:86	1		
	Avn Age, F '52:36-37	1		S*
	Avn Wk, Mr 1, '48:20-21	1		
	Mr 8, '48:29	2		
	Ag 30, '48:20-27	1		
	Can Avn, O '46:100	1		
	Flight, S 5, '46:248	1		
	Flying, D '46:51-53	1		
	Interavia, No. 6, '46:20-27	1		P, S
	O '48:555			C
	Ap '49:191-193	4		
	Nav Avn News, O '46:11	1		
	Ja '47:9	2		
	S '47:16	1		
	My '48:23	1		C
	Ap '49:20-21	4		C
	Skyways, Ja '49:24-25	1		C
	US Air Svcs, D '46:25	2		
	Je '48:15	2		
	N '48:15	2		

LOCKHEED (cont.)

1893.	"Constitution" (cont.)		
	West Fly, O '46:70	1	
	Je '51:16	1	
1894.	"Excalibur"		
	Pop Avn, N '39:33-34+	1	
1895.	"Hudson"		
	Civil designation: 14 "Super Electra"		
	US military designation: AT-18		
	Air BP, No. 22:25-28	4	C, P
	Can Avn, Ag '40:56	1	
	CAHS, Win '69:116-120 (RCAF-)	1	
	:121 (-I to -VI:RCAF)		Pr†
	Comm Avn, Jl '39:30	1	
	Mr '42:76-80	4	
	N '43:138+	4	
	Flight, Jl 9, '42:40b (-V)	1	
	My 16, '46:492a-493	4	
	Flying, F '42:20-23+	1	
	Pop Avn, Ap '39:19	1	S
	Roundel, Ja/F '63:33 (Canadian-)	4	P
	RAF Fly Rev, D '60:26-28	4	
	West Fly, My '42:28-31+	4	
1896.	"Jet Star"		
	Military designation: C-140		
	Aeroplane, N 29, '57:815-817	1	C
	Ja 16, '59:72-74	1	C
	Avn Wk, S 30, '57:108-113	1	
	My 5, '58:78-99	1	C
	Ja 8, '62:67-74 (Pt 1)	4	C
	Ja 15, '62:70-75 (Pt 2)	4	
	D 16, '68:31-33	4	C
	Bee-Hive, Fall '60:1	1	P
	Flight, S 20, '57:490	1	S*
	Mr 29, '62:484-485	1	C
	Mr 19, '64:424-425	1	C
	Ag 20, '64:297-299	4	
	Flying, O '57:38-39	1	
	O '58:32-34+	4	
	Mr '61:24-27+	1	
	Mr '65:94-95	1	
	Ap '68:60-64	4	
	Fly Rev Intl, Jl '69:50-51 (TWA-)	4	
	Interavia, No. 7, '58:711-715	1	C, S
	RAF Fly Rev, Ja '63:30-33 (C-140A & C-140B)	1	Pr, S*

LOCKHEED (cont.)

1897.	"Orion"		
	See also P-3 and P3V-1		
	Flight, My 13, '32:427	4	
	Pop Avn, Je '32:365+	4	
	F '34:92	1	
1898.	"Saturn"		
	Air Force, S '46:28-29	1	S
	Aviation, Ja '45:169	1	
	Ag '46:93	1	
	Can Avn, S '46:85	1	
	Comm Avn, Ja '45:45-46	1	S
	Interavia, No. 6, '46:41-47	1	C
	Skyways, F '45:92	1	
	West Fly, D '44:76	1	
	Ag '46:56	1	
1899.	"Sirius" Landplane		
	Aero Digest, Jl '31:60-64	1	S*
	Ja '34:14		P
	AAHSJ, Win '65:266-280	1	
	:275 (Fierro's "Anahuac"-)		Pr†
	Aviation, O '30:232-240	1	
1900.	"Sirius" Seaplane		
	Aeroplane, Ag 19, '31:468		S*
	Acft Eng, S '31:225-226	6 (Floats)	
	AAHSJ, Win '65:269-271 ("Tingmissartoq"-)	1	
	Aviation, F '34:33-36 ("Tingmissartoq"-) (Pt 1)	4	C, S*
	Ap '34:102-105 ("Tingmissartoq"-) (Pt 2)	4	
	Flight, Mr 1, '34:187-188 ("Tingmissartoq"-)	4	S*
	US Air Svcs, S '33:4 ("Tingmissartoq"-)	4	P
1901.	"Super Constellation" L-1049		
	Aeroplane, Ag 28, '53:256-259 (-E)	1	
	Air Pict, N '51:326, 337	1	P
	Air Prog, '52 Annual:38-39 (-C)		S*
	Avn Age, Je '51:32-33	1	
	O '53:30-31 (-B)	1	S*
	S '55:32-33 (-G)	6 (Flap Actuator)	S*
	Avn Wk, D 24, '51:20-25 (-B-55)	1	
	N 3, '52:85-86 (-C)	1	

LOCKHEED (cont.)

1901. "Super Constellation" (cont.)
Flying, Je '53:28-29+ 1 C
Interavia, No. 11, '51:628-629 (-C) 1
No. 2, '54:97-101 3
Skyways, Ja '54:12-13+ (-C:KLM) 1
My '55:8-10 (-G) 6 (Radar)
West Avn, N '51:13 (-B) 3
O '52:33-36+ (-B) 6 (Cargo)
West Fly, Mr '51:24 1

1902. "Super Constellation" L-1249
Air BP, No. 8:12 (Trans-Canada-) Pr*
Avn Wk, S 21, '53:92-93 (-B) 1
Can Avn, Ap '54:26-27+ 1

1903. "Super Constellation" L-1449
Avn Wk, D 6, '54:17 1
D 27, '54:16-18 1

1904. "Super Constellation" and "Starliner" L-1649A
Air Cl, Je '69:36 S*
Avn Age, O '55:26-31 1
Avn Wk, O 10, '55:35-42 6 (Wing) S*
Flight, Jl 6, '56:55-58 1 S*
Je 14, '57:800 4
Interavia, No. 11, '56:878 1
West Avn, Mr '56:20 6 (Wing)

1905. "Vega"
Air Prog, F/Mr '63:35-37+ ("Century of Progress"-) 4 Pr
N '68:56-57+ (Last flyable-) 1 C
AAHSJ, Ja/Mr '57:1-35 1
Jl/S '58:172-176 ("Century of Progress"-) 4
Fall '69:199-201 ("Stanavo"-) 1 P
Bee-Hive, Ag '31:1-6+ ("Winnie Mae"-) 4
S '33:1-5 ("Winnie Mae"-) 4
:8 ("Winnie Mae"-) 6 (Engine)
Spg '61:24-28 ("Winnie Mae"-) 4
Flight, Ap 3, '31:288-289 1
My 27, '32:469 (Earhart's "Friendship"-) 4
Jl 27, '33:750 ("Winnie Mae"-) 4
Flying, S '65:74-79 1 C, P*
Hist Avn, Mr '69:18-21 ("Winnie Mae"-) 1

LOCKHEED (cont.)

1905.	"Vega" (cont.)			
	US Air Svcs, Ag '33:20-23 ("Winnie Mae"-)	4		P
	F '35:19-20 (Earhart's-)	4		
1906.	"Ventura"			
	Derived from 18 "Lodestar"			
	USAAF designation: B-34 "Lexington"			
	US Navy designation: PV-1			
	Flight, D 31, '42:716a	1		
	Ja 21, '43:76	1		
	My 16, '46:492a-493	4		
	Flying, D '43:72-73+	4		
	Roundel, O '64:33 (RCAF-)	1		
1907.	A-11			
	See also YF-12A and SR-71			
	Air Force, Ap '64:33-37+	1		
	Avn Wk, Mr 9, '64:16-19	1		
	Flight, Mr 12, '64:377-379	1		S*
	Fly Rev Intl, My '64:6-7+	1		
	Interavia, No. 4, '64:421-422	1		
1908.	AH-56A "Cheyenne"			
	Development of XH-51A			
	Air Force, F '68:52-55	1		
	Air Prog, Jl '67:44-45	1		
	Ap '68:24, 58-59	1		C
	Avn Wk, My 8, '67:25-26	1		
	My 15, '67:28-31	1		
	Je 5, '67:68-69	1		C
	O 30, '67:50-55	6	(Armament)	S
	D 18, '67:20-21	1		
	:56-57	3		
	D 25, '67:64-66	1		C
	Ja 15, '68:40-41	1		
	Mr 24, '69:23	2		
	Ap 21, '69:18-19	2		
	My 26, '69:24-26	3		
	Flight, Je 8, '67:958-959	1		
	Je 12, '69:979			C*
	Fly Rev Intl, Jl '67:715+	1		
	D '68:37			P*
	Interavia, No. 10, '67:1562-1563	1		
	US Army Avn, Jl '67:64-65	1		
	Ja '68:2-4	1		
	D '68:42-50	3		

LOCKHEED (cont.)

1909. AT-18 "Hudson"
Civil designation: 14 "Super Electra"
RAF designation: "Hudson"
Can Avn F '43:112 1
Flight, Mr 18, '43:283 1
West Fly, F '43:112 1

1910. B-34 "Lexington"
Lend-Lease designation of "Ventura"

1911. C-5A "Galaxy"
Civil version: L-500
Aeroplane, D 9, '65:26-28 1
Ja 31, '68:18-21 1
Aerosp Intl, S/O '67:24-25 1
Mr/Ap '68:42-43 1
Air Force, D '65:33-40 1 D
O '67:70-72 1
Ap '68:64-80 1
:81-83, 102-104 2
:84-85 6 (Engine)
:86-87 6 (Tips)
:88-93 6 (Nav & Radar)
:94-98 4
:99-100 3
O '68:54-60 4
Ap '69:79-80 1
S '69:51-55 1
AF/Sp D Intl, Mr '66:28-33 1 D
Air Prog, Ja '66:41-44+ 1 S
Je '68:42-45 1 C
O '68:84 C
Acft Eng, Je '68:8-21 1
Avn Wk, Ap 26, '65:52-62 2
My 3, '65:22-23 6 (Engine)
Je 28, '65:50-57 1
O 11, '65:26-28 1
D 5, '66:52-58 6 (Navigation)
Ag 14, '67:40-41 3
N 20, '67:164-166, 228-251 3
:167-187 1
:192-202 6 (Avionics)
:207-211 6 (Engine)
Mr 11, '68:20-21 1
Mr 25, '68:40-41 3
Jl 1, '68:24-27 3

LOCKHEED (cont.)

1911.	C-5A (cont.)			
	Avn Wk, Jl 1, '68:62-66	2		
	Jl 8, '68:22-25	2		
	Jl 15, '68:17	2		
	Mr 3, '69:40-41	2		
	Jl 21, '69:41-49	3		
	S 8, '69:59-63	1		P
	Flight, Jl 15, '65:93-95	1		
	F 10, '66:223-228	1		S*
	Je 30, '66:1116-1117	1		
	Mr 14, '68:390-391	1		
	Ag 22, '68:291-293+	6	(Navigation)	
	S 26, '68:510	2		
	Flying, N '68:46-47			C
	Fly Rev Intl, Je '66:607-611	1		
	S '68:74-79	1		P*
	Interavia, No. 3, '65:424	1		
	No. 11, '65:1686-1687	1		
	No. 8, '67:1262-1265	1		C, S*
	No. 5, '68:578-580	1		
	No. 9, '68:1094-1095	2		P
	Sp/Aeron, Mr '65:40-45	1		
	My '69:88-90	6	(Brakes)	
1912.	XC-35			
	Special variant of 10E "Electra"			
	Aero Digest, F '38:54-56	1		
	Pop Avn, N '37:49	1		C*
	Jl '38:54-56+	4		C*
	US Air Svcs, O '38:17	4		P
	West Fly, F '38:12	1		C
1913.	C-56 and C-57 "Lodestar"			
	Civil designation: 18 "Lodestar"			
	Military variant: C-60			
	Flight, D 2, '43:614	1		
1914.	C-60 "Lodestar"			
	Civil designation: 18 "Lodestar"			
	Military variants, C-56 and C-57			
	Aero Digest, Ap 15, '44:79+	4		
1915.	C-69 "Constellation"			
	Civil designation: "Constellation," L-049			
	Air Cl, Je '69:31, 33-35, 42-43	1		C, P
	Air News, F '43:20-22+	1		
	D '45:47-49+	4		C*

LOCKHEED (cont.)

1915.	C-69 (cont.)		
	Avn Wk, S 21, '53:29-34 (X-:"Beast of Burbank")	1	
	Flying, Mr '43:35+	1	
	Skyways, O '45:34-41+	1	C, P*, S*
1916.	C-121 "Super Constellation" and "Warning Star"		
	Improved C-69		
	Civil designation: "Constellation," L-749		
	US Navy version: WV-2		
	Avn Wk, Ag 24, '53:17-18 (R-/-C)	1	
	Skyways, F '52:26-27+ (V-:"Columbine II")	4	
	West Avn, S '53:18 (R-/-C)	1	
1917.	C-130 "Hercules" or "Herky Bird"		
	Civil version: L-100		
	Aeroplane, Ag 20, '54:253-254 (Y-)	1	
	My 20, '56:16-22	1	
	Ap 5, '67:4-7 (-K)	4	
	Air Pict, N '67:398-399 (-K)	4	
	Air Prog, Ap '65:12-17+	4	C
	F '68:33-36+ (H-/-H)	1	S*
	S '69:48-51+	1	
	Avn Age, N '55:30-31 (-A)	1	S*
	Avn Wk, S 6, '54:26-28 (-A)	3	
	Jl 30, '56:72	2	
	D 3, '56:50-67 (-A)	1	
	Ag 12, '57:37 (-B)	1	
	My 19, '58:60-77 (-A& -B)	1	C
	D 18, '61:83-86	6	(Engine Inst)
	Je 4, '62:87-89 (J-/-A)	4	
	Can Avn, N '64:27+ (-B)	1	
	Flight, Ag 13, '54:205 (Proto-)	1	
	D 10, '54:842-844 (-A)	1	
	S 14, '56:491-494 (-A)	1	S*
	Je 11, '64:985	1	
	My 6, '65:705-711+ (-E)	4	C
	Jl 6, '67:22-23 (-K)	1	
	O 5, '67:585-586 (-K)	1	
	Flying, Ja '58:28-30+	1	
	Interavia, No. 2, '57:125-127 (-A)	1	
	Nav Avn News, Mr '64:24-26	4	
	N '64:20-22	2	
	Roundel, My '60:27 (-B:RCAF)	1	D
	Mr '61:8-10 (-B:RCAF)	1	
	My '64:12-15 (-B:RCAF)	4	
	Ja/F '65:2-3 (-E:RCAF)	1	

LOCKHEED (cont.)

1917. C-130 (cont.)

RAF Fly Rev, My '57:34-37 (-A)	1		Pr, S*
Jl '63:42-43 (-E)	1		P*
US Air Svcs, Ag '55:21	1		
D '56:17	2		
West Avn, S '54:13 (Y-)	1		

1918. C-140

See LOCKHEED "Jetstar"

1919. C-141 "Starlifter"

Civil version: L-300

Aeroplane, Ja 17, '63:9-10	1		
D 9, '65:20-26	1		
Air Pict, Ap '67:107-110	1		C
:110-111	4		
Air Prog, Jl '69:27+	4		
Avn Wk, O 9, '61:50-65	1		
Ag 26, '63:30-33	3		C
S 2, '63:68-77	1		C
Ag 31, '64:38-43	2		
Ap 3, '67:80-93	6	(Landing)	C*
Bee-Hive, Ja '64:37	1		P
Spg '65:14-17	2		
Can Avn, Mr '67:30-31	4		P
Flight, N 16, '61:762-763	1		
S 20, '62:485-487+ (-A)	1		
S 12, '63:462-465 (-A)	1		S
Je 11, '64:979-985 (-A)	1		C
N 11, '65:804-805 (Super-)	2		
Flying, D '62:40+	1		C
Fly Rev Intl, Ja '64:33-35+ (-A)	1		Pr, S*
Interavia, No. 1, '62:87-89	1		
No. 9, '62:1145-1147	1		
No. 11, '62:1488-1490	6	(Nav Comp)	C
No. 3, '63:312-313	3		
No. 12, '65:1912-1913	6	(Navigation)	C
West Avn, S '63:19	1		

1920. CL 289

Avn Wk, F 20, '67:58-59	2		

1921. CL 475

Avn Wk, Jl 31, '61:26-27	1		
Flight, Mr 1, '62:323-325	2		
Interavia, No. 6, '62:727-729	2		

LOCKHEED (cont.)

1922.	CL 760 Counterinsurgency (COIN) aircraft		
	Avn Wk, My 25, '64:74-82	1	C
1923.	CL 840		
	Lock Hor, Sum '65:43		D*
1924.	F-5 "Lightning" Photo-reconnaissance version of P-38		
	Flying, N '43:54-55 (-A)	4	
	Fly Rev Intl, Ap '68:224, 227 (-A, -B, -E, -F)		Pr*
1925.	YF-12A See also A-11 and SR-71		
	Air Force, N '64:46-47	1	
	Je '65:50-51	1	
	S '69:34+	1	
	AF/Sp D Intl, Jl '65:36-37	2	
	Air Prog, Ja '65:50-51	1	
	O '65:60-62	1	
	Avn Wk, O 5, '64:16-17	1	
	O 12, '64:30-33	1	
	O 19, '64:54-63	1	
	Mr 15, '65:181		P*
	Ag 11, '69:65-68	1	S*
	D 22, '69:22	2	
	Bee-Hive, Fall '64:10-12	1	
	Flight, O 15, '64:667-669	1	
	Fly Rev Intl, Ja '65:16-21	1	Pr
	Interavia, No. 12, '64:1806-1807	1	
1926.	F-80 "Shooting Star" Earlier designation: P-80		
	Aeroplane, My 7, '48:526-527 (T-/-C)	1	
	O 21, '49:557 (T-&F-)		S*
	Airpwr Hist, Jl '62:180-183	1	
	Avn Age, D '50:17-18	4	
	Avn Wk, Ag 2, '48:13 (T-/-C)	1	
	Flight, My 6, '48:495 (T-/-C)	1	
	Ag 17, '50:197	4	
	Flying, D '48:33-35+ (-A)	1	C, S*
	Interavia, My '48:276-277 (T-/-C)	1	P
	N '48:610-613 (T-/-C)	4	C, P
	Skyways, Jl '48:30-31 (T-/-C)	1	
	Ja '49:30-32+ (T-/-C)	4	

LOCKHEED (cont.)

1926.	F-80 (cont.)		
	US Air Svcs, My '48:25 (T-/-C)	1	
	Jl '48:21 (T-/-C)		D
	Jl '53:18 (Q-)	1	
	D '54:22 (-R)	4	
	West Fly, My '48:8 (T-/-C)	1	
	O '48:13+ (T-/-C)	1	
	O '49:22+	1	
1927.	XF-90		
	Only two built		
	Aeroplane, Jl 1, '49:18-21	1	
	Avn Wk, My 23, '49:12-13	1	
	D 25, '50:27-29	1	
	Flying, Je '50:14-15+	1	
	Nav Avn News, S '49:1		P
	F '50:12		P
	RAF Fly Rev, Ja '57:13	1	
	S '60:21+	1	
	Skyways, Ag '49:30-31	1	
	My '50:14-15+	4	
	West Fly, Je '49:9	1	
	Ag '50:22	2	
1928.	F-94 "Starfire"		
	Development of T-33		
	Aero Digest, Ag '51:20-23+ (-C)	1	S*
	N '54:32-33 (-C)		S*
	Air Pict, Ap '56:126-127 (-C)		S*
	Avn Wk, Je 26, '50:13 (-A)		S*
	Je 25, '51:20-23	1	
	Jl 7, '52:16-17 (-C)	1	
	:61-64	4	
	N 23, '53:19 (-C)	1	S*
	Ag 16, '54:380-381 (-C)	1	
	Bee-Hive, Sum '52:31-32	1	P
	Flight, Jl 11, '52:46 (-C)	1	
	Flying, Ag '51:22-23+ (-A to -C)	1	
	O '52:29 (-C)	1	
	F '54:37 (-C)	1	S*
	RAF Fly Rev, O '62:57+ (To -D)	1	
	Skyways, O '49:19	1	
	Ap '53:10-11+	4	P
	West Avn, N '51:12	3	
	Jl '52:11 (-C)	1	
	D '53:26 (-C)	1	S*
	West Fly, O '50:18+	1	

LOCKHEED (cont.)

1929. F-104 "Starfighter"

See also Canadian-licensed CANADAIR CF-104

Aeroplane,	Ap 20, '56:284-285	1	
	Je 20, '58:841	4	
	My 18, '61:555-558 (-G)	1	S*
Air Force,	Ag '60:60-64	4	
Air Pict,	Jl '56:232-233	1	
	Ag '58:277-279	1	Pr
Air Prog,	F/Mr '65:17-23+ (-D)	4	C*, S*
Avn Age,	Jl '56:26-31	1	
	My '58:60-67	6 (Engine Intake)	
Avn Wk,	Jl 12, '54:15 (X-)	1	
	Ag 9, '54:21-24 (X-)	1	
	D 3, '56:32-33 (X-)	1	C
	F 18, '57:81 (-B)	1	
	F 10, '58:66-84 (-A)	1	
	O 27, '58:73-85	1	
	Je 15, '59:71-72 (-9)	1	
	N 16, '59:34 (-G)	1	
	Je 20, '60:167-173 (-G)	1	
	O 1, '62:74-75 (-G)	3	
	N 12, '62:90-91 (-J)	3	
	F 11, '63:72-79 (T-/-G)	4	
	Ag 9, '65:80-85 (N-/-A)	1	
Can Avn,	Jl '56:45-46	4	
	Ag '59:20-23	1	
Flight,	Ap 20, '56:440-443 (-A)	1	
	N 23, '56:824	1	
	My 30, '58:739-743 (-A)	1	S*
	O 19, '61:622-628 (-G)	1	S*
	Mr 14, '63:368-377 (-G)	3	
	Je 17, '65:979-983+ (-G)	4	P
Flying,	Ja '56:34-35+	1	
Fly Rev Intl,	Je '64:21-25 (-G)	1	
	O '65:120, 122 (-G&C-)		Pr*
Interavia,	No. 5, '56:360-361 (-A)	1	
	No. 6, '56:439 (-A)	4	
	No. 4, '58:326-327	6 (Grnd Support)	
	No. 1, '59:43-46	1	
	No. 4, '60:483 (-G)	1	
	No. 1, '61:81-82 (-VTOL)	1	
	No. 1, '63:88-90 (-G)	6 (Radar)	
	:104 (-G)	1	P
	No. 8, '63:1192-1195 (-G)	3	
	:1196-1200 (-G)	1	C, S*
	No. 12, '67:1917-1920	1	

LOCKHEED (cont.)

1929. F-104 (cont.)
Lock Hor, Sum '65:20 (N-) P*
RAF Fly Rev, Ap '56:17-18+ (-A) 1
F '57:31-32 (-A) 4 S*
RAFs Qtr, Vol. 1, No. 2:100 (-G) D
Sp/Aeron, N '58:132-133 6 (Electronics)
West Avn, My '56:18 (X- to -B) 1
S '59:18-19 (-A) 2

1930. XH-51A "Aerogyro"
Developed into AH-56A "Cheyenne"
Avn Wk, My 13, '63:90-97 1
D 28, '64:44-45 1
Interavia, No. 9, '63:1338-1340 1
Lock Hor, Spg '65:59 P*
Sum '65:4-19 1 C, P*
Sp/Aeron, Ag '63:105+ 2 S*
US Army Avn, My '69:26-31 2
West Aerosp, N '63:18 1

1931. L-100
Military version: C-130 "Hercules"
Avn Wk, S 11, '67:85-88 1
Can Avn, N '66:24-25+ 4

1932. L-286
Air Prog, My '67:10-15+ 4 C
Flying, O '67:61-63 4 C
US Army Avn, My '69:26-31 2

1933. L-300
Military version: C-141
Avn Wk, O 29, '62:44-47 1
Interavia, No. 10, '64:1493-1495
(-A&-B) 1 C

1934. L-500
Military designation: C-5A
Aeroplane, Ja 31, '68:18-21 1
Aerosp Intl, S/O '67:24-25 1
Mr/Ap '68:42-44 1 D
AF/Sp D Intl, S '66:16-24 1
O '66:41-43 1
Avn Wk, Je 13, '66:82-85 1
N 20, '67:213-214 1
:214-227 4
Mr 24, '69:32-33 6 (Loading)

LOCKHEED (cont.)

1934.	L-500 (cont.)			
	Flight, F 8, '68:192-193	1		
	Interavia, No. 4, '67:491	1		
	No. 10, '69:1670-1672	1		
1935.	L-1011 "Tristar"			
	Aeroplane, Jl 5, '67:16-19	1		
	N 1, '67:4-6	1		
	Ap 3, '68:34	1		
	Ap 10, '68:4-9	1		
	Je 26, '68:26-27	1		S*
	Acft Eng, N '69:25-30	1		
	Avn Wk, Je 26, '67:30-33	1		
	S 18, '67:36-40	1		
	D 18, '67:22	1		
	Ap 8, '68:30-34, 36-37	4		
	Ap 15, '68:40-41	6	(Interior)	
	Ap 22, '68:62-72	1		C
	:75-84	6	(Avionics)	S*
	Ap 29, '68:59	1		
	Ag 26, '68:45-52	6	(Pneumatic)	
	Ja 20, '69:55-57	6	(Food Service)	
	Mr 31, '69:67-69+	6	(Avionics)	
	S 8, '69:25	3		
	O 20, '69:48-50	3		
	N 17, '69:55-56 (-8.4A&D)	1		
	Can Avn, Mr '69:12-16	1		C
	Flight, Jl 13, '67:42-43	1		
	N 2, '67:711-714	1		
	Ap 11, '68:503-504	1		
	Je 27, '68:945-947	1		S*
	Ag 29, '68:327-330	4		
	F 20, '69:289-300	1		C*, Pr, S*
	S 18, '69:440-441 (-8.4)	1		
	Fly Rev Intl, N '67:955	1		
	Ag '68:429	1		P*
	Ja '69:22-23	1		
	N '69:21 (-8.4)	1		
	:29-30	1		
	Interavia, No. 12, '67:1876-1879	1		
	No. 8, '68:949-953	1		S*
	No. 9, '69:1515-1519	2		S*
	Sp/Aeron, Ag '68:42-54	1		S*

LOCKHEED (cont.)

1936.	LASA 60		
	See also MACCHI-LOCKHEED AL. 60		
	Flying, F '61:42-43+	4	C
	West Avn, My '60:58	1	
1937.	YO-3A		
	Quiet observation aircraft based on glider		
	Avn Wk, My 12, '69:25	1	
	O 20, '69:193	1	P
1938.	P2V "Neptune"		
	Japanese licensee: KAWASAKI P2V-Kai		
	Aero Digest, Ap '53:28-48	1	S*
	Air Pict, F '52:55	1	
	Air Prog, '54/'55 Annual:42-43 (-5)		S*
	Acft Eng, Je '50:156-162	1	S*
	Am Helicop, F '46:26-27	1	
	Avn Wk, Ja 5, '48:14 (-2)	1	
	Jl 25, '49:20-28	1	S*
	F 15, '54:28-34	3	
	Can Avn, N '46:56+	1	
	Ap '55:36-37 (-7)	1	
	Comm Avn, Ja '46:48-50	1	
	Flight, O 10, '46:398-399 ("Truculent Turtle")	4	
	Ap 3, '47:297 ("Truculent Turtle")	4	
	O 30, '53:589-594	4	
	Flying, Mr '52:30-31 (-5)	1	
	Fly Rev Intl, F '66:335		P*
	Nav Avn News, F '46:29	1	
	D '47:22 (-1&-2)	1	
	Je '48:1-5	1	C
	S '48:20-21	4	
	Ja '50:13 (-w. skiis)	1	
	Mr '52:24 (-5)	1	
	Ap '54:9 (-7)	1	
	Je '61:12-13 (-1 to -7)	1	P
	RAF Fly Rev, Ap '55:30-33 (-1 to -7)	1	Pr, S*
	Jl '55:10+ ("Truculent Turtle")	1	
	Mr '61:26	1	
	:27 (-5:Australian)		P*
	Ag '61:50 (-5:RAF M.R.1)	5	
	Sentinel, Jl/Ag '68:6-9 (-7:RCAF)	4	P
	Skyways, Jl '50:10-11+ (-4)	1	

LOCKHEED (cont.)

1938. P2V (cont.)

US Air Svcs, Ja '46:20	1		
O '46:18+ ("Truculent Turtle")	4		
F '48:28	2		
West Avn, O '51:34+ (-5)	1		

1939. P-3 "Orion"
Earlier designation: P3V-1

Avn Wk, Mr 27, '67:72-88 (-C)	6	(Anti-sub)	
Flight, O 31, '68:708-709 (Y-/-C)	1		
Flying, Je '63:24-25+ (-A)	1		
Fly Rev Intl, Mr '64:9-14 (-A)	1		Pr, S*
Interavia, No. 9, '63:1392-1396	1		
Lock Hor, No. 7, Ja '69:54 (Y-/-C)			P*
Nav Avn News, Je '63:32-34 (-A)	3		
Jl '69:24-25 (-C)	1		C

1940. P3V-1 "Orion"
Later designation: P-3

Avn Wk, S 17, '62:63-65	1		
Nav Avn News, O '62:12-13	1		

1941. P-38 "Lightning"
See also F-5 and XP-58

Aero Album, Sum '68:70-71 (-J:Bong's)	1		Pr†
Sum '69:3-11 (French-)	4		Pr†
Aero Digest, Mr '41:88-92	2		
Je '43:202-203			S
N '43:244-247	6	(Electrical)	C*
F 1, '44:114-116	6	(Hydraulic)	S*
My 1, '45:84-89+ (-L) (Pt 1)	1		S*
My 15, '45:80-85+ (-L) (Pt 2)	1		S*
Aeronautics, Ag '44:46-49	1		
Air News, O '43:56-57			S*
Ag '45:24-26+	4		
My '46:18-23+	1		
Air Prog, Win '62/'63:91			C
O/N '64:42-43 (-J)	1		S*
N '69:49-51+	1		
Air Tech, Je '43:21-39	1		C*, S*
Acft Eng, Ja '44:26-28	3		
AAHSJ, Sum '60:126-131	1		P

LOCKHEED (cont.)

1941.	P-38 (cont.)			
	Aviation, Ap '41:72-73	1		S*
	Je '41:50-51+	1		
	Ap '42:74-77+	6	(Struts)	
	Ja '43:101+	1		
	S '43:121-134+	3		
	Ag '44:123-147	1		C*, S*
	Comm Avn, O '40:38-40	1		
	Mr '45:44-46 (-L)	1		
	Flight, S 24, '42:336b	1		
	N 4, '43:497-500	1		
	Je 22, '44:665-670+ (-J)	1		S*
	Mr 29, '45:337-338 (-L)	1		
	Flying, S '41:57-58			P*
	Mr '42:53-54+	2		
	My '43:24-25+	4		
	Je '43:44-45+	4		
	Mr '44:49			C
	Ap '45:39-46 (-J)	1		C, S
	Ag '45:26-28+	1		
	Fly Rev Intl, Ap '68:224, 226 (-I, -F, -H, -J, -L)			Pr*
	Pop Avn, Ap '38:58 (X-)	1		
	Pop Fly, Ap '39:31 (X-)	1		
	Skyways, Ap '44:34-41+	1		P*, S*
	US Air Svcs, My '43:16-17	6	(Wing Tanks)	
	West Fly, Ag '43:40-41	3		
	Mr '44:118 (-J)	1		
	Mr '45:116 (-L)	1		
	Ja '46:36-37	1		
1942.	XP-49			
	Modified P-38			
	Air Pict, My '60:176-177	1		
1943.	XP-58 "Chain Lightning"			
	Two-seat version of P-38			
	Air Pict, Je '59:230-232+	1		
1944.	P-80 "Shooting Star"			
	Later designation: F-80			
	USAF Trainer: T-33			
	US Navy Trainer: T2V-1			
	Aero Digest, S 1, '45:60-61+	1		
	Air Cl, Ag '69:50-65	1		C*, P*
	:63 (-C)			S*
	:64 (Y-/-A)			C

LOCKHEED (cont.)

1944.	P-80 (cont.)		
	Air News, O '45:30+	1	
	Air Prog, Ap/My '63:75 (-R)		C
	Acft Eng, Mr '48:75-86	1	C, P, S*
	Aviation, Jl '45:149-151	1	
	Avn Wk, Jl 19, '48:22-26	1	
	Can Avn, S '45:58-59+	1	
	Comm Avn, Ag '45:78-80	1	
	Flight, Ap 12, '45:392-394	1	
	Ag 9, '45:152	1	
	Je 26, '47:596-597	4	
	Flying, O '45:31+	1	
	Ap '46:44-45	4	
	Mr '56:28-29+	1	
	Interavia, No. 7, '47:15-17	1	P
	Skyways, Jl '45:20-21+	4	
	Ag '46:20-21+	4	
	US Air Svcs, S '45:22		P
	F '46:26-28	4	
	Jl '47:21-22 (-R)	1	
	West Fly, Mr '45:68+	1	
	S '45:48-49+	1	
	Mr '46:34	4	
	My '46:25	4	
1945.	PO-1W "Po' One"		
	Later designation: WV-1		
	Commercial designation: "Constellation," L-749		
	Air Cl, Je '69:39		P
	Avn Wk, Je 27, '49:14	1	
	US Air Svcs, F '50:16		P
1946.	PV-1		
	B-34s requisitioned for US Navy		
	US Air Svcs, Ap '43:12	1	P
1947.	R7V- "Super Constellation"		
	Commercial equivalents: -1 = L-1049, -2 = L-1249		
	Air Cl, Je '69:38 (-2)		P
	Avn Wk, S 13, '54:18 (-2)	1	
	West Avn, Ja '53:19 (-1)	2	
	O '54:46 (-2)	1	
1948.	S-3A		
	Air Prog, N '69:54	1	
	Avn Wk, S 15, '69:96-102+	1	

LOCKHEED (cont.)

1948.	S-3A (cont.)			
	Flight, Ja 30, '69:172	1		D
	Ag 14, '69:260	1		
	Fly Rev Intl, D '69:25-26	1		
	Nav Avn News, S '69:7	1		P
1949.	SR-71			
	See also A-11 and YF-12A			
	Avn Wk, D 14, '64:20-21	1		
	Ja 17, '66:33-34	1		
1950.	SST L-2000			
	Aeroplane, S 22, '66:4-7	1		S*
	Air Force, Jl '66:74-89	1		
	AF/Sp D Intl, Jl '66:6-15	1		
	Acft Eng, Ja '67:25-38 (-7A&-7B)	1		S*
	Avn Wk, Mr 23, '64:38-44	1		C
	Mr 29, '65:38-44	1		
	Ja 10, '66:43-50	1		
	F 21, '66:99	6	(Antenna)	
	Mr 28, '66:40	1		
	S 5, '66:42-47 (-7)	1		C
	O 31, '66:190-204	6	(Avionics)	
	N 21, '66:76-90 (-7)	1		
	D 5, '66:74-86 (-7)	3		
	Flight, Ja 20, '66:95-100 (Pt 1)	1		S*
	Ja 27, '66:144-147 (Pt 2)	1		
	Interavia, No. 8, '66:1156-1159 (-7)	1		S
	No. 12, '66:1900-1903 (-7A)	1		S*
	Lock Hor, Spg '65:4-19	1		D*, S
	:28-47	2		
	No. 5, '66:3-33	1		D*, S*
	:34-53	3		D*, S*
1951.	T2V-1, TV-2 and TO-2 "Seastar"			
	USAF version: T-33			
	Air Cl, Ag '69:61, 63			P
	Avn Age, My '56:20-25	1		S*
	Avn Wk, Ap 11, '55:15	1		
	Ap 23, '56:62-66	1		S*
	Fly Rev Intl, Ja '68:50			Pr*
	RAF Fly Rev, F '58:26-29	1		Pr, S*
	US Air Svcs, Ja '56:18	1		P
	West Avn, Mr '56:32	1		

LOCKHEED (cont.)

1952. T-33 "Shooting Star" or "T-Bird"
US Navy version: T2V-1
Canadian-built version: CANADAIR CL-30
Japanese-built version: KAWASAKI T-33

Aeroplane, Ag 21, '53:225-228	4	
Fly Rev Intl, Ja '68:50 (-A:Netherlands' "Whisky 4," China, Germany)		Pr*

1953. U-2

Aeroplane, My 27, '60:639	1	P
Air Prog, Fall '60:58-59	1	P
Avn Wk, F 11, '57:30	1	
My 2, '60:158	1	
My 16, '60:28-29	1	P
My 23, '60:32-37	1	P
Ag 12, '63:72-77 (W-)	1	
N 18, '63:53-57	4	
Can Avn, Je '60:55	1	
Flight, N 23, '56:809	1	
Flying, Jl '60:49	1	
Fly Rev Intl, F '66:331-332+	1	
Lock Hor, Sum '65:21		P*
RAF Fly Rev, Jl '57:32	1	
Jl '60:18-19+	4	
West Avn, Je '60:37	1	
F '61:Cover, 6-8+	4	P*

1954. VZ-10 "Hummingbird"
Later designation: XV-4

Avn Wk, D 11, '61:30-31	1	
Ag 20, '62:35	1	
Flight, Ap '62:510-512	2	

1955. WV-2 "Warning Star" or "Willy Victor"
Commercial designation: "Super Constellation," L-1049

Flight, N 26, '54:762-763	1	
Interavia, No. 2, '56:120-121	1	P
Nav Avn News, My '58:9	1	
US Air Svcs, Ag '56:21	1	
West Avn, Mr '58:12	1	

LOCKHEED (cont.)

1956.	XFV-1		
	Air Power, Aut '54:6-10	1	P
	Avn Wk, Mr 22, '54:14-18	1	
	Mr 29, '54:16-17	4	
	Interavia, No. 5, '54:295-297	1	
	Nav Avn News, My '54:20-21	1	
	RAF Fly Rev, Je '62:43+	1	
	US Air Svcs, Ap '54:8-9	1	P
1957.	XV-4 "Hummingbird"		
	Earlier designation: VZ-10		
	Air Prog, Je/Jl '64:12-13 (-A)	1	
	Avn Wk, D 16, '63:82-83 (-A)	1	
	Je 10, '68:17 (-B)	1	
	Ag 5, '68:78-84 (-B)	1	
	Flight, Je 11, '64:986-987 (-A)	1	
	Ja 5, '67:36 (-A)	1	
	Fly Rev Intl, N '64:27-28+ (-A)	1	
	N '68:31-32 (-B)	1	
	Interavia, No. 12, '61:1685-1686	1	
	No. 5, '64:652-655 (-A)	1	S*
	No. 11, '68:1399-1400 (-B)	1	
	US Army Avn, S '64:6-7 (-A)	1	S
1958.	10 "Electra"		
	See also US military XC-35		
	Aero Digest, Ap '34:52-53	1	
	Ag '35:42		C
	Aeroplane, Ap 11, '34:653-654	1	
	Ap 7, '37:412-413	4	C
	D 27, '57:950-955	1	
	N 21, '58:742-744	4	C
	Air Cl, D '68:51-63	1	C, P
	Air Prog, Fall '61:6-12	1	C
	F '68:37+ (-A:Koepke-Pellegreni)	1	
	Ap '68:40-41+ (-A:Koepke-Pellegreni)	1	
	Aviation, Ap '34:127	1	
	Avn Age, D '55:22-27	1	
	Ja '57:32-33	6 (Antenna)	S*
	Je '58:40-45	6 (Lndg Gear)	S*
	Avn Wk, Je 13, '55:12+	1	
	O 31, '55:24-31	1	C*
	Mr 26, '56:92-93	6 (Flight Inst)	C*
	Jl 2, '56:58-60	3	S*

LOCKHEED (cont.)

1958.	10 (cont.)			
	Avn Wk, O 8, '56:71	1		
	Mr 3, '58:107			P*
	Mr 31, '58:46-63 (Pt 1)	1		
	Ap 7, '58:79-84 (Pt 2)	1		
	D 22, '58:56-63 (Pt 1)	4		
	D 29, '58:53-57 (Pt 2)	4		C*
	F 23, '59:37	6	(Prop)	
	O 24, '60:37-39	4		
	Bee-Hive, Ap '34:3-4+ (-A)	1		
	CAHS, Fall '68:73+ (-A:Canadian)	1		
	Flight, Ja 25, '34:74-75 (-A)	1		
	Ap 22, '37:384-386	6	(Cockpit)	C*
	N 4, '55:712-716	1		S*
	N 11, '55:740	3		
	Jl 6, '56:39-42	1		S*
	S 21, '56:523	2		
	N 1, '57:693-696	1		S*
	Jl 25, '58:135-137	1		S*
	N 14, '58:761-762+	4		
	O 16, '59:404	1		Pr
	Flying, S '58:39	6	(Fuel)	
	F '61:46-47+	4		
	Interavia, No. 7, '56:519-521	1		S*
	No. 4, '57:311-316	1		C, S*
	Pop Avn, Ap '34:225-226	1		
	Pop Fly, My '37:92 (-A)	1		C
	Sp/Aeron, N '58:62-63	4		
	D '59:81-82+	6	(Anti-ice)	S*
	West Avn, Jl '55:20	1		
	D '57:33	1		
	West Fly, Ap '34:20+	1		
1959.	12 "Electra" or "Small Electra"			
	Air Cl, D '68:51-63	1		P, S*
	Air Prog, F/Mr '63:84-85			Pr†
	Aviation, F '36:36+	1		
	S '36:33-35	1		C
	Flying, D '55:33+	1		
	Pop Avn, O '37:21-22+ (Mattern's "The Texan"-)	4		
	Ag '39:35 (Military-)	1		
	West Fly, Mr '36:13	1		

LOCKHEED (cont.)

1960. 14 "Super Electra" or "Sky Zephyr"
Military versions: "Hudson" and AT-18
Aero Digest, My '37:54 1
S '37:45-46 1
Aeroplane, O 12, '39:456-459 (Pt 1) 6 (Slots)
O 19, '39:486-488 (Pt 2) 6 (Slots)
Air Cl, D '68:50 (Netherlands-) Pr*
:63 P
Aviation, Mr '37:39-41 1
Jl '37:32-34+ 1
Avn Wk, Mr 6, '61:38-40 1
Can Avn, S '37:12+ 1
Flight, Jl 21, '38:58 (Hughes' world flight) 4
Ag 18, '38:138-140 4 C
O 6, '38:294-296 1 S*
Pop Avn, Ap '38:66 4
West Fly, Ag '37:36 1
F '38:32 4
Ag '38:12-14 (Hughes'-) 4

1961. 18 "Lodestar"
Military designations: C-56, C-57, C-60
Aero Digest, D '39:49 1
Can Avn, D '47:71 6 (Lube Chart)
Comm Avn, Ja '41:34-35 1
Flight, Ja 16, '41:48e-f 1
US Air Svcs, Ap '40:26 ("Gilmore Oil"-) P*
D '40:34 4
West Fly, Ap '41:68 D*

LOENING USA

1962. "Air Yacht"
CAHS, Fall '63:98-99 1 Pr
Win '66:97-101 1 Pr

1963. M-1
AAHSJ, Jl/S '57:176-178 1

1964. OA-1
AAHSJ, Win '67:265-267 1 Pr

LOHNER Austria

1965. Flying Boat (1916-1917)
Avn Wk, F 18, '63:36 1 D*

LOHNER Canada

1966. No. 1 (1910)
Canada's first aircraft
CAHS, Spg '69:16-20 4 Pr

LOIRE France

1967. 43, 45, 46
Fly Rev Intl, Ja '64:63 1
Ap '66:493-495 (-46) 1 Pr*
Pop Avn, O '36:30 (-46) 1

1968. 210
RAF Fly Rev, Jl '61:54 1

LOIRE-NIEUPORT France

1969. LN-10
Fly Rev Intl, S '65:44 1
RAF Fly Rev, O '59:9+ 1

1970. LN-40, LN-401, LN-402, LN-411
Fly Rev Intl, Ag '66:768-769 1
RAF Fly Rev, My '59:46 1
O '59:9 1
Jl '60:9+ 1

1971. LN-42
Fly Rev Intl, Ag '66:768-769 1

LONGREN USA

1972. "Air Flivver"
Air Prog, O '69:48-49+ 1

LTV USA
See LING-TEMCO-VOUGHT

LUALDI Italy

1973. L. 55
Avn Wk, Je 23, '58:57 1 C

LUSCOMBE			USA
1974.	"Observer" Model 90		
	Flying, S '48:36-37+	4	C
LVG (Luft-Verkehrs Gesellschaft)			Germany
1975.	C. II		
	Flying, Ap '60:81		D
1976.	C. V		
	Cr & Cockade, Sum '69:181	1	P
1977.	D. 10 "Flying Flounder"		
	Cr & Cockade, Aut '68:240-242	1	P
1978.	D. 12		
	Cr & Cockade, Aut '68:242-243	1	P
LWD (Lotnicze Warsztaty Doswiadczaine)			Poland
1979.	"Mis" (Bear)		
	RAF Fly Rev, Mr '56:15	1	
LWF (Lowe, Willard and Fowler)			USA
1980.	"Owl" (1920)		
	Flying, N '42:64-65	1	
LWS (Lubelska Wytwórnia Samolotow)			Poland
1981.	"Zubr" (Bison)		
	Fly Rev Intl, Je '65:43	1	
MACCHI			Italy
	RAF Fly Rev, N '56:50-51 (WW II fighters)		
1982.	M. 39 and M. 52 Schneider Seaplanes		
	RAF Fly Rev, Jl '62:47	1	
1983.	M. 41		
	Acft Eng, Je '29:120	1	
1984.	M.B. 320		
	Air Pict, N '52:341	1	
	Flight, O 20, '49:518	1	

MACCHI (cont.)

1985.	M.B. 326		
	Aeroplane, S 28, '56:489	1	S
	Ja 8, '60:38-39	4	
	Air Pict, Mr '58:98-99	1	S
	Acft Eng, My '64:130	1	S
	Avn Wk, O 15, '56:97	1	S
	Ag 19, '68:56-62 (-G)	1	C
	Bris Sid Jnl, Spg '61:66-69	1	
	Vol. 7, No. 3, '66:26-29	4	C
	Flight, S 21, '56:521	1	
	F 7, '58:175	1	S*
	S 20, '62:492-494	1	
	N 18, '65:852-856	1	C*, S*
	Fly Rev Intl, Mr '65:21-23	1	
	F '69:25-27	1	
	Interavia, No. 1, '57:65 (-A)	1	S
	No. 10, '59:1242-1243	1	C, S*
	No. 2, '62:226 (-C)	6 (Elecrn)	C*, S*
	RAF Fly Rev, My '61:32-35	1	Pr, S*
	D '62:34-35	1	
1986.	M.C. 72		
	Holder of 1934 world speed record		
	Aero Digest, S '33:44-45	1	
	Air Prog, O '65:16+	1	
	Acft Eng, Je '33:129-131	1	S*
	Flight, Ap 20, '33:367, 383	4	P
	N 1, '34:1152	4	
1987.	M.C. 200 "Saetta" (Thunderbolt)		
	Can Avn, O '40:24-25	1	
	Fly Rev Intl, Ag '64:56 (168 Squadriglia)		Pr*
	RAF Fly Rev, S '55:22	1	
	N '56:51-52	1	
	Ap '59:22	4	
	O '60:32-33	1	
1988.	M.C. 202 "Folgore" (Lightning)		
	Flight, Ja 7, '43:12a	1	
	Fly Rev Intl, Jl '64:62 (-III)		Pr*
1989.	M.C. 205 "Veltro" (Greyhound) V Model		
	"Orione" (Orion) N Model		
	RAF Fly Rev, N '58:16 (-V:1° Gruppo Caccia)		Pr*
	:17	1	

MACCHI-LOCKHEED Italy/USA

1990. AL. 60 "Santa Maria"
See also LOCKHEED LASA 60
Can Avn, O '61:27-28 4
Flight, O 5, '61:548-549 4

MCCULLOCH USA

1991. MC-4
Enlarged version of JOV 3
Developed into JOVAIR 4E and 4ES
Air Pict, Ap '53:106 P
Am Helicop, Ap '51:6-7+ 1 C
Ap '53:10 1 C
Avn Age, My '53:30-31 1 C
Avn Wk, Ap 30, '51:32-36 1 C
Flight, My 18, '51:596-597 1 C

MCCURDY Canada
See AERIAL EXPERIMENT ASSOCIATION "Silver Dart"

MCDONNELL USA
Nav Avn News, Ja '57:14-16 (US Naval aircraft)

1992. XF2D-1 "Banshee"
Am Helicop, Jl '47:16-17+ 1
Nav Avn News, Ag '47:22-23 1

1993. F2H "Banshee"
Aero Digest, S '52:22-38 1 S*
Aeroplane, Ap 5, '57:471-474 (-3) 1 P
Air Pict, Mr '51:78 (-2) 1
Avn Opns, F '50:29+ 1
Avn Wk, O 31, '49:22-25 1
Can Avn, S '56:36-38 (-3) 4
F '63:18-19 4
CAHS, Fall '66:75-76 (-1 to -3) 1
Flight, Jl 17, '47:63-64 1
Nav Avn News, Ap '48:1-5 (X-/-1) 1
RAF Fly Rev, Jl '60:16 1
:17 (-4:Canadian) P*

MCDONNELL (cont.)

1994. F3H "Demon"
Avn Wk, Ja 18, '54:9, 19 (-1N) 1
N 7, '55:16 (-1) 1
Ja 23, '56:39 (-1) 1
Nav Avn News, Ap '54:8 (-1N) 1
RAF Fly Rev, Mr '54:18-19 (-1N) 1

1995. F-4 "Phantom"
Earlier designations: XFD-1 and FH-1
Later version: MCDONNELL DOUGLAS FGR. 2 "Phantom"
Aerosp Intl, N '66:16-25 (-II) 1
Ja/F '69:48-50 (-E:Japan) 1
Jl/Ag '69:28-30+ (-M: RAF) 6 (Recon Pod) S*
Air Force, My '62:36-37) (-H-II) 1
Ag '66:30-36 (-II) 1
Air Pict, Mr '69:78-79 (RAF-) 4
Air Prog, Fall '62:8-9 (-H-II) S*
My '67:30-33 (R-/-C) 4
Mr '68:44-45+ 4
Ag '69:23-26+ (-J:Blue Angels) 4
AAHSJ, Sum '67:114-119 (-II) 1 Pr
Avn Wk, Je 9, '58:43-44 (-H-1) 1
O 13, '58:66-77 (-H-1) 1
Mr 9, '59:71 P*
Je 20, '60:156-165 (-H-1) 1
Jl 30, '62:38-46 (-H-1) 1
Mr 16, '64:71 (-C/-II and -B) P*
Jl 4, '66:24-25 (-K/-II) 1
Jl 25, '66:78-79 (-K/-II) 1
Jl 10, '67:31 (-E) 6 (Armament)
My 5, '69:102-109+ (-J:Blue Angels) 1
S 1, '69:36-39+ (-E:Thunderbirds) 1
Flight, O 3, '58:547 (-H-1) 1 Pr
My 27, '65:832-835 1 D
Ja 20, '66:118 (FV-) 1
Je 30, '66:1091-1099 (-K and -M) 1 S*
:1100-1101 (-B) 4
Jl 14, '66:57-58 (-K&-M) 1 S*
S 21, '67:483-487 (-K&-M) 1
O 12, '67:629 (-K&-M) 6 (Afterburner)
:630 (-E) 1
Ap 4, '68:484 (RAF-) 1 P*

MCDONNELL (cont.)

1995. F-4 (cont.)			
Flight, Ag 15, '68:253-258 (-K)	4		P
N 20, '69:768-771 (R-/-C)	4		
Flying, My '62:62-63+ (-H-1/-II)	1		
Fly Rev Intl, Jl '64:21-25+ (-II)	1		
:23 (-B)			Pr
:26-28	4		C*
N '65:153 (To -M)	1		
Mr '66:430 (YR-/-C)			Pr*
S '66:5-10 (-B&-C)	1		Pr
My '67:531-532 (-K&-M)	1		P*
Mr '69:50-51+ (-K&-M)	1		
Je '69:68-69 (-J:Blue Angels)	1		P*
Jl '69:34-40	1		
S '69:82 (Blue Angels & Thunderbirds-)			P*
O '69:48 (Thunderbirds-)			P*
:68 (-M)	6	(Recon Pod)	S*
Interavia, No. 10, '60:1235 (-H/-II)	1		
No. 11, '63:1698-1701 (-A to -C)	1		S*
No. 2, '65:259-263 (-C)	1		C*, S*
No. 5, '66:737-741 (-A to -K)	1		S*
Nav Avn News, My '60:19-21 (-H-II)	2		
N '60:20-21 (-H-II)	4		
Mr '61:24-26 (-H-II)	3		
Ja '62:20-22 (-H-II)	4		
Je '62:12 (-H-II)	4		
My '63:12-13 (Air Force-)	1		
S '65:6 (-B&-C)	1		
Ag '66:17 (-K)	1		
S '68:15 (-K&-M)	1		
:16-19	3		
RAF Fly Rev, Ap '61:38-39 (-H-1/-II)	1		P*
Ja '62:30-33 (-II)	1		Pr, S*
RAFs Qtr, Spg '67:xii (-K:RAF)			P

1996. XF-85 "Goblin" Only two built			
Aero Digest, D '49:42-43+	6	(Landing Skid)	
Air Prog, Ap/My '63:73			C
Am Helicop, N '48:11	1		
Avn Wk, D 1, '47:14	1		
My 17, '48:10	1		
Can Avn, Jl '48:74	1		
Flying, Ap '49:30-31+	1		

MCDONNELL (cont.)

1996.	XF-85 (cont.)			
	Fly Rev Intl, D '67:1060-1062	1		
	RAF Fly Rev, My '55:12	1		
	Skyways, Ag '48:30-31	1		
	Mr '49:33-34	1		
1997.	XF-88			
	Developed into F-101 "Voodoo"			
	Am Helicop, N '48:11	1		
	Avn Wk, S 4, '50:27-33	1		
	O 29, '51:16-17	1		
	Jl 6, '53:17	2		
	Nav Avn News, F '49:33			P
	RAF Fly Rev, Ag '61:48	1		
	Skyways, D '48:17	1		
1998.	F-101 "Voodoo"			
	AAHSJ, Win '69:282-289	4		D
	Avn Wk, Ag 15, '55:18 (-A)	1		
	Jl 16, '56:67	1		
	Je 7, '65:57-58 (-B)	6	(Radar)	
	Flight, D 3, '54:809 (-A)	1		
	Mr 9, '56:268	1		P
	Ap 6, '61:441-445	4		
	N 5, '64:805-806 (C-/-B)	4		
	Fly Rev Intl, Mr '66:430 (R-/-C)			Pr*
	Roundel, S '61:8-9 (C-/-B)	1		D
	RAF Fly Rev, F '55:40-41 (-A)	1		
	N '58:37-38	1		
	F '61:17+	4		
	West Avn, O '55:28-29+ (-A)	2		
1999.	XFD-1 "Phantom"			
	Later designations: FH-1 and F-4			
	Am Helicop, Mr '46:33	1		
	Aviation, N '46:40-48	1		S*
	Flight, Ap 10, '47:314-316	1		S
	RAF Fly Rev, Jl '58:11	1		
	US Air Svcs, Ag '46:10-12	1		
2000.	FH-1 "Phantom"			
	Earlier designation: XFD-1			
	Later designation: F-4			
	Avn Wk, Je 21, '48:21-25	1		
	Nav Avn News, Jl '56:31	2		
	RAF Fly Rev, Jl '58:11	1		

MCDONNELL (cont.)

2001.	XH-20 or J-1 "Little Henry"		
	Am Helicop, D '47:10-11	1	P
	Avn Wk, N 24, '47:14	1	
	Interavia, F '48:84-86	1	P
2002.	XHJD-1 "Whirlaway"		
	Am Helicop, O '49:22	1	
	Jl '50:6-9+	1	C
2003.	XP-67 "Bat"		
	Air Pict, Ap '59:139-141	1	
2004.	XV-1 "Convertiplane"		
	Air Pict, S '55:276		P
	Air Power, Win '57/'58:88		P
	Am Helicop, F '54:11	1	
	Mr '54:8+	2	
	My '55:16	1	
	N '55:3-4	4	
	Avn Wk, F 15, '54:17	1	
	D 12, '55:17	1	P
	Nav Avn News, Ap '54:8	1	P
	West Avn, Mr '54:16	1	
	Je '55:27	1	
	S '56:26	4	
2005.	119 UCX		
	Avn Wk, F 9, '59:32-34	1	
	F 16, '59:95	1	
	Flying, My '59:27	1	
2006.	120		
	Am Helicop, S '58:6	1	
	Avn Wk, N 10, '58:66-73	1	
	Nav Avn News, Ap '60:31 (-D)	1	
	Sp/Aeron, D '58:4	1	

MCDONNELL DOUGLAS USA

Aerosp Intl, S/O '68:41-45 (Merger)

2007.	C-9A "Nightingale"		
	Aerosp Intl, Ja/F '68:29-32	1	
	Air Force, Mr '68:102-105	1	
	Ag '68:82	1	
	Ja '69:38-39	4	
	Bee-Hive, Sum '69:22-29	4	

MCDONNELL DOUGLAS (cont.)

2008.	F-15		
	Air Force, Jl '69:28-34	1	
	Avn Wk, S 2, '68:16-17	1	
	Ja 13, '69:110	1	
	N 24, '69:32	1	D
	D 8, '69:88-89+	1	
	Sp/Aeron, F '67:74-85	1	
	Ag '69:18-20	1	
	D '69:36-47	1	S*
2009.	FGR. 2 "Phantom"		
	Original design: MCDONNELL F-4		
	Flight, O 30, '69:665-669	4	P
2010.	188		
	Original STOL design: BREGUET 941		
	Air Prog, Ja '69:18-19+ (-E)	4	
	F '69:48-49 (-E)		C
	Avn Wk, Ag 12, '68:45	1	
	S 30, '68:28-29	2	
	N 4, '68:79-99	6 (Auto Flight Control)	C
	Flight, O 17, '68:612-613 (-E)	1	C
	Jl 17, '69:97-99	4	
	Interavia, No. 7, '69:903-905	1	C*, P

MCKINNON USA

2011.	"Goose" G-21-C		
	Original design: GRUMMAN G-21 "Goose"		
	Flying, Je '58:50	1	

MAKHONINE France

2012.	123 Variable Wing		
	Flight, My 13, '32:417-418	1	S
	Flying, N '50:27+	1	
	Pop Avn, D '31:104	1	
	RAF Fly Rev, Ap '59:48	1	

MANSYU Manchuria

2013.	"Hayabusa" (Falcon)		
	Air Pict, My '61:141-142+	1	

MARCEL BESSON France

2014. MB-411
Fly Rev Intl, Ag '67:812 1

MARTIN USA
Nav Avn News, Ag '56:9-12 (Naval aircraft)
Skyways, Ap '43:12-15+ (Flying boats)
US Air Svcs, F '48:26 (Operations in 1947)
Ag '49:8-10+ (40th Anniversary)

2015. "Baltimore" Model 187
Aero Album, Fall '68:12-15 (-V:French) 1
Aeroplane, Ag 24, '45:220-222 4
Comm Avn, Jl '41:70 1
Flight, Jl 31, '41:65 1
Jl 30, '42:120a 1
S 24, '42:333 1
Fly Rev Intl, Je '69:73 (-V:Greece) Pr*
RAF Fly Rev, S '61:42 (-I to -V) 1
RAF Qtr, Mr '44:76 P
US Air Svcs, D '41:28 4

2016. Biplane (1909)
US Air Svcs, Ag '49:8-9 1 D

2017. "Clipper" Model 130
Only three built
Aero Digest, Ja '35:34-36 1
Ag '35:41 C
D '35:34-36 6 (Sound Proofg)
Air Pict, Ap '67:133 1
AAHSJ, Fall '65:172-183 1
Aviation, Ja '35:11-14 1 S
Flight, Ja 24, '35:99 1
Flying, Mr '44:24-25+ ("China"-) 4
Pop Avn, F '35:82 1
F '36:86-88+ ("China"-) 4 S*
D '37:22-24+ ("China"-) 4
US Air Svcs, Ja '35:28-32 4 P
D '35:14-15 ("China"-) 4 P
N '36:15-16+ ("China"-) 4 P

2018. COIN (Counterinsurgency Aircraft)
Interavia, No. 8, '64:1121-1123 1 S

MARTIN (cont.)

2019.	"Maryland" Model 167		
	Aero Digest, D '39:71-74 (-W)	1	
	AAHSJ, Spg '65:34-41	1	
	Aviation, Ap '42:112-113		S*
	Flight, D 28, '39:525 (-W)	1	
	My 30, '40:494a-c (-F)	1	
	D 10, '42:632b	1	
	Flying, D '40:24-25+	4	
	Pop Avn, Ap '40:43 (-F)	1	
	RAF Fly Rev, Jl '55:9	1	
	My '61:51+	1	
2020.	"Midget"		
	Flight, N 19, '54:735-736	1	S*
2021.	AM-1 and BTM "Mauler"		
	Flight, F 3, '49:144	1	
	Nav Avn News, O '47:33		P
	US Air Svcs, Mr '46:12-13	1	P
	My '49:10	1	P
2022.	B-10 Models 123 and 139		
	See also MARTIN 139 and XB-907		
	Aviation, Ag '35:27-30	1	
	Flight, F 15, '34:156	1	
	S 19, '35:305-306	1	
	Fly Rev Intl, Ag '65:49-51	1	Pr*
	Pop Avn, S '35:150	1	
	RAF Fly Rev, Ja '63:45	1	
	US Air Svcs, Jl '34:12-14 (Alaska flight)	4	
	S '34:12-15 (Alaska flight)	4	
2023.	B-12		
	B-10 with engine change		
	Bee-Hive, O '35:2-3	4	
	Fly Rev Intl, Ag '65:49-51	1	
2024.	B-26 "Marauder," "Martian" and "Widow Maker"		
	US Navy designation: JM-1 and JM-2		
	Aero Digest, Ap '41:133-134+	3	
	Ja '43:244-245		S
	Jl 15, '44:79-82	1	S*
	Aeronautics, O '44:28-32	1	
	Aeroplane, Mr 26, '43:360-361 (-A or -I)	1	
	Ag 3, '45:130-133	1	
	Ag 10, '45:157-160	1	

MARTIN (cont.)

2024. B-26 (cont.)

Aerosp Hist, Sum '69:41-43+ (-B50:"Terre Haute Tornado") 1
Air Tech, N '42:28 (-B) P
Ja '45:17-27+ 1 S*
AAHSJ, Fall '66:190-196 4
Aviation, S '44:183 S*
Can Avn, My '41:48+ 3
O '41:47-48+ 3
D '41:66+ 6 (Plexiglas)
Je '44:105-106 6 (Landing Gear)
Comm Avn, Ja '41:51+ 3
S '41:44-46 3
My '44:76 6 (Landing Gear)
Flight, Ja 23, '41:68c-d 1
Jl 30, '42:120b 1
Flying, O '44:58-59 4 (Bail-out) S*
Fly Rev Intl, O '63:52 (-B, -C, -I, -II, JM-1 & JM-2) Pr*
RAF Fly Rev, S '63:24-26+ 1
:24-25, 27 (-B) Pr, S*
Skyways, N '44:18-22+ 1 S*
US Air Svcs, Ap '43:18-20 1 P
Jl '47:31 (X-/-H:"Stump Jumper") 6 (Tandem Gear)
West Fly, My '47:48 (X-/-H:"Stump Jumper") 6 (Landing Gear)

2025. XB-48

Am Helicop, F '48:11-12+ 1
Aviation, Je '47:60 1
Avn Wk, S 8, '47:26-33 3
Can Avn, Ag '47:79 1
Interavia, No. 8, '47:51-52 1 P
Nav Avn News, O '48:33 P
US Air Svcs, Je '47:22 1
Mr '48:10 1 P

2026. XB-51

Aeroplane, S 30, '49:479 P
Air Pict, Jl '65:247 1
Avn Wk, O 3, '49:12-13 1
Flying, F '51:22-23+ 1
Skyways, Ja '50:18-19 1
US Air Svcs, O '49:14 1 P

MARTIN (cont.)

2027. B-57 "Canberra"
Original design: ENGLISH ELECTRIC "Canberra"

Aero Digest,	N '52:22-25	3	
Aeroplane,	Ag 31, '51:246-247 (-A)	1	
Avn Wk,	Jl 23, '51:19-31 (-A)	1	
	Ag 31, '53:17-18 (-A)	1	
	Ja 4, '54:28-37 (-A)	1	
	Mr 14, '55:74 (-B)		P*
	N 28, '55:13-15	4	
	Jl 13, '64:22-24 (R-/-F)	1	
	Jl 20, '64:51-57 (R-/-F)	1	
Flight,	Ja 15, '54:60-61 (-A)	1	
	Ap 9, '54:424	1	
	Ja 11, '57:37-41 (R-)	4	
Flying,	Mr '55:58 (-B)	1	
Fly Rev Intl,	F '69:45 (-B: Pakistan)		Pr*
RAF Fly Rev,	O '57:47 (R-/-D)	1	
RAFs Qtr,	Sum '69:139-141 (RAAF-)	4	

2028. XB-907 Bomber
Variant of B-10

RAF Fly Rev,	Ja '63:45	1	
US Air Svcs,	S '32:26	1	P
	Mr '33:23	1	
	Ap '33:27	1	

2029. MB-1 and MB-2

AAHSJ,	Fall '62:187-194	1	P
Aviation,	Ja '43:91 (-2)		P

2030. XP2M-1

Flight,	Mr 22, '34:286-287	1	
US Air Svcs,	Ag '31:52	1	

2031. P4M-1 "Mercator"

Am Helicop,	F '48:11-12+ (X-)	1	
Nav Avn News,	D '46:10 (X-)	1	D
	O '50:19	4	
	F '51:19	1	
RAF Fly Rev,	N '61:47+	1	
US Air Svcs,	Mr '47:31 (X-)	1	
	Ag '47:12-13 (X-)	1	P
	Jl '48:8	1	P
West Fly,	Mr '47:50 (X-)	1	

MARTIN (cont.)

2032.	P5M "Marlin"			
	Aero Digest, Jl '52:24-34 (-1)	1		S*
	Avn Wk, Jl 24, '50:27-28 (-1)	6	(Water Flaps)	
	Ja 17, '55:28-45 (-2)	1		S
	Interavia, No. 6, '55:436+	2		
	Nav Avn News, S '48:11 (X-/-1)	1		
	S '50:32	6	(Water Flaps)	
	N '53:18 (-2)	1		
	Ja '69:20-21	1		
	US Air Svcs, My '51:22 (-1)	1		
	My '52:11 (-1)	1		
2033.	XP6M-1 and YP6M-1 "Seamaster"			
	Aeroplane, Ja 14, '55:38-39	1		P
	Air Pict, Mr '55:72-73	1		
	Avn Wk, Ja 10, '55:16-18	1		
	O 3, '55:21	1		P
	Mr 5, '56:44-47	3		
	My 14, '56:34	2		
	Je 4, '56:58-59	2		P
	N 19, '56:34	2		
	My 6, '57:31-37	2		
	F 3, '58:38-39	1		
	S 22, '58:30-31	6	(Beach Gear)	
	Flight, Ja 14, '55:36-37	1		
	Mr 30, '56:350-352	3		
	Fly Rev Intl, My '64:61	1		
	RAF Fly Rev, Mr '55:52-53	1		
2034.	XPB2M-1 "Mars"			
	Air_Tech, Je '44:24-27+	1		
	Can Avn, O '41:100	1		
	F '42:94	1		
	CAHS, Spg '68:18-21	1		
	Spg '69:25 ("Caroline"-)			P
	Comm Avn, O '41:44	1		
	Flight, Mr 4, '43:228a	1		
	Flying, Ag '45:70			C
	Nav Avn News, Mr 15, '44:8	1		C, S
	O '46:32	1		
	Ap '49:13 ("Caroline"-)	4		
	Je '50:13 ("Marshall"-)	4		
	S '50:13 ("Caroline"-)	4		
	Ap '52:14-15	4		
	Mr '55:27	4		
	O '56:10	1		
	Mr '69:28-29	1		

MARTIN (cont.)

2034. XPB2M-1 (cont.)

US Air Svcs,	N '41:29	1	
	Ja '44:12		P
	O '45:35 ("Philippine"-)	1	
	S '48:12-14 ("Caroline"-)	1	P
West Fly,	Ag '42:92	1	
	S '45:112 (Commercial-)	1	

2035. PBM-1 "Mariner"

Aviation,	Ap '39:30-31 (X-)	1	
Flight,	N 19, '42:552b	1	

2036. PBM-3 "Mariner"

Aero Digest,	Je 1, '44:86-91+	3	
Comm Avn,	O '44:58-62 (-R)	4	
Flying,	Ja '42:113		P
West Fly,	Je '44:146	6 (Fuel Tank)	

2037. PBM-5 "Mariner"

Fly Rev Intl,	O ' 68:53 (Uruguayan-)		Pr*
US Air Svcs,	Ja '46:18-19 (X-/-A)	1	
	O '47:31 (-A)	4	

2038. PM-2

Nav Avn News,	Ap '48:33		P

2039. SV-5 Lifting Body
Military designation: X-24A

Avn Wk,	My 16, '66:64-75 (-P)	1	S*
	Jl 10, '67:99-101 (-D&-J)	2	

2040. XT5M-1 "Diving Bomber"
Produced as BM-1

US Air Svcs,	Ap '30:35	4	
	S '30:56	2	

2041. X-24A Lifting Body
Prototype: SV-5

Air BP,	No. 45:2-6	1	P
Air Force,	Ap '68:177	2	
Avn Wk,	Jl 24, '67:46	1	

2042. 139
Export version of B-10B

RAF Fly Rev,	Ja '63:45	1	

MARTIN (cont.)

2043.	156			
	Aviation, D '37:36-40+	1		
	Can Avn, Ap '38:12	1		
	Pop Avn, F '38:18-19 ("Russian Clipper"-)	1		
2044.	202 "Martinliner"			
	Acft Eng, F '49:39-41	3		
	AAHSJ, Sum '62:120-125+	1		
	Am Helicop, Mr '46:32+	1		
	Aviation, Ja '47:38-43	1		S*
	Je '47:72, 75			S*
	Avn Opns, Ap '50:21	6	(Anti-ice and Defrost)	S*
	Can Avn, Je '47:76-78	1		
	Comm Avn, S '44:76-78	1		
	Flight, N 2, '44:481	1		
	Interavia, No. 3, '46:68-72	1		C
	No. 4, '47:51-53	1		
	Skyways, Ja '49:44+	6	(Fire Ext)	S*
	US Air Svcs, O '45:16-17	1		S
	D '46:20	4		
	My '47:14-16	1		
	Jl '47:17	2		
	O '47:21	4		
	D '47:15	1		
2045.	270			
	Avn Wk, Je 23, '52:34-38	1		
2046.	303			
	US Air Svcs, F '46:22	1		
2047.	404			
	Avn Age, S '51:26-27	6	(Temp Cntrl)	S*
	N '51:16-18	6	(Auto Trim)	
	Ja '52:38-39	1		S*
	Avn Wk, Ag 14, '50:55-56	1		C
	Je 25, '51:14-16	1		
	Interavia, No. 11, '50:588-590	1		
	Skyways, O '52:14-17+	1		C
	US Air Svcs, S '50:24	1		
	D '50:10	1		P
	Ja '51:21-22 (Airlift-)	1		

MARTIN-BAKER			GB
	RAF Fly Rev, S '57:43-45 (Fighter designs)		
2048.	M.B. 1		
	Flight, D 20, '34:1346-1348	3	S
2049.	M.B. 2		
	Flight, Je 1, '39:563	1	
	D 19, '52:753-754	1	
2050.	M.B. 3		
	Fly Rev Intl, F '66:364	1	
2051.	M.B. 5		
	Aeroplane, N 30, '45:615-618	1	
	Ja 9, '48:48-51	1	C*
	Flight, N 29, '45:588-590	1	C
	D 18, '47:679-681	4	
MARTINSYDE			GB
2052.	A		
	CAHS, Sum '66:37 (-II)		Pr
	:38-41 (-I&-II)	1	
2053.	F. 4 "Buzzard"		
	Air Pict, Ja '58:7	1	
2054.	F. 6		
	Air Pict, Je '69:225	1	
	CAHS, Sum '66:40	1	
2055.	G. 100 and G. 102 "Elephant"		
	Air Pict, S '55:269	1	
2056.	S. 1 "Scout"		
	Air Pict, Ja '63:16-19	1	
	Avn Wk, O 17, '60:36		D*
MATRA-CANTINIEAU			France
2057.	MC-101		
	Am Helicop, D '52:10-11+	1	P
MAUBOUSSIN			France
2058.	M. 11		
	Flight, Ja 15, '32:48-49	1	

MAULE USA

2059. M-4 "Rocket"
Flying, Ap '65:32-37 4

MAX HOLSTE France

2060. "Broussard" MH-1521
Aeroplane, F 19, '60:234-236 4 C
Avn Wk, Ag 10, '59:104-109 4
Flight, Mr 23, '56:329-330 4

2061. "Super Broussard" MH-250, MH-260, MH-262
Avn Wk, Ja 25, '60:87-91 (-260) 1
Flight, Mr 27, '59:439 (-250&-260) 1
N 13, '59:542-544 (-260) 4 C
Je 22, '61:860-861 (-260) 4 C
Flying, D '61:33+ (-260) 4 C
Interavia, No. 8, '61:1113-1115 (-260 & -262) 1 C
RAF Fly Rev, Jl '61:42-43 (-260) 1 P*

MAYO COMPOSITE GB
See SHORT "Mayo Composite"

MERCKLE Germany

2062. SM-67
Avn Wk, Jl 17, '61:119-121 1

MERCKLE-KIEBITZ Germany

2063. LF 2 Model 501
Avn Wk, Mr 14, '60:106-107 1

MERIDIONALI Italy
See IMAM

MESSERSCHMITT Germany

2064. Bf. 108 or Me. 108 "Taifun" (Typhoon)
Air Pict, O '57:368-369 (-B1) 1
Air Prog, Ag/S '64:66-67 (SNCAN-built) 1 C
Can Avn, Jl '35:28 1
Flight, Ap 8, '37:357 1
Je 29, '39:661 1 C
Jl 29, '43:118a (-B) 1

MESSERSCHMITT (cont.)

2064. Bf. 108 or Me. 108 (cont.)

RAF Fly Rev, S '57:23	1		
F '63:43+	1		
West Fly, O '47:24-26	1		

2065. Bf. 109 or Me. 109
Czech-built: AVIA S. 199
Spanish-built: HISPANO Ha-1109, Ha-1110, Ha-1112

Aero Album, Spg '69:44-45 (2-seater)	1		
:46 (-G)	6	(Cockpit)	C
Aeronautics, My '45:42-44	2		
Aeroplane, O 19, '39:483			D
O 26, '39:512-513	1		S*
O 4, '40:372-373	1		C
Ag 1, '41:120-122 (-F)	1		S*
F 6, '42:152-153 (-E&-F)	1		
Air Cl, Mr/Ap '65:69 (-G6:J.G. 54)			Pr*
D '68:26-27, 34-35 (-K-4: Hartmann's)			D*, Pr*
:28-29 (-G-6)			P
:31 (-F)			C
AF News Ltr, Je '42:37-39	1		S*
Air Pict, Je '58:201-203	1		Pr
Ap '60:109-111+ (Pt 1)	1		
My '60:149-153 (Pt 2)	1		Pr
Je '60:211-214 (Pt 3)	1		Pr
Jl '60:251-255 (Pt 4)	1		Pr
Ag '60:281-285 (Pt 5)	1		Pr
D '63:407 (-E:Heydrich's)	5		
S '64:303 (-E:Wick's)	5		Pr†
:303 (-E-3:'41 USA exhibit)	5		
Ja '66:27 (North African-)	5		
S '68:348 (-F-2:Mölder's)	5		
S '69:324-329 (-E)	4		C*
N '69:429 (-F-4:Marseille's)	5		
Air Prog, Ja '43:54 (-E)			C*
'54/'55 Annual:44-45 (-E)	1		Pr†, S*
Ja '67:28 (Belenye's-)			Pr†
D '68:42-43 (-E)			Pr*
Ja '69:46-47+ (-F-4: Marseille's)	1		Pr†
Ag '69:46-47+ (-G)	4		
Air Tech, O '44:60-61+	2		

MESSERSCHMITT (cont.)

2065. Bf. 109 or Me. 109 (cont.)

Acft Eng, D '41:346-348	6 (Radio)	P
Mr '42:70-72 (-E)	1	
:72-75 (-E&-F)	2	
Je '42:164-167	6 (Radio)	
Ag '44:239-241	6 (Fuel Tank)	
Aviation, Ag '41:100-102+	6 (Radio)	
O '41:50-51 (-F)	1	S*
Mr '43:169-171 (-E)		S*
Can Avn, S '41:31+ (-F-2)	1	S*
CAHS, Sum '64:50 (-E-3:Captured "1190")	1	
Comm Avn, Ja '41:54		C
Mr '42:44-52+ (-E&-F)	2	
Ag '43:98-106 (-E&-F)	6 (Engine)	
Flight, O 5, '39:286a-d	1	
D 14, '39:490e-h	1	
O 24, '40:347-350	1	C*, S
Jl 31, '41:62 (-F)	1	
F 5, '42:115-120 (-E&-F)	1	C*
N 11, '43:530b (-G)	1	
D 18, '69:Inset		P*
Flying, Mr '41:16-17	1	
Fly Rev Intl, O '63:48 (-E:Swiss)		Pr*
:49-51 (Swiss-)	4	
F '64:56 (-B&-E:Spain, Rumania, Slovak, Hungary)		Pr*
Ap '64:54 (-E&-G)		Pr*
O '65:125 (-TL)	1	
D '65:242, 244, 246 (12 German aces)		Pr*
N '67:1002 (-E)		Pr*
Pop Avn, Mr '40:27	1	S*
Jl '40:40	3	
RAF Fly Rev, S '57:24 (-Z)	1	
O '58:53 (-H)	1	
Ja '59:30	5	Pr†
F '59:6 (-E-3:'41 USA exhibit)	1	
N '59:22-24 (-E-3:USA)	1	Pr*
:46 (-E-7/U2)	1	
F '60:9+ (-E-3:USA)	1	
Ap '60:50 (-E-3:USA)	5	
Je '60:54 (-E-3:USA)	1	
O '60:9 (Swiss-)	1	

MESSERSCHMITT (cont.)

2066. Bf. 110 or Me. 110 "Zerstörer" (Destroyer)

Aero Digest, My '41:155-167	1	C
F 15, '44:124-125 (-C)		S
Mr 15, '44:120-121 (-C)		S*
Aeroplane, Ja 12, '40:36-41	1	S*
O 11, '40:406-407	1	C
Air Prog, O '66:34-37+ (Jabs'-)	1	Pr†
Acft Eng, N '41:308-310	1	
Aviation, Je '41:44-45+	1	S
S '41:79		S*
Ag '42:197	6 (Gun)	S*
Can Avn, Jl '40:18-21	1	Pr
Je '41:32-34+	1	S
Comm Avn, My '41:27-30+	1	S
Flight, Ja 18, '40:54a-b	1	
N 14, '40:412g	6 (Radio)	C*
S 10, '42:282a	1	
Flying, Mr '41:17	1	
Fly Rev Intl, O '64:86-93	1	Pr*, S*
D '65:246 (Streib, Gildner, Drewes, Schnaufer's-)		Pr*
Ag '67:806 (-C to -F)		Pr*
My '68:297 (-G-2&-4)		Pr*
RAF Fly Rev, Jl '55:43-44	1	
S '60:36-37 (-C)	4	

2067. Bf. 161 or Me. 161 and Bf. 162 or Me. 162 "Jaguar"

Aeroplane, Ap 12, '40:511 (-161)		D
Fly Rev Intl, Ag '67:813 (-162)	1	
RAF Fly Rev, N '62:41 (-161&-162)	1	

2068. Me. 163 "Komet"

Air News, D '45:19	1	
Air Prog, Je/Jl '63:43		Pr†
Aviation, Ja '46:77-80	1	
Flight, Mr 20, '47:231-232	6 (Engine)	S
Flying, Ja '50:12-15 (-A to -C)	1	
Fly Rev Intl, Mr '65:52 (-B-1a)		Pr*
S '65: (Vol. 21):27-29	1	Pr
My '67:585 (-S)	1	S
RAF Fly Rev, Je '54:11-13	1	
D '56:30-32+	1	Pr
Ja '59:17-19+	4	
:16 (-B-1)		Pr*
My '63:27-31+ (-B)	1	Pr, S*

MESSERSCHMITT (cont.)

2069.	Me. 209		
	Air Pict, Je '60:211-214	1	
	Fly Rev Intl, Mr '64:59 (-V1 to -V3)	1	
	RAF Fly Rev, O '58:50+	1	
	D '61:51-52	1	
2070.	Me. 210		
	Later development: Me. 410		
	Aero Digest, Mr '43:267-268+	1	
	Aeroplane, F 12, '43:194-197 (-A1)	1	S*
	Air News, Ap '43:7-9+	1	S*
	Air Prog, Jl '43:54-55 (-A1)		S*
	Air Tech, Ap '43:48-51 (-A1)	1	S*
	Aviation, Ap '43:235-237 (-A1)	1	S
	Can Avn, My '43:60 (-A1)		S*
	Flight, F 11, '43:143-146 (-A1)	1	S*
	Mr 25, '43:310a	1	
	Flying, My '43:45 (-A1)		S*
	RAF Fly Rev, Ag '55:33-34	1	
2071.	Me. 261 "Adolphine" (Little Adolph)		
	RAF Fly Rev, D '59:39+	1	
2072.	Me. 262 -A-1 = "Schwalbe" (Swallow) or -A-2 = "Sturmvogel" (Stormbird)		
	World's first operational jet fighter		
	Aeroplane, My 4, '45:516	1	
	Air Cl, Je '69:50 (-A-2A:Jagd Geschwader 7)		Pr*
	:51-61	1	P, Pr†
	Air News, D '45:20-21	1	
	Air Pict, Ja '59:15-17+	1	Pr
	Air Prog, Win '62/'63:84		C*
	Acft Eng, O '46:347-348	6 (Hydraulic)	S
	AAHSJ, Fall '63:183		Pr
	Aviation, O '45:115-135 (Pt 1)	1	C, S*
	N '45:115-130 (Pt 2)	1	S*
	CAHS, Win '64:110-112	4	
	Flying, F '53:20-21+	1	
	My '55:32-34+	1	
	Fly Rev Intl, O '66:152, 154 (-A-1&-A-2)		Pr*
	RAF Fly Rev, F '54:11-13	1	
	Je '55:27-29+	1	
	F '56:26-30	1	Pr, S*
	Je '60:35	5	
	Ja '63:26-29+	4	P

MESSERSCHMITT (cont.)

2073.	Me. 264			
	RAF Fly Rev, F '59:45-46	1		
2074.	Me. 309			
	RAF Fly Rev, N '61:37-38	1		
2075.	Me. 323 "Gigant" (Giant)			
	Air Pict, O '65:367	1		
	Flight, Jl 8, '43:49	1		
	S 16, '43:310a	1		
	Flying, Ja '44:30-31	1		S
	RAF Fly Rev, Ja '55:29	1		
2076.	Me. 328			
	Suicide jet			
	RAF Fly Rev, My '55:25-27+	1		
	F '58:32-33	1		
2077.	Me. 329			
	RAF Fly Rev, O '55:11	1		
2078.	Me. 410 "Hornisse" (Hornet)			
	Development of Me. 210			
	Acft Eng, O '46:347-349	6	(Hydraulic)	S
2079.	P. 1101			
	RAF Fly Rev, O '62:63	1		

MESSIER France

2080.	Monoplane (1936)			
	Pop Avn, Ja '36:23	1		

MEYERS USA

2081.	OTW Trainer			
	Aviation, Je '41:97	1		

MFI (Malmo Flygindustri) Sweden

2082.	9 "Mili-Trainer"			
	Air Prog, S '66:40-44	1		C, Pr

MFI (cont.)

2083.	10 "Vipan" (Peewit)		
	Flight, My 7, '64:779-780	4	
	Fly Rev Intl, Ag '64:40-41	4	
	RAF Fly Rev, Ap '61:42-43	1	
2084.	15		
	-A = Air Force -B = Army		
	Avn Wk, F 3, '69:61 (-A&-B)	1	
	O 6, '69:19	1	P
	Fly Rev Intl, S '69:31 (-A&-B)	1	
	D '69:30-31 (-A&-B)	1	

MIG (Mikoyan and Gurevich) Russia

Fly Rev Intl, N '65:155-161+

2085.	[Faithless]		
	STOL fighter		
	Avn Wk, Ap 7, '69:21	1	P
	Ap 14, '69:32	1	P
	Fly Rev Intl, Je '69:33	1	
2086.	[Flogger]		
	Variable geometry fighter		
	Avn Wk, Jl 17, '67:26-27	1	
2087.	E-166		
	Avn Wk, Jl 31, '67:58-59	1	
2088.	I-270 Rocket Fighter		
	Fly Rev Intl, F '68:109	1	
2089.	1		
	RAF Fly Rev, N '55:11	1	
2090.	3		
	Flight, D 16, '43:666	1	
	Fly Rev Intl, Ja '67:327-329	1	Pr*
	RAF Fly Rev, N '55:11	1	
2091.	8 "Utka" (Duck)		
	Flying, Ap '49:58	1	
	Fly Rev Intl, F '65:43	1	
2092.	9 [Fargo]		
	Fly Rev Intl, Ag '66:774		Pr*
	Nav Avn News, Mr '52:20-21	1	
	RAF Fly Rev, D '56:15	1	

MIG (cont.)

2093.	15 [Fagot]		
	First code name: [Falcon]		
	Aero Digest, Jl '51:17-19+	2	
	Aeroplane, Ag 1, '52:160-162	1	S*
	N 13, '53:656-657		P
	Air Force, Je '51:25	1	S*
	Air Prog, Fall '62:73		C
	Ap/My '63:71		C
	Avn Age, F '51:21-25	1	Pr, S*
	Avn Wk, N 13, '50:36	1	
	Ap 2, '51:32	1	S
	Jl 7, '52:10-15	1	
	Ap 13, '53:17-19	1	S*
	Je 1, '53:14	1	
	N 2, '53:46-48	3	
	N 30, '53:16-17	1	P
	Flight, Jl 25, '52:102-104	1	D
	Ja 28, '55:112-113	1	C*, S*
	Fly Rev Intl, Ja '65:29-32+ (Pt 1)	1	Pr, S*
	F '65:31-33 (Pt 2)	1	
	Interavia, No. 8-9, '50:432	1	
	Nav Avn News, Ja '51:14-15	1	
	Ag '53:12 (-2)	1	
	RAF Fly Rev, Je '53:11-14	4	S*
	D '53:21+ (Suk's-)	4	
	Ja '54:12-13	1	S*
	US Air Svcs, My '52:8	1	
2094.	15UTI [Midget]		
	Two-seat trainer version		
	Flight, Ap 24, '69:669-672	4	C
2095.	17 [Fresco]		
	Air Pict, Jl '59:260-261	1	Pr
	Air Power, Ap '56:174		P
	Avn Wk, Ja 9, '56:28-29	1	
	Fly Rev Intl, F '64:42-43 (-C:Poland)	1	
	S '64:20-23 (-A to -D)	1	Pr
	S '69:51 (-PF:Hungary)		Pr*
	RAF Fly Rev, Ag '58:8-9	1	Pr*
2096.	19 [Farmer]		
	Air Pict, S '58:315-317	1	Pr
	Ag '63:244-245	1	Pr
	Avn Age, O '57:36-45	1	

MIG (cont.)

2096. 19 (cont.)

Avn Wk, Ja 7, '57:28-29	1	
D 2, '68:42-52 (Pakistani-)	4	
Flying, F '54:21	1	S
Fly Rev Intl, F '64:22-25+	1	Pr
F '69:45 (Pakistani-)		Pr*

2097. 21 [Fishbed]

Earlier code name: [Faceplate]

Air Force, Ap '59:54-55	1	
Air Pict, D '62:384-385	1	Pr
Air Prog, My/Je '66:6-8+	1	
Avn Age, F '58:44-49	2	
Avn Wk, D 16, '57:67	1	
S 5, '66:31 (-C)	1	
Jl 31, '67:50-56 (-C)	4	
:60 (-F)	1	
Ap 7, '69:21 (-G)	1	P
Ap 14, '69:32 (-G)	1	P
Flight, My 9, '63:662	1	
Fly Rev Intl, N '63:20, 22	1	
:21 (-C&-D)		Pr
N '66:195+	1	
N '68:43 (-F:India)		Pr*
RAF Fly Rev, Mr '61:15-16	1	

2098. 21 PF [Fishbed-D]

Air Prog, D '67:34		Pr
Avn Wk, Jl 24, '67:22-23	1	
Flight, Ap 24, '69:664		P
Fly Rev Intl, Ja '68:11		P*
Interavia, No. 5, '69:603	1	P
RAFs Qtr, Win '67:271		P

2099. 23 E-266 [Foxbat] and [Flipper]

Air Pict, S '62:280-281 [Flipper]	1	Pr
Avn Wk, D 18, '67:54-55	1	
Je 3, '68:22	4	
Fly Rev Intl, Je '64:28-29+ [Flipper]	1	Pr
Ja '68:26	4	P
O '68:27-28	1	

MIL — Russia

Fly Rev Intl, O '65:69-73 (Helicopters)

MIL (cont.)

2100.	Mi-1 [Hare]		
	Aeroplane, Ag 17, '56:204-205	4	
	Jl 12, '67:26	1	
	Air Pict, Ap '52:114	1	
	N '63:358-359	1	Pr
	Avn Wk, Jl 8, '57:57-71	1	C*, S*
	Ap 7, '58:45	1	
	Jl 18, '60:76	1	
	Flight, Ap 20, '56:444-445	1	S*
	RAF Fly Rev, My '56:36-37	1	S*
	Skyways, Je '52:13+	1	
2101.	Mi-2 [Hoplite]		
	Polish manufacturer: WSK/MIL Mi-2		
	Aeroplane, Jl 12, '67:26	1	
	Avn Wk, Ja 16, '67:73	1	
2102.	Mi-6 [Hook]		
	Aeroplane, N 22, '57:760	1	
	Air Pict, Mr '64:76-77	1	Pr
	Air Prog, N/D '65:24-25	1	C
	Avn Wk, N 18, '57:34	1	
	Je 22, '59:91-92	1	
	Ag 7, '61:75	1	
	Ag 24, '64:86-89	1	C
	Jl 26, '65:36-47	1	
	Fly Rev Intl, My '66:549	1	
	D '68:34-35		P*
	RAF Fly Rev, N '59:34-35	1	S*
	US Army Avn, N '65:30-32	1	
2103.	Mi-8 [Hip]		
	Avn Wk, My 18, '64:99	1	
	Jl 26, '65:36-47	1	C
	Fly Rev Intl, My '66:549	1	
	Interavia, No. 3, '69:253-257	1	C
2104.	Mi-10 [Harke]		
	Air Pict, Mr '64:77	1	Pr
	Air Prog, N/D '65:18-20	1	C
	Avn Wk, O 9, '61:25	1	
	My 31, '65:24	1	
	Je 14, '65:84-86	1	
	Jl 26, '65:36-47	1	
	My 16, '66:29 (-K)	1	
	Ap 10, '67:48-49	1	

MIL (cont.)			
2104.	Mi-10 (cont.)		
	Fly Rev Intl, My '66:550	1	
	US Army Avn, N '65:30-32	1	
MILES			GB
	Comm Avn, D '44:161-163+		
2105.	"Aerovan" M. 57		
	Aeronautics, S '46:32-35	1	P
	Ag '47:44-45	4	C
	D '47:45		P*
	Aeroplane, Je 8, '45:658-660	1	C, P
	Air News, S '45:73-75	1	
	Acft Eng, F '47:68-70 (Pt 1)	3	S
	Mr '47:101-107 (Pt 2)	3	S*
	Can Avn, Ag '45:86	1	
	Comm Avn, Jl '45:86+	1	
	Flight, Je 7, '45:604-605	1	
2106.	"Falcon" M. 3		
	Aeroplane, S 11, '35:317		S*
	Flight, Ag 23, '34:870-871	1	S
	Ja 10, '35:37-38	1	S*
	F 21, '35:203	1	
	Ag 1, '35:135	4	
	Ag 21, '69:Inset	1	P*
	Pop Fly, S '36:316	1	
2107.	"Graduate"		
	Flight, O 15, '64:665-666	1	
2108.	"Hawk" M. 2		
	Aeroplane, F 14, '34:270	1	
	F 27, '42:251	1	
	Can Avn, N '69:41 (Canadian-)	1	
	Pop Fly, Je '33:123	1	P
	Ja '34:536-537	4	
2109.	"Hawk Major" M. 2E and M. 2F		
	Aeroplane, Ag 1, '34:144-145	4	
	Ag 22, '34:227-230	1	S*
	Ag 7, '35:188-189 (De luxe-)	4	C*
	Flight, Jl 12, '34:717	1	
	D 13, '34:1336	6 (Flaps)	
	Ag 1, '35:135	4	
	Pop Avn, My '37:19+	1	

MILES (cont.)

2110.	"Libellula" M. 39B		
	Aviation, S '44:181+	1	
	RAF Fly Rev, Mr '58:52	1	
2111.	"Magister" M. 14		
	Aeronautics, N '46:91-92	4	C*
	Can Avn, Mr '42:31	1	
	Pop Fly, D '37:490, 514		P
	RAF Fly Rev, O '62:23	4	
2112.	"Magister II" M. 18		
	Aeroplane, F 5, '43:151	1	
	Flight, F 4, '43:115-118	1	C
	RAF Fly Rev, F '59:6	1	
	O '61:69	1	
2113.	"Marathon" M. 60		
	Later version: HANDLEY PAGE "Marathon"		
	Aeroplane, My 31, '46:630-633	1	
	Jl 19, '46:68-71	1	S*
	O 28, '49:585-587	4	
	Flight, My 30, '46:545-546	1	
	S 4, '47:248a-249	1	C, S
	Interavia, No. 4, '46:44-45	1	
2114.	"Martinet" M. 25		
	Tow-target aircraft		
	Can Avn, N '42:43	1	
	Flight, Jl 8, '43:39	1	
	Ap 11, '46:366a-367	4	
	Fly Rev Intl, D '66:277	1	
2115.	"Master I" M. 9A and M. 9B		
	Aeronautics, N '39:15-18	3	
	Aeroplane, Jl 26, '39:128-132	1	C, S*
	Air BP, No. 30:20-23	4	C
	Can Avn, N '40:41		C*
	Flight, Jl 27, '39:77-80	1	C*, S
	Ap 18, '40:348-349	4	C*
	Ap 11, '46:366a-367	4	
	Skyways, D '42:42+	1	C*, S*

MILES (cont.)

2116.	"Master II" and "Master III" M. 19 and M. 27		
	Air BP, No. 30:20-23	4	
	Flight, Je 26, '41:434 (-II)	1	
	O 22, '42:444a (-III)	1	
	Ap 11, '46:366a-367 (-II&-III)	4	
	Skyways, D '42:42+	1	C*
2117.	"Merchantman" M. 71		
	Flight, Ag 28, '47:221-225	1	S
2118.	"Mercury" M. 28		
	Aeronautics, Ag '46:38-39	4	
	Aeroplane, F 5, '43:152	1	
	Flight, F 4, '43:115-118	1	
	Ap 29, '43:444a	1	
2119.	"Merlin" M. 4		
	Aeroplane, Ap 3, '35:388-390	1	S*
	Flight, Ap 4, '35:351-353	1	S
2120.	"Messenger" M. 38		
	Aeronautics, D '45:43-46	1	S*
	Ja '47:44-45	4	C, P
	D '47:44		P*
	Aeroplane, Jl 29, '49:145-146	4	
	Can Avn, Je '45:84-85	2	
	Comm Avn, Je '45:94+	1	
2121.	"Messenger" M. 48		
	Flight, Ap 12, '45:396-398	1	
2122.	"Mohawk" M. 12		
	Aviation, F '37:47	1	S*
	Flight, N 5, '36:473-475	1	S
	US Air Svcs, D '36:38 (Lindbergh's-)	1	
	Ja '37:14 (Lindbergh's-)	1	P
2123.	"Monarch" M. 17		
	Aeroplane, Je 22, '38:779-781	1	
	:781	4	S*
	Flight, Je 23, '38:612a-d	1	C*

MILES (cont.)

2124.	"Monitor" M. 33		
	Aeroplane, O 5, '45:398-399 (-T.T.2)	1	
	Ja 4, '46:16-17 (-T.T.2)		P
	Flight, O 4, '45:376	1	
	RAF Fly Rev, Ap '61:47	1	
2125.	"Nighthawk" M. 7		
	Flight, F 20, '36:202c-d	1	C*, S
2126.	"Peregrine" M. 8		
	Aeroplane, My 6, '36:550		S*
	D 2, '36:721-722	1	C*
	Air Pict, F '58:66	1	
	Flight, N 14, '35:501	1	
	D 3, '36:594-597	1	C*, S
2127.	"Satyr" M. 1		
	Flight, Ag 5, '32:733	1	P
	Ag 12, '32:766		P
2128.	"Sparrowjet" M. 77		
	Aeroplane, O 1, '54:512-515	1	S*
	O 8, '54:549-551	4	C*
	Air Pict, Mr '54:83	1	
2129.	"Student" M. 100		
	Aeroplane, D 16, '55:937-939	1	S*
	My 24, '57:725-726	1	S*
	S 6, '57:390		S*
	S 5, '58:373-374	4	C
	Avn Wk, O 21, '57:127	1	C
	Jl 27, '59:84-93	4	
	Ag 5, '63:95-100	4	
	Flight, D 16, '55:915-917	1	S*
	S 18, '59:279	4	
	O 15, '64:665-666 (-2)	1	
	RAF Fly Rev, Ja '56:25	1	
	Jl '57:33-34	1	
2130.	"Whitney Straight" M. 11		
	Flight, F 4, '37:112-114	1	S*
	Ap 5, '62:521-522	4	
	Pop Fly, Mr '37:619	1	

MILES (cont.)

2131. "X Minor" M. 30
RAF Fly Rev, Mr '58:11+ 1
West Fly, D '43:76 1

2132. HDM. 105
Aeronautics, Je '57:48-49 1 P
Flight, Ap 19, '57:504-505+ 1 S*
Je 6, '58:776-777 4

2133. M. 20
Flight, Jl 6, '44:15 1
RAF Fly Rev, Ja '58:11 1

2134. M. 35
Tandem-wing experiment
RAF Fly Rev, Mr '58:52 1

2135. M. 52 E. 24/43
Interavia, No. 6, '46:47 1
RAF Fly Rev, Ja '57:9 1

2136. M. 56
Can Avn, Jl '45:58+ 1

2137. M. 61
Flight, Ap 12, '45:396-398 1

2138. M. 68
Acft Eng, O '47:335 1
Avn Wk, S 29, '47:22 1
Can Avn, O '47:72 1
Flight, S 4, '47:252 1

2139. M. 218
See BEAGLE B. 218

2140. X. 1 to X. 14 Projects
RAF Fly Rev, Mr '58:11+ 1

MILITARY AIRCRAFT FACTORY Argentina

2141. "Bombi" Ae. M.B. 1 and Ae. M.B. 2
RAF Fly Rev, Je '63:49 1

MILLET-LAGARDE France

2142. ML-10
Flying, F '51:25+ 1
RAF Fly Rev, Ag '57:40 1

MIP Poland

2143. "Smyk" (Chit)
Flight, F 11, '43:152-154 1 S

MISSISSIPPI STATE UNIVERSITY USA

2144. XV-11A "Marvel"
Avn Wk, Jl 15, '68:54-69 1
Fly Rev Intl, F '67:351-352 1 S*
US Army Avn, Je '68:14-17 1 S*

MITSUBISHI Japan

2145. A5M1-4 Type 96 [Claude]
Flight, Jl 15, '43:70 (-2a) 1
RAF Fly Rev, Ja '59:8+ (-1 to -4) 1

2146. A6M1-8 "Zero-sen" [Zeke] or [Zero]
See also clipped wing A6M3 [Hamp]
Air Force, Ap '45:20+ (-52) 4
AF News Ltr, Jl '42:32-33 1
Air Pict, N '64:371 5
Air Prog, Ja '43:8-10+ 4
S '43:26-27+ 1
'56/'57 Annual: 33-35 1
F/Mr '64:18-19+ 1 S*
AAHSJ, Fall '69:207 1
Aviation, Ap '44:223 D*
Flight, My 20, '43:526b 1
Flying, N '42:25-26+ 1
D '43:65-66+ 1
Hist Avn, N '69:27-29 1
RAF Fly Rev, F '55:35 1
D '56:16-22 (-5C) 1 Pr*
Skyways, N '42:24-27 1
S '44:24-28+ 1 S*
West Fly, N '42:27+ 1

MITSUBISHI (cont.)

2147. A6M3 [Hamp] or [Hap]
Aviation, My '45:119-140 1 C*, S*
Flight, Ag 9, '45:146-148 1 S*
Flying, S '45:69 C
Hist Avn, N '69:27-29 1
RAF Fly Rev, D '56:20-21 S*

2148. A7M1 and A7M2 "Reppu" (Violent Wind) [Sam]
Fly Rev Intl, My '65:41+ 1
Nav Avn News, F '65:22 1
RAF Fly Rev, D '56:15 1

2149. C5M "Karigane II" (Wild Goose II) Type 98 [Babs]
Command reconnaissance version: Ki-15
Flight, D 16, '43:667 1

2150. F1M1-4 "Zero-kan" [Pete]
RAF Fly Rev, O '62:49 1

2151. G3M1-3 Type 96 [Nell]
RAF Fly Rev, O '58:41-42 1
O '59:36 1 Pr
:37-38 4

2152. G4M1-3 "Flying Lighter" [Betty]
RAF Fly Rev, D '57:25-26 1

2153. J2M1-5 "Raiden" (Thunderbolt) [Jack]
Air Cl, Mr/Ap '65:63 (-3) Pr
Air Pict, S '59:331-334 (-3) 1 Pr
RAF Fly Rev, My '62:18-20 1
:18-19 (-3) S*

2154. J8M1 "Shusui" (Rigorous Sword)
Japanese version of MESSERSCHMITT Me. 163B
RAF Fly Rev, Mr '56:39-40 P, S
My '61:51 1

2155. K3M1 and K3M2 Type 90 [Pine I]
RAF Fly Rev, Ja '60:40 1

2156. Ki-15 "Karigane" (Wild Goose) [Babs]
Land reconnaissance version: C5M
Aeroplane, Ap 14, '37:439 ("Kamikaze") 1
Flight, Ap 15, '37:374+ ("Kamikaze") 4 C,
Pop Avn, Jl '37:40+ ("Kamikaze") 4
RAF Fly Rev, Mr '63:47+ 1

MITSUBISHI (cont.)

2157.	Ki-21 Type 97 [Sally]		
	Earlier code names: [Gwen] and [Jane]		
	Flight, Je 17, '43:634b	1	
	RAF Fly Rev, Ja '56:17	1	
2158.	Ki-30 [Ann]		
	Air Prog, F '68:46-47+	1	
	RAF Fly Rev, F '63:39-40	1	
2159.	Ki-46 "Shin-Shitei" (Command Reconnaissance) [Dinah]		
	Fly Rev Intl, Ja '66:314, 316 (-II&-III)		Pr*
	RAF Fly Rev, Jl '57:35-36	1	
2160.	Ki-51 [Sonia]		
	RAF Fly Rev, F '63:39-40	1	
2161.	Ki-67 "Hiryu" (Flying Dragon) [Peggy]		
	Fighter variant: Ki-109		
	RAF Fly Rev, My '58:37-38	1	
2162.	Ki-83		
	RAF Fly Rev, N '57:42	1	
2163.	Ki-109		
	RAF Fly Rev, My '58:37-38	1	
2164.	MU-2		
	Avn Wk, Mr 3, '69:80-81 (-G)	1	
	Can Avn, Mr '69:24-26 (-B&-F)	4	
	Flying, O '64:26-31	1	C
	D '65:55-59	1	
	O '66:60-61	1	
	Jl '67:40-47 (-B)	4	C
	Fly Rev Intl, Jl '69:25-26 (-G)	1	
	Interavia, No. 6, '66:932-934	1	S*
2165.	XT-2		
	Aerosp Intl, Jl/Ag '69:46	1	
	Air Force, Jl '69:25	1	
	Avn Wk, Ag 25, '69:20-22	1	
	Flight, F 20, '69:302	1	
166.	Type 92		
	Flight, Ja 5, '33:13	1	

ML (R. Malcolm, Ltd.) GB

2167. "Flying Jeep" or "Rotaplane 10/42"
Air Pict, Ja '66:17 1

2168. "Utility" Mk. 1
Flight, Ag 17, '56:246 1
My 31, '57:751-752 1

MONOSPAR GB
See also GENERAL AIRCRAFT

2169. S.T. 4
Pop Fly, F '35:576-577 3

2170. S.T. 6
Aeroplane, Mr 14, '34:464 4
Pop Fly, Je '34:126-127 4

2171. S.T. 10
Aeroplane, O 31, '34:523-524 4
Acft Eng, S '34:240 1
Flight, Jl 12, '34:708-709 1 S
Jl 19, '34:744-747 1 Pr, S

2172. S.T. 11
S.T. 10 with retractable gear
Aeroplane, S 5, '34:286-287 1 S*
O 31, '34:523-524 4

2173. S.T. 12
Flight, Ja 3, '35:12-13 1 S

2174. S.T. 18 "Croydon"
Only one built
Aeroplane, My 6, '36:551 S*
Flight, My 16, '35:535-536 1
My 7, '36:478-480, 486 1 C, S

2175. S.T. 25 "De Luxe," "Universal" and "Jubilee" (3 versions)
Flight, Je 20, '35:666h 1
Je 27, '35:718-719 1
Ap 2, '36:344-346 1 C*,
Je 4, '36:607 1
Pop Avn, F '38:37 ("Florence Nightingale") 1

MONSTED-VINCENT USA

2176. MV-1 "Starflight"

Avn Wk, O 18, '48:15	1	
Flying, Ja '49:56	1	
Skyways, F '49:26-27+	1	

MONTE-COPTER USA

2177. 14

Am Helicop, Je '58:9+	1	

2178. 15 "Triphibian"

Air Pict, O '63:339	1	

MOONEY USA

2179. MU-2
See MITSUBISHI MU-2

MORANE-SAULNIER France

Air Pict, My '55:132-134 (Pt 1) (Aircraft types)
Je '55:168-170 (Pt 2) (Aircraft types)
Jl '55:212-213 (Pt 3) (Table of all acft types)

2180. A. I

Air Prog, O '65:56		Pr†
Cr & Cockade, Aut '63:264-271	1	C, S, P
:272-273	4	

2181. L "Parasol"

Aerosp Hist, Win '69:56		D*
Avn Wk, N 12, '62:36-37 (-2)	1	D*
Cr & Cockade, Aut '64:294		P
Flying, N '60:75 (Warneford's-)	1	D
Pop Fly, Ja '35:531	1	D
RAF Qtr, Jl '39:308-311	1	

2182. N "Bullet" and "Avenger"

Avn Wk, Jl 11, '60:90 (Kasakov's-)		D*
RAF Fly Rev, Je '58:16 (Russia's 19 Sq)		Pr*
RAF Qtr, Jl '39:309	1	

2183. 138-E2

Aeroplane, My 29, '42:623 (RAF-)	3	

MORANE-SAULNIER (cont.)

2184.	225 and 275		
	Air Prog, N '66:47		Pr†
2185.	325		
	Fly Rev Intl, N '68:71 (-C1)	1	
	Pop Avn, Jl '35:25+	1	
2186.	405		
	Trainer version: 430		
	Pop Avn, Ag '37:22 (-C1)	1	
2187.	406		
	Trainer version: 435		
	Swiss-built version: DORNIER D-3800		
	Flight, Ap 4, '40:304a-c	1	D, S
	Fly Rev Intl, Je '64:64		Pr*
	RAF Fly Rev, Ag '56:11	1	
	Je '60:24, 26-27	1	
	:25 (Groupe de Chasse 1/7-)		Pr*
	Mr '61:19-21 (Mörko Morane-)	1	
2188.	410		
	Variant of 406		
	Fly Rev Intl, My '69:73	1	
2189.	430 and 435		
	Fighter version of 435: 405		
	Fly Rev Intl, O '69:73	1	
2190.	450		
	RAF Fly Rev, S '62:43+	1	
2191.	700, 703 and 704		
	Avn Age, Ag '53:30-31	1	
2192.	730 to 733		
	Air Pict, My '52:150	1	
2193.	755 "Fleuret"		
	Avn Age, Je '54:22-23	1	S
	Flight, Ja 9, '53:44	1	

MORANE-SAULNIER (cont.)

2194. 760 "Paris"
Aeroplane, Ag 21, '59:32-33 4 C
Air Prog, Ap '67:19 (-III) 1 C
Avn Age, Mr '55:74-77 1
Flying, Ja '56:24-26+ 1 C
RAF Fly Rev, S '61:28-30 (-I to -III) 1 S*
:31 (-I) Pr
Skyways, Ag '55:12-14 1 C
West Avn, My '55:18 1
O '55:86-87 1

MORROW USA

2195. 1-L "Victory Trainer"
Aviation, F '42:116-117 S*
Can Avn, D '41:95-96 1

MUEGYETEMI SPORTREPULO EGYESULET Hungary
See TECHNICAL UNIVERSITY

MUREAUX France

2196. 113. R2
Flight, Ag 16, '34:846-847 1

2197. 130. A2
Aero Digest, Ap '30:108 1
Acft Eng, Mr '30:70 1

2198. 170. C1
Flight, My 10, '34:469 1

MURRAYAIR New Zealand

2199. MA-1
Converted STEARMAN "Kaydet"
Fly Rev Intl, My '69:32 1
N '69:30 1

MYASISHCHEV Russia

2200. MYA-4 [Bison]
Air Force, My '55:31-32+ 1
Avn Wk, Ag 8, '55:16-17 1
Jl 31, '67:63 (-201M) 1

MYASISHCHEV (cont.)

2201.	[Bounder]		
	Air Force, Ja '60:46-47	1	
	Air Pict, Jl '62:208-209	1	Pr
	Avn Wk, Ag 7, '61:72-73	1	
	Fly Rev Intl, D '64:32-34	1	Pr
	RAF Fly Rev, Je '60:13-14	1	
2202.	DVB-102		
	Fly Rev Intl, My '66:581	1	

NAGLER USA

2203.	NH-160	
	Avn Wk, Mr 26, '56:55-58	1

NAGLER & ROLZ Germany

2204.	NR. 54 and NR. 55 One-man helicopters	
	RAF Fly Rev, S '57:28+	1

NAKAJIMA Japan

RAF Fly Rev, Je '55:43-44 (WW II interceptors)

2205.	B5N1-2 [Kate]	
	RAF Fly Rev, Mr '57:49-50	1
2206.	B6N1-2 "Tenzan" (Heavenly Mountain) [Jill]	
	RAF Fly Rev, Mr '57:49-50	1
2207.	C6N1-2 "Saiun" (Painted Cloud) [Myrt]	
	RAF Fly Rev, Ap '55:10	1
	My '59:35	1
2208.	G8N1 "Renzan" (Mountain Range) [Rita]	
	Fly Rev Intl, My '64:63	1
	RAF Fly Rev, Ag '55:10	1
2209.	G10N1 "Fugaku" (Mt. Fuji)	
	Fly Rev Intl, Ap '64:57	1
	RAF Fly Rev, My '57:13	1
2210.	J1N1 "Gekko" (Moonlight) [Irving]	
	RAF Fly Rev, Ap '61:44 (-S)	1

NAKAJIMA (cont.)

2211.	J5N1 "Tenrai" (Heavenly Thunder)		
	RAF Fly Rev, D '55:11	1	
2212.	J8N1 "Kikka" (Chrysanthemum) or "Kitsuka" (Orange Blossom)		
	Air BP, No. 43:18-21	4	P
	RAF Fly Rev, Ap '55:41-42	1	
2213.	Ki-12		
	Air Pict, Je '65:203	1	
2214.	Ki-27 [Nate]		
	RAF Fly Rev, Ja '59:8+	1	
2215.	Ki-43 "Hayabusa" (Peregrine Falcon) [Oscar]		
	Air Pict, O '66:387 (Indonesian-)	5	
	Air Prog, Je/Jl '63:44		Pr†
	Aviation, Ag '45:191		S*
	Fly Rev Intl, Ag '65:52 (-I to -III)		Pr*
	RAF Fly Rev, N '57:16 (-II:Thailand)		Pr*
2216.	Ki-44 "Shoki" (Demon) Type 2 [Tojo]		
	Air Cl, Mr/Ap '65:69 (-IIb)		Pr*
	Fly Rev Intl, D '66:272 (-Ic, -II & Proto-)		Pr*
2217.	Ki-49 "Donryu" (Dragon Swallower) Type 100 [Helen]		
	RAF Fly Rev, N '60:61-62	1	
2218.	Ki-115 "Tsurugi" (Sabre)		
	RAF Fly Rev, My '61:40-41	1	
2219.	Type 91		
	Flight, Jl 6, '33:679	1	

NAMC (Nihon Aeroplane Manufacturing Co., Ltd.) Japan

2220.	XC-1A		
	Avn Wk, Ja 27, '69:46-47	1	
	Fly Rev Intl, Mr '69:26	1	
2221.	YS-11		
	Aeroplane, Mr 6, '59:284-286	1	S
	S 13, '62:54-55	1	
	F 23, '67:20-26	1	

NAMC (cont.)

2221.	YS-11 (cont.)		
	Avn Wk, D 26, '60:60-63	1	
	Ag 21, '61:87	1	
	O 10, '66:47-49	1	
	O 31, '66:176-177	1	C
	Ag 11, '69:60 (-A-207)	1	
	Flight, D 28, '61:984-985	1	
	Fly Rev Intl, D '66:239-240	1	Pr
	N '68:28-30	1	
	Interavia, No. 1, '60:100-101	1	
	No. 4, '62:427	1	

NAPIER-HESTON GB

2222.	Racer (1940)		
	Aeroplane, Ap 16, '43:447-449	1	S*
	Air Prog, S '43:58-59		S*
	Jl '66:56+	1	
	Flight, Ap 15, '43:387-391	1	S
	Fly Rev Intl, S '66:77-78	1	
	RAF Fly Rev, F '58:9	1	

NARDI Hungary

2223.	F.N. 315		
	RAF Fly Rev, Ja '63:49	1	

NASA (National Aeronautics and Space Administration) USA

2224.	"Hyper 3" Lifting Body		
	Avn Wk, S 29, '69:52-55+	1	

NATIONAL USA

2225.	NA-75		
	Modified STEARMAN 75		
	Flight, O 19, '56:639	1	
	West Avn, N '56:24	1	

NAVAL AIRCRAFT FACTORY Norway

2226.	M.F. 11		
	Fly Rev Intl, Ap '69:52		Pr*

NAVAL AIRCRAFT FACTORY USA

Nav Avn News, Ag '57:1-5
O '67:22-24 (Aircraft in WW I)

2227. F5L
Nav Avn News, Jl '55:24 2

2228. N3N "Yellow Peril" or "Yellow Bird"
AAHSJ, Fall '62:236-239 (-3) 1 Pr
Win '68:288-290 4 P
Nav Avn News, D '59:24-25 (X-/-1 to X-/-3) 1
Jl '66:6 (-1) P

2229. PBN-1 "Catalina"
Aero Digest, Ja 15, '44:88-94 1 S

2230. PN-9
Pop Avn, N '37:28, 30+ 4 P

NAVARRO GB

2231. "Chief"
Flight, Mr 6, '31:198-200 1 S

NAVY-CURTISS USA

2232. NC-1 to NC-4
AAHSJ, Spg '69:74 (-4) P
Boeing Mag, Jl '69:10-12 1
Nav Avn News, My '69:6-36, 45 1 D, S
Pop Avn, D '27:27-31+ (Atlantic crossing) 4
Jl '38:52-53+ 1 Pr
US Air Svcs, My '44:12-14+ 1
My '49:9-12+ (Atlantic crossing) 4

NAVY-WRIGHT USA

2233. NW-1 and NW-2
Air Pict, N '62:361-362 1

NB (Nicholas-Beazley Airplane Co., Inc.) USA

2234. "Plymocoupe"
AAHSJ, Fall '66:184-185 1 C, P
Pop Avn, Ag '35:96-97 1 C
N '35:292 1

NB (cont.)

2235.	3 "Barling"		
	Can Avn, F '30:14-15	1	
2236.	8		
	Air Prog, D '66:48-49+	4	C

N.E. (Designed by Royal Aircraft Factory) GB

2237.	1		
	Air Pict, Jl '61:202-206+	1	

NHI (Nederlandse Helicopter Industrie N.V.) Netherlands

2238.	H-3 "Kolibrie" (Hummingbird)		
	Air BP, No. 9:34-36	1	C
	Air Pict, F '59:39-41	1	C*
	Am Helicop, Mr '58:6-7+	1	
	Avn Age, D '57:184-185	1	
	Avn Wk, Ap 28, '58:85-86	1	S*
	Flight, N 14, '58:760	1	

NICHOLAS-BEAZLEY USA
See NB

NIEUPORT France
Air Pict, Jl '57:234-236 (Scouts)
Cr & Cockade, Aut '66:237-253 (V-struters)

2239.	Monoplane (1919)		
	Pop Avn, Jl '36:30	1	Pr
2240.	4		
	Air Prog, F/Mr '65:33 (-G)		Pr†
2241.	11 "Bébé" (Baby)		
	Air Cl, Ap '69:28	1	D
	Air Prog, F/Mr '65:34 (-C-1)		Pr†
	Avn Wk, O 9, '61:86-87	1	D*
	Cr & Cockade, Spg '61:86		Pr†
	Fall '69:246		P
	Flying, Ap '60:81 (Guynemer's-)	1	D
	Pop Avn, Ja '36:22		P
2242.	12		
	Air Prog, F/Mr '65:35 (-C-2)		Pr†
	AAHSJ, Win '61:Supp		P*

NIEUPORT (cont.)

2243.	14 Triplane		
	Air Prog, F/Mr '65:36		Pr†
	Pop Avn, My '36:344	1	Pr†
2244.	17		
	Air Pict, F/Mr '65:37 (-C-1)		Pr†
	Avn Wk, F 16, '59:147	1	D*
	Cr & Cockade, Spg '61:85		Pr†
2245.	23		
	Pop Fly, N '32:445	1	D
2246.	24		
	Air Pict, Jl '57:234-236 (Nungesser's-)	5	Pr
	Air Prog, Ja '65:57 (Nungesser's-)		Pr
	F/Mr '65:38		Pr†
	Cr & Cockade, Sum '66:193-194 (-bis: Restoration)	1	P
	Fly Rev Intl, Je '66:72 (-bis:Greece)		Pr*
2247.	27		
	Air Prog, F/Mr '65:38 (-C-1)		Pr†
2248.	28		
	Air Prog, F/Mr '65:39 (-C-1)		Pr†
	AAHSJ, Win '61:Supp		P*
	Sum '65:108-109	1	P
	Avn Wk, Je 8, '59:123		D*
	Flying, F '61:77 (-C-1:Rickenbacker's-)	1	D
	Jl '67:64-67	4	
	Fly Rev Intl, Mr '69:42 (Swiss-)		Pr*
2249.	29 "Vitesse" (Speed)		
	See also NIEUPORT-DELAGE 29		
	Aero Album, Sum '68:43-50	4	Pr
	Air Prog, F/Mr '65:40 (-C-1)		Pr†
2250.	121		
	Pop Avn, Ag '36:18 (Peruvian-)	1	P
2251.	140, 160 and 161		
	RAF Fly Rev, O '59:9+	1	

NIEUPORT-DELAGE			France
2252. "Sesquiplan" Racers			
Aero Album, Spg '68:21			P
Win '68:16-21		1	P, Pr
Air Prog, F/Mr '65:41			Pr†
2253. 29			
See also NIEUPORT 29			
Aero Album, Sum '68:43-48, 51		4	Pr
2254. 37			
Aero Album, Win '68:16-21		1	
2255. 52 and 62 Series			
RAF Fly Rev, N '61:44+		1	
S '63:32, 34		1	
:33 (-622:4e Escadrille)			Pr*
2256. 590 "Colonial"			
Pop Avn, F '33:78		1	
NIEUPORT & GENERAL			GB
2257. "Nighthawk"			
Air Pict, Ag '63:248-251 (Pt 1)		1	
S '63:298-301 (Pt 2)		1	
O '68:380-381		1	
Cr & Cockade, Spg '67:19-28		1	P, Pr, S
NIHON			Japan
See NAMC			
NIKITIN-SHEVCHENKO			Russia
2258. IS-1			
Air Pict, Je '66:227		1	
NOORDUYN			Canada
2259. "Norseman"			
Air Prog, O/N '64:33-36 (-IV)		1	C*, Pr†
Can Avn, Ja '36:10		1	
Mr '37:25-28		1	
Ag '45:50-53+ (-V)		1	C
My '65:1		1	D*
CAHS, Spg '67:9		1	

NOORDUYN (cont.)

2259.	"Norseman" (cont.)		
	Comm Avn, Jl '39:32 (-IV)	1	
	Je '45:30-33 (-V)	1	
	Flight, Je 3, '43:580b	1	
	Roundel, Je '64:33 (RCAF-)	1	P
2260.	C-64 "Norseman"		
	USAAF designation		
	Can Avn, Ap '43:52-53	1	
	Comm Avn, Mr '44:72+ (-A)	1	
	US Air Svcs, F '45:31-32	1	
	West Fly, Mr '44:120 (U-/-A:Aeronca)	1	

NORD France

2261.	260 and 262		
	260 = Unpressurized 262 = Pressurized		
	Earlier designation: MAX HOLSTE 260		
	Aeroplane, S 27, '67:17-20	4	
	Air Pict, S '62:282-283 (-262)	1	
	Avn Wk, Ag 20, '62:104-106	4	C
	Je 28, '65:44-45	3	
	N 8, '65:57-62	4	
	Can Avn, Mr '67:20-22	4	
	Flight, F 21, '63:254-255	1	
	Fly Rev Intl, My '65:28-30	1	S*
	O '66:106-107	1	C, Pr, P*
	Interavia, No. 2, '68:175-177	6 (Auto Landing)	C
2262.	500 "Cadet"		
	Avn Wk, Ag 5, '68:45	1	P
2263.	600		
	Interavia, No. 12, '65:1954-1955	1	S
2264.	1110		
	Flying, Je '61:46-47+	4	C
2265.	1400+ "Noroît"		
	Air Pict, F '51:39	1	
	Acft Eng, Jl '50:202-203+	1	
2266.	1402 "Gerfauti" (Gyrfalcon)		
	Aeroplane, Ag 13, '54:211		P
	Air Pict, Mr '54:68, 91	1	P
	Flight, Ja 29, '54:118	1	
	Ap 13, '56:413-416	1	C

NOORDUYN (cont.)

2267. 1500+ "Griffon" (Griffin)
Originally named "Guépard" (Cheetah)
Avn Wk, Ja 26, '59:52 — 1
Sp/Aeron, My '59:61-62+ (-II) — 6 (Engine)

2268. 1700+ "Norelic"
Flight, Ap 8, '48:393 — 1

2269. 2100+ "Norazur"
Fly Rev Intl, D '64:52 — 1
Mr '65:43 — 1

2270. 2200
RAF Fly Rev, F '62:45 — 1

2271. 2500+ "Noratlas"
Aeronautics, O '52:56 — 1
Avn Age, My '52:43 — 1
Avn Wk, Jl 17, '67:89-96 — 4
Interavia, No. 1, '54:35-39 — 3
RAF Fly Rev, Je '56:27-29 — 1 — S*

2272. 3400
Air Pict, My '58:171 — 1
Flight, Mr 11, '60:333-334 — 4 — C

NORTH AMERICAN — USA
Comm Avn, O '40:36-37+
Flying, S '61:28-29+ (Pt 1)
O '61:52-53+ (Pt 2)
N '61:58-59+ (Pt 3)
Nav Avn News, N '56:16-19 (US Navy aircraft)
Pop Avn, Ap '37:41+
Ja '39:42-45 (Operations in 1939)
Skyline, Vol. 7, No. 1, '37:4-9
Vol. 3, No. 1, '42:22-27
Vol. 25, No. 4, '67:2-52 (Entire issue)

2273. "Sabreliner"
Civil version
USAF designation: T-39
US Navy designation: T3J
Avn Age, Je '58:30-31 — 1 — S
Avn Wk, D 9, '57:53-65 — 1 — D
Flight, Mr 19, '64:422-423 — 4

NORTH AMERICAN (cont.)

2273.	"Sabreliner" (cont.)		
	Flying, O '60:92	1	
	My '64:48+	4	
	My '66:42-47	1	C
	RAF Fly Rev, Mr '60:18-19	1	P*, S*
	Skyline, Vol. 17, No. 1, '58/'59:50-55	1	C
	Vol. 23, No. 1, '65:2-15	4	
2274.	XA2J-1		
	Aero Digest, F '52:20-27	1	
	Air Pict, D '59:453-454	1	
	Avn Wk, Ja 14, '52:16	1	
2275.	A3J "Vigilante"		
	Later designation: A-5		
	Avn Wk, My 26, '58:30-31	1	
	S 8, '58:33 (-1)	1	
	S 22, '58:36		D*
	My 18, '59:140-141	3	
	Ag 6, '62:62-63 (-2)	1	
	Ag 27, '62:88-96	1	
	Flight, My 30, '58:731-732	1	
	Ja 30, '59:156 (-1)	1	Pr
	Interavia, No. 7, '58:739	1	
	No. 10, '60:1233-1234	1	
	Nav Avn News, Je '58:18	1	
	F '61:13	4	
	N '61:19-22	3	
	RAF Fly Rev, Ja '60:16-17	1	P*
	Ag '62:32-35 (-1)	1	Pr, S*
	Skyline, Vol. 16, No. 3, '58:2-3, 48	1	P
	Vol. 17, No. 2, '59:40-43	1	P
	Vol. 18, No. 1, '60:18-25	2	
	Sp/Aeron, F '59:58-60	1	
2276.	A-5 "Vigilante"		
	Earlier designation: A3J		
	Air Prog, Ja '65:39 (R-/-C)		Pr
	Je '69:38-41+ (R-/-C)	1	P*
	Flight, Ag 13, '64:251-157 (-A, -B, R-/-C)	1	S*
	Fly Rev Intl, Jl '69:27-28 (R-/-C)	1	
	Interavia, No. 9, '65:1410-1411 (R-/-C)	6 (Cameras)	S*
	Nav Avn News, Mr '63:17 (-A to -C)	1	P

NORTH AMERICAN (cont.)

2277.	A-36 "Invader"		
	Original design: P-51		
	Flight, N 11, '43:530a	1	
	Skyline, Vol. 4, No. 6, '43:15	4	
	Vol. 5, No. 2, '44:8-9	4	
	Skyways, D '43:104		D*
2278.	AJ-1 "Savage"		
	Aero Digest, F '52:20-27	1	
	Can Avn, N '48:98 (X-)	1	
	Flight, S 23, '48:380 (X-)	1	
	Nav Avn News, O '48:22 (X-)	1	
	Jl '50:1	1	P
	Skyline, Vol. 7, No. 2, '49:20-21	2	
	Vol. 9, No. 3, '51:4-9	4	
	Skyways, Ja '49:52 (X-)	1	
	West Fly, O '48:20 (X-)	1	
2279.	AJ-2 "Savage"		
	Nav Avn News, Ja '55:1-4	4	
2280.	AT-6 "Texan" and "Harvard"		
	See also SNJ, BT-9 and NA-16		
	Aeroplane, S 7, '45:276-277	1	
	Air Cl, Mr/Ap '65:36-39+	4	P*
	Air Pilot, S '44:4-6	4	S
	Air Prog, Ap/My '63:68 (-G)		C
	My '68:26-29+ (-IV:Canada)	1	C
	Aviation, Ag '42:144-148		S*
	Can Avn, Je '40:30-31+	3	
	O '40:19-22+ (Canadian-)	3	
	Mr '41:24-26+	4	C
	Ja '48:51	6 (Lube Chart)	
	Comm Avn, O '40:25-29 (Canadian-)	3	
	Ap '42:24-27 (-A)	3	
	Flight, Ag 22, '46:192a-b	4	
	Flying, Je '69:60-65	4	
	Fly Rev Intl, N '67:996 (Brazilian-)	4	P*
	Roundel, Ap '52:39-41	2	
	My '65:6-10 (RCAF-)	4	
	RAF Fly Rev, My '61:39	4	
	Skyline, Vol. 1, No. 3, '40:4-5	4	
	:12 (-II)	1	
	Vol. 2, No. 1, '41:8 (-A)	3	
	Vol. 2, No. 3, '41:6-9	3	
	Vol. 3, No. 5, '42:2-5+	4	

NORTH AMERICAN (cont.)

2280.	AT-6 (cont.)		
	Skyline, Vol. 4, No. 2, '43:20-21	1	
	Vol. 7, No. 2, '49:14-16	1	
	US Air Svcs, Je '40:12 (-A)		P
2281.	YAT-28E		
	Avn Wk, Mr 16, '64:187		P*
2282.	XB-21 "Dragon"		
	Aviation, F '37:41	1	
	RAF Fly Rev, My '60:50	1	
	West Fly, Ja '37:16+	1	
2283.	B-25 "Mitchell"		
	See also XB-28		
	Aeronautics, O '43:34-37		P
	Aerosp Hist, Win '68:50 (MATS-)		P*
	Air Cl, Mr/Ap '65:67 (-J)		P, S*
	Air News, Je '44:32-33+	6 (Gun)	S
	Air Pict, Ag '53:238	1	
	Air Pilot, Jl '43:44	6 (Wing)	
	Air Prog, Ap/My '63:70 (-B)		C
	Air Tech, O '42:33		P
	N '43:21-33+	1	C, S*
	Aviation, Mr '43:106-111+	3	
	Mr '45:119-142	1	S*
	Avn Wk, Mr 27, '50:15 (-J)	1	
	Comm Avn, Mr '45:84-86 (T-)	4	
	Ag '45:116 (Eisenhower's-)	1	
	Flight, Jl 9, '42:40 (-C)	1	
	S 17, '42:308a (-C)	1	
	Flying, Jl '44:64 (-G)		P*
	S '65:16 (-B&-H)	1	D
	Fly Rev Intl, O '67:934, 936 (-B, -C, -D, -H, -J, -II)		Pr*
	O '68:53 (-J:Uruguay)		Pr*
	Hist Avn, D '69:8-10+	1	Pr, S*
	Roundel, Jl/Ag '64:33 (RCAF-)	1	P
	Skyline, Vol. 2, No. 1, '41:12-13	1	
	Vol. 3, No. 3, '42:2-4+, 18-19	4	
	Vol. 5, No. 1, '44:10-11	4	
	Skyways, My '44:18-23+	1	C, S*
	West Fly, Ja '43:44-45+	4	
	N '43:40-41+	3	
	Mr '44:46-47+ (-G)	6 (75mm Cannon)	

NORTH AMERICAN (cont.)

2284.	XB-28		
	High altitude B-25		
	RAF Fly Rev, D '62:47	1	
2285.	B-45 "Tornado"		
	Air Pict, My '52:135 (R-/-C)	1	P
	Avn Wk, Ag 11, '47:24-25	2	S
	Flight, O 13, '49:498	6 (Cockpit)	C
	Ap 22, '55:523-525 (-A)	4	C
	Interavia, No. 4, '47:14-16 (X-)	1	P
	Skyline, Vol. 7, No. 2, '49:17-19	1	
	Vol. 8, No. 3, '50:10-11 (R-/-C)	1	
	Skyways, Je '52:10-12+ (R-/-C)	4	
	US Air Svcs, Ap '47:16 (X-)	1	
	West Fly, O '50:16-17+	1	
2286.	XB-70A "Valkyrie"		
	Aeroplane, N 4, '60:613-616	4	
	Air Force, Ja '60:29-33	1	
	AF/Sp D Intl, Ag '66:43-44	4	
	Air Prog, Ag/S '64:26-29	1	
	My/Je '66:18-19	1	
	Avn Wk, F 22, '60:87-90	6 (Ejection Seat)	
	Je 27, '60:30-31	1	
	Ag 8, '60:102-113	1	
	Mr 6, '61:83-85	6 (Aux Power)	
	Ag 12, '63:85-86	1	
	My 18, '64:26-30	1	
	Je 1, '64:84-91	1	
	S 28, '64:24-26	2	
	My 24, '65:60-65	1	C
	Je 6, '66:65-73 (Pt 1)	2	
	Je 13, '66:60-70 (Pt 2)	2	
	F 13, '67:38-39	1	
	Boeing Mag, Mr '59:10-11	6 (Wings)	
	Can Avn, My '67:16-18	4	
	Flight, Je 25, '64:1055-1062 (Pt 1)	1	S*
	Jl 2, '64:18-24 (Pt 2)	1	S*
	O 1, '64:577	2	
	O 7, '65:615	1	
	Flying, Ap '60:34-36	1	D
	Fly Rev Intl, Ja '64:22-25+	1	Pr
	Ap '64:22	1	
	My '66:511-512	1	
	Interavia, No. 6, '61:800-803	1	
	No. 7, '64:982-984	1	

NORTH AMERICAN (cont.)

2286. XB-70A (cont.)

RAF Fly Rev, Ag '58:13-14	1		
Ap '60:9+	1		
Skyline, Vol. 23, No. 3, '65:2-13	3		
Sp/Aeron, My '60:75+	6	(Hydraulic)	
Je '60:110-113	3		S*
Ja '62:85+	6	(Hydraulic)	S*
West Aerosp, Je '64:12-14+	1		
West Avn, N '60:26-28+	6	(Escape)	
N '61:28-31+	1		

2287. BT-9 "Harvard"

Aeroplane, Ja 25, '39:118-119 (-B)	1		
Flight, Ja 26, '39:84-85	1		
US Air Svcs, Ap '39:18	4		P
West Fly, Je '36:21-22	1		P

2288. F2J

USAF designation: F-86

Avn Wk, Ja 29, '51:17	3		

2289. F-51 "Mustang"

Earlier designation: P-51

Later development: CAVALIER "Mustang"

Flying, D '48:30-31 (-H)	1		C*
Fly Rev Intl, O '68:53 (-D:Uruguayan)			Pr*

2290. F-82 "Twin Mustang"

See P-82

2291. F-86 "Sabre"

Foreign-built: COMMONWEALTH CA-21 and CANADAIR CL-13

US Navy designations: FJ- "Fury" (early) and F2J (later)

Interim designation for -D model: F-95

Aero Digest, Ap '51:22-23+	6	(Wing Tanks)	
Je '52:22-31	1		S*
Aeroplane, Je 26, '53:832-833 (-E)			S*
Jl 3, '53:20-23	1		
Air Force, Ja '53:36-39	1		
Air Prog, '54/'55 Annual:18-19 (-F)			S*
Ag '60:60-64	4		
F/Mr '63:82-83 (-D&-K)			Pr†
Ap/My '63:69 (-A)			C
:76 (-D)			C
Am Helicop, Ap '53:13 (-D)	6	(Rockets)	

NORTH AMERICAN (cont.)
2291. F-86 (cont.)

Avn Age, Jl '56:64-69 (-K)	6	(Fire Control)	
Avn Wk, O 6, '47:14 (XP-)	1		
Je 14, '48:11-12 (X-)	2		
S 27, '48:15 (-A)	2		
N 21, '49:18-26 (-A)	1		
Mr 20, '50:20-22 (E-/-A)	1		S*
Je 30, '52:52-54	4		
N 3, '52:13-15	4		
F 16, '53:59-60	6	(Yaw Damper)	
S 7, '53:15-16	4		
S 21, '53:48-50 (-D)	6	(Auto Engine Control)	
O 19, '53:21-26 (-F)	4		
Ja 11, '54:31 (-D)	1		
Ag 16, '54:380 (-D)	1		
D 6, '54:30-35 (-H)	1		
Flight, N 18, '48:592-593	1		
Ap 13, '50:455	4		
F 8, '51:148-149	1		
Ja 30, '53:136-146	1		C*, S*
:146-147	4		
Ja 15, '54:63-66 (-A to -K)	1		
Mr 25, '55:396 (-K)	6	(Fire Control)	
Ap 15, '55:476 (-D)	1		
Ja 25, '57:111-114 (-K)	3		
Flying, N '48:15-17+	1		
My '51:35			P*
Fly Rev Intl, O '64:94 (Aerobatic teams-)			Pr*
S '68:63 (-D:Denmark)			Pr*
F '69:45 (-F:Pakistan)			Pr*
Ap '69:53 (-F:Norway)			Pr*
Nav Avn News, F '48:33 (X-)			P
RAF Fly Rev, Je '55:52-53 (-D)	1		
Jl '55:30-33 (-F)	1		Pr, S*
Ap '57:16-20+ (Pt 1)	1		
My '57:38-41+ (Pt 2)	1		
Ap '59:23 (-F:Chinese Nat AF)			P*
Jl '63:30 (-K:Germany)			Pr*
:31+ (-K)	1		
Ag '64:56 (-F-40:Japan's "Blue Impulse")			Pr*
Skyline, Vol. 8, No. 1, '50:17-18	2		S*
Vol. 10, No. 2, '52:4-7 (-D)	1		
Vol. 11, No. 1, '53:4-7	1		S
:11-15 (-D)	4		

NORTH AMERICAN (cont.)

2291.	F-86 (cont.)			
	Skyways, Ap '50:14-16+	4		
	Je '51:10-12+	4		
	US Air Svcs, D '47:21 (XP-)	1		
	Mr '51:10-11	1		P
	West Avn, D '52:8 (-D)	4		
	Ag '53:21-23 (-D)	4		
2292.	YF-93A			
	Cancelled in favor of F-86D			
	Avn Wk, Ja 30, '50:13	1		
	N 3, '52:18	1		
	Flight, F 8, '51:148-149	1		
	Interavia, Nos. 1-2, '50:58	1		P
	Skyways, Ap '50:17	1		
2293.	F-95A			
	Interim designation: F-86D			
	Aeroplane, Jl 6, '51:15	6	(Air Cond)	S*
2294.	F-100 "Super Sabre"			
	See also F-107			
	Aeronautics, S '57:140-141	4		
	Aeroplane, Je 1, '56:474-476	1		
	Je 28, '57:905-906	4		
	Air Pict, Jl '56:230 (To -D)	4		
	Air Prog, F '68:18-21+	4		C
	Am Helicop, D '53:26	1		
	Avn Age, Ap '57:54-57 (-F)	1		
	Avn Wk, O 26, '53:12-13	1		
	D 19, '55:14-16	4		
	Bee-Hive, Fall '53:3-6	1		
	Flight, N 6, '53:612-614	1		
	Jl 30, '54:137-138 (-A)	1		S*
	F 25, '55:251	2		
	My 20, '55:680-684 (-A)	1		S*
	Je 8, '56:696-697, 721 (-C)	1		P
	Interavia, No. 12, '53:727 (Y-)	1		
	RAF Fly Rev, D '54:26-28 (-A)	1		Pr
	D '55:38-41 (-A to -C)	1		Pr, S*
	F '58:17-19+ (-A to -J)	1		
	:16 (-C-10-NH: "Thunderbirds")			Pr*
	O '60:19 (-F)	4		
	Skyline, Vol. 12, No. 2, '54:16-19	1		
	Vol. 13, No. 5, '55:20-25 (-C)	4		
	Vol. 20, No. 4, '62:26-31	3		
	West Avn, Ag '55:32 (-C)	1		

NORTH AMERICAN (cont.)

2295. F-107
Fighter-bomber version of F-100; cancelled for REPUBLIC F-105
Avn Wk, N 25, '57:27 (-A) 1
D 16, '57:80-81 (-A) 1 P
Ja 20, '58:69 (-A) 1
RAF Fly Rev, My '57:11 1
F '58:46 1

2296. F-108 "Rapier"
Avn Wk, D 29, '58:58-62 6 (Comm)

2297. FJ "Fury"
US Navy designation
USAF designation: F-86 "Sabre"
Aero Digest, O '48:28-30 (-1) 1 S*
N '52:38 (-2) 6 (Nose Gear)
Aeroplane, Ja 17, '47:74-75 (X-/-1)1
AAHSJ, Sum '68:116-122 (-1) 1 C, Pr†
Avn Wk, Jl 7, '47:35 (X-/-1) 6 (Generator)
Mr 22, '48:13 (-1) 4
Can Avn, F '47:21 (X-/-1) 1
Flight, Ja 15, '54:65-66 (-2&-3) 1
Flying, D '48:30-31 (X-/-1) 1 C*
Nav Avn News, O '47:25 (X-/-1) 1
F '48:1-5 (-1) 1
Je '49:9 (-1) 2
S '54:13-15 (-2) 4
Ja '55:27 (-4) 1
Ap '55:8 (-3) 2
Jl '55:32-33 (-3) 4
S '57:26-27 (-4B) 1 P
RAF Fly Rev, My '57:38-41+ 1
D '61:30 (-1:Calif Naval Air Reserve) Pr*
:31 (-1) 1
Skyline, Vol. 10, No. 3, '52:14-15 (-2)4
Vol. 13, No. 1, '55:22-27 (-3) 2
Vol. 13, No. 5, '55:12-19 (-1 to -4) 1
Skyways, Je '48:28-29+ (-1) 1
West Avn, Mr '52:28 (-2) 1

2298. L-17 "Navion"
Flying, Ag '51:24-25+ 4

NORTH AMERICAN (cont.)

2299.	NA-16 See also: AT-6, BT-9, SNJ, NA-50A, NA-68 and P-64		
	Aviation, Jl '36:38-41	1	
	S '36:44	2	
	Comm Avn, Jl '39:34 (-E)	1	
	Flight, F 27, '36:226	1	
2300.	NA-35 Primary Trainer		
	Aero Digest, F '40:113-114	1	
	Skyline, Vol. 1, No. 2, '40:10-11	1	S
	West Fly, F '40:23-24	1	
2301.	NA-50A Peruvian import version of NA-16		
	RAF Fly Rev, O '58:53	1	
2302.	NA-68 Thailand import version of NA-16		
	RAF Fly Rev, O '58:53	1	
2303.	NA-300 Military designation: OV-10A		
	Avn Wk, Jl 13, '64:38-42	1	S*
	Interavia, No. 10, '64:1502	1	
2304.	O-47		
	Fly Rev Intl, Je '64:71	1	
	Pop Avn, Jl '37:22	1	
2305.	OV-10A "Bronco" COIN (Counterinsurgency aircraft) Manufacturer's model: NA-300		
	Air Prog, Ag '66:30-32+	1	S
	Jl '69:48-50+	4	P*
	Avn Wk, F 8, '65:20-22	1	
	Jl 12, '65:32-34	1	
	D 13, '65:28-29	1	
	D 20, '65:59	1	
	Mr 10, '69:135		P*
	N 17, '69:67-69+	4	
	Flight, Mr 4, '65:337	1	
	Fly Rev Intl, Je '65:31-35	1	S*
	Ap '66:448-449	1	P*
	Je '68:311-313	1	P*, S*
	Interavia, No. 9, '65:1413-1416	1	S*
	No. 11, '69:1815		P*

NORTH AMERICAN (cont.)

2305. OV-10A (cont.)

Nav Avn News,	Ap '65:12	1	
	S '65:20-21	1	
	Je '66:28	1	
	S '68:24-25	1	
	S '69:22-33	4	
Skyline,	Vol. 23, No. 2, '65:28-39	1	C
	Vol. 26, No. 1, '68:40-47	3	
	Vol. 27, No. 4, '69:28-36	4	

2306. P-51 "Mustang"

Later designation: F-51

Dive-brake variant: A-36

See also CAVALIER "Mustang" and JURCA "Gnatsum"

Aero Album,	Fall '68:26-27 (-D:Eagleston's)	4	Pr†
	Win '68:34-35 (-D:Preddy's)	4	Pr†
	Spg '69:24-25 (-X)	1	P
	Sum '69:3-11 (French-)	4	Pr†
	:20-21 (-D:Carr's)	4	Pr†
Aero Digest,	Ja 1, '44:94-95		S
	Jl 15, '45:64-68+	1	S*
Aeronautics,	N '44:43-46	1	S
Aeroplane,	N 5, '43:527-531 (-I)	1	C*, S*
	Ja 28, '44:101 (-B)	1	P
Air Cl,	Mr/Ap '65:65 (-A)		P, S*
	D '68:6-7 ("Flying Undertaker"-)	5	P
Air Force,	O '46:29-36	1	S
	Mr '64:80-86	1	
Air Prog,	My '43:54-57		S*
	O '43:24		D
	'54/'55 Annual:46-47 (-B)		C*, S*
	Win '62/-63:86		C*
	Ja '68:42-43		S
	Jl '68:18-21+ ("Calif. Airmotive")	1	C
	Je '69:28-29+ (-D:Meyer's)	4	Pr†
Air Tech,	My '43:19-23+	1	S*
Airpwr Hist,	O '62:197-206	1	
AAHSJ,	Sum '64:128-131 (8th AF-)	4	
	:152		D
	Win '65:304 (-B&-D)		D
	Spg '68:31-33 (Swedish-)	4	
	Win '69:250-258 (RCAF-)	1	Pr†
Aviation,	F '43:147		S*
	Jl '43:162-167+	3	
	Jl '44:127-147	1	C*, S*
	Jl '45:197-201	1	

NORTH AMERICAN (cont.)

2306.	P-51 (cont.)		
	CAHS, Fall '69:67-71 (RCAF-)	4	
	Comm Avn, Je '43:44	1	
	Ag '45:117 (-H)	1	
	Flight, Jl 23, '42:92b	1	
	Jl 30, '42:122-123	1	
	Ag 20, '42:209-211	1	
	D 10, '42:627-630	4	S*
	Ja 27, '44:91 (-B)	1	
	Je 20, '46:613-615 (-F)	1	
	Flying, Ap '44:27 (-II)	1	
	S '44:32-38 (-A to -D)	1	S*
	S '45:86 (-H)	1	
	Fly Rev Intl, S '67:874 (-III&-B)		Pr*
	My '68:272 (-IV:19 Sq)		Pr*
	Roundel, Ap '64:33 (RCAF-)	1	P
	RAF Fly Rev, Mr '59:53 (British-)	1	
	Mr '60:26	1	
	:27 (-D:Gilier's)		Pr*
	Skyline, Vol. 3, No. 4, '42:5+	1	
	Vol. 4, No. 1, '43:16-17		D
	Skyways, Ja '44:22-26+	1	S*
2307.	P-64		
	Used only as trainer		
	Export versions: NA-50 and NA-58		
	RAF Fly Rev, N '57:9	1	
2308.	P-82 "Twin Mustang"		
	Later designation: F-82		
	Air Cl, O '69:14-18+	1	P, S
	Air Pict, Ap/My '63:78 (-B)		C
	Can Avn, Mr '46:40	1	
	Comm Avn, D '45:62-66	1	
	Flying, Ja '46:49	1	
	Ap '47:35	4	
	Fly Rev Intl, N '65:188-189	1	
	RAF Fly Rev, Ja '56:17	1	
	US Air Svcs, D '45:20	1	P
	Mr '47:7 ("Betty Jo"-)	4	
2309.	XSN2J-1		
	Nav Avn News, O '47:25	1	
	West Fly, My '47:46	1	C

NORTH AMERICAN (cont.)

2310. SNJ
USAAF designation: AT-6
Flying, Ja '42:101 P
Nav Avn News, S '47:17 4 C
Skyline, Vol. 4, No. 1, '43:10-11 4

2311. T2B "Buckeye Trainer"
Nav Avn News, My '64:3 1
Skyline, Vol. 25, No. 1, '67:40-47 4

2312. T2J-1 "Buckeye"
Air Prog, F/Mr '63:86-87 Pr†
Avn Wk, My 13, '57:52-59 1 C, S
D 30, '57:60-61 3
Skyline, Vol. 15, No. 3, '57:24-27 1 S*
Vol. 17, No. 3, '59:36-41 4
Sp/Aeron, D '58:68-69 1 P

2313. T3J "Sabreliner"
See also "Sabreliner"
USAF designation: T-39
Nav Avn News, Ja '62:33 1

2314. T-6
See AT-6

2315. T-28 "Trojan"
Utility conversion: PAC "Nomad"
Aero Digest, My '54:31-36 (-B) 1 S*
Air Pict, My '50:134+ 1
Avn Age, N '50:24-25 1 S*
Avn Wk, F 4, '63:27 (Y-/-E) 1
Flight, D 22, '49:798-799 1
Fly Rev Intl, My '64:56-57 (-A&-B) 1 Pr*
Skyline, Vol. 7, No. 3, '49:11-13 3
Vol. 20, No. 3, '62:28 P
Skyways, D '50:14-15+ 4 C

2316. T-39 "Sabreliner"
See also "Sabreliner"
US Navy designation: T3J
Aeroplane, N 28, '58:785-786 1 S*
Avn Wk, F 6, '61:102-108 1
Je 5, '61:54-57 4 C
Flying, Jl '61:24-26+ 4
Interavia, No. 5, '61:702-703 1 S*

NORTH AMERICAN (cont.)

2316.	T-39 (cont.)		
	RAF Fly Rev, Mr '62:26-29	1	Pr, S*
	Skyline, Vol. 18, No. 2, '60:44-49	2	
	Vol. 20, No. 3, '62:14-18	4	
	Vol. 22, No. 1, '64:38-43	4	
2317.	X-15		
	Aeroplane, N 14, '58:730-733	1	
	My 15, '59:579		C
	N 13, '59:488-490	1	S*
	Air Force, N '58:84-88	1	D
	O '60:42-47	4	
	Je '68:62-71	1	
	Air Pict, Mr '59:90-91+	1	Pr
	My '64:142-143 (-A-2)	1	Pr
	Air Prog, Ag/S '64:50-51 (-A-2)	1	Pr
	Avn Age, Jl '58:22-23+	1	
	Avn Wk, My 6, '57:26-27	1	
	F 3, '58:26-28	1	D
	Je 16, '58:117-126	1	
	O 20, '58:26-29, 101	1	D*
	O 27, '58:88-98	1	D
	Ja 12, '59:53-59	6 (Flight Inst)	
	F 16, '59:67	6 (Wing Pylon)	
	F 23, '59:28-30	2	
	Mr 9, '59:97		D*
	Mr 30, '59:59-65	6 (Engine)	
	My 4, '59:29		C
	Ag 31, '59:88-89	2	S*
	N 16, '59:32-33	2	
	N 20, '61:95-99	6 (Flight Cntrl)	
	:52-67 (Pt 1)	1	
	N 27, '61:57-63 (Pt 2)	1	
	N 4, '63:38 (-A2)	1	
	S 27, '65:79-87	6 (Re-entry)	C*
	O 18, '65:22	1	
	Ap 17, '67:77-83	1	
	Jl 24, '67:74-81 (-A2)	1	
	Ag 26, '68:85 (-3)		C*
	Flight, Ja 24, '58:100-101	1	
	F 14, '58:196-197	1	S*
	My 8, '59:639-643	1	C, S*
	Ja 24, '63:136-140 (-A)	1	S*
	Ja 16, '64:106	2	
	D 3, '64:982-983 (-A-2)	1	
	N 11, '65:833-835	1	
	F 15, '68:219-220	1	

NORTH AMERICAN (cont.)

2317. X-15 (cont.)

Flying, Je '60:24-27+	2		
Interavia, No. 1, '60:57-60	1		C
No. 1, '61:101-103	4		S*
No. 5, '64:706-707 (-A-2)	1		S*
No. 1, '66:56-57	4		
Nav Avn News, Ap '59:12-13	3		
D '60:7-11	4		
My '62:6-10	1		
Roundel, S '59:10-11	1		
RAF Fly Rev, D '58:24-26	1		
D '59:16-18	1		Pr*, S*
Ag '60:13-15 (Pt 1)	4		
S '60:53-54 (Pt 2)	4		
Skyline, Vol. 16, No. 1, '58:14-23	1		
Vol. 17, No. 1, '58/'59:2-23	1		
Vol. 17, No. 3, '59:18-21	4		
Vol. 19, No. 3, '61:12-17	1		
Vol. 25, No. 2, '67:10-15	1		
Sp/Aeron, Jl '59:57-59	6	(Eng Cntrl)	S*
F '61:55-58	2		S*
N '61:61-63	6	(Cooling)	
West Aerosp, N '63:7 (-No. 2)	1		
Mr '64:2 (-A-2)	1		
West Avn, Je '58:14	1		
S '58:4	3		
N '58:8-13	1		
Ag '59:4	4		
O '59:14-15	1		

NORTHERN AERO COMPANY USA

2318. "Pheasant"

CAHS, Sum '64:39+ (Canadian-) 1

NORTHROP USA

West Fly, Ag '34:23-24+

2319. "Alpha"

Aviation, D '30:361-362 1

Flight, S 11, '31:915-916 1

2320. "Beta"

Flight, S 11, '31:916-917 1

Pop Avn, Jl '32:10 1

NORTHROP (cont.)

2321. "Delta"
Passenger version of "Gamma"
Aero Digest, D '33:28-29+ 1
S '35:45 C
CAHS, Sum '65:33-39 (-I to -III:Canada) 1 C*, Pr†
:35 (-II:Canada) S*
:37 (Canadian-) 5
Comm Avn, Jl '39:34 1
Flight, F 8, '34:123-124 1
Fly Rev Intl, S '65 (Vol. 21):61 1
West Fly, O '34:42 ("Richfield Eagle"-) 4

2322. "Flying Wing" (1929)
Aero Digest, Mr '30:128 1
Aviation, Mr 29, '30:645-648 1
Flight, Ap 11, '30:411-412 1
Fly Rev Intl, D '69:75 1
West Fly, Mr '30:46-49 2

2323. "Flying Wing" (1940) N1M and N9M
Experimental design for XB-35
Aero Digest, N '41:116-121 1
Aviation, D '41:82-83+ 1
Can Avn, Ja '40:21+ 1
Comm Avn, N '41:52+ 1
Flight, N 13, '41:337-338 1
RAF Fly Rev, Ap '63:41 1
US Air Svcs, N '41:15-16 1
West Fly, N '41:40-42+ 1

2324. "Gamma"
Other versions: "Delta" and "Victoria"
Bomber development: 2E
Aero Digest, Ja '33:40-41 1
S '35:45 C
AAHSJ, Win '62:311 D
Bee-Hive, D '32:7 (Lincoln-Ellsworth-) 1
Can Avn, Jl '33:13 (Hawks' "Sky Chief"-) 1
Flight, Ja 19, '33:56-58 (Hawks' "Sky Chief"-) 1
Pegasus, Ap '46:2 P
Pop Avn, Mr '33:162 ("Sky Chief"-) 1
S '33:184+ ("Sky Chief"-) 6 (Auto Pilot) C
D '36:35-36 (TWA Hi-Alt Research-) 1
RAF Fly Rev, Jl '58:11 1

NORTHROP (cont.)

2324. "Gamma" (cont.)

US Air Svcs, Ja '33:20-22 ("Sky Chief"-)	4	P
Jl '33:32 ("Sky Chief"-)	6 (Instruments)	C*
West Fly, N '32:32 (Ellsworth's-)	1	

2325. "Nomad"

Canadian name

USAAF designation: A-17

CAHS, Spg '67:16-18	1	Pr

2326. "Pioneer"

Military designation: YC-125

Air Force, F '47:42-43+	1	
Aviation, N '46:67	1	
Can Avn, F '47:42-44	1	C
Interavia, No. 8/9, '46:67-71	1	C
West Fly, Ja '47:22-23+	1	

2327. "Victoria"

Private version of "Gamma"

Flight, F 8, '34:123-124	1	

2328. A-17

Canadian name: "Nomad"

Export development: DOUGLAS DB-8A

Fly Rev Intl, Mr '65:41+	1	
Pop Avn, Mr '36:174	1	
O '37:41 (-A)	1	
D '39:19 (-A)		P
US Air Svcs, Ja '36:12-13	1	

2329. XB-35 "Flying Wing"

Developed from "Flying Wing" (1940)

Aero Digest, Je '46:44-45+	1	
My '48:59-60+	6 (Lndg Gear)	S*
Air Force, Jl '46:29-32	1	
Aviation, My '46:92-93	1	
Ag '46:55-61	1	
Ap '47:37-41	1	
Avn Wk, O 13, '47:33	1	C*
Can Avn, Je '46:117	1	
Flight, My 9, '46:469-470	1	
Mr 20, '47:240a-241	1	C, S*
Flying, S '46:34-35+	4	

NORTHROP (cont.)

2329.	XB-35 (cont.)		
	Interavia, No. 2, '46:18-22	1	
	No. 4, '47:26-31	1	P
	Pegasus, Ag '46:2		P
	Skyways, Je '46:42-43+	1	
	West Fly, My '46:22-24	1	
	D '46:20-21+	3	
2330.	YB-49 and YRB-49		
	Am Helicop, D '48:3	2	
	Avn Wk, O 13, '47:14	1	
	Ag 2, '48:11-12	1	
	Flight, Mr 20, '47:240a-241	1	
	Flying, F '48:19-21+	1	
	Interavia, Ja '48:47	1	
	F '49:103	3	
	Skyways, Ja '48:36-37+	1	
	US Air Svcs, Ap '47:34	1	
	Jl '48:11	1	P
2331.	YC-125 "Raider"		
	Military designation of "Pioneer"		
	Air Pict, Ap '51:113		P
	AAHSJ, Spg '61:38-41+	1	
	Avn Opns, My '50:26-27	1	
	Ag '50:36	4	
	Avn Wk, N 8, '48:21-26	1	
	Can Avn, Ag '49:30-32	1	
	Flight, My 19, '49:603	1	
	RAF Fly Rev, My '62:39	1	
	US Air Svcs, F '50:8	1	P
	West Fly, O '48:15+	1	
2332.	F-5 "Freedom Fighter" and "Skoshi Tiger"		
	USAF fighter development of N-156F		
	Spanish-built version: CASA SF-5		
	See also T-38		
	Air Force, D '64:34-40	1	
	Ag '66:45-48	4	
	Ap '69:23-24 (-21)	1	S*
	AF/Sp D Intl, F '65:11-15	1	
	Air Prog, My '69:23-25 (-A&-B)	1	S*
	Avn Wk, S 23, '63:105		D*
	O 12, '64:48-53	4	
	Ja 17, '66:28-32	4	
	Mr 11, '68:79 (R-/-A)	1	S*
	Mr 3, '69:42+ (-21)	1	S*

NORTHROP (cont.)

2332.	F-5 (cont.)		
	Can Avn, Ag '65:14-17+ (-A-15)	1	C, P
	D '65:22-24	1	
	F '66:16-19 (-A)	4	
	Flight, D 17, '64:1046-1049	1	S*
	Jl 1, '65:15	1	
	Jl 29, '65:174-179 (-B)	4	C
	Fly Rev Intl, Ag '64:21-27 (-A&-B)	1	Pr, S*
	S '65 (Vol. 21):13-16 (-B)	4	C
	D '65:197-199	4	
	Mr '66:430 (-A)		Pr*
	Ap '66:475-476	4	
	Interavia, No. 11, '62:1462-1464 (-A)	1	C*
	No. 7, '69:912-915 (-21)	1	S*
	Sentinel, Mr '66:5-7 (C-)	4	
	N/D '67:18-19 (C-)	2	
2333.	F-15A "Reporter" Reconnaissance version of P-61 Redesignated RF-61C in 1948		
	Aero Digest, Je '46:44-45+	1	
	AAHSJ, Fall '69:186-188	1	P
	Flight, My 9, '46:469-470	1	
	Flying, Ag '46:25		P
	Interavia, No. 3, '46:75-76	1	P
	RAF Fly Rev, Je '63:45	1	
	West Fly, My '46:69	1	
2334.	F-89 "Scorpion"		
	Aero Digest, Ja '51:30-32+	1	
	Ja '55:28-31 (-D)	1	S*
	Avn Wk, Ag 20, '51:21-26	3	
	D 3, '51:15 (-D)	1	
	Can Avn, Ap '49:81 (X-)	1	
	Flight, F 23, '51:215-216 (-A)	1	
	F 24, '56:216-220 (-C)	4	C
	Flying, Ja '50:14-15+ (X-)	1	
	Interavia, N '48:614-615 (X-)	1	[illegible]
	No. 5, '52:269-271 (-D)	1	
	RAF Fly Rev, My '56:48-50	1	P
	Skyways, D '48:23 (X-)	1	
	Je '50:14-15+	4	
	D '51:14-15+	1	
	US Air Svcs, O '48:25 (X-)	1	
	West Avn, Jl '51:24-25	3	
	S '52:12-13 (-D)	6 (Rockets)	
	O '52:20-22	6 (Deceleron)	
	West Fly, N '50:12-13+	3	

NORTHROP (cont.)

2335.	XFT		
	RAF Fly Rev, Ap '60:9 (-1&-2)	1	
2336.	HL-10 Lifting Body		
	Air BP, No. 45:2-6	1	
	Air Prog, F/Mr '65:24		S*
	Avn Wk, Jl 13, '64:21	1	S*
	F 21, '66:61-65	1	
	Jl 25, '66:105-112	1	
	Ja 16, '67:68-73	1	
	Flight, F 3, '66:197-198	1	S*
	Ja 5, '67:38	2	
	D 5, '68:957	2	
	Interavia, No. 4, '66:529	1	S*
2337.	M-2 Lifting Body		
	Air BP, No. 45:2-6	1	
	Air Prog, F/Mr '65:25		D, S*
	D '66:10-13+	4	
	Avn Wk, S 9, '63:34	1	
	Jl 13, '64:20-21	1	S*
	Je 28, '65:83	1	
	O 11, '65:83	1	
	O 18, '65:26	2	
	Jl 25, '66:105-112	1	
	Ja 16, '67:68-73	1	
	Flight, S 12, '63:478-479	2	
	Jl 8, '65:70-71	1	
	My 5, '66:770	2	
	Jl 21, '66:115	2	
	West Aerosp, F '64:18-21+	2	
2338.	MX-324		
	RAF Fly Rev, D '55:15	1	
	Jl '61:57	1	
2339.	N-3PB		
	Aero Digest, Mr '41:250-252	1	
	Aviation, Jl '41:65		S*
	Comm Avn, Je '41:48+	1	
	Flight, O 28, '43:479	1	
	Fly Rev Intl, Ap '69:53 (Norwegian-)		Pr*
	RAF Fly Rev, Ja '57:11	1	
	Skyways, My '43:99 (Norwegian-)		D*
2340.	N-102 "Fang"		
	Air Pict, My '64:154	1	

NORTHROP (cont.)

2341. N-156F "Freedom Fighter"
US military designations: F-5 and T-38
Aeroplane, D 4, '59:573-578 1 S*
Avn Wk, Jl 28, '58:54-66 1
Je 15, '59:65-69 1
Je 27, '60:80-83 6 (Weapons)
Flight, Ja 8, '60:43-50 1 C*, S*
Interavia, No. 8, '58:800-802 1
No. 8, '59:972-974 3
No. 7, '60:828-835 6 S*
RAF Fly Rev, F '60:26-29 1 Pr, S*
S '62:34-35 1 S
Sp/Aeron, My '59:52-55 1 S*
Ag '59:58-59 6 (Access Drive)
West Avn, Je '59:25 1

2342. XP-56
Air Force, F '45:36 P
RAF Fly Rev, D '55:15 1

2343. P-61 "Black Widow"
Reconnaissance version: F-15A
Aero Digest, N 1, '45:54-58 (Pt 1) 1 S*
D 1, '45:50-54+ (Pt 2) 1 S*
Aeronautics, N '44:50-51 1
Air News, S '45:50-52+ 4 C
:57-61 1
Air Pict, N '66:427 1
Air Tech, S '44:25 P
Flight, O 5, '44:368-369 1
Flying, Je '45:32-34+ 4
Skyways, S '44:18-20+ 1
S '45:34-41+ 1 S*
West Fly, S '44:53+ 1
D '44:54-55 6 (Ailerons)
O '45:154 (-C) 1

2344. XP-79 "Flying Ram"
Air Pict, My '59:180-181 (-B) 1
Pegasus, Ag '46:3 (-B)
RAF Fly Rev, D '55:15 (-B) 1

NORTHROP (cont.)

2345.	T-38 "Talon"			
	USAF trainer version of N-156			
	Aeroplane, My 15, '59:576-577	1		
	Air Force, Mr '61:99-102	1		
	Air Prog, Fall '61:73 (-A)			C*, S*
	F/Mr '64:4-11+	4		C*
	F '69:40-44+	4		P*
	Avn Wk, O 28, '57:32-33	1		
	Ja 13, '58:50-53	1		
	Ag 29, '60:96-103	4		
	Ap 16, '62:59-63	4		
	Flight, Mr 16, '61:333-334	4		C
	Flying, My '61:30-31+	4		C, S*
	Ag '62:48-49+	4		
	Interavia, No. 2, '58:143	1		S
	No. 4, '61:463-464, 469-470	4		
	:465-468	1		C*, S*
	No. 8, '64:1102-1104	4		C*
	RAF Fly Rev, Ja '59:8-9	1		P*
	N '60:32-38 (-A)	1		Pr, S*
	West Avn, F '58:20-22	1		
	S '58:7	1		S*
	My '61:22-24	1		
2346.	X-4 "Bantam"			
	Flying, S '50:28-29	1		
	Interavia, Mr '49:136-137	1		P
	RAF Fly Rev, N '57:44	1		
	Skyways, Mr '49:19	1		
	US Air Svcs, O '48:25	1		
2347.	X-21			
	Air Force, Ag '63:31-36	1		S*
	Air Prog, O/N '63:37 (-A)	1		S*
	Avn Wk, F 11, '63:80-81	3		
	Ap 29, '63:30 (-A)	1		
	Je 24, '63:52-63 (-A)	1		
	S 27, '65:98-103	2		
	Flying, Ap '64:38-41+	6	(Wing)	S*
	Sp/Aeron, O '62:60-64	2		S*
	West Aerosp, F '63:14-15+	1		S*

NORTHROP (cont.)

2348. 2-E
Bomber development of "Gamma"
Aeroplane, F 13, '35:185-190 3 S*
Flight, F 7, '35:138-142 1 S

OKL (Osrodek Konstrukcji Lotniczych) Poland
See also PZL

2349. TS-8 "Bies" (Dare-devil or Fiend)
RAF Fly Rev, Mr '58:30-33 1 Pr, S*

2350. TS-11 "Iskra" (Spark)
Fly Rev Intl, D '64:28-31 1 Pr, S*

OKO (Opytny Konstruktorsky Otdel) Russia

2351. 1
Fly Rev Intl, D '69:75 1

OMEGA USA

2352. BS-12
Earlier designation: SB-12
Avn Wk, Je 12, '61:123-125 1
Flying, Ag '61:32-33+ (-D-1) 4

2353. BS-14
Avn Wk, Je 12, '61:123-125 1

2354. BS-17A "Airliner"
Avn Wk, Je 12, '61:123-125 1

2355. SB-12
Later designation: BS-12
Am Helicop, Je '56:6-9+ 1
Ja '57:11 2
Avn Age, Ag '56:32-33 1
Avn Wk, My 7, '56:50-55 1
Mr 11, '57:77 1
Can Avn, S '59:53-54 (-D) 1

OMNIPOL Czech
See CZECHOSLOVAKIA

ON MARK USA

2356. "Counter Invader" YB-26K
Avn Wk, Mr 18, '63:66 1
My 18, '64:96-97 1
N 2, '64:43 6 (Photography)
Air Prog, Ag/S '64:48-49 1 C

2357. "Marksman"
Transport conversion of DOUGLAS B-26
Flight, Jl 11, '63:58-60 4 C
Flying, Mr '61:36-37+ 1 C
West Avn, Je '60:44 1

ON MARK/AERO SPACELINES USA

2358. "Mini Guppy" B-377MG
Avn Wk, Jl 3, '67:72-76 1
Boeing Mag, F '69:7-9 1
Flight, F 15, '68:233-234 1
Fly Rev Intl, Ap '69:28 (Advanced-)1

2359. "Pregnant Guppy" B-377PG
Avn Wk, Je 24, '63:80-84 1
Boeing Mag, F '69:7-9 1
Flight, F 4, '65:188-189 1
F 15, '68:233-234 1

2360. "Super Guppy" or "Very Pregnant Guppy" B-377SG
Avn Wk, Ap 19, '65:43-45 1
Ag 23, '65:42-43 1
Boeing Mag, F '69:7-9 1 P
Flight, S 9, '65:457 1
S 16, '65:489 1
F 15, '68:233-234 1 C
Fly Rev Intl, Ap '69:28 (Advanced-)1

ORENCO USA

2361. PW-3
Three built; never flown
AAHSJ, Fall '61:170 C

PAC (Pacific Airmotive Corp.) USA

2362. "Nomad"
Development of NORTH AMERICAN T-28A
Flying, D '61:26-27+ 4

PAC (cont.)

2363. "Tradewind"
Development of BEECHCRAFT C-45
Flying, Ap '63:26-27+ 4 C

2364. 1 Airliner
Fly Rev Intl, Jl '69:30-31 1

PANDER Netherlands

2365. S-4 "Postjäger" (Post Chaser)
Air Pict, Mr '59:98-99 4

PARNALL GB
Air Pict, O '65:336-341 (Pt 1)
N '65:402-404 (Pt 2)

2366. "Hamble Baby"
Flight, N 29, '57:845-848 1

2367. "Panther"
Aeronautics, Jl '41:35 C*
Aeroplane, O 15, '43:453 1
Air Pict, Je '62:180-183 1
Pop Fly, Ag '32:268 P
RAF Qtr, Ap '38:Supp 1 D

2368. "Parasol"
Aeroplane, Ap 15, '30:683-684 1 S
Flight, Ap 17, '31:329-331 1 S

2369. "Pixie"
Aeroplane, Mr 26, '43:371 (-I&-II) 1

2370. "Possum"
Aeroplane, N 6, '42:550 1

2371. "Puffin"
Aeroplane, Ja 8, '43:54 1

2372. G. 4/31
Air Pict, S '63:286-287 1 P
Flight, Ja 23, '36:85, 88

2373. 382
Flight, Mr 23, '39:284-286 1

PARTENAVIA Italy

2374. P. 68
Fly Rev Intl, F '69:27 1

PAYEN France

2375. PA. 22 and PA. 112 "Fléchair"
PA. 112 reached mock-up stage only
Air Pict, My '66:191 1
Fly Rev Intl, F '67:396 1

2376. PA. 49
Avn Wk, Ag 3, '53:39 1

PAZMANY Taiwan
See CHINESE NATIONALIST AIR FORCE

PEMBERTON-BILLING GB

2377. "Night Hawk"
See SUPERMARINE "Night Hawk"

2378. P.B. 23 and P.B. 25
Air Pict, S '58:313 1

PERCIVAL GB
See also HUNTING PERCIVAL

2379. "Gull Four" and "Gull Six" P. 1 and P. 3
Aeroplane, Ap 25, '34:733 1
Air Prog, F/Mr '66:57 (-Four, Mk 2) Pr†
Acft Eng, N '33:259-260 1 S
O '34:276 1
Flight, Jl 29, '32:708-710 1 Pr
Mr 30, '33:292-293 1
O 19, '33:1041-1042+ ("Miss Southern Cross") 4
Ap 26, '34:411-412 1
Mr 21, '35:300-302 1 C, S

2380. "Merganser"
Aeronautics, O '47:32-33 1
Aeroplane, My 23, '47:537-539 1 C
Can Avn, Jl '46:46-48 1
Flight, My 9, '46:459-460 1 S
My 22, '47:472-474 1 S*

PERCIVAL (cont.)

2381.	"Mew Gull" P. 2 and P. 6		
	Aeroplane, Mr 21, '34:497-498	1	P
	S 23, '36:389		S*
	Acft Eng, O '34:276	1	
	Flight, Mr 22, '34:270-272	4	
	Ag 3, '51:135-136 (Henshaw's-)	1	
	:137-138	4	
	Pop Avn, Ap '36:258	1	
	Pop Fly, N '37:457	1	
	Ap '39:30	4	
	RAF Fly Rev, Jl '60:9	1	
2382.	"Prentice" P. 40		
	Aeroplane, Ap 26, '46:496-497	1	
	D 19, '47:788+	1	C*, S*
	My 7, '48:536-537	4	
	N 29, '57:818-820	4	C
	Acft Eng, My '48:137-147	3	S
	Aviation, Jl '46:91	1	
	Can Avn, N '46:52	1	
	Flight, Ap 25, '46:427-428	1	
	D 18, '47:689-695	1	C, S*
	Je 17, '48:651-653	4	C*
	Ag 26, '48:227-228 (-I&-II)	1	
	Ja 13, '49:33-34	4	
2383.	"Prince" P. 50 See also Mark V version: HUNTING PERCIVAL "President"		
	Aeroplane, Ag 20, '48:215-219	1	S*
	O 14, '49:534-535	1	
	My 19, '50:579-582	4	C
	Ja 25, '52:102-107	1	S*
	Air Pict, N '51:337	1	P
	Mr '52:82-83	1	
	Avn Age, Mr '51:26-27	1	S*
	Avn Wk, My 1, '50:44-45	1	
	Can Avn, My '47:33	1	
	F '53:30-31 (Survey-)	1	
	Flight, Ag 19, '48:209-213	1	C, S*
	Ag 26, '48:227-228	1	
	S 1, '49:264-265	2	
	O 6, '49:476-477	1	
	F 9, '50:196-197 (Survey-)	1	
	My 11, '50:572-574	4	C
	RAF Qtr, Jl '52:216		P

PERCIVAL (cont.)

2384. "Proctor" P. 28, P. 30, P. 31, P. 34, and P. 44
Marks I-V

Aeronautics, S '46:48-49	4	C
N '46:95-96 (-V)	4	C*
Aeroplane, Mr 3, '44:243-246	1	S
N 1, '46:512-513 (-V)	1	C
My 19, '50:854		P*
Acft Eng, D '49:391-394	1	S*
Can Avn, My '44:62-64+ (-IV)	1	S
F '46:40+	1	
Comm Avn, Ap '44:56-58 (-IV)	1	
Flight, S 2, '43:251	1	
F 24, '44:210 (-V)	1	

2385. "Provost" P. 56
Later development: HUNTING PERCIVAL "Jet Provost"

Aeronautics, S '52:94-95		P
Aeroplane, Mr 2, '51:262-263		P
My 4, '51:536-542	1	C, S*
Ja 25, '52:102-107	1	
F 5, '54:151-153	4	C
Air Pict, Jl '51:211, 213 (-I&-II)	1	P
Air Prog, F/Mr '66:54	1	Pr
Flight, My 18, '50:620	1	P
Mr 2, '51:265	1	
Mr 23, '51:340	1	
My 4, '51:527-533 (-II)	1	C*, S*
Ag 31, '51:251	1	
S 14, '51:353-355 (-II)	4	C*
Jl 30, '54:141 (-I)	6 (Armament)	
Flying, S '52:34-35+	1	C
RAF Fly Rev, O '53:33-34	4	S
RAF Qtr, O '52:313 (-I)		P

2386. "Sea Prince" P. 57

Aeronautics, Ap '54:44-45 (-C. Mk 2)		P
Aeroplane, Ja 25, '52:102-107	1	S*
Air Pict, Mr '52:82-83 (-T. Mk 1)	1	S*

2387. "Vega Gull" P. 10

Aeroplane, S 23, '36:388		S*
Flight, Ja 2, '36:14b-c	1	S
O 8, '36:360c-d	1	C, S
S 30, '37:328-329	4	C

PERCIVAL (cont.)

2388.	"Q-4"		
	Flight, D 16, '37:596-598	1	
2389.	"Q-6" P. 16		
	Aeroplane, D 22, '37:785-788	1	C, S*
	Ag 3, '38:151-152	4	C
	Can Avn, Ja '38:7+	1	
	Flight, D 16, '37:596-598	1	C, S
	Pop Fly, Mr '38:685	1	
	S '38:295 ("Bird of Eden"-)		C

PETLYAKOV Russia

2390.	Pe-2 [Buck]		
	Acft Eng, My '44:124-127	1	S*
	Flight, Mr 18, '43:276-277	1	
	Jl 22, '43:96	1	
	F 10, '44:142-144	1	S*
	Fly Rev Intl, D '64:50+	1	
	Mr '67:453 (-I&-VI)	1	
	Interavia, N '49:643		P
2391.	Pe-3		
	Three-seat version of Pe-2		
	Fly Rev Intl, D '64:52	1	
2392.	Pe-8		
	Fly Rev Intl, Ap '64:57-58	1	

PFALZ Germany

Air Prog, Ap/My '64:9-24+ (Many aircraft photos)

2393.	D. III		
	Air Cl, Ap '69:51	1	D
	Air Prog, Je/Jl '64:70-71		Pr
	Ap '67:38 (Berthold's-)		Pr†
	Cr & Cockade, Win '60:28-53	1	Pr, S*
	Fall '69:277 (-A: Köhler's)	1	P
	Pop Fly, F '34:589	1	D
2394.	D. XII		
	Cr & Cockade, Aut '61:291		Pr*
	Fly Rev Intl, D '65:254 (Replica-)		P*
	Pop Fly, F '35:588	1	D

PFALZ (cont.)

2395. D. XV
Cr & Cockade, Sum '62:178-179 1 Pr

2396. Dr. I
Pop Fly, D '34:468 1 D

2397. E. IV
Only 24 built
Cr & Cockade, Aut '68:313 P

PHOENIX GB

2398. P. 5 "Cork"
Air Pict, Ag '62:232-236 1
Ja '66:26 1

PHÖNIX Austria

2399. C. 1 "Dront"
Air Prog, Ja '66:46 Pr†

2400. D. I
Air Prog, Ja '66:47 Pr†

2401. 122
Foreign development: ARMY AIRCRAFT FACTORY S. 21
Fly Rev Intl, Ja '66:318 1

2402. 222
Foreign development: ARMY AIRCRAFT FACTORY S. 21
Air Pict, Ap '57:129 P
Fly Rev Intl, Ja '66:318 1

PIAGGIO Italy
Air Pict, Ag '68:290-293 (Pt 1)(Many aircraft photos)
S '68:334-338 (Pt 2)

2403. P. 7
RAF Fly Rev, My '58:6+ 1

2404. P. 23M and P. 23R
Air Pict, Jl '56:253 1
Fly Rev Intl, Ja '67:333 1
RAF Fly Rev, O '61:65 1

PIAGGIO (cont.)

2405.	P. 50		
	RAF Fly Rev, O '61:65 (-II)	1	
2406.	P. 108		
	Flight, Je 17, '43:634a (-B)	1	
	O 21, '43:458	1	
	RAF Fly Rev, S '57:32	1	
	Jl '60:9	1	
	S '61:32-33	1	
2407.	P. 119		
	Air Pict, Mr '60:93-94+	1	
	RAF Fly Rev, Jl '57:43	1	
2408.	P. 136		
	Aeroplane, Ap 4, '58:487-489	4	C
	Air Pict, Ap '52:116	1	
	Acft Eng, Ap '52:107-111	3	
	Avn Wk, Ag 15, '49:42	1	
	Flight, Ap 11, '58:486-488 (-L)	4	
2409.	P. 148		
	Aeroplane, N 2, '51:586-588	4	C
	Avn Age, Ag '51:31	1	C
2410.	P. 149		
	Aeroplane, Ap 10, '59:438-439	4	C
	Flight, Jl 18, '58:73 (-D)	4	
2411.	P. 150		
	Flight, D 5, '52:705	1	
2412.	P. 166 "Portofino" (B Model)		
	Aeroplane, Ja 30, '59:128	1	
	Ap 3, '59:407-409	4	
	Air Pict, Ag '58:264-265	1	
	Je '61:163-166	4	C, S
	My '65:164 (-C)	1	
	Flight, My 23, '58:709-711	1	
	S 5, '58:415-416	4	
	My 1, '59:614	1	
	Ap 18, '63:567-569 (-B)	4	C

PIAGGIO/DOUGLAS Italy/USA

2413.	PD 808 "Vespa Jet"		
	Air Prog, Ap '67:10-13	1	C
	Avn Wk, Je 14, '65:268-279	1	
	S 26, '66:97-105	4	
	Ja 15, '68:103-105	1	
	Flight, Mr 8, '62:367	1	
	Jl 1, '65:21-28	1	S*
	S 29, '66:557-559	4	
	Flying, O '68:38-42	4	
	Fly Rev Intl, Ja '66:287-289+	1	

PIASECKI USA

Flying, Jl '49:24-25+

2414.	YH-16A "Transporter"		
	Only two built		
	Aeroplane, Je 10, '55:776-777	1	P
	Am Helicop, O '53:12+	1	
	Avn Wk, S 21, '53:16-18	1	
	D 12, '55:16	2	
	Can Avn, O '53:86	1	
	Fly Rev Intl, Ja '68:53	1	
	West Avn, D '54:33	1	

2415.	H-21 "Workhorse" or "Flying Banana"		
	Based on Navy HRP-2		
	Civil designation: PH-42		
	Later design: VERTOL H-21		
	Aero Digest, D '53:22-29	1	S*
	Am Helicop, F '50:14+	1	D
	F '55:17 (-A:"Flying Fireman")	1	
	D '55:4 (-A:"Flying Fireman")	4	
	Army Avn D, Je '55:5-14 (-C)	1	S
	O '56:10-14	4	
	Avn Wk, Jl 21, '52:22-32	3	S*

2416.	XHJP-1		
	Am Helicop, Ap '49:9-10+	1	C
	Can Avn, Ap '49:23+	1	C
	Flight, Mr 24, '49:341	1	
	Nav Avn News, Je '49:9	1	

PIASECKI (cont.)

2417.	HRP-1 "Rescuer" and "Flying Banana"		
	See also PV-3		
	Am Helicop, O '49:10-11+	4	
	Mr '52:6-8+	2	
	Nav Avn News, Ap '47:20 (X-)	1	
	Skyways, O '49:26-27+	4	C
2418.	HRP-2		
	Civil designation: PH-42		
	USAF designation: H-21		
	Am Helicop, N '49:8+	1	
2419.	PH-42		
	USAF designation: H-21		
	US Navy designation: HRP-2		
	Am Helicop, Ag '55:13+	1	
	Avn Age, Jl '55:26-30	1	
	Can Avn, Ag '55:59	1	
2420.	PV-2		
	Aviation, N '43:229	1	
	West Fly, D '43:140	1	
2421.	PV-3		
	Later development: HRP-1		
	Aero Digest, O 1, '45:62-63+	1	
	Comm Avn, Ag '45:110+	1	
	West Fly, S '45:96	1	
2422.	VZ-8P "Flying Jeep"		
	Am Helicop, N '58:18	1	
	Avn Wk, Jl 13, '59:129	1	
2423.	59K		
	Am Helicop, Mr '58:9	1	P

PILATUS Switzerland

	Air Pict, Jl '65:228-229		
2424.	P-3		
	Flight, Jl 2, '54:13	1	
	Interavia, No. 3, '54:161	1	
	RAF Fly Rev, D '59:30-33	1	Pr, S*

PILATUS (cont.)

2425. PC-6 "Porter"

Aeroplane, D 18, '59:632-633	4	C
Air Pict, Ap '64:115-117	4	
Avn Wk, Je 15, '59:101-102	1	
Flight, Ja 1, '60:13	1	
My 13, '60:653-654	4	
Jl 22, '65:158 (Astazou X-)	1	C
Flying, Ap '62:38-39+	4	C, P
Interavia, No. 8, '60:956-958	4	
No. 11, '62:1494-1496	1	

2426. PC-6/-- "Turbo-Porter"

See also FAIRCHILD HILLER/PILATUS "Heli-porter"

Aeroplane, My 27, '65:22-23	1	
Jl 22, '65:29-30	4	
Avn Wk, D 17, '62:92-97	1	
Flight, O 20, '66:674-675	4	C

2427. PC-7 "Turbo-Trainer"

Air Prog, S '66:83		P
Avn Wk, Jl 17, '67:109-111	1	

2428. PC-8 "Twin Porter"

Aeroplane, Je 28, '67:26	1	
Avn Wk, O 9, '67:77-85	1	
Flight, My 9, '68:712	1	S
Interavia, No. 7, '67:1101	1	

PILGRIM USA

2429. 100A Transport

Aero Digest, Ap '32:82-83	1	
Aviation, N '31:654-655	1	

PIPER USA

Flight, Mr 24, '66:479-483
Flying, Ap '54:23-25 (Aircraft genealogical chart)
Interavia, No. 4, '68:483-486
Nav Avn News, O '60:20-21
Pop Avn, Ap '40:14-17+

2430. "Cub"

Military designation of J-3: L-4
Glider version: TG-8

Aero Digest, N '42:224-225 (TG-8)		S*

PIPER (cont.)

2430.	"Cub" (cont.)		
	Air Prog, Fall '60:46-47 (J-3)	1	Pr
	D '69:32-34+ (J-3)	4	C
	Air Tech, Mr '45:15-20+	1	S*
	Aviation, D '37:43	1	
	F '41:131		S
	Flying, Mr '45:24-25+	4	
	S '65:30-35	1	
	Skyways, O '49:20-22+	4	
	F '51:18-19+	4	
	US Army Avn, D '67:65 (L-4)		P
	West Fly, D '37:32+ ("Sport"-)	1	
	O '38:36	1	
2431.	"Cub Coupe"		
	Flight, Mr 2, '39:207-208	4	
	Flying, Ag '40:45	1	S*
	Pop Avn, N '38:48	1	
2432.	L-59		
	Flight, Ap 29, '43:444b (-A)	1	
2433.	PA 18 "Super Cub"		
	Avn Age, Jl '50:63	1	
	Avn Wk, Ap 24, '50:18	1	
	Ja 7, '52:15 (-A)	1	
	Can Avn, Je '54:122 (-A)	1	
	Flight, Ja 11, '52:54 (-A)	1	
	Flying, My '50:28-29+	4	
	Ap '52:12-13+ (-A)	4	
	Ag '54:42 (-A)	1	
	Skyways, O '50:14-15+	4	
	Mr '52:25 (-A)	1	
2434.	PA 25 "Pawnee"		
	Avn Wk, D 15, '58:91-95	1	
	S 14, '59:99	1	C
	D 25, '61:74-75 (-235)	1	
	Can Avn, Ag '66:13-14 (-235 "B")	1	
	Flight, D 19, '58:934	1	
2435.	PT-1		
	Air Prog, Jl '43:30-33+	1	S*
	Aviation, Jl '43:233+	1	
	Comm Avn, Ag '43:108-112	1	
	Flight, S 9, '43:282	1	
	Flying, Jl '43:110+	1	
	S '45:64		P*

PITCAIRN USA

2436. PA-5 "Mailwing"
Air Prog, Fall '60:48-49 1 Pr

2437. PA-19
Aero Digest, F '33:48-50 1
US Air Svcs, F '33:15-16 1

2438. PA-36 "Whirl Wing"
Aeroplane, N 1, '40:491-492 1
Aviation, S '40:46-47+ 1 S*

2439. PA-39
Air Pict, Je '59:228-229 1 P

2440. PAA-1
Aviation, Jl '31:408-415 1

2441. PCA-2
Aviation, Jl '31:408-415 1
US Air Svcs, D '30:36-37 4

PITCAIRN-CIERVA USA

2442. C-8
AAHSJ, Win '66:278-279 1 Pr

PLAGE I LASKIEWICZ Poland

2443. Lublin R-X
Acft Eng, Ag '29:191 1

POLIKARPOV Russia
Fly Rev Intl, Jl '68:404-407 (Pt 1)
Ag '68:455-456 (Pt 2)

2444. DI-6
RAF Fly Rev, F '62:41 1

2445. I-5
Air Pict, Ag '64:251 1
Fly Rev Intl, S '64:45+ 1

POLIKARPOV (cont.)

2446. I-15 "Chato" (Little One)
Variant: I-153
I-152 = I-15 bis
Fly Rev Intl, Ja '68:44-45, 47 (Incl. -bis) 1 Pr*
RAF Fly Rev, My '57:13+ 1

2447. I-16 "Rata" (Rat), "Yastrebok" (Young Eagle), "Ishak" (Little Donkey), and "Mosca" (Fly)
Later development: I-180
Air Pict, D '58:425-427 1 Pr
Air Prog, Je/Jl '63:45 Pr†
Fly Rev Intl, My '65:54 (Type 10 & 24) Pr*
N '69:58-65 (Pt 1) 1 P, Pr*, S*
D '69:60-64 (Pt 2) 1 Pr
Cont. in Ja '70
RAF Fly Rev, Ja '56:13 1
S '59:47-48 (Japanese captured-) 1
My '62:23-24 (Spanish War-) 4

2448. I-153 "Chaika" (Gull)
Development of I-15
Fly Rev Intl, Ag '66:769+ 1
Ja '68:46, 48 (China, Russia & Finland-) 1 Pr*
RAF Fly Rev, My '57:52 1
Mr '62:34-35 1 Pr*

2449. I-180
Basic design: I-16 "Rata"
Fly Rev Intl, F '69:73 1

2450. IL-400
Designation of I-1 prototype
Fly Rev Intl, Je '66:640 1

2451. NB (T) (1944)
Fly Rev Intl, S '67:877 1

2452. Po-2 [Mule]
RAF Fly Rev, Ap '60:34-36 4 P

2453. R-5 "Natasha"
RAF Fly Rev, Je '58:6 1

POLIKARPOV (cont.)

2454.	R-10		
	RAF Fly Rev, D '57:50	1	
2455.	VIT-1 and VIT-2		
	Fly Rev Intl, Mr '66:419	1	
	RAF Fly Rev, My '63:41+	1	

PONNIER — Belgium

2456.	M-1 and M-2		
	Cr & Cockade, Sum '66:132-133		Pr†

PORT VICTORIA — GB

2457.	P.V. 7 "Grain Kitten"		
	Hist Avn, O '69:27-28	1	
2458.	P.V. 8 "Eastchurch Kitten"		
	Hist Avn, O '69:27-29	1	

PORTSMOUTH — GB

2459.	"Aerocar"		
	Aeroplane, D 10, '48:730-731	4	
	Aviation, Mr '46:52	1	
	Avn Wk, Ag 18, '47:33		S*
	Can Avn, Ja '46:36	1	

POTEZ — France

2460.	25		
	Fly Rev Intl, Mr '69:42 (-A. 2:Swiss)		Pr*
	RAF Fly Rev, Jl '62:43	1	
2461.	37-R2		
	Acft Eng, Ap '32:87	1	P
2462.	39		
	Pop Avn, Ag '36:18	1	P
2463.	41		
	Aeroplane, F 21, '34:310	6 (Lndg Gear)	S*
	Air Pict, Je '66:216-217	1	
2464.	45		
	Pop Avn, Ja '34:14+	1	

POTEZ (cont.)

2465.	49			
	Pop Avn, D '33:379		1	
2466.	50			
	Flight, O 19, '33:1043		1	
2467.	53			
	Pop Avn, S '33:166		1	
2468.	54			
	Acft Eng, F '35:43		1	
	RAF Fly Rev, Ja '58:11+		1	
	RAF Qtr, Ja '35:16			P
2469.	62			
	Pop Avn, S '35:172		1	
2470.	63			
	Flight, O 26, '39:329			D
	Mr 28, '40:285-286b		1	C, D, S
	RAF Fly Rev, Ap '57:13		1	
	Ja '63:20-23+	(-.11)	1	Pr*
2471.	65			
	Air Pict, F '68:65		1	
2472.	75			
	Air Pict, N '53:325			P
	Flight, My 14, '54:614-616		4	
	RAF Fly Rev, My '61:53		1	
	US Army Avn, Jl '57:30-31		1	
2473.	141			
	Pop Avn, Je '38:31			P
2474.	840			
	Aeronautics, Jl '61:20-22		1	C
	Avn Wk, Ap 3, '61:34		1	
	My 29, '61:71		1	
	D 25, '61:70-73		4	
	Can Avn, D '62:26		1	
	Flight, Jl 22, '60:131		1	
	Je 21, '62:988-989		4	
	Flying, Ag '61:25+		1	C
	O '62:24-25+		4	
	Interavia, No. 3, '60:343		1	
	No. 12, '61:1687-1689		1	

POTEZ-HEINKEL			France/Germany
2475.	CM. 191 Development of FOUGA "Magister" Avn Wk, Ap 23, '62:102 RAF Fly Rev, Ap '63:26-29	 1 1	 Pr, S*
POWELL			USA
2476.	P.H. 2 Air Prog, Ag/S '64:18-19	 1	 S*
PRAGA			Czech
2477.	E-51 Fly Rev Intl, Jl '65:59	 1	
2478.	XE-55 RAF Fly Rev, My '63:41	 1	
PRUDDEN			USA
2479.	TM-1 and SE-1 AAHSJ, Sum '67:130-135	 1	 P
PRUDDEN-WHITEHEAD See ATLANTA			USA
P-V (Piasecki, Venzi) See PIASECKI			USA
PWS (Podlaska Wytwórnia Samolotow)			Poland
2480.	16 Pop Avn, N '37:17	 1	
2481.	26 Fly Rev Intl, Ag '68:465		 P*
2482.	41 Air Pict, O '63:328-329	 1	
2483.	54 Flight, Ag 31, '33:865-866	 1	

PZL (Panstwowe Zaklady Lotnicze) Poland

2484. P. 1
Acft Eng, Jl '30:183 1

2485. P. 6
Development of P. 1
Aero Digest, N '31:43 1
RAF Qtr, Ap '31:287-288 1

2486. P. 11 "Jedenastka" (Eleventh)
Later development: P. 24
Air Pict, D '55:380-381 (-C) 1 C, Pr
:380 (-C) 5
Fly Rev Intl, My '66:574, 576 (-B&-C) Pr*
Ag '68:464 (-C) P*
Pop Fly, Ap '37:21 P
RAF Fly Rev, S '56:24 1
O '60:23-26+ 1
:25 (-C:121 Sq) Pr*

2487. P. 23 "Karas" (Crucian)
Air Pict, S '64:279-283 1 C, Pr
Fly Rev Intl, F '65:42 1

2488. P. 24
Prototype: P. 11
Flight, My 3, '34:444 1
Fly Rev Intl, My '66:576 (-C&-F) Pr*
Je '69:72-73 (-F:Greece) Pr*
Pop Avn, S '37:48 1
RAF Fly Rev, My '55:9-10 1
Jl '55:9 1
S '56:24 1
N '56:15 1

2489. P. 37 "Loŝ" (Elk)
Air Pict, F '68:53-58 1 Pr
Acft Eng, Jl '38:225-226 1
Fly Rev Intl, N '63:49 1
RAF Fly Rev, S '56:23 1
N '56:15 1

2490. P. 38 "Wilk" (Wolf)
RAF Fly Rev, O '60:11 1

2491. P. 44 "Wicher" (Whirlwind)
Acft Eng, Ag '38:251 1

PZL (cont.)

2492.	P. 45 "Mars"		
	Air Pict, O '63:328-329	1	
2493.	P. 46 "Sum" (Electric Eel)		
	RAF Fly Rev, S '56:24	1	
2494.	P. 50 "Jastrzab" (Hawk)		
	Air Pict, D '62:386-388	1	
	RAF Fly Rev, Jl '55:9	1	
	S '56:23-24	1	
	N '56:15	1	
2495.	101 "Gawron" (Rook)		
	Flight, Ag 3, '61:158	4	
2496.	104 "Wilga" (Thrush)		
	Air Pict, Ag '65:278	1	
	Ag '68:295-296 (-34)	1	C
	Air Prog, Je '69:26-27+ (-32&-35)	4	

RAMOR Austria

2497.	K.E. 14		
	Flight, N 16, '33:1160	1	P

R.E. (Designed by Royal Aircraft Factory) GB

2498.	7		
	Air Pict, O '57:385	1	
2499.	8 "Harry Tate"		
	Air Prog, My '67:84		D*
	CAHS, Win '64:6-9 (Canadian-)	4	
	Cr & Cockade, Aut '64:284		P
	Spg '69:97-98		P
	Flight, O 15, '54:575-581	1	
	Pop Fly, S '32:356	1	D
	RAF Qtr, Ap '37:Supp	1	D
	Ap '39:207-211	1	

REARWIN USA

Air Prog, Ja '68:44-48+

REGGIANE Italy

2500.	Re 2000 "Falco I" (Falcon)		
	Fly Rev Intl, Jl '66:703-707	1	Pr*
	S '67:870	1	
	S '69:51 (Hungarian "Héja II"-)		Pr*
	RAF Fly Rev, Ja '59:6	1	
	F '59:16 (-Series I: Hungary)		Pr*
2501.	Re 2001 "Falco II" (Falcon II)		
	Flight, N 12, '42:526b	1	
	Fly Rev Intl, S '67:870-873	1	Pr*
	RAF Fly Rev, Ag '63:41	1	
2502.	Re 2002 "Ariete" (Ram)		
	Air Pict, Ja '69:25	1	
	Fly Rev Intl, S '67:872		Pr*
	O '67:937-939	1	Pr*
	RAF Fly Rev, O '59:50+	1	
2503.	Re 2007		
	Fly Rev Intl, My '66:580	1	

REICHENBERG Germany

See FIESELER Fi-103

REID & SIGRIST GB

2504.	R.S. 1 "Snargasher"		
	Aviation, Ag '41:111	1	
	Flight, My 11, '39:476c-e	1	S
	Fly Rev Intl, Ap '67:516	1	
2505.	R.S. 3 "Desford"		
	Aeroplane, My 10, '39:593-595	1	S*
	D 28, '45:751-753	4	C
	Flight, Ja 24, '46:79-81	1	
	Jl 22, '48:87-89	4	C
	Fly Rev Intl, Ap '67:516	1	
2506.	R.S. 4 "Bobsleigh"		
	Fly Rev Intl, Ap '67:516	1	

RENARD Belgium

2507.	R-31		
	Acft Eng, F '33:41-42	1	

RENARD (cont.)

2508.	R-36 to R-40		
	Fly Rev Intl, Mr '67:452-453	1	

REPUBLIC USA

Air News, Je '45:52-55+

2509.	"Seabee"		
	Modification: UCC "Twin Bee"		
	Aviation, My '46:45-62	1	C*, S*
	Can Avn, S '46:47-48+	4	C
	Comm Avn, Ja '45:48+	1	
	O '45:76-78	1	
	Flight, S 5, '46:255-257	1	S
	Ag 14, '47:160-162	4	
	Flying, O '46:38-39+	4	C
	Mr '59:52-53+	4	
	My '59:49+	4	
	Ag '64:65-67	1	
	Skyways, Mr '46:34-41+	1	S*
2510.	EP-1		
	Built for Sweden		
	Similar to REPUBLIC P-35		
	Aviation, Je '40:52-53	1	S
2511.	XF-12 and XR-12 "Rainbow"		
	Only two built		
	First designation: XF-12		
	Aero Digest, Ap '46:44-45	1	
	Aeroplane, D 6, '46:686-691	1	
	D 13, '46:718-722	1	S*
	Air Cl, Ag '69:38-42	1	P
	Acft Eng, Ja '48:2-10+	1	
	Avn Wk, N 10, '47:28-30	1	
	Can Avn, F '46:88	1	
	Comm Avn, S '45:62-63	1	
	D '45:74	1	
	Flight, Ag 22, '46:192	1	
	Interavia, No. 7, '46:45-51	1	S*
	O '48:576	6 (Control)	S*
	Pegasus, Jl '46:4		P
	RAF Fly Rev, Ap '57:50	1	
	Ja '60:40	1	
	Skyways, D '48:28-29	4	
	US Air Svcs, F '48:18		P
	D '48:12-13	4	P
	West Fly, F '46:62	1	

REPUBLIC (cont.)

2512. F-84 "Thunderjet"
Earlier designation: P-84
See also F-84F, RF-84F and YF-96

Aeroplane, My 14, '54:604-605 (-E)			P
Air Pict, My '52:133 (-G)	1		
AAHSJ, Sum '69:128-129	4		P
Avn Opns, Mr '50:24-25+	1		
Avn Wk, S 5, '49:20-25	1		
Je 18, '51:15 (-G)	1		
Ag 1, '55:16 (X-/-H)	1		
Flying, Mr '49:22-26+	4		C
F '50:20-21	3		
Ja '52:24-25 (-E)	4		
My '52:34-35+	1		S
Fly Rev Intl, S '68:62 (-G:Denmark)			Pr*
Je '69:73 (-G:Greece)			Pr*
Pegasus, Jl '48:13			P
RAF Fly Rev, Je '57:8-9 (-K)	1		
Je '59:6 (X-/-H)	1		
Ag '61:22-24	1		S*
:25 (-G)			Pr
Skyways, O '49:14-15+ (To -E)	1		
:23 (-E)	1		
Jl '50:18-19+ (-E)	4		
US Air Svcs, Je '48:15	6	(Rockets)	
Mr '49:24 (-D)	1		
Jl '49:20 (-E)	1		
Ap '52:18-20	1		
Ag '55:9 (X-/-H)	1		
West Avn, S '55:32 (X-/-H)	1		

2513. F-84F "Thunderstreak"

Aero Digest, S '51:25-34	1	S*
Aeroplane, D 24, '54:917-920	1	
Air Force, F '53:29-30	1	
Air Pict, Ja '60:15-17	1	Pr
Avn Wk, Ap 7, '52:16-17	1	
D 15, '52:31-39	3	
Mr 30, '53:17		S*
Flight, D 24, '54:890-892	1	
RAF Fly Rev, Ja '55:23-25	1	S*
Ag '60:38	1	
:39 (German-)		Pr*

REPUBLIC (cont.)

2514.	RF-84F "Thunderflash"		
	FICON = "Fighter Conveyer" project		
	Air Prog, Ap/My '63:68 (-FICON)		C
	Avn Wk, Ap 7, '52:16-17	1	
	Ag 31, '53:12-13 (-FICON)	1	
	S 7, '53:16 (-FICON)	1	
	Ag 1, '55:22-23 (-FICON)	1	P
	Fly Rev Intl, S '68:63 (Danish-)		Pr*
	Ap '69:53 (Norwegian-)		Pr*
	RAF Fly Rev, Je '57:8-9	1	
	US Air Svcs, F '55:22	1	
2515.	XF-91 "Thunderceptor"		
	Only two built		
	Air Force, Ap '49:25		P
	Air Power, Win '55:100		P
	Flight, Mr 10, '49:275	1	
	RAF Fly Rev, Je '56:13	1	
	US Air Svcs, Ja '50:23	2	
	F '53:9-10	1	
2516.	YF-96		
	Became F-84F "Thunderstreak"		
	Avn Wk, My 15, '50:17 (-A)	1	
2517.	XF-103		
	Development of XF-91		
	RAF Fly Rev, S '59:15	1	
	O '59:11	1	
2518.	F-105 "Thunderchief" or "Thud"		
	Air Force, O '66:37-43	1	C, S*
	Air Pict, O '57:354-355	1	Pr
	Air Prog, F '67:50-51	1	
	:84		C
	O '67:38-40+	1	C
	Avn Age, Ap '58:54-59 (-B)	1	
	Avn Wk, My 20, '57:27 (-B)	1	
	Jl 29, '57:28-29	1	
	Ag 5, '57:74-75	1	P
	S 9, '57:101-106 (X-)	2	
	N 25, '57:66-67	1	
	D 28, '59:37-45	4	
	F 25, '63:105-111 (-D)	4	
	Mr 20, '67:116		D*
	My 22, '67:16-17	4	

REPUBLIC (cont.)

2518.	F-105 (cont.)			
	Avn Wk, S 23, '68:78-81	4		
	Ja 20, '69:64-65+ (-D)	6	(Bomb)	
	Flight, My 31, '57:724 (-A&-B)	1		
	Je 12, '59:799-802 (-B)	1		S*
	F 6, '64:204-212 (-D)	1		S*
	F 20, '64:284-285 (-D)			C*
	Flying, N '66:60			P*
	Fly Rev Intl, Ja '64:21-23	1		
	Mr '66:430 (-D)			Pr*
	Jl '67:699-702 (-D)	1		Pr
	Interavia, No. 8, '58:817 (-B&-E)	1		
	No. 3, '60:311-317 (-D)	1		S*
	RAF Fly Rev, My '57:11	1		
	F '58:35-36 (-B)	1		
	My '59:18 (-B)	1		P*
	O '60:28-31 (-B)	1		Pr, S*
	Sp/Aeron, F '60:119-121	6	(Water Injec)	S*
2519.	P-43 "Lancer"			
	Aero Digest, Ja '41:168+	1		
	RAF Fly Rev, Ap '56:9	1		
	Mr '59:45	1		
2520.	P-47 "Thunderbolt"			
	See also XP-72			
	Aero Album, Fall '68:28-29 (-D:Holt's)	1		
	Win '69:40-41 (-D: Truluck's)	4		Pr†
	Aero Digest, N '43:282-283			S
	Ja 1, '45:74-78+	1		C, S
	Aeronautics, N '44:28-33	1		
	Aeroplane, My 21, '43:591-592	1		
	Air Force, D '46:25-29	1		
	D '47:100-105	4		
	Air Pict, O '61:315-316+ (X-/-J)	1		
	F '67:61 (German-)	1		
	Air Prog, Fall '61:80			C
	Ap/My '63:78 (-N)			C
	My/Je '65:47 (-D:Johnson's)			Pr†
	F '68:57 (Mahurin's-)			Pr†
	Air Tech, O '42:14-15	1		
	D '43:21-35	1		S*
	AAHSJ, Spg '65:75 (-D)			D
	Aviation, Ag '42:100-105+	3		
	Mr '44:193, 195			S*
	Ap '44:191, 193			S*

REPUBLIC (cont.)

2520. P-47 (cont.)

Aviation, My '44:199		S*
Je '44:201		S*
Jl '44:201, 205		S*
F '45:223-225	6 (Dive Flaps)	S*
F '46:160-162		S*
Bee-Hive, My/Je '41:6 (X-/-B)	1	
Can Avn, D '45:128 (-J)	1	
Comm Avn, O '44:44-48	1	
Flight, Jl 30, '42:120	1	
My 20, '43:521-524	4	
Flying, Ag '43:47-48+	1	
Ap '44:42-46	4	
F '45:34-38	1	C, S
Fly Rev Intl, D '63:52 (-D:Dominica & USA)		Pr*
:52 (-II:73 OTU & RAF 81 Sq)		Pr*
RAF Fly Rev, Ja '57:9 (X-/-H)	1	
N '57:7	5	
Mr '60:33+	4	
D '62:23	1	
Ag '63:24-26+	1	
:24-25 (-D)		S*
:27 (-N-5-RE)		Pr
Skyways, Je '44:34-41+	1	S*
Mr '51:30-31+	1	
US Air Svcs, O '44:11	1	
Mr '45:30 ("Burma Yank")		P
West Fly, My '45:94 (-N)	1	

2521. XP-69

Two ordered and cancelled

RAF Fly Rev, My '55:10	1	
Mr '60:9	1	

2522. XP-72

Development of P-47; two built

RAF Fly Rev, My '55:10	1	

2523. P-84 "Thunderjet"

Later designation: F-84

Aero Digest, F '48:44-48	1	C, S
Air News, Ap '46:46-47 (X-)	1	P
Am Helicop, Je '46:13 (X-)	1	P
Avn Wk, My 24, '48:18-21	1	
Interavia, No. 6, '47:50-53	1	C, P
Skyways, Jl '48:14-16+	4	

REPUBLIC (cont.)

2523. P-84 (cont.)
US Air Svcs, N '46:13 (X-) 1
West Fly, Jl '46:58 (X-) 1

2524. 2PA "Guardsman"
USAAF designation: AT-12
Comm Avn, S '40:50 1
Flight, O 17, '40:321 1

REY France

2525. R. 1
Variable wing aircraft
Air Pict, Ap '51:112 1 P
RAF Fly Rev, O '55:11 1

RILEY USA

2526. 400
Remanufactured DE HAVILLAND D.H. 104 "Dove"
Flight, D 31, '64:1127-1128 1
S 30, '65:578-580 4 C

ROBBINS & SCHIEFER USA

2527. Pursuit (1917)
AAHSJ, Win '60:308-309 1 P, Pr

ROBERTSON USA

2528. SRX-1 "Skylark"
Avn Wk, Je 7, '54:32-39 1

2529. 210
CESSNA 210 STOL conversion
Flying, O '68:44-47 1 P

ROBEY & CO. GB
Air Pict, N '61:349-351 1

ROBINSON GB

2530. "Redwing"
Flight, Je 27, '30:674-675 1
D 19, '30:1457-1460 (-II) 1 S

ROGERS USA

2531. "Pacific Standard" (1921)
Air Prog, Ap/My '64:25+ 1

ROGOZARSKI Yugoslavia

2532. SIM-XIV-H
RAF Fly Rev, Mr '63:49 1

ROHR USA

2533. M.O. -1
Flight, Ja 16, '47:55-56 1

ROLAND Germany
Air Pict, F '57:62-63 (D. I to D. 29 with photos)

2534. C. II "Walfisch" (Whale)
Avn Wk, Ag 5, '63:64-65 1 D*

2535. D. VI
Air Prog, Ja '64:12-16+ 1 P, Pr†

ROLLASON GB

2536. "Luton Beta" B. 2
Air Pict, Je '69:217-219 1 C

ROMEO Italy
See IMAM

ROTOR-CRAFT USA

2537. X-2A "Dragonfly"
Avn Wk, Ag 9, '48:21-25 1
West Fly, O '45:156 1

2538. XR-11
Interavia, S '48:520-522 1

2539. XRH-1 "Pinwheel"
West Avn, My '57:20 1
D '60:60+ 1

ROTORCRAFT GB

2540. "Grasshopper"
Acft Eng, N '62:319 1

ROUSSEL France

2541. R. 30
RAF Fly Rev, N '58:11 1

ROYAL AIRCRAFT FACTORY GB
See A.E., B.E., C.E., F.E., N.E., R.E., and S.E.
Flight, Jl 1, '55:17-19 (List of aircraft types built)

RUMPLER Germany

2542. "Eindecker" or "Taube"
Avn Wk, Ap 23, '62:36-37 1 D*
Ja 25, '65:58-59 1 D*
Flying, My '59:65 1 D

2543. C. III
Cr & Cockade, Sum '69:174 (Zorer's-)1 P

2544. C. IV
Avn Wk, Je 8, '59:123 D*
Cr & Cockade, Sum '69:179 1 P

2545. D. I
Cr & Cockade, Spg '67:73 1 P

2546. 7D
Prototype of RUMPLER D. I
Cr & Cockade, Spg '67:71-73 1 P

RWD (Rogalski, Wigura and Drzewiecki) Poland
Air Pict, Je '56:218 (Types of aircraft)

RYAN USA
Air News, My '45:54-58+
Flying, Jl '54:23-25 (Aircraft genealogical chart)
Nav Avn News, My '58:21
West Avn, S '52:16-17

2547. "Brougham" (1928)
M-1 with enclosed cockpit
Flying, O '64:41

RYAN (cont.)

2548. "Flex-wing"

Flying, Jl '61:22-23+	1	
Interavia, No. 11, '61:1520	4	

2549. XF2R-1 "Dark Shark"

Turboprop variant of FR-1 "Fireball"

AAHSJ, Win '63:240-242	1	Pr
Nav Avn News, Ja '47:12	1	
S '47:16	2	
RAF Fly Rev, Ja '58:41	1	
West Fly, F '47:50	1	

2550. FR-1 "Fireball"

Aero Digest, N 1, '45:62-63+	1	
Air News, D '45:28-29	1	
Mr '46:37-39+	4	
Acft Eng, Ja '49:25-27	1	S*
AAHSJ, Win '63:231-247	1	Pr
Am Helicop, F '46:18-19+	1	
O '46:16-17+	1	
Aviation, N '45:171+	1	
Jl '46:95		S
Ag '46:101		S
Comm Avn, O '45:80-82	1	
Flight, Ja 10, '46:29-31	1	
Interavia, Ag '48:446		P
Nav Avn News, N '45:14	1	S
RAF Fly Rev, Ja '58:11+	1	
Skyline, Ja '46:18-19+	1	P*
US Air Svcs, O '45:30	1	
West Fly, N '45:40-41	1	S
F '46:34-35+ (X-)	1	

2551. FR-2 to FR-4 "Fireball"

Nav Avn News, S '47:16 (-4)	1	
RAF Fly Rev, Ja '58:11+ (-2, -3, -4)	1	

2552. M-1

Open cockpit

Modifications: "Brougham" and NY-P

Flying, Mr '63:63	1	D

RYAN (cont.)

2553.	NY-P "Spirit of St. Louis"		
	Modified M-1		
	Aero Digest, My '52:26-40	1	S*
	Aerosp Intl, My/Je '67:21-27	1	
	Air Prog, '52 Annual:54-55	1	S*
	Je '67:15-18+	1	S
	Interavia, No. 5, '67:753	4	
	Nav Avn News, S '60:12-15	1	
	Pegasus, My '57:7-9	1	
2554.	YO-51 "Dragonfly"		
	Aviation, Ap '40:64-65	1	S
	Flight, Ap 17, '41:284g	1	P
	Fly Rev Intl, F '64:64	1	
2555.	XPT-16		
	US Air Svcs, Ag '39:20		P
2556.	PT-20		
	Development of XPT-16		
	Pop Avn, Mr '40:23		P
	US Air Svcs, Mr '40:4		P
2557.	PT-22 "Recruit"		
	Civil designation: ST-3		
	Flying, D '60:57+	4	
2558.	YPT-25		
	Civil designation: ST-4		
	Plastic-bonded wood construction		
	Aero Digest, O '42:198-202	1	
	Air Tech, N '42:22-24+	1	
	Aviation, O '42:235+	1	
	Can Avn, O '42:107+	1	
	Comm Avn, S '42:90-94	1	
	Flight, N 12, '42:531	1	
	D 30, '43:722	1	
	Flying, Ja '43:50	3	
2559.	S-C		
	Acft Eng, O '41:288	1	
	Aviation, S '37:42-46	1	
	D '37:32-33+	1	
	D '38:38	1	
	West Fly, S '37:28	1	
	D '37:37 (-145W&-150)	1	
	N '38:23	1	

RYAN (cont.)

2560.	ST- Military designations: ST-3 = PT-22, ST-4 = YPT-25		
	Aero Digest, Ap '42:112-116 (-S: Seaplane)	1	
	Acft Eng, O '41:286-287 (-3)	1	
	Aviation, Ag '36:35-37 (-3)	1	
	Can Avn, Ja '41:58 (-3)	1	
	Comm Avn, Ja '41:40-42 (-3)	1	
	S '42:90-94 (-4)	1	
	Flight, Ag 9, '34:817	1	
	F 6, '41:118-120 (-3)	1	
	Pop Avn, Ag '36:53-54	1	
	West Fly, Jl '34:27-28	1	
2561.	STM-2		
	Acft Eng, O '41:286-287	1	
	Aviation, Je '40:60	1	S
	West Fly, F '40:30+	1	
2562.	VZ-3RY "Vertiplane"		
	Air Prog, Ag/S '64:55-56	1	C, P
	Am Helicop, D '57:10	1	
	Avn Wk, F 24, '58:77	1	
	Fly Rev Intl, N '63:47	1	
	Sp/Aeron, O '59:150-152	1	
	US Army Avn, My '58:49	1	
2563.	VZ-11 Later designation: XV-5A and XV-5B		
	Avn Wk, Ag 20, '62:32-33	1	
	Flight, S 27, '62:526-527	1	S*
	Interavia, No. 10, '62:1347-1349	1	S*
	Sp/Aeron, O '62:39+	1	
2564.	X-13 "Vertijet" Only two built		
	Am Helicop, My '57:6	2	
	Ag '57:8-9+	1	
	Avn Wk, My 13, '57:29-31	1	
	My 20, '57:64-65	2	
	Ag 26, '57:120-121	1	
	Flight, D 13, '57:928	1	
	Interavia, No. 8, '57:816	1	
	RAF Fly Rev, Jl '61:59	1	

RYAN (cont.)

2565.	XV-5A		
	Earlier designation: VZ-11		
	Air Prog, F/Mr '63:10-11		S*
	Avn Wk, Ja 21, '63:64-75	1	C
	S 23, '63:120-121	3	
	Mr 2, '64:30	1	
	N 23, '64:32-33	1	
	D 14, '64:38-45	1	S
	Je 6, '66:74-80	4	
	Jl 18, '66:65-66	1	
	Flight, S 30, '65:567	2	
	Fly Rev Intl, N '64:26-27, 47	1	S*
	S '65:5-8	1	
	Interavia, No. 11, '63:1733	1	
	No. 5, '64:655-656	1	S
	No. 4, '65:520-523	1	C, S
	US Army Avn, S '64:11-12	1	S
	West Aerosp, Mr '64:6-7	1	
2566.	XV-5B		
	Air Prog, O '68:50-53+	1	C
	Fly Rev Intl, O '68:27	1	

SAAB (Svenska Aeroplan Aktiebolaget) Sweden

Air BP, No. 5:2-6 (Operations in 1957)
Air Pict, F '63:34-38 (Pt 1)
Mr '63:72-75 (Pt 2)
Ap '63:114-117 (Pt 3)
Flight, My 26, '48:575-580
Nav Avn News, My 27, '58:17-19

2567.	17		
	SAAB's first design		
	Fly Rev Intl, N '67:997-1000	1	Pr*
2568.	18		
	Air Power, Sum '55:258 (T-/-B)		P
	Fly Rev Intl, D '64:52-53	1	
2569.	21		
	Jet version: -R		
	Flight, Je 24, '48:689-692 (-R)	4	C
	RAF Fly Rev, Je '59:16 (-A)	1	Pr*
	S '59:17 (-R)	1	

SAAB (cont.)

2570.	29 "Tunnan" (Barrel or Flying Barrel)		
	Aeronautics, F '54:44-45		P
	Aeroplane, N 2, '51:581		P
	Air Pict, O '52:294-295	1	S
	Avn Age, S '50:24-25	1	S*
	Avn Wk, Ag 11, '52:32-33	1	S
	Flight, Je 17, '48:654	1	
	My 4, '50:556-558	1	S
	O 26, '51:537	1	
	O 9, '53:495	1	
	O 16, '53:531-533	4	P
	Interavia, No. 7, '51:372-375	1	S*
	:376-378	2	
	Nav Avn News, Jl '54:24	4	
	Roundel, My '55:25 (-C)	4	
	RAF Fly Rev, N '55:30-33 (To -F)	1	Pr, S*
2571.	32 "Lansen" (Lance)		
	Air Pict, D '56:425-427	1	
	:426-427 (-A)	5	Pr
	Air Power, Sum '55:260		P
	Avn Wk, F 6, '56:37	1	
	Flight, F 3, '56:140-141	1	
	D 30, '60:1017-1020	1	S
	Fly Rev Intl, O '69:42-47	1	Pr*
	Interavia, No. 12, '52:691	1	P
	No. 3, '54:160	1	
	Nav Avn News, Mr '53:41		P
	RAF Fly Rev, O '58:30-33 (-A)	1	Pr, S*
	O '59:16-17	1	P*
2572.	35 "Draken" (Dragon)		
	See also 210		
	Aeroplane, O 26, '56:633		P
	Air Pict, Je '59:217-220+ (-A)	1	Pr
	Ja '67:4-6 (-F)	4	
	Air Prog, Ag/S '64:58-59 (-F)	1	S
	Avn Wk, O 29, '56:60-62	1	
	Mr 24, '58:46-65 (Pt 1)	1	
	Mr 31, '58:64-67 (Pt 2)	3	
	Ag 17, '59:99	3	
	My 13, '68:58		P
	Can Avn, Mr '58:28-29+	1	

SAAB (cont.)

2572. 35 (cont.)

Flight, N 11, '55:745 1
D 30, '60:1017-1020 1 S
Ag 8, '63:207 (-D) 6 (Ejection Seat)
Ja 21, '65:102-107 (-A) 4
Ap 28, '66:701-702 (-F) 1
Fly Rev Intl, N '63:33 (-D) 6 (Ejection Seat)
My '64:58 (-A) Pr*
Ag '64:17-19 1 Pr
Ag '66:761-764 (-A to -F) 1 P*
F '67:392 (-A, -B, -D, -E, -F) Pr*
Interavia, No. 12, '55:932 1
No. 12, '56:978-979 1
No. 1, '58:120 (-A) 1
No. 3, '58:239-242 1
No. 9, '60:1117-1121 (-A to -C) 1
No. 3, '64:371-373 (-F) 6 (Fire Control)
No. 5, '68:634 P
RAF Fly Rev, Ag '59:24-27 (-A) 1 Pr, S*
Mr '63:34 (-A) Pr*
:35 1
Sp/Aeron, Ap '61:59-61 (-A&-B) 3 S

2573. 37 "Viggen" (Thunderbolt)

Aerosp Intl, My/Je '68:42-44 1
Air Force, Mr '68:63-67 1
Air Pict, F '63:43 1
Ap '67:116-118+ 1
Air Prog, N '66:46 Pr†
Ap '68:18-23+ 1 Pr, S*
Avn Wk, O 21, '63:57-61 1
My 17, '65:117 1
Je 14, '65:257-259 1
My 29, '67:234-258 1
My 20, '68:55-60 2
F 3, '69:64-66 6 (ILS) S*
Je 2, '69:226-238+ 6 (Avionics) C*, S
Flight, O 26, '56:670-671 1 P
D 27, '62:1010-1012 1 S
F 16, '67:270 4
Ap 13, '67:548 1
Ap 20, '67:635-642 1 S*
Fly Rev Intl, Jl '64:6+ 1
Ag '65:27-29 1
Ap '67:493 1
Je '67:595-600 1 P

SAAB (cont.)

2573. 37 (cont.)

Interavia, No. 1, '63:109-111	1		
No. 11, '64:1687-1689	1		D
No. 4, '65:524-525	6	(Engine)	
No. 6, '65:886-889	1		S*
No. 11, '65:1742-1743	6	(Engine)	
No. 12, '66:1920-1921 (Proto-)	2		
No. 5, '67:710-713	1		C, S
No. 6, '68:749-751	6	(Electronics)	C
No. 3, '69:276-277	6	(Thrust Rvrsr)	
:278-279	6	(Landing)	
RAF Fly Rcv, Mr '63:15-16	1		

2574. 90 "Scandia"

Aeroplane, Ag 27, '48:239-240	4	
Avn Age, F '53:30-31	1	C, S
Flight, Je 3, '48:607-611	1	S
:611-613	4	C
Ag 26, '48:243	4	
D 1, '49:719	1	
Interavia, No. 4, '46:30-33	1	
F '49:99-102	1	C*

2575. 91 "Safir"

Air Pict, N '55:362 (-C:Australia)	1	
Avn Wk, O 29, '51:30 (-B)	1	
Ag 17, '59:89-97 (-D)	4	C
Flight, O 23, '47:459-462	1	C, S
N 7, '52:586 (-B)	3	
Ja 1, '54:2-3+ (-C)	1	C
Fly Rev Intl, D '69:53 (-D:Austria)		Pr*

2576. 105

Aerosp Intl, S/O '69:20-23	1	
Air Force, O '69:65-68	1	
Avn Wk, Je 3, '63:29	1	
Mr 23, '64:83	1	
S 25, '67:40-59 (-XT&-SK60)	1	C, S
:60-65 (-XT)	4	C
Flight, Je 29, '67:1083 (-XT)	1	
Fly Rev Intl, Je '65:28-30	1	S*
Interavia, No. 8, '61:1090-1091	1	C, S*
No. 7, '63:1046	1	
No. 2, '64:214	1	C
No. 12, '64:1802-1805	1	C
No. 4, '68:458-460 (-XT)	1	
No. 7, '68:880-881 (-XT)	4	C
RAF Fly Rev, Ag '63:16-17	1	P*

SAAB (cont.)

2577. 210
Small scale replica of 35 "Draken"
Flight, S 5, '52:324 1
Flying, Je '56:34-35 1

2578. 1071
Avn Wk, Jl 1, '68:29-32 1 S*
Interavia, No. 2, '68:199-200 1
No. 8, '68:981 1

2579. 1073
Avn Wk, Je 3, '68:36-39 1
Interavia, No. 2, '68:199-200 1
No. 8, '68:981 1

SAAC (Swiss American Aviation Corp.) USA
See LEAR JET

SABCA (Société Anonyme Belge de Constructions Aéronautiques) Belgium

2580. S XI
Flight, S 4, '31:885-886 1

2581. S XII
Flight, D 18, '31:1237-1238 1

SAI (Societa Aeronautica Italiana) Italy
See AMBROSINI

ST. LOUIS USA

2582. XPT-15 and YPT-15
Pop Avn, My '36:324 1

SALMSON France

2583. D. 6 and D. 7 "Cri-Cri"
Fly Rev Intl, Je '64:71 1

2584. 2 A2
AAHSJ, Fall '69:217 (Sykes & Miller's-) D
Avn Wk, N 7, '66:62-63 1 D*
Cr & Cockade, Spg '64:70-73 6 (Lewis Guns) P
Spg '68:103 P

SANGER-BREDT Germany

2585. Rocket Bomber
Air Pict, O '63:339 1

SANTOS-DUMONT France

2586. "Demoiselle" (Dragonfly)
Air News, Ap '46:27+ 1
Flying, O '60:51+ 1
Hist Avn, Ag '69:20-23 1 S*

SARO GB
See SAUNDERS-ROE

SAUNDERS Canada

2587. ST-27
Stretched DE HAVILLAND D.H. 114 "Heron"
Can Avn, Ag '69:17 1
Flight, Ag 7, '69:200-201 1
Fly Rev Intl, O '69:26-27 1

SAUNDERS GB

2588. "Kittiwake" (1920)
Aeroplane, N 27, '42:634 1

SAUNDERS-ROE GB

2589. "Skeeter"
Earlier development: CIERVA W. 14 "Skeeter"
Aeroplane, D 17, '54:887 S*
My 11, '56:376-380 1 S*
Jl 5, '57:19-21 (-12) 4 C
Am Helicop, Ap '55:9-12+ (Pt 1) 1
Ag '55:8-11+ (Pt 2) 1
O '55:12-13+ (Pt 3) 1 C
Ja '57:6-8 6 (Rocket Bst) S*
Jl '58:6-8 3 S*
Flight, Mr 30, '56:355-359+ (-6) 1 C, S*
Ja 18, '57:83-84 (-5) 6 (Rocket Bst) S*
Jl 5, '57:15-17 (-6) 4 C
Fly Rev Intl, O '65:124-125 1

2590. A. 7 "Severn"
Flight, Ja 9, '31:25-26 1

SAUNDERS-ROE (cont.)

2591.	A. 10		
	Air Pict, My '55:118-119	5	Pr
2592.	A. 17 "Cutty Sark"		
	Aeroplane, N 8, '33:816	4	
	Flight, Ja 2, '31:3		C
	Jl 10, '31:658-660	4	
2593.	A. 19 "Cloud"		
	Acft Eng, Ag '30:198-200	1	
	Flight, Jl 25, '30:831-832	1	Pr
2594.	A. 21 "Windhover"		
	Acft Eng, Ag '30:199	1	
	Flight, Jl 22, '32:680-681	4	
2595.	A. 27 "London"		
	Flight, F 8, '34:129	1	
	Mr 26, '36:332-335	1	S
	Jl 7, '38:12a (-I&-II)	1	
2596.	A. 37 "Shrimp"		
	RAF Fly Rev, Ja '62:49	1	
2597.	P. 531 "Wasp"		
	Aeroplane, Ag 8, '58:197-201	1	S*
	Ja 1, '60:13-16 (-2)	1	S*
	Am Helicop, S '58:7+	1	P
	Avn Wk, S 8, '58:59	1	
	Flight, Ag 8, '58:199-203	1	C, S*
	D 18, '59:745	1	
2598.	S. 36 "Lerwick" or "Flying Pig"		
	Fly Rev Intl, Mr '68:165	1	
	RAF Fly Rev, Ag '61:48	1	
2599.	S.R./A. 1		
	Aeronautics, N '47:88-92	1	P
	Aeroplane, Ag 27, '48:243-247	1	S*
	Air Pict, N '67:417	1	
	Avn Wk, Ag 25, '47:27	1	
	Can Avn, Mr '47:25	1	
	Flight, N 7, '46:497	1	
	Ag 7, '47:127-128	1	
	Jl 29, '48:123-129	1	S*
	Flying, D '47:35	1	

SAUNDERS-ROE (cont.)

2599.	S.R./A. 1 (cont.)			
	Fly Rev Intl, My '67:584	1		
	Interavia, No. 9, '47:48-49	1		
	Nav Avn News, Ja '48:32	1		
2600.	S.R. 45 "Princess"			
	Aero Digest, Jl '50:22-23+	1		
	Aeronautics, S '51:142-155	1		
	Ja '53:30-37	1		P
	Aeroplane, F 6, '48:160-164	3		S*
	Ap 7, '50:405-410	3		S*
	S 5, '52:353-361	1		D, P
	Acft Eng, O '52:300-318 (Pt 1)	1		S*
	N '52:326-334 (Pt 2)	1		
	Avn Wk, My 22, '50:19-29	2		
	D 10, '51:22-23	1		
	Can Avn, O '49:73	1		
	Flight, Ja 15, '48:63-68	1		S*
	Jl 15, '48:75-76	6	(Control)	
	Ag 4, '49:118-119	3		
	N 3, '49:594-595	6	(Engine)	
	N 9, '51:582-584	1		
	S 26, '52:411-422	1		S*
	F 20, '53:224	6	(Cocooning)	
	N 13, '53:637	1		
	Fly Rev Intl, Ja '66:318	1		
	Interavia, F '48:105	1		
	O '48:576-578	6	(Control)	S*
2601.	S.R. 53 Rocket-Jet Fighter			
	Avn Wk, My 27, '57:33-34	1		
	D 23, '57:56-63	1		
	Flight, My 24, '57:697-700	1		
	RAF Fly Rev, O '57:4-5	1		Pr*
	Mr '58:34-35	1		
2602.	S.R. 177			
	Prototype: S.R. 53			
	Avn Wk, D 23, '57:56-63	1		
	Fly Rev Intl, S '66:77	1		
	Interavia, No. 2, '58:145	1		
	RAF Fly Rev, Mr '58:34-35	1		

SAVIELIEV-ZALEWSKI Russia

2603.	S.Z. "Quadruplanes"			
	Air Pict, Mr '62:82-83	1		

SAVOIA-MARCHETTI Italy

2604. S. 31
See BUDD BB-1

2605. S. 55
Aero Digest, F '31:44-45 (Rome-to-Rio) 4
Ag '33:56-57 1
Aeroplane, D 24, '30:1464 (Rome-to-Rio) 4
Acft Eng, S '33:197-199 1
Flight, Ag 17, '33:819-820 (Balbo to USA) 4
Pop Avn, O '33:231-232+ (Balbo to USA) 1
RAF Fly Rev, N '62:39+ 1
US Air Svcs, F '31:35-38 (Rome-to-Rio) 4
Ag '33:10-14 (Balbo to USA) 4 P

2606. S. 65
Air Pict, N '64:368-369 1
RAF Fly Rev, N '62:41 1

2607. S. 66
Flight, O 13, '32:958-959 1
Pop Avn, Mr '33:170 1
RAF Fly Rev, N '62:41 1

2608. S. 71
Flight, My 29, '31:476-477 1

2609. S.M. 74
Aero Digest, Mr '35:48 1
Flight, Ja 3, '35:17 1

2610. S.M. 79 "Sparviero" (Sparrow Hawk) or "Gobbo Maledetto" (Hunchback)
Aeroplane, D 20, '40:704 1
Flight, Jl 11, '40:24b-d 1 D
RAF Fly Rev, Jl '55:35-36 (Flying Bomb-) 1
My '58:11 (-B) 1

2611. S.M. 80
Flight, D 21, '33:1286 1 S
Pop Avn, My '35:284 1

2612. S.M. 81 "Pipistrello" (Bat)
Flight, Jl 11, '40:24d-25 1

SAVOIA-MARCHETTI (cont.)

2613. S.M. 82 "Canguro" (Kangaroo) or "Marsupiale" (Marsupial)
Aeroplane, Ja 31, '41:142-144 1 S*
Air Pict, S '54:273 1
Flight, Jl 1, '43:12a 1

2614. S.M. 83
Flight, Jl 1, '43:12b 1

2615. S.M. 84
Original designations: S.M. 79 bis and S.M. 94
Flight, N 5, '42:498b 1
Fly Rev Intl, Jl '66:709 1

2616. S.M. 85
Flight, Jl 11, '40:24d-25 1
RAF Fly Rev, S '62:47 1

2617. S.M. 89
Fly Rev Intl, O '69:73 1

2618. S.M. 91 and S.M. 92
RAF Fly Rev, N '59:46+ 1

2619. S.M. 93
Fly Rev Intl, O '69:73 1
RAF Fly Rev, My '63:43 1

2620. S.M. 94
See S.M. 84

2621. S.M. 95
Interavia, No. 2, '46:46-47 1

2622. S.M. 102 "Veltro" (Greyhound)
Interavia, Ag '49:482-483 1 C, P

2623. SV-20
Fly Rev Intl, Ag '69:28-29 (-A&-C)1

SAVOIA-MARCHETTI/F.F.A. Italy/Switzerland

2624. A.S.-202 "Bravo"
Interavia, No. 4, '69:398-399 1 C

SCA (Stabilimento Costruzioni Aeronautiche)			Italy
2625.	S.S. 3 "Anitra" (Duck) (1937)		
	Fly Rev Intl, D '69:75	1	
SCHEMP-HIRTH			Germany
2626.	Gö-9		
	RAF Fly Rev, Ja '31:44	1	
SCOTTISH AVIATION			GB
2627.	"Prestwick Pioneer"		
	Aeroplane, O 29, '48:559-561	4	
	Avn Wk, O 2, '50:24	1	
	Flight, O 2, '47:389-392	1	S*
	Ag 18, '49:205-206 (-II)	1	
2628.	"Twin Pioneer"		
	Aeronautics, Ja '58:32-41	1	S*
	Aeroplane, Mr 18, '55:349-354	1	S*
	Je 1, '56:489-492	2	C
	D 7, '56:849-852	4	
	Flight, Mr 18, '55:357-361	1	S*
	N 25, '55:805-808	4	C
	Jl 6, '56:63	1	
	Ap 11, '58:474-475	1	
	Ag 21, '59:34-35 (CC. 1)	4	P
	Jl 1, '60:7-9 (Series 3)	4	C
	RAF Fly Rev, Mr '61:25	4	
S.E. (Designed by Royal Aircraft Factory)			GB
2629.	4		
	Air Pict, My '56:169	1	
	Flight, Ap 9, '42:344a	1	
2630.	5		
	Aeronautics, Je '55:53 (-A)		P
	Aeroplane, D 25, '59:657-659 (-A)	4	C
	Air BP, No. 22:2-5 (-A)	1	P*
	Air Cl, Ap '69:50 (-A)	1	D
	Air Prog, Ag/S '65:83 (-A)		P
	AAHSJ, Win '61:Supp (-A)		P*
	Spg '62:32-35 (-A)		C, Pr, S
	Avn Wk, Mr 16, '59:123 (-A)		D*

S. E. (cont.)

2630. 5 (cont.)

Flight, Jl 17, '53:85-90+	1		
Ja 15, '60:86	6	(Armament)	
F 5, '60:194	6	(Armament)	
F 27, '69:Inset (-A)	1		P*
Flying, Jl '59:19 (-A)	1		D
Ja '64:34-35+	4		C, P
Fly Rev Intl, My '68:265 (74 Sq-)			Pr*
Pop Fly, Jl '32:225a	1		D
RAF Fly Rev, Je '62:34 (-A:McCudden's)			Pr*
RAF Qtr, Ap '37:Supp (-A)	1		D
Ja '39:Plate			D

SEDDON GB

2631. "Mayfly" (1910)

Air BP, No. 17:2-5	1		P

SEIBEL USA

Avn Wk, F 18, '52:21-24

2632. S-3

Am Helicop, D '47:12-13+	6	(Control)	
Avn Wk, N 3, '47:22-27	1		
Skyways, Jl '48:35	1		

2633. S-4 "Skyhawk"

USAF designation: YH-24

Am Helicop, Jl '49:12+	1		
My '50:6-7	1		
O '51:6-8+ (-A&-B)	1		S*
Avn Age, S '50:28-29	1		
Avn Wk, O 23, '50:25-27	1		

SEPECAT (Société Européenne de Production de l'Avion École de Combat et d'Appui Tactique) Anglo-French

2634. "Jaguar" BR-121

Aerosp Intl, My/Je '69:9-10+	1		
Air Force, F '69:44-48	1		
Air Pict, Je '68:209-211	1		S
Air Prog, O '68:56-60+	1		Pr, S*
Acft Eng, Ag '68:8-15, 23-29	1		S*
:16-19	6	(Brake Chute)	
:20-22	6	(Lndg Gear)	S*

SEPECAT (cont.)

2634. "Jaguar" (cont.)

Avn Wk, F 13, '67:39	1		
Ap 22, '68:20-22	1		
Ap 29, '68:28-30	1		
My 13, '68:100-101	1		
Je 2, '69:322-323	6	(Range Finder)	
Flight, My 27, '65:815-816	1		
S 8, '66:419-421+	1		
Mr 9, '67:353-354	1		
Ap 18, '68:575-579	1		S*
Ap 25, '68:609-610	1		
Jl 11, '68:39 (Proto-)			C*
D 5, '68:959	6	(Ejection Seat)	
Ap 17, '69:641	1		
O 16, '69:600-604	1		
Fly Rev Intl, My '68:277-280+	1		
Interavia, No. 4, '68:455-457	1		
No. 6, '68:752-756	1		S*
No. 5, '69:598-601	1		C*

SERVICIOS AEREAS DE AMERICA Mexico

2635. "Cuauhtemoc M-1"
See also MAULE M-4
Avn Wk, My 4, '64:91 1

SEVERSKY USA

Aero Album, Spg '69:18-23 (Pt 1)
Fall '69:33-37 (Pt 2)
RAF Fly Rev, My '61:22-24

2636. "Executive"
Pop Avn, O '38:31 (Doolittle's-) 1 C

2637. EP-106
US military designation: P-35
RAF Fly Rev, My '61:23 (Sweden: Flygflottilj 8) Pr*

2638. P-35
Derived from SEV-2
Swedish import version: EP-106

Aero Album, Win '68:24-33	1	P, Pr†
Air Prog, '55/'56 Annual:84-85		S*
Pop Avn, N '36:22+	1	
Jl '37:33	1	
D '37:38	1	
RAF Fly Rev, Ja '58:9+	1	

SEVERSKY (cont.)

2639.	XP-41		
	Turbo-supercharged P-35		
	RAF Fly Rev, N '58:6	1	
2640.	SEV-2		
	Aero Album, Spg '69:20-21 (-XP)	1	P
	Fall '69:33 (-A&-B3)	1	P
	Aero Digest, D '37:78-80 (-PA-L)	1	
	Fly Rev Intl, S '65 (Vol. 21):61	1	
	US Air Svcs, O '37:36 (-PA-L)	1	P
2641.	SEV-3		
	Aero Album, Spg '69:18-19, 21 (-M)	1	P
	Aviation, Jl '33:224-225	1	
	Flight, N 16, '33:1154-1155	1	
	Fly Rev Intl, O '63:62	1	
	S '65 (Vol. 21):61	1	
	Pop Avn, O '33:222	1	
	S '36:26	6 (Fuel Tanks)	
2642.	SEV-3XAR		
	Aero Album, Spg '69:18-19	1	P
	Aero Digest, F '35:51	1	
	Aviation, F '35:70-71	1	
	Fly Rev Intl, S '65 (Vol. 21):61	1	
	US Air Svcs, F '35:30	1	

SFECMAS (Société Francaise d'Etudes et de Constructions de Matériels Aéronautiques Spéciaux) France
See NORD

SG (Sznycer Gottlieb) Canada

2643.	VI		
	First Canadian helicopter		
	Later design: CANAMERICAN SG VI-E		
	Am Helicop, Mr '47:8-9	1	
	O '47:8-9+	1	P
	Mr '48:16+ (-D)	1	
	Mr '51:10-11	2	
	Avn Wk, Ap 5, '48:21-22 (-D)	1	
	Can Avn, F '50:28+ (-DM)	1	
	Flight, Ap 15, '48:420-421	1	

SHCHERBAKOV Russia

2644. 2 "Shchuka" (Pike)
Fly Rev Intl, Jl '69:65 1

SHIN MEIWA Japan

2645. PS-1
Avn Wk, Mr 24, '69:53 2

2646. PX-S
Flight, S 19, '68:437-438 1
Fly Rev Intl, S '66:64 1

2647. UF-XS
Rebuilt GRUMMAN UF-1 "Albatross" for PX-S development
Air Pict, Je '64:176-177 1 Pr

SHORT GB
Air Pict, Mr '57:74-76 (Pt 1)
Ap '57:112-114 (Pt 2)
My '57:148-151 (Pt 3)
Je '57:185-188+ (Pt 4)
Flight, Ap 20, '39:404e-412
Interavia, No. 11, '49:656-659
No. 12, '69:1914-1917 (Operations in '69)
Pop Avn, O '38:32-35+ (Many aircraft photos)

2648. "Belfast" SC. 5
Earlier name: "Britannic"
Aeronautics, Jl '61:48-49 1
Aeroplane, D 12, '63:28-33 1
Acft Eng, S '63:249-259, 265-274, 281-284 1 C, S*
:260-264 6 (Engine) S*
:275-278 6 (Controls) S*
:279-280 3
F '64:46-47 1
My '64:138-139 1
O '67:33-37 2
Avn Wk, My 15, '61:47 1
Ap 16, '62:64-65 3
Ja 13, '64:39 1
Flight, Ja 3, '63:31 6 (Landing Gear)
S 19, '63:499-508 1 C, S*
Ja 9, '64:40-41 2
Ja 16, '64:94-97 2

SHORT (cont.)

2648.	"Belfast" (cont.)		
	Flight, Mr 9, '67:370-371	6 (Auto Landing)	C
	Ag 3, '67:179+ (-C.1)	1	
	Fly Rev Intl, Je '64:8	1	
	O '64:50-55	1	C, Pr
	Interavia, No. 1, '62:63-65	1	
	No. 3, '64:380-383	1	C
2649.	"Britannic" SC. 5		
	Later name: "Belfast"		
	Aeroplane, S 25, '59:261-264	1	
	Avn Wk, Mr 2, '59:37	1	
	Flight, F 20, '59:240 (-3)	1	
	S 4, '59:107 (-3A)	1	
2650.	"Calcutta" S. 8		
	Flight, O 3, '30:1081 (Service type-)	1	
	O 23, '69:11, 15-16	4	P
2651.	"Crusader" (1927)		
	Air Pict, N '64:366-368	1	
	RAF Fly Rev, Mr '58:50+	1	
2652.	"Empire" Class S. 23		
	Aeroplane, O 28, '36:530-542	1	C*, S*
	Jl 7, '37:4	1	S*
	Aviation, N '36:39 ("Canopus")		S*
	Can Avn, Ap '36:6-7	1	
	Flight, F 20, '36:202a-b	3	
	Mr 5, '36:258-260	1	S*
	Jl 9, '36:54b-d	1	
	S 10, '36:262, 264-265	1	C
	O 29, '36:430-434, 444	1	C*, S*
	:440d-h	3	S
	Jl 15, '37:68-72	4	
	Pop Avn, Ja '37:19-20 ("Canopus")	1	S*
	O '37:31 ("Caledonia")		P
	Pop Fly, Ag '37:262-265	4	C
	:268-269+	4	S*
	US Air Svcs, Ap '37:17-20+	4	
2653.	"Folder" (1913)		
	Flight, D 14, '56:925-926	1	
	Ja 4, '57:23-24	1	
2654.	"Gurnard"		
	RAF Qtr, Ja '30:96 (-II)		P

SHORT (cont.)

2655. "Hythe" Class
See "Sunderland"

2656. "Kent" S. 17
Flight, O 17, '30:1130-1132 1 Pr
F 27, '31:175 P
Ag 19, '32:780-781 1
Pop Avn, My '33:304 1

2657. "Mayo Composite" S. 20/S. 21
See also "Mercury" for upper component
Aeroplane, O 6, '37:418-420 1
F 16, '38:195-201 1 C, S*
Mr 2, '38:266-267 2 P
Air Prog, Ja '43:5-7+ 1
Bris Sid Jnl, Aut '63:20-21 1 P
Flight, N 7, '35:476-478 (Pt 1) 2 D
N 14, '35:505-506 (Pt 2) 2
Ag 19, '37:180-181 2
F 10, '38:124-125 2
F 17, '38:Supp:a-h, 159 1 C, S*
Mr 3, '38:209 2
Pop Avn, Ja '38:35 1
Pop Fly, Je '39:104-107+ 1 P
US Air Svcs, Mr '38:31 4

2658. "Mercury" S. 20
Upper component of "Mayo Composite"
Flight, S 16, '37:290-292 1
Jl 28, '38:79-80 (East-west Atlantic crossing) 4
O 13, '38:321-322 (Africa flight) 4

2659. "Sandringham" S. 25
Developed from "Sunderland"
Aeroplane, D 7, '45:671-673 1
Flight, Jl 19, '45:63-65 1

2660. "Sarafand"
See also R6/28
Aeronautics, F '46:82 C*
Flight, Je 13, '35:631-634 1 C

SHORT (cont.)

2661.	"Scion" S. 16		
	Aeroplane, F 14, '34:262-264	4	
	Mr 6, '35:272-276	1	S*
	Air Pict, Mr '56:100-101	1	Pr
	S '64:303 (G-AEZF)	1	
	Acft Eng, O '34:275	1	
	Flight, F 15, '34:147-148	1	
	F 28, '35:220	1	
	US Air Svcs, F '36:34	1	P
2662.	"Scion Senior" S. 22		
	Aeroplane, O 30, '35:533-535	1	S*
	Flight, O 31, '35:450, 452-455	1	P, S
2663.	"Scipio" Class		
	See "Kent"		
2664.	"Scylla" S. 17		
	Land version of "Kent"		
	Aeroplane, Ap 4, '34:601-608	1	C*, P, S*
	Acft Eng, Ap '34:97-99	1	
	Flight, Ap 5, '34:327-334	1	Pr, S
2665.	"Seaford" S. 45		
	Original designation: "Sunderland IV"		
	Aeroplane, Ja 4, '46:27-29	1	
	Flight, Ja 3, '46:8-10	1	
2666.	"Sealand" S.A. 6		
	Aeroplane, D 5, '47:732-734	1	S
	Je 22, '51:784-786	4	C
	S 7, '51:299		P
	Acft Eng, F '50:49-52+ (Pt 1)	1	S*
	Mr '50:87-88 (Pt 2)	1	
	Flight, N 27, '47:611 (Proto-)	1	
	D 4, '47:633-638 (Proto-)	1	S
	Je 22, '51:733-735+	4	C*, S
	Interavia, Ap '49:216-217	1	
2667.	"Seamew" S.B. 6		
	Aeroplane, Ja 20, '56:86-91 (-1)	1	S*
	Air Pict, N '53:324		P
	Avn Wk, D 21, '53:41	1	
	Flight, Ag 6, '54:164	1	
	F 18, '55:209	6 (Armament)	
	Jl 29, '55:153	1	P

SHORT (cont.)

2667.	"Seamew" (cont.)			
	Flight, Ja 20, '56:81-86 (-1&-2)	1		S*
	:86 (-1)	4		C*
	RAF Fly Rev, N '53:42	1		P
	Ap '56:52-53	1		P*
	F '63:47	1		
2668.	"Sherpa" S. B. 4			
	Aeroplane, D 18, '53:843-846	1		S*
	Air Pict, F '54:39	6	(Wing)	S*
	Ja '55:14-15	5		Pr
	Acft Eng, Ja '54:18-20	1		S*
	Avn Age, F '54:6-9	1		S*
	Flight, O 16, '53:535	1		
	Interavia, No. 2, '54:107	1		S*
2669.	"Shetland" S. 35			
	Two built			
	Aeronautics, Jl '45:38-39	1		
	S '45:32-35	1		P
	Aeroplane, My 18, '45:568			C
	My 25, '45:602-603			P
	Je 1, '45:628-631	6	(Aux Power)	S*
	Jl 27, '45:102-105	1		
	:106-108	2		
	D 7, '45:661-667	1		S*
	Air Pict, Jl '67:249	1		
	Can Avn, N '45:66-69	1		
	Comm Avn, Jl '45:71-72	1		
	Flight, My 17, '45:529-534	1		C, S*
	D 6, '45:607-608 (-II)	1		S
	Flying, Je '46:44			P*
	RAF Fly Rev, Jl '62:45	1		
	RAF Qtr, Je '46:136			P
2670.	"Shirl"			
	Air Pict, N '59:404+	1		
2671.	"Silver Streak"			
	Flight, Jl 16, '42:67-69	1		S
2672.	"Singapore" S. 19			
	Aeroplane, My 31, '33:991-992 (-II)	1		C
	Flight, S 5, '30:988-989 (-II)	1		
	Je 1, '33:509-511 (-II)	1		C
	Je 25, '36:692-693 (-III)	1		S*
	RAF Qtr, O '35:ix			P

SHORT (cont.)

2673.	"Skyvan" and "Turbo Skyvan" SC. 7			
	Aeroplane, Je 16, '66:17-20	1		
	:22-24	4		
	S 18, '68:36 (Series 3)	1		
	Air Pict, D '68:451	2		
	Air Prog, Ap '67:16-17	1		C, S
	Acft Eng, Mr '63:73	1		
	Ja '65:1-12+, 17-21, 32-34	1		S*
	:13-16	6	(Engines)	S*
	:22-24	6	(Interior)	C*
	:25-29+	3		
	:30-31	2		
	Ap '68:31 (Series 3)	1		
	Je '69:6-19 (Series 3)	1		S*
	:19-20 (Series 3)	6	(Flight Deck)	C
	Avn Wk, S 7, '64:146-149	4		C
	Jl 1, '68:76-78 (Series 3)	1		C
	Ap 21, '69:22 (-3M)	1		
	S 29, '69:51	4		
	Can Avn, F '69:16-18	4		
	Flight, Ag 26, '60:310-311	1		
	S 2, '60:326	1		
	My 28, '64:897-901	1		S*
	My 19, '66:845-849	1		S*
	:849-850	4		C
	Ap 18, '68:580	1		P*
	Ag 1, '68:164-165+ (Series 3)	1		S*
	:166-169 (Series 3)	4		C
	Jl 10, '69:66-67			P
	Ag 28, '69:321-322	1		
	S 4, '69:366			P
	Flying, N '69:74-78	4		C
	Fly Rev Intl, My '65:31-35	1		S
	Interavia, No. 3, '63:377	1		
	No. 1, '64:106-107	1		S
2674.	"Solent" S. 45			
	Aeroplane, Je 11, '48:702-703	4		C
	Jl 9, '48:47-48			D
	Flight, D 12, '46:650a-b	1		P
	Ap 10, '47:320-321	6	(Radio)	
	Ag 21, '47:193-197	1		S

SHORT (cont.)

2675. "Sperrin" S.A. 4

Aeronautics, Mr '52:34-35	1	P
Aeroplane, F 18, '55:204-209	1	S*
Air Pict, Ja '69:25	1	
Flight, D 17, '54:869-873	1	C*, S*
Ja 21, '55:79-83	1	S*

2676. "Stirling" Bomber S. 29

Aero Digest, Ap '42:147-154	1	C
My '42:156-160		S*
Jl '42:174-176		S*
Aeroplane, O 24, '41:452-453	1	
Ja 30, '42:133-135	1	
F 27, '42:228		C
:239-242 (-I)	1	S*
S 11, '42:305-310	1	C, S*
Air News, Ap '43:23-25	1	S
Air Prog, O '42:56-57		S*
Acft Eng, S '41:250-251	1	P
My '42:141-148		S*
Aviation, My '42:147		S*
Flight, O 23, '41:276b-278	4	
Ja 29, '42:94-101	1	C, S*
Ag 6, '42:148b	1	
Flying, Jl '43:72		C
Fly Rev Intl, My '68:268 (-I:214 Sq)		Pr*
Roundel, My '61:18-22	4	D
RAF Fly Rev, Ap '58:32-34+	4	
West Fly, F '43:49	1	

2677. "Stirling V" Transport S. 29

Flight, Ag 23, '45:205-207	2	

2678. "Sturgeon" S.A. 2

Aeronautics, My/Je '47:48-51	1	P
Aeroplane, O 18, '46:443-448	1	S*
Ap 27, '51:506-509 (-II)	4	C
Flight, O 17, '46:422-426	1	S
Ap 20, '51:442-444 (-II)	4	
My 4, '51:526 (-II)	1	

2679. "Sunderland" or "Flying Porcupine" S. 25

Aeronautics, Je '48:52		P*
Aeroplane, Ja 25, '39:106-107, 115-117	1	S*

SHORT (cont.)

2679. "Sunderland" or "Flying Porcupine" (cont.)

Air Pict,	Jl '67:238-240 (Pt 1)	4	
	Ag '67:280-282 (Pt 2)	4	
Flight,	Ja 26, '39:75-80 (-I)	1	C, S*
	O 1, '42:362a	1	
	My 18, '51:580-581	1	
Pop Avn,	S '38:35		P
Roundel,	Ja/F '64:33 (-III)	1	P
RAF Fly Rev,	F '57:33-36	4	
	D '58:18-20	1	

2680. "Valetta" Landplane S. 11

One built

Flight,	My 20, '32:436-437	1	

2681. "Valetta" Seaplane S. 11

Aeroplane,	Jl 23, '30:252-254	1	
	O 1, '30:770-772	4	
Acft Eng,	Ag '30:207-208	1	
Flight,	Jl 25, '30:825-830	1	Pr, S
	O 3, '30:1092-1093	4	

2682. "C" Class Flying Boat

See "Empire" Class

2683. "G" Class S. 26

Developed from S. 23 "Empire" Class

Aeroplane,	Jl 19, '39:98-100	1	S*
	Je 8, '45:652-653	1	
Flight,	Jl 20, '39:58d-60	1	C, S*
	S 21, '44:315+ ("Golden Hind")	1	
	My 14, '54:623 ("Golden Hind")	1	
Pop Avn,	Ag '39:37 ("Golden Hind")	1	

2684. N. 2B

Air Pict,	Ag '60:273	1	
Flight,	D 28, '56:1004	1	
	Ja 4, '57:23-24	1	

2685. PD. 16

Interavia,	No. 2, '57:146-147	1	

2686. R6/28

See also later "Sarafand"

Aeronautics,	Jl 13, '32:90-94	1	P
Flight,	Jl 15, '32:643, 666	1	P

SHORT (cont.)

2687.	R24/31 "Knuckleduster"		
	Aeroplane, Jl 10, '35:60-61	1	
	Flight, F 8, '34:120-121	4	
	Jl 11, '35:56-58	1	C
2688.	S. 1 to S. 74 Seaplanes		
	Flight, D 14, '56:921-926 (Pt 1)	1	
	Ja 4, '57:23-24 (Pt 2)	1	
2689.	S. 38		
	Aeronautics, F '46:78		C*
2690.	S. 81 "Gun-carrier" and S. 82		
	Flight, D 14, '56:925	1	
	D 21, '56:965	1	
	Ja 4, '57:23-24	1	
2691.	SB. 3		
	RAF Fly Rev, Mr '59:6	1	
2692.	SB. 5		
	Aeronautics, Ag '53:40-41		P
	Ap '54:46	1	
	Avn Wk, N 28, '60:61	1	
	Can Avn, Ap '53:98-100	1	
2693.	SC. 1		
	VTOL aircraft		
	Aeroplane, D 13, '57:876-879	1	
	D 12, '58:855-856	4	C
	Je 10, '60:701-709	1	S*
	S 9, '60:366-368	4	C
	F 14, '68:26	1	
	Avn Wk, O 12, '59:84-89	1	
	S 26, '60:64-65	1	
	O 17, '60:78-83	4	
	Flight, D 21, '56:949	1	
	My 10, '57:621	1	
	Je 10, '60:793-799	1	S*
	Ag 29, '63:358		P
	Interavia, No. 8, '60:982-984	1	S*
	No. 1, '61:78	1	S*
	RAF Fly Rev, S '57:12-13	1	P*
	D '60:22-23+	4	
	:24-25	1	P*, S*
	Sp/Aeron, S '60:59-62	1	S*

SHORT (cont.)

2694.	184 Type (1914)		
	Flight, D 21, '56:968	1	C
	D 28, '56:999-1003	1	
	Ja 4, '57:23-24	1	Pr
2695.	225		
	Air Prog, My/Je '65:46		Pr†
2696.	320		
	Air Pict, Mr '58:103	1	
	Flight, D 28, '56:1003	1	
	Ja 4, '57:23-24	1	
2697.	364		
	Flight, D 28, '56:1003-1004	1	
	Ja 4, '57:23-24	1	
2698.	827/830 (1915)		
	Flight, D 21, '56:967-968	1	
	Ja 4, '57:23-24	1	

SIDDELEY GB

Interavia, Ag '49:464-466

2699.	R.T. 1		
	Air Pict, Mr '59:85	1	

SIEBEL Germany

2700.	Fh. 104 "Hallore" (Halle-born)		
	Acft Eng, N '38:354-355 (-A)	1	S*
	Fly Rev Intl, Ja '69:68	1	
2701.	Si 201		
	RAF Fly Rev, O '55:11	1	
2702.	Si 204		
	Air Pict, D '53:371	1	
	Flight, Ap 6, '44:369-370	1	
	RAF Fly Rev, O '58:8+	1	

SIEMENS-SCHUCKERT Germany

Air Pict, F '58:40-42 (1909-1919)

RAF Fly Rev, F '60:34-36 (D. I to D. VI)

SIEMENS-SCHUCKERT (cont.)

2703.	D. I		
	Cr & Cockade, Aut '61:250-254	1	P, Pr
	RAF Fly Rev, F '60:34-36	1	
2704.	D. II		
	RAF Fly Rev, F '60:34-36	1	
2705.	D. III		
	RAF Fly Rev, F '60:34-36	1	
2706.	D. IV		
	Cr & Cockade, Win '63:409 (Udet's-)		Pr
	Pop Avn, Je '32:375-376	1	Pr
	RAF Fly Rev, F '60:34-36	1	Pr*
2707.	D. VI		
	Cr & Cockade, Sum '62:151-152	1	Pr
	Win '63:398-400	1	Pr
	RAF Fly Rev, F '60:36	1	
2708.	DR. I		
	Fly Rev Intl, Ja '69:68	1	P
2709.	L. I		
	Cr & Cockade, Win '66:377-380	1	P, Pr

SIKORSKY Russia

2710.	"Ilya Mourometz"		
	Aeroplane, Ap 17, '42:451	1	
	Cr & Cockade, Sum '63:168-179	1	P
	Fly Rev Intl, D '68:60-63	1	
	Pop Fly, Ag '38:239-241	4	
	RAF Fly Rev, My '57:20-22	1	
2711.	S 16		
	Air Pict, Ap '64:111-113	1	
2712.	S 20		
	Developed from S 16		
	Air Pict, Ap '64:113	1	

SIKORSKY USA

Flight, O 10, '63:607-617 (Helicopters)
Nav Avn News, Mr '58:32-33 (US Navy helicopters)

SIKORSKY (cont.)

2713. CH-3
USAF designation
See also RH-3

Avn Wk, N 27, '67:81 (-E)	6 (Armament)	

2714. CH-37 "Mojave"
Army designation
US Marine designation: HR2S
Civil designation: S-56

Avn Wk, My 10, '65:108-112	4	
US Army Avn, Jl '68:32-34	4	

2715. CH-53A "Sea Stallion"
US Navy designation
USAF designation: HH-53B and HH-53C
Civil designation: S-65

Air Prog, Ap '66:83	1	S
Avn Wk, Ap 15, '63:85-97	1	C
N 18, '68:100-105	2	
Bee-Hive, Spg '68:30-32	6 (Avionics)	C
Flight, N 19, '64:862	1	
Interavia, No. 2, '65:266	1	
No. 9, '68:1091-1093	1	C
Nav Avn News, Ja '65:3	2	
Mr '65:16-17	3	
Ap '65:9	2	

2716. CH-54A "Flying Crane" or "Skycrane" or "Tarhe"
US Army designation
Civil designation: S-64

Avn Wk, Ja 1, '68:42-45	4	
Ja 8, '68:23	6 (Military Pod)	
Bee-Hive, Sum '65:28-32	4	
Win '69:33		P
US Army Avn, Ag '63:49	1	
N '64:12-15	1	
Jl '66:30-31	4	
F '68:2-3	4	
N '69:2-6	2	

2717. H-5
Earlier designation: R-5
US Navy designation: HO3S
Civil designation: S-51

Air Pict, Jl '51:195 (-H)	1	

SIKORSKY (cont.)

2717. H-5 (cont.)

Air Prog, Ap/My '63:72 (Y-/-A) C
Interavia, No. 3, '51:159-160 6 (Rotor Anti-ice)

2718. H-34 "Choctaw"
US Army designation
Civil designation: S-58
US Marine designation: HUS-1
US Navy designation: HSS-1
US Army Avn, Mr '57:5-9 4

2719. H-19 "Chickasaw"
USAF designation
Civil designation: S-55
US Marine designation: HRS Series
US Navy designation: HO4S
Am Helicop, N '49:11+ 1 P
Bee-Hive, Ja '50:4-8 1
Sum '51:26-27 4
Sum '68:9-13 (U-/-D) 4
Flight, Ag 15, '52:187-188 (Atlantic crossing) 4
US Air Svcs, D '49:21 1
Ja '51:12 1 P

2720. HH-3
Avn Wk, O 30, '67:40 (-F) 1

2721. HH-53B and HH-53C "Super Green Giant"
USAF designation
US Navy designation: CH-53A
Civil designation: S-65
Air Force, N '68:95 (-C) 1
Avn Wk, S 12, '66:37 (-B) 1 D
Jl 3, '67:64-69 (-B) 1
Je 17, '68:56-58 (-B) 4

2722. HNS-1 "Hoverfly I"
US Navy designation
Civil designation: VS-316
US Army designation: R-4
Nav Avn News, My 1, '44:6 1

SIKORSKY (cont.)

2723. HO4S US Navy designation Commercial designation: S-55 USAF designation: H-19 US Marine designation: HRS-Series			
US Air Svcs, O '50:19	1		P
2724. HOS-1 "Hoverfly II" US Navy designation US Army designation: R-6 British name: "Hoverfly II"			
Nav Avn News, S '46:24	1		
2725. HR2S US Marine designation US Army designation: H-37 Civil designation: S-56			
Avn Wk, Ja 25, '54:18-19 (-1)	1		
2726. HSS-1 US Navy designation Civil designation: S-58 US Army designation: H-34 US Marine designation: HUS-1			
Am Helicop, Jl '58:16-17	1		
Ag '58:10-11	6	(Radar)	C
Avn Wk, Je 23, '58:54	1		C
Jl 14, '58:36			D*
Nav Avn News, Jl '58:24	1		
2727. HSS-2 US Navy designation Civil designation: S-61 Later designation: SH-3			
Am Helicop, S '58:9+	1		
F '59:10	1		
Avn Wk, S 1, '58:56-57	1		
F 9, '59:87	3		
Mr 30, '59:30-31	1		P
Ja 16, '61:52-63	1		C
Can Avn, F '63:20-21	1		
Flight, F '61:34-35+	4		C
Flying, Ap '62:106	4		
Nav Avn News, Jl '61:22-23	4		
Jl '62:31	4		

SIKORSKY (cont.)

2727.	HSS-2 (cont.)		
	RAF Fly Rev, Ag '61:20-21	1	P*
	West Avn, O '61:21	1	
2728.	XPBS-1		
	Bee-Hive, S '37:1-4	1	
	Pop Avn, N '37:21	1	P
	N '38:52+	1	
	West Fly, S '37:38+	1	
2729.	R-4 "Hoverfly I"		
	Civil designation: VS-316		
	US Navy designation: HNS-1		
	British name: "Hoverfly I"		
	Aeroplane, S 15, '44:310-313 (Y-)	1	S*
	Jl 27, '45:100-101 (-B)		C, P
	Air Pilot, D '44:4-5+	4	C*
	Air Prog, Ap/My '63:70		C
	Bee-Hive, F/Mr '43:22-23 (X-)	4	
	N '44:6-7 (-B)	1	
2730.	R-5		
	Later designation: H-5		
	US Navy designation: HO2S-1		
	Am Helicop, Je '46:24-25+	2	
	My '49:8-9+ (-D)	2	
	US Air Svcs, F '46:24	4	
2731.	R-6 "Hoverfly II"		
	US Army designation		
	US Navy designation: HOS-1		
	British name: "Hoverfly II"		
	Aero Digest, Jl 1, '45:92-94+	1	
	West Fly, N '45:39+	4	
2732.	RH-3 "Sea King"		
	Mine-sweeper version		
	Nav Avn News, My '67:6-8 (-A)	4	P
2733.	S-35		
	AAHSJ, Ap/Je '58:119-128	4	
2734.	S-38		
	AAHSJ, Fall '59:155-185, 232	1	C, P, Pr
	Flying, Mr '64:53	1	D

SIKORSKY (cont.)

2735.	S-39		
	Air Pict, Mr '66:115 ("Osa's Ark"-)	1	
	AAHSJ, Spg '62:3-16	1	C*, P, Pr
	Win '63:267-271	1	
2736.	S-40		
	Aero Digest, F '30:126-127	1	
	S '31:66-67	1	
	AAHSJ, Spg '67:79		P
	Aviation, O '31:594-598	1	
	Bee-Hive, S '31:1-4	1	P
	D '31:3-6	4	
	Flight, Ja 24, '30:154-155	1	
	N 13, '31:1135	1	
	Pop Avn, O '31:22	1	
	D '31:89	1	S*
	US Air Svcs, N '31:16-18	1	P
	D '31:30 ("American Clipper"-)		P
	Ja '34:28		P
2737.	S-41		
	Aero Digest, N '30:70	1	
	US Air Svcs, O '30:41	1	
2738.	S-42		
	Aero Digest, Ap '34:56-57+	1	
	My '34:18		P
	Ag '34:54-56 (Pt 1)	1	
	S '34:24-26 (Pt 2)	1	
	O '34:50-52 (Pt 3)	1	
	Ag '35:41		C
	Aeroplane, Ap 25, '34:723-724	1	
	Ap 1, '36:409-412 (-A)	1	S*
	Jl 14, '37:43-44 (-B)	1	
	Aviation, Mr '34:96	1	
	Ap '34:106-108	1	
	Ag '34:246-249	1	
	:273	3	
	Bee-Hive, Mr '34:7	1	
	My '34:4-5	4	
	S '34:1-7	4	P
	Flight, Ag 30, '34:908	2	
	Jl 15, '37:68-72 (-B)	1	C
	Pop Fly, Je '36:131 (-A)	1	

SIKORSKY (cont.)

2738. S-42 (cont.)
US Air Svcs, Ap '34:26 1 P
My '34:12-14 1
S '34:20-21 ("Brazilian Clipper") 4
My '35:16-18 ("Pioneer") 4 C
Jl '35:27-28 4
Jl '37:15-16+ (-B:"Bermuda Clipper") 4
S '38:20-22+ (-B:"Bermuda Clipper") 4 P

2739. S-43
Aero Digest, Jl '35:54-58 1
Ag '35:40 C
My '36:18 P
Aeroplane, Ag 25, '37:229-231 1
Aviation, D '34:412 1
Jl '35:34-36 1
Bee-Hive, Ja '35:8-9 1
Jl '35:4-5 1
Flight, Je 20, '35:665-666 1
Pop Avn, Ag '35:93-94 1
Ap '39:18 6 (Tail)
US Air Svcs, Jl '35:27 1

2740. S-51
Civil designation
USAF designation: H-5
US Navy designation: HO3S
Am Helicop, My '46:25+ 1
Ag '46:12-13+ 4 C
Bee-Hive, Spg '51:30-31 1
Flight, Mr 10, '49:283-285 4 C
US Air Svcs, Mr '46:40 1
Je '46:24 C
Je '47:7 4

2741. S-52
Civil designation
USAF designation: H-18
US Navy designation: HO5S-1
Am Helicop, Ja '47:28+ 1
Ja '48:12+ 1
Avn Age, Jl '53:30-31 (-2) 6 (Engine) S
Avn Wk, Je 28, '48:14 2
F 22, '54:22-26 (-5) 1

SIKORSKY (cont.)

2741. S-52 (cont.)

Bee-Hive, Ja '47:23	1		
Flight, Ag 7, '53:178	1		

2742. S-55

Civil designation
USAF designation: H-19
US Marine designation: HRS-Series
US Navy designation: HO4S
British-built: WESTLAND "Whirlwind"

Aero Digest, F '53:26-40	1		S*
Air BP, No. 13:14 (N.Y. Airways-)			Pr*
Air Power, Sum '55:276			P
Air Prog, '54/'55 Annual:34-35			S*
Am Helicop, Jl '56:8	6	(Float Gear)	
Avn Wk, D 12, '49:15	1		
O 20, '52:49-58	1		
S 24, '56:88-92	4		
Bee-Hive, Ja '50:4-8	1		
Can Avn, S '56:33-35	4		

2743. S-56

Civil designation
US Army designation: H-37
US Marine designation: HR2S

Bee-Hive, Ja '54:11	1		
Ja '57:20-21	2		
Flight, Ap 30, '54:536-537	1		

2744. S-58

Civil designation
US Army designation: H-34
US Marine designation: HUS-1
US Navy designation: HSS-1

Am Helicop, Ag '58:6-7+ (Eisenhower's-)	4		
N '58:15	4		
Avn Age, O '55:34-39	1		
Avn Wk, S 17, '56:28-31	4		
S 24, '56:88-92	4		
O 29, '56:63	1		
Flight, Ap 30, '54:536-537	1		
N 11, '55:737-738	1		S*

SIKORSKY (cont.)

2745. S-60 "Flying Crane"

Am Helicop,	Mr '59:11+	1	P
Avn Wk,	O 13, '58:29-30	1	
	Ap 20, '59:29-30	1	
	O 19, '59:122-123	1	P
	Mr 28, '60:89-97	4	
Bee-Hive,	Fall '59:28-29	1	
US Army Avn,	N '59:37-38	1	

2746. S-61 "Jolly Green Giant"

-N version: Amphibious; -L version: Land;
-R version: "Jolly Green Giant"
See also H-3, HSS-2, and WESTLAND SH-3

Aeroplane,	Ja 8, '60:36-37	1	C
Avn Wk,	D 19, '60:42-43 (-L)	1	
	My 7, '62:67-79 (-A)	4	
	D 4, '67:6779- (-A)	4	
Bee-Hive,	Fall '64:22-25 (-A)	6 (Armament)	
Flight,	O 10, '63:609-611, 614-616 (-N)	1	S*
	Jl 27, '67:136-138 (-N)	4	C
Interavia,	No. 1, '64:105 (-R)	1	
	No. 4, '64:511-513 (-R)	1	S*
	No. 7, '65:1082-1084 (-L&-N)	1	

2747. S-62

Civil designation
US Navy designation: HH-52

Am Helicop,	Ja '58:8	1	D
	My '58:6-7+	1	
Avn Wk,	Je 2, '58:74-77	1	C
	Ag 25, '58:64-71	4	C
	Mr 9, '59:229		P*
	Ap 3, '61:40-41	4	
Bee-Hive,	Spg '62:22-23	6 (Rescue Pltfm)	
Flight,	Je 6, '58:778	1	
Flying,	Ag '58:49+	1	

2748. S-64 "Skycrane" or "Flying Crane"

Civil designation
US Army designation: CH-54A

Avn Wk,	D 25, '61:38-44	1	C
	Je 11, '62:32-33	1	
	Je 24, '63:72-73	1	
Bee-Hive,	Fall '63:8-10	1	
Flight,	O 10, '63:612, 616	1	S*
Flying,	Mr '62:28-29+	1	

SIKORSKY (cont.)

2749. S-65
Civil designation
USAF designation: HH-53B and HH-53C
US Navy designation: CH-53A

Flight, O 10, '63:613, 616-617	1	S*
F 20, '69:278 (-A)	1	
Interavia, No. 7, '65:1084	1	D, S
No. 3, '69:253-257	1	

2750. SH-3 "Sea King"
US Navy designation
British-built: WESTLAND SH-3D

Nav Avn News, Je '66:41 (-A)		P
Sentinel, N/D '67:20-23 (-A:RCAF)	4	

2751. VS-44 "Excalibur"

Aero Digest, D '41:224-227	1	
Air Prog, F/Mr '64:44-48 (-A)	1	C
Aviation, Ja '42:60-63+ (-A)	1	C
Mr '42:93 (-A)		S*
O '42:154-158 (-A)		S*
Bee-Hive, D '41:20-21	1	
Fall '60:33-37	4	
Flight, My 14, '42:479-480 (-A)	1	
Fly Rev Intl, Je '64:67+	1	
US Air Svcs, Mr '42:12		P

2752. VS-300

Aeroplane, F 21, '41:233-234	1	S*
Air Prog, '54/'55 Annual:24-25		S*
Aviation, Jl '40:65+	1	S*
Bee-Hive, Je '40:6-7	1	
D '40:5-6	1	
Mr/Ap '42:21-22	1	
Flight, My 29, '41:374 (-A)	2	
S 3, '42:261-263	1	
US Air Svcs, Jl '40:15	1	
Je '41:41	1	
N '43:39	1	
Je '46:24-25	4	

SILVERCRAFT Italy

2753. SH-4

Avn Wk, Mid-D '68:41	1	
Fly Rev Intl, D '68:30	1	

SIPA (Société Nouvelle Industrielle Pour l'Aéronautique) France

2754.	S-12		
	Flight, D 14, '51:752-753	4	C
2755.	S-200 "Minijet"		
	Air Pict, My '54:130-132	4	
2756.	S-251 "Antilope" (Antelope)		
	Avn Wk, Je 24, '63:96-99	4	C

SKODA KAUBA Czech

Fly Rev Intl, N '65:169-172

2757.	SK-257 and SK-V4		
	RAF Fly Rev, Ap '63:43	1	

SKYCRAFT USA

2758.	"Skyshark"		
	Earlier name: "Skylark"		
	Avn Wk, My 28, '62:89-91	1	
	Flying, My '62:49	1	

SLINGSBY GB

2759.	"Hengist"		
	Troop glider		
	RAF Fly Rev, F '58:9+	1	
	Ag '63:39	1	

SM (Shackleton & Murray) GB

2760.	1 (1932)		
	Flight, My 11, '33:440-445	1	C, Pr, S

SNCAC (Société Nationale de Constructions Aeronautiques du Centre) France

See AEROCENTRE

SNCAO (Société Nationale de Constructions Aeronautiques de l'Ouest) France

2761.	C.A.O. 200		
	Fly Rev Intl, Je '67:679	1	
	RAF Fly Rev, Jl '57:39	1	
	N '58:11	1	

SNCAO (cont.)

2762. C. A. O. 700
RAF Fly Rev, Ja '58:41 1

SNCASE (Société Nationale de Constructions Aéronautique de Sud-Est) France
See LIORE ET OLIVIER and SUD-EST

SNCASO (Société Nationale de Constructions Aéronautiques du Sud-Ouest) France
See SUD-OUEST

SNECMA (Société Nationale d'Etude et de Construction de Moteurs d'Aviation) France

2763. C-400 "Atar Volant" (Flying Atar)
Avn Wk, Ag 5, '57:50-59 1
Interavia, No. 5, '57:409 P
No. 7, '57:713-715 (-P. 2)4 C*

2764. C-450 "Coléoptère" (Coleoptera)
Air Pict, Ap '55:98-99 4
O '66:387 1
Avn Age, N '55:32-37 1
Flight, Je 24, '55:861-865 1
Fly Rev Intl, F '65:42-43 1
Interavia, No. 1, '55:35-38 1 S*
:53-55 6 (Wing)
RAF Fly Rev, Ag '58:47 1

SNOW USA

2765. S-2C
Later development: AERO COMMANDER 600 "Snow Commander"
Air Prog, Ja '67:16-19+ 4 C
Avn Wk, Mr 23, '64:87 1
Flying, F '62:48-49+ 4 C
F '65:38-43 4

SOKO Yugoslavia

2766. "Galeb" (Seagull) G2-A
Flight, N 28, '68:901-905 1 C, S*
Fly Rev Intl, S '68:43-44 1
Ja '69:36-41 1 S*
Interavia, No. 7, '69:884-886 1

SOKO (cont.)

2767. "Jastreb" (Hawk) J-1
Avn Wk, S 16, '68:96-97 1
Flight, N 28, '68:901-905 1 S*
Fly Rev Intl, S '68:45 1
Ja '69:36-41 1 P*, Pr*
Interavia, No. 7, '69:884-886 1

2768. "Kraguj" P-2
Fly Rev Intl, S '68:45 1 P
Ja '69:41 1
Interavia, No. 7, '69:886 1 C

SOLAR USA
West Fly, D '38:10-12+

2769. MS-1 Sesquiplane
Aero Digest, D '30:52 1
AAHSJ, Sum '67:130-135 1 P, Pr
Aviation, D '30:363-364 1

SOLDENHOF Switzerland

2770. Tailless
Flight, D 26, '30:1478 1

SOMERS-KENDALL GB

2771. SK-1
Aeroplane, D 31, '54:952-953 1 S*

SOPWITH GB
Aero Album, Sum '69:12-17 (Early racers)
Air Pict, Jl '60:222-225 (Civil aircraft)
Flight, N 30, '51:677-696 (Many aircraft photos)
Interavia, Je '49:334-337 (Many aircraft photos)
Pop Avn, D '35:371-372+ (WW I aircraft types)

2772. "Baby"
Cr & Cockade, Spg '63:110 P
Flight, N 22, '57:821-822 1
N 29, '57:845-848 1 Pr
Fly Rev Intl, Ap '69:52 (Norwegian-) Pr*

2773. "Buffalo"
Cr & Cockade, Sum '61:150 Pr

SOPWITH (cont.)

2774.	"Camel"		
	Air Cl, Ap '69:31 (-F.1)	1	D
	Air Force, O '69:82	4	
	Air Pict, Ja '64:9-12 (-F.1)(Pt 1)	1	Pr
	F '64:46-48 (-F.1)(Pt 2)	1	
	Mr '64:81-84 (-F.1)(Pt 3)	1	
	Ag '64:242-248 (-2F.1)	1	Pr
	Air Prog, F/Mr '64:53 (-1F.1)		Pr†
	AAHSJ, O/D '58:281-286	4	
	Win '61:Supp		P*
	Avn Wk, My 11, '59:155+	1	D*
	CAHS, Fall '66:73		D
	Cr & Cockade, Aut '62:256		Pr
	Spg '66:74-92, 104	1	Pr, P
	Fall '69:Back Cover (German Insignia)		P
	Flight, Ap 22, '55:527-532 (Pt 1)	1	
	Ap 29, '55:560-563 (Pt 2)	1	
	Ap 4, '58:454		D
	Ap 18, '68:581 (-2F.1)	1	P*
	My 2, '68:680-681a (-F.1)	4	
	Flying, Ap '59:19 (-F.1:Brown's)	4	D
	Fly Rev Intl, My '68:265 (201 Sq-)		Pr*
	Pop Avn, Ag '35:91+	1	
	Pop Fly, My '32:98	1	D
	S '34:300	4	D
	O '37:404 (-F.1)		P
	RAF Fly Rev, Ap '58:16		Pr*
	Ag '60:41	5	
	RAF Qtr, Ap '37:Supp (-F.1)	1	D
2775.	"Circuit Seaplane" (1913)		
	Aeroplane, F 6, '42:166	1	
2776.	"Cuckoo" T. 1		
	Air Pict, Ja '61:21-23	1	
	RAF Qtr, Jl '37:Supp	1	D
2777.	"Dolphin"		
	Air Pict, My '61:132-135+	1	
	Air Prog, F/Mr '64:54		Pr†
	Cr & Cockade, Sum '64:176-178		Pr, S*
	Pop Avn, Ja '35:11-12	1	
	Pop Fly, Mr '33:681	1	D

SOPWITH (cont.)

2778. "Grasshopper"
Air Pict, O '65:348-349 1 Pr

2779. "1 1/2 Strutter"
Air Prog, F/Mr '64:50 Pr†
Flight, S 28, '56:542-546 (Pt 1) 1
O 5, '56:586-591 (Pt 2) 1
Pop Fly, N '33:429 1 D

2780. "Pup"
Naval fold-wing version: BEARDMORE S.B. 3D
Air Cl, Je '69:44-47 Pr†
Air Prog, F/Mr '64:51 Pr†
F '69:2, 35, 83 (Replica-) P*, C
Cr & Cockade, Win '62:328 D, Pr
:385 (Replica-) P
Flight, Jl 20, '50:90-91 4
Ja 1, '54:8-12 1
Flying, Jl '69:100-103 4
Pop Fly, Je '32:168 1 D
RAF Fly Rev, Ja '57:11 1

2781. "Salamander"
Pop Fly, Ap '34:8-9 3 P

2782. "Schneider"
Aero Album, Sum '69:14-15 1 P
Flight, N 15, '57:765-766 1
N 29, '57:845-848 1

2783. "Snipe" 7F-1
Air Prog, F/Mr '64:55 Pr†
Avn Wk, Je 1, '64:36-37 1 D*
CAHS, Fall '64:70 (Barker's-) 1 D
Win '65:108-111 (Barker's-)1 D
Cr & Cockade, Aut '60:28-29 4
Sum '61:179 S
Flight, F 21, '63:265-267 1
Flying, Ja '60:19 (Barker's-) 1 D
Fly Rev Intl, S '65 (Vol. 21):62 P*
My '68:265 (208 Sq-) Pr*
Pop Fly, Ap '34:8-9 3 P
S '34:303 1 D

SOPWITH (cont.)

2784.	"Tabloid"		
	Air Prog, F/Mr '64:49		Pr†
	Flight, N 8, '57:733-736	1	
	N 29, '57:845-848	1	Pr
	RAF Qtr, Jl '39:314-316	1	
2785.	Tractor Biplane (1913)		
	Fly Rev Intl, D '66:276	1	
2786.	"Transport"		
	Air Pict, Ja '65:22-23 ("Atlantic")	1	Pr
2787.	"Triplane," "Tripe" or "Tripehound" PC-10		
	Air Cl, D '68:42-43 (Collishaw replica-)	1	P*
	Air Prog, F/Mr '64:52		Pr†
	AAHSJ, Fall '66:228		D
	Avn Wk, O 21, '63:36-37	1	D*
	Can Avn, Ap '68:138	1	D*
	CAHS, Fall '65:73	1	
	Cr & Cockade, Sum '67:158-161	1	
	Flight, Ap 19, '57:507-510 (Pt 1)	1	
	Ap 26, '57:553-557 (Pt 2)	1	
	Pop Fly, My '33:92	1	D
	RAF Qtr, Ap '37:Supp	1	D
	Sentinel, Mr '67:27		P

SOUTHERN GB

2788.	"Martlet"		
	Flight, My 8, '31:399	1	

SOUTHERN USA

2789.	"Roadable"		
	West Fly, Mr '46:66	1	
2790.	BM-10		
	Aero Digest, Je '41:162-165	1	
	Aviation, Je '41:83, 95	1	S*

SPAD France

Pop Avn, N '37:34-36+ (WW I aircraft types)
See also BLERIOT-SPAD

SPAD (cont.)

2791.	7		
	Air BP, No. 11:35 (USAF Museum's-)		P*
	Air Pict, Jl '56:231	1	
	Mr '66:100-103 (British-built) (Pt 1)	1	Pr
	Ap '66:144-146 (British-built) (Pt 2)	1	Pr
	Avn Wk, Ag 31, '64:54-55	1	D*
	Cr & Cockade, Spg '61:87		Pr†
	Sum '66:Back Cover (Guynemer's-)		D
	Aut '66:262-263 (Guynemer's-)		Pr†
	Win '69:345-349 (Baracca's-)	1	C*, P
2792.	11		
	Air Cl, Mr/Ap '65:40-41	1	D*
	Air Prog, Mr '68:60-61		Pr†
2793.	13		
	Air Cl, Ap '69:32	1	D
	Air Pict, Mr '66:100-103 (Pt 1)	1	
	Ap '66:144-146 (Pt 2)	1	Pr
	O '66:383 (Drewitt's-)	1	Pr
	AAHSJ, Win '61:Supp		P*
	Avn Wk, Ag 3, '59:131+		D*
	Ja 15, '62:36		D*
	Cr & Cockade, Aut '60:38-39 (Nungesser's-)		Pr†
	Sum '64:160 (Rockwell's-)		P
	Aut '68:244-256 (Rickenbacker's-)	1	P
	Flying, Ap '55:22+	4	
	Fly Rev Intl, Jl '64:64 (Belgian-)		Pr*
	Je '69:73 (Greek-)		Pr*
	Pop Avn, Je '31:19	1	
	Pop Fly, D '32:492	1	D

SPARMANN Sweden

2794.	S-1		
	Pop Avn, Ja '37:38	1	

SPARTAN GB

2795. "Arrow"
Flight, N 7, '30:1207-1210 1 S

2796. "Cruiser"
Flight, Jl 22, '32:689-692 1 Pr, S
S 9, '32:851-852 1
F 23, '33:172-173 (Iraq-) 1

2797. "Executive"
Aviation, N '37:34-35+ 1 C

2798. "Three Seater" (1930)
Air Pict, Ap '64:119 1

SPORTARUTERMELÖ NV Hungary

2799. R-18 "Kanya" (Kite)
Glider tug
Air Pict, D '50:364 1 P

STAAKEN Germany

2800. R "Giant"
Cr & Cockade, Sum '63:169 (-IV) Pr
RAF Fly Rev, O '60:50+ 1

STANDARD USA
Pop Avn, Ag '37:26+ (Aircraft type descriptions)

2801. E-1 (1918)
Air Prog, Ap/My '63:69 C

STATE AIRCRAFT FACTORY Finland

2802. "Myrsky" (Storm)
RAF Fly Rev, Je '59:51-52 1

2803. "Pyry"
Fly Rev Intl, D '65:253 1

STEARMAN USA

2804. XA-21
Pop Avn, F '40:29 P
RAF Fly Rev, F '60:9 1

STEARMAN (cont.)

2805.	N2S "Yellow Peril" US Navy version Army versions: PT-13 and PT-17		
	Nav Avn News, S '47:17	4	C
2806.	NS-1 "Yellow Peril"		
	Pop Avn, My '36:322	1	
	D '36:37-38	1	
2807.	PT-13 "Kaydet" See also MURRAYAIR MA-1 US Navy designation: N2S		
	Air Prog, Win '61/'62:29-34	1	Pr
	Ap/My '63:79 (-D)		C
	Flying, Je '44:42-47	4	
	Pop Avn, D '36:37-38	1	
	US Air Svcs, Je '39:22 (-A)		P
2808.	PT-17 "Kaydet" See also MURRAYAIR MA-1		
	Air Prog, D '68:18-21+	4	
	Flying, Mr '43:48+	4	
	Je '44:42-47	4	
2809.	PT-27 Winterized PT-13 with canopy for Canada		
	Flight, Ja 14, '43:41-43	1	
2810.	4-EM Mailplane		
	CAHS, Win '64:1 (Canadian-)	1	
2811.	73		
	Pop Avn, Ap '35:220	1	
	US Air Svcs, D '39:22 (-B1:Cuba)		P
2812.	76D1 For Argentine Navy Similar to NS-1 and PT-13		
	Pop Avn, D '36:37-38	1	P
2813.	81		
	Bee-Hive, Jl '33:7	1	
	Pop Avn, D '33:374-375	1	Pr

STEARMAN-HAMMOND USA

2814.	Y Earlier version: HAMMOND Y		
	Flight, N 18, '37:482-484	1	C*, S
	F 3, '38:100-102	4	C
	Pop Avn, My '38:53	4	
	Pop Fly, Ja '38:561+	1	
	N '38:410-412+	4	
	West Fly, Je '37:12-13+	3	
	F '39:13-14	1	

STEWARD-DAVIS USA

2815.	"Stolmaster" Jet-assisted FAIRCHILD C-119		
	Air Prog, F '68:49+, 84	1	P

STINSON USA

Air Prog, O/N '63:16-31 (Aircraft photos and specifications)

2816.	"Detroiter"		
	Flying, Ag '63:73		D
2817.	"Reliant"		
	AAHSJ, Spg '63:52 (-V-77B)	1	
	Sum '69:79-80 (Catlos'-)	1	
	Aviation, F '37:35-36 (-SR9)	1	C, S*
	Can Avn, Ap '39:47	1	
	Flight, O 17, '35:Supp:d	4	C*
	Jl 29, '37:129	6 (Inst)	C*
	Jl 13, '39:47 (-SR10C)	1	
	Pop Avn, Ja '36:28+ (-SR6-A)	1	C
	N '38:26-28+ (US Forest Svc-)	1	
2818.	"Senior Trainer" Model O Development of STINSON "Reliant"		
	Pop Avn, Mr '35:156	1	
2819.	A Airliner		
	Aero Digest, Mr '34:44-45	1	
	Flight, Mr 29, '34:303	1	
	Pop Avn, Mr '34:162-163	1	Pr
2820.	AT-19 "Reliant"		
	Flight, N 18, '43:557	1	

STINSON (cont.)

2821. L-1 "Vigilant"
Earlier designation: O-49
Flight, Ap 22, '43:418b (-A) 1

2822. L-5 "Sentinel"
Aeroplane, Mr 11, '49:278-279 4 C
RAF Fly Rev, Mr '63:45 1
Skyways, Je '47:28-31+ 4

2823. U Airliner
Aero Digest, My '30:148 1
My '32:66-67 1
Pop Avn, Jl '33:31 1 Pr, S*
US Air Svcs, D '30:57 1

STIPA Italy
See CAPRONI STIPA

STOUT USA
See also FORD

2824. "Sky Car"
Flight, Ap 17, '31:334 1
Flying, Jl '48:33+ 1

STROUKOFF USA

2825. YC-134
Development of FAIRCHILD C-123 "Provider"
Avn Wk, F 27, '56:60 1
D 24, '56:34-37 1
Ja 7, '57:58-60 1
RAF Fly Rev, Mr '62:43 1

SUD-AVIATION France
See also SUD-EST and SUD-OUEST
Air Pict, F '67:36-39

2826. "Caravelle"
See SUD-EST "Caravelle"

2827. "Caravelle Super"
Avn Wk, Je 10, '63:48-49 (-A) 1
Interavia, No. 4, '63:451-453 (-A&-B)1 C

SUD-AVIATION (cont.)

2828. "Caravelle 12"
Stretched variant of "Caravelle Super"

Flight, F 13, '69:238+	1		
Fly Rev Intl, Ap '69:27	1		

2829. "Magister"
See FOUGA "Magister"

2830. SA 321 "Super Frelon" (Super Hornet)

Avn Wk, Je 26, '61:55-59 (-0)	1		
Ap 29, '63:53 (-0)	1		
Je 10, '63:96-99 (-0)	1		
Ja 22, '68:54-61 (-F&-J)	1		
Flight, Jl 9, '64:55-59 (-0)	1		C, S*
Fly Rev Intl, D '68:37			P*
Je '69:36-37	1		
Interavia, No. 1, '62:90-92 (-0)	1		S*
No. 6, '68:768	1		
No. 1, '69:68-69 (-G:ASW)	1		C
No. 3, '69:253-257	1		C

2831. SA 330 "Puma"

Avn Wk, Jl 29, '63:46-47	1		
My 3, '65:57	1		
F 5, '68:57	6	(Landing Gear)	
Fly Rev Intl, Jl '66:680-682	1		
Ap '69:29-30	1		
Interavia, No. 8, '64:1134	1		S
No. 6, '65:883-885	1		
No. 6, '68:769	1		S*

2832. SA 340

Flight, F 16, '67:265	1		
Interavia, No. 11, '66:1802-1803	1		
No. 6, '68:770	1		S

2833. SA 341

Avn Wk, Je 2, '69:246-247	1		C

2834. SA 3164
Armed version of SUD-EST SE 3160 "Alouette III"

Avn Wk, Jl 13, '64:47	6	(Armament)	

SUD/BAC France/GB
See BAC/SUD

SUD-EST France
See also SUD-AVIATION

2835. SE 100
RAF Fly Rev, F '57:11 1
Mr '62:47 1

2836. SE 200
RAF Fly Rev, Ja '62:49 1

2837. SE 210 "Caravelle"
Later development: SUD-AVIATION "Caravelle Super" and "Caravelle 12"
Aeronautics, My '58:24-34 1 S*
My '59:66-71 4
Aeroplane, F 5, '54:154-159 1
My 13, '55:629-633 1 S*
F 3, '56:144-145+ 1
F 15, '62:170-171 (-VIR) 4 C
D 6, '62:30 (-10A) 1
Air BP, No. 7:6-9 1
No. 15:16 (Finnair-) Pr*
Acft Eng, Mr '62:86-87 3
Avn Age, Jl '55:32-33 1 S*
Mr '56:18-23 1 S*
Ja '57:62-67 6 (Engine)
My '57:66-73 3
Jl '57:70-79 2
Avn Wk, Ap 29, '57:41 C
My 13, '57:41-42 1
My 26, '58:60-79 3
My 2, '60:119-131 1
Ag 8, '60:53-54 (-VIR) 1 C
My 1, '61:118-129 (-III&-VII) 4
Flight, S 24, '54:472-473 (Proto-) 1
My 20, '55:689-694 (Proto-) 1 S*
Jl 6, '56:31-33 1 S*
N 1, '57:705-707 1 S*
D 20, '57:944-945 4 C
Jl 25, '58:101-103 1 S*
N 13, '59:549 (-1) 1 Pr
Je 8, '61:797-798 (-XA) 4
S 12, '63:456-458 1
Fly Rev Intl, Jl '67:737-738 (-VIR) 1 Pr
Interavia, No. 10, '55:770-773 1 C
No. 4, '57:308-309 2 C
No. 8, '57:810-814 (Proto-) 4
No. 2, '61:199-201 1

SUD-EST (cont.)

2838.	SE 700			
	Am Helicop, Ap '46:28+	1		
2839.	SE 1010			
	Acft Eng, Jl '48:200-205	3		S
	D '48:368-369	3		C
	Fly Rev Intl, Mr '68:164-165	1		
2840.	SE 2010 "Armagnac"			
	Acft Eng, Jl '47:233-238+ (Pt 1)	3		S
	Ag '47:265-271+ (Pt 2)	3		
	N '50:318-322	2		
	Interavia, No. 1, '46:49-52	1		
	No. 3, '50:135-136, 147	1		
	:136-139	2		
	:140-143	3		S*
	:144-146	6	(Lndg Gear)	S*
2841.	SE 2410+ "Grognard" (Grumbler)			
	Air Pict, Mr '51:83			P
	Avn Wk, Ap 16, '51:36-38	1		
	Ja 14, '52:33	6	(Flight Cntrl)	S
	N 2, '53:28-30	1		S*
	Flight, O 16, '53:536+	6	(Armament)	
	Fly Rev Intl, S '68:110	1		
	RAF Fly Rev, F '61:39+	1		
2842.	SE 3101			
	Am Helicop, F '49:13	1		
2843.	SE 3120 "Alouette" (Lark)			
	Am Helicop, S '52:9-12	1		S
2844.	SE 3130 "Alouette II" (Lark II)			
	Am Helicop, Ag '55:6-7+	1		Pr
	Ja '58:6-7+	1		
	Ap '58:8+	2		
	Avn Age, Jl '58:30-31	1		
	Avn Wk, My 6, '57:50-53	1		
	Flight, N 30, '56:841	1		
	Je 21, '57:843-844	4		
	Flying, N '58:37+	4		
	O '60:32-33+ (Republic's-)	4		
	Interavia, No. 6, '68:766	1		
	US Army Avn, My '59:13-15	1		

SUD-EST (cont.)

2845. SE 3160 "Alouette III" (Lark III)
Later version: SUD-AVIATION SA 3164
Avn Wk, Ap 29, '68:111 — 6 (Armament)
Interavia, No. 6, '68:766-767 — 1 — S

2846. SE 3200 "Frelon" (Hornet)
RAFs Qtr, Vol. 1, No. 3:240 — P

2847. SE 5000+ "Baroudeur" (Fighter)
Aeroplane, Ja 1, '54:11-15 — 1
Air Pict, F '54:38-39 — 1
N '54:328-329 — 5 — Pr
Avn Wk, Jl 27, '53:29-36 — 1
S 2, '57:74-75 (-5004) — 1
Flight, Ag 6, '54:182-183 — 1
Interavia, No. 3, '54:150-153 — 1
Nav Avn News, Mr '54:14-15 — 1
RAF Fly Rev, F '55:22-25 — 1 — Pr, S*

SUD/NORD — France/Germany

2848. SN 600 "Diplomate" or "Corvette"
Avn Wk, My 27, '68:117-119 — 1
Fly Rev Intl, S '68:46-47 — 1
Interavia, No. 8, '69:996-997 — 1 — S

SUD-OUEST — France

2849. "Deltaviex" (Delta VX)
RAF Fly Rev, F '60:9 — 1

2850. M. 1 and M. 2
Half-scale flying model for SO 4000
RAF Fly Rev, Jl '60:50 — 1

2851. SO 30 "Nene"
Avn Wk, F 12, '51:33-34 — 1

2852. SO 30P "Bretagne" (Brittany)
Flight, Ag 22, '52:204 (Auriol's-) — 1
O 10, '52:480-481 (-I&-II) — 1 — C
Interavia, No. 1, '50:13-17 — 3
:43-47 — 1 — S*
:47-48 — 2
:48-50 — 6 (Press)

SUD-OUEST (cont.)

2853.	SO 30R "Bellatrix"		
	Aviation, Je '46:93	1	
	Flight, F 19, '48:213	1	
	Interavia, No. 5, '46:20-25	1	C, P
2854.	SO 1100 "Ariel I"		
	Acft Eng, O '50:292-295	1	C
2855.	SO 1110 "Ariel II"		
	Air Pict, O '50:296-297	1	
	Acft Eng, O '50:292-295	1	S*
2856.	SO 1120 "Ariel III"		
	Air Pict, O '50:296-297	1	
	O '51:303	1	
	Am Helicop, Ap '52:10-12+	1	C, S
	Avn Wk, Jl 2, '51:25-26	1	
2857.	SO 1221 "Djinn"		
	Aeroplane, N 26, '54:760-764	4	C
	Am Helicop, O '54:6-8+ (Pt 1)	1	S*
	N '54:6-8+ (Pt 2)	1	C*, S*
	N '58:6-7+	1	P, S*
	Avn Age, S '54:32-33	1	C, S
	Avn Wk, Je 21, '54:43-44	1	S
	US Army Avn, My '59:15+	1	
2858.	SO 4000		
	Half-scale flying model: M. 1 and M. 2		
	Fly Rev Intl, Ap '68:228	1	
	RAF Fly Rev, F '58:45	1	
2859.	SO 4050 "Vautour" (Vulture)		
	Aeroplane, D 11, '53:795	1	
	Avn Age, N '54:126-131	1	
	D '54:38-45	3	
	My '56:42-47	1	
	Avn Wk, Ag 12, '57:50-59	1	
	Interavia, No. 3, '54:154-155	1	
	:158	6 (Lndg Gear & Hydr)	
	:158-159	4	
	No. 6, '55:397		P
	RAF Fly Rev, My '58:26-29+	1	Pr, S*

SUD-OUEST (cont.)

2860. SO 6000 "Triton" (Merman)
Fly Rev Intl, S '67:876 1
RAF Fly Rev, My '58:42 1

2861. SO 6020 to SO 6026 "Espadon" (Swordfish)
RAF Fly Rev, Jl '60:50 1

2862. SO 8000 "Narval" (Narwhal)
RAF Fly Rev, Mr '63:49 1

2863. SO 9000 "Trident I"
Air Power, Aut '55:16 P
Avn Wk, Mr 16, '53:15 1
RAF Fly Rev, Ag '55:40-41 1
Mr '57:35-37 1 S*

2864. SO 9050 "Trident II"
Aeronautics, Je '57:36 P
Avn Age, S '56:48-53 1
Avn Wk, Ap 9, '56:55 1
Ja 7, '57:33 2
Ap 22, '57:30 1
Ag 19, '57:50-61 1
RAF Fly Rev, Mr '57:34-37 1 Pr

SUKHOI Russia
Fly Rev Intl, F '66:351-354
:373-375 (Pt 1)
Mr '66:435-437 (Pt 2)
Ap '66:490-492+ (Pt 3)

2865. Su-2
Fly Rev Intl, Ap '65:43 1
F '66:374-375 1 Pr*

2866. Su-7 [Fitter]
Air Prog, S '67:23 (-MB) Pr
D '67:35 Pr
Avn Wk, Jl 17, '67:28-29 1
Jl 24, '67:24 1
Jl 31, '67:61 1
Ag 21, '67:75 1
Ap 7, '69:22 (-B) 1
Ap 14, '69:33 1 P
Flight, D 30, '65:1120-1121 1

SUKHOI (cont.)

2866.	Su-7 (cont.)		
	Fly Rev Intl, D '63:22-25	1	Pr
	Interavia, No. 5, '69:603 (-B)	1	P
2867.	Su-9 [Fishpot]		
	Air Pict, My '63:140-141	1	Pr
	Avn Wk, Ja 20, '58:38-39	1	
	Fly Rev Intl, D '63:22-25	1	Pr
	RAF Fly Rev, Ja '58:9	1	
2868.	[Flagon A]		
	Twin-jet interceptor		
	Fly Rev Intl, Je '69:33-35	1	P
	Interavia, No. 5, '69:602	1	P
2869.	[Flagon B]		
	STOL fighter		
	Avn Wk, Jl 17, '67:36-37	1	
	Ap 7, '69:20	1	P
	Fly Rev Intl, Je '69:33-35	1	P
	Interavia, No. 5, '69:602	1	P

SUNDORPH USA

2870.	Cabin Racer		
	Aviation, O '37:45	1	

SUNRISE USA

2871.	S-1600 "Areion"		
	Fly Rev Intl, Mr '66:406	1	

SUPERMARINE GB

Air Pict, S '53:269-271 (Pt 1) (S. 4 to Spitfire 4)
O '53:306-308 (Pt 2) (Spitfire 5 to Spit 14)
N '53:333-335 (Pt 3) (Spitfire 16 to Spiteful)
D '53:362-363 (Pt 4) (E. 10/44 to 525)
Flight, Ap 18, '40:350e-354
O 2, '53:461-480 (With aircraft types and photos)

2872.	"Air Yacht"		
	Flight, F 28, '30:250-252	1	

SUPERMARINE (cont.)

2873.	"Attacker"		
	See also "Sea Attacker"		
	Aeroplane, Je 12, '53:766-775	1	S*
	Air Pict, D '50:357 (-F.1)	1	
	Ap '66:138-140	1	
	Je '66:219 (Nene-)	1	
	Flight, My 15, '47:446-450	1	C, S*
	N 13, '47:541-542	1	
	Mr 4, '48:247-248	4	
	Fly Rev Intl, F '69:45 (-Pakistan: No. 11 Sq)		Pr*
	RAF Fly Rev, N '58:11 (-F.B.2: Pakistan)	5	
2874.	"Night Hawk"		
	Pop Fly, F '33:647		P
2875.	"Scapa"		
	Aeroplane, Ap 25, '34:722-723	1	
	F 26, '36:260-261	1	S*
	Flight, Ap 26, '34:416-418	1	
	Je 27, '35:711	1	S*
	F 27, '36:220-223	3	S
	RAF Qtr, Ap '35:Supp	1	
2876.	"Scimitar"		
	Aeroplane, Jl 26, '57:114-119	1	
	Ap 4, '58:478-479	1	
	F 20, '59:217-222	1	S*
	Air Pict, Ap '60:121-123+	1	Pr
	Avn Wk, Je 2, '58:60	1	
	Flight, S 20, '57:482-485	1	
	Ap 4, '58:449-451	4	
	Je 6, '58:769		Pr
	Ap 15, '60:538	4	
	Nav Avn News, Jl '60:34-35	4	
	RAF Fly Rev, Ap '61:22-25	1	Pr, S*
2877.	"Sea Attacker"		
	Basic design: "Attacker"		
	Avn Wk, Mr 15, '48:25		S*
2878.	"Sea Otter"		
	Flight, S 7, '44:254-255 (-I)	1	
	O 10, '46:383-385 (Civil-)	1	S

SUPERMARINE (cont.)

2879. "Seafang"
Royal Navy version
RAF version: "Spiteful"
Flight, Je 13, '46:598a-b 1 P

2880. "Seafire"
Aeroplane, Ag 24, '45:216-217 (-XV) P
Ap 16, '48:446-447 (-47) 6 (Wing
(-47) Fold) C*, S*
Flight, S 7, '44:254-255 (-III) 1
D 26, '46:707-710 1
Nav Avn News, O '50:41 (-47) P
Skyways, Ja '45:25+ (-I to -III) 1

2881. "Seagull"
RAF Qtr, Ap '38:Supp (-III) 1 D

2882. "Seagull" (1947)
Aeroplane, D 12, '47:757-761 1 S*
Avn Wk, S 20, '48:33 1
Flight, My 1, '47:383-384 1
D 11, '47:663-667 1 S*
N 18, '48:600-601 1
RAF Fly Rev, Ag '58:6 1

2883. "Seagull V"
See "Walrus"

2884. "Southampton"
Mk V renamed "Stranraer" in 1935
Air Pict, Ap '68:132-134 (Far East Flight,
'27-'28) 4
Pop Fly, D '33:488-490 (Far East Flight,
'27-'28) 4 D*
RAF Qtr, Ap '32:168 P

2885. "Spitfire"
Twenty-four mark numbers built
Developed from: 224
Aero Album, Spg '69:40-41 (-Vc:Hanks') 4 Pr†
Aero Digest, O 1, '44:128 (-14) 1
Aeronautics, F '41:26-31 1
Jl '41:33 C*
Ag '44:76 D*
S '44:38-39 (-12) 1 P
Ja '45:44 1 P*
F '47:45 (-VIII) P*

SUPERMARINE (cont.)
2885. "Spitfire" (cont.)
Aeroplane, N 30, '38:702 (-I) S*
Ap 12, '40:517-524 1 P, S*
Ap 21, '44:442-443 (-XII) P
Je 23, '44:700-701 (-VIII) P
S 15, '44:304-307 (-XIV) 1 P
Jl 13, '45:43-44 D*
O 5, '45:390-391 (-21&-22) P
N 2, '45:508-509 (-XIV) P
S 20, '46:332-333 (-XVII) S*
N 22, '46:632-633 (-VIII) S*
Air BP, No. 19:2-5 (-VIII) 4 C
No. 34:30-32 (-XII) 1
(-XIV) C
Air Pict, S '53:269-271 (To -4) (Pt 1) 1
O '53:306-308 (-5 to -14) (Pt 2) 1
N '53:333-335 (-16 to Spiteful) (Pt 3) 1
D '53:362-363 (Table) (Pt 4) 1
Ap '65:135 (-N.17) 1
Je '68:217 5
Air Prog, Je '43:45 P
'54/'55 Annual:48-49 (1940) 1 C, Pr†, S*
Ap/My '63:67 (-XVI) C
Ap/My '64:56-57 (-IXE) Pr†
S '66:47 (-VIII:Caldwell's) Pr†
Ag '68:34-39+ (-16) 4 C, P*
Acft Eng, Ap '39:167-180 (-I) 3 S*
N '41:306, 318 (-V) 1
AAHSJ, Fall '67:220-221 (-IX) C
Aviation, Jl '43:183 (-IX) S*
Can Avn, Mr '41:31 C*
Flight, Je 30, '38:636-636b P
D 1, '38:499-502 1 S*
F 9, '39:134a-c 2
Mr 28, '40:290-293+ 4
S 26, '40:246a-f 1
:247-250 4 S*
Ap 10, '41:271-273 3
N 12, '42:526a (-V) 1
Ap 1, '43:336a (-IX) 1
S 23, '43:340 (-VB) 1
Ap 20, '44:414-415 (-XII) 1
My 4, '44:471 1
Jl 13, '44:38-39 (-XI) 1

SUPERMARINE (cont.)

2885. "Spitfire" (cont.)

Flight,	Ag 17, '44:172-173 (-XI)	1	
	S 14, '44:282-284 (-XIV)	1	
	O 12, '44:402 (-XIV)	1	
	Ag 16, '45:169-171 (-XIV)	4	
	O 4, '45:371 (-21&-22)	1	
	Ja 24, '46:90a-91 (-I to -IX) (Pt 1)	1	
	Ja 31, '46:116a-117 (-IX-XXII) (Pt 2)	1	
	D 26, '46:707-710 (All Mks)	1	
	Je 13, '58:803 (-VB)		D
	Mr 21, '68:409 (-LF.16)	1	P*
	My 16, '68:755-756 (-8)	4	C
	O 16, '69:Inset (-9)	1	P*
Flying,	Jl '42:30-32+	1	
	N '44:68 (-Vb)		P*
Fly Rev Intl,	My '66:577-578 (-V:Germany)	1	P
	O '66:134-135+ (-24)	4	
	My '68:272 (-II:41 Sq)		Pr*
	S '68:62 (-IX:Denmark)		Pr*
	N '68:42 (-XIV:India)		Pr*
	Ap '69:53 (-IX:Norway)		Pr*
	Je '69:73 (-XVI:Greece)		Pr*
Pop Avn,	O '38:43	4	
Roundel,	Je '62:33 (-V:Canada)	4	P
RAF Fly Rev,	S '54:28	1	
	Ap '58:12+	1	
	:13 (-XVI)		P*
	Mr '60:9+ (-VB)	1	
	Ja '63:19	4	
RAF Qtr,	Je '46:137 (-XXII)		P
Skyways,	Mr '44:18-23+	1	C*, S*
West Fly,	D '44:142 (-XIV)	1	

2886. "Spiteful"

RAF designation

Royal Navy designation: "Seafang"

Flight,	N 1, '45:478-480 (-XIV)	1	
Fly Rev Intl,	S '67:876-877	1	

2887. "Stranraer"

Development of "Southampton"

Can Avn,	F '40:9-13 (Canadian-)	3	
Comm Avn,	Jl '39:36	1	
Flight,	Ag 1, '35:129-131	1	

SUPERMARINE (cont.)

2887. "Stranraer" (cont.)

Fly Rev Intl, Mr '63:47	1		
RAF Qtr, Ja '38:xxiii			P

2888. "Swift"

Air Prog, '53/'54 Annual:62-63+	1		S*
Avn Wk, Je 29, '53:18	1		
Flight, Ag 17, '51:195	1		
Fly Rev Intl, Mr '64:58-59	1		
Interavia, No. 2, '52:92-96	1		S*
No. 11, '53:648 (-F.4)			P
RAF Fly Rev, S '53:13-15 (-F.4)	1		
S '54:17-18 (-F.4)	4		
Ag '61:37	4		

2889. "Walrus" or "Shagbat"

Aeroplane, My 15, '42:551	1		
Flight, Mr 29, '34:297-300	1		
Ja 21, '43:68a (-II)	1		
Roundel, Je '65:33 (RCAF-)	1		P
RAF Qtr, O '34:xxiii			D

2890. S. 6

Aeroplane, D 16, '31:1376-1379 (-B)	3		
Air Pict, My '57:162-163 (-B:Boothman's)	5		Pr
:163	1		
Acft Eng, O '31:247-248 (-B)	1		S*
Ap '32:91-95	6	(Engine)	S*
Can Avn, N/D '31:10-11+	1		
Flight, Ap 18, '30:448-449 (Pt 1)	4		
Ap 25, '30:467-468 (Pt 2)	4		
My 2, '30:494-495 (Pt 3)	4		
Ag 14, '31:801 (-B)			P
S 18, '31:933-950	4		
O 2, '31:981-982 (-B)	1		
:989-994	6	(Engine)	P
Fly Rev Intl, Jl '67:743-746	1		C*, S*
Pop Avn, N '31:23 (-B)	1		
D '31:85+	4		
RAF Qtr, Ja '30:78-90	4		D*

2891. 224 and F. 7/30

Developed into "Spitfire"

Fly Rev Intl, F '64:64	1		
RAF Fly Rev, Ap '58:17+	1		

SUPERMARINE (cont.)

2892.	322 and S. 24/37 "Dumbo"		
	Aeroplane, Ap 26, '46:488-491	2	
	Flight, Ap 25, '46:409-413	1	S*
	Je 20, '46:625	1	
	Fly Rev Intl, Ap '64:61	1	
2893.	508, 525, and 529 "Scimitar" prototypes		
	Aeroplane, My 14, '54:595 (-525)		P
	RAF Fly Rev, O '62:61	1	
2894.	510 Swept-wing "Attacker"		
	Aeronautics, Ag '50:45		P
	Aeroplane, D 16, '49:816-819	1	S*
	Flight, Jl 28, '49:102-103	1	P
	Mr 2, '50:290-294	1	S
2895.	535 Development of 510		
	Aeroplane, Je 8, '51:692-693		P
	Avn Wk, S 18, '50:18	1	
	My 7, '51:32	1	S
2896.	544 N. 113		
	Flight, F 3, '56:127, 129	1	
2897.	545		
	Air Pict, Jl '57:233	1	

SVENSKA Sweden

2898.	J6 "Jaktfalk" (Gyrfalcon)		
	RAF Fly Rev, S '59:17	1	
	Je '60:9	1	
2899.	"Tigerschwalbe" (Tiger-Swallow)		
	Pop Avn, Je '33:366	1	

SWEARINGIN USA

2900.	FS-226 "Metro"		
	Avn Wk, S 29, '69:76	1	P
	Flight, O 30, '69:664	1	
	Fly Rev Intl, Ja '69:24-25	1	
	Interavia, No. 11, '69:1792-1793	1	

SWEDEN (Kungl. Flygförvaltningens Flygverkstad) Sweden

2901. J 22
Air Prog, Ja '65:42-43 (-B) Pr†
Flight, Ag 26, '43:221-222 1

SWITZERLAND Switzerland

2902. D-3802 and D-3803
RAF Fly Rev, S '58:23+ 1

SZNYCER GOTTLIEB Canada
See SG

TACHIKAWA Japan

2903. Ki-36 [Ida]
Fighter version
Air Prog, My '67:40-41 1
My '69:47+ 1

2904. Ki-55 [Ida]
Trainer version
Air Prog, My '67:40-41 1

2905. Ki-74 [Patsy]
RAF Fly Rev, Je '60:38 1

2906. Ki-77 Asahi:A-26
Fly Rev Intl, Mr '64:43-45 1
RAF Fly Rev, Jl '61:54 1
West Fly, Ja '46:28-29+ 1 C

2907. Ki-94
RAF Fly Rev, Mr '56:12+ 1

TARRANT GB

2908. "Tabor"
Aeronautics, Ap '60:71-72 1
Aeroplane, Ap 3, '42:394 1
Air Pict, Ag '56:267 1

TAYLORCRAFT GB
See also later AUSTER

TAYLORCRAFT (cont.)

2909. "Auster"
Aeroplane, S 24, '43:356-357 (-I&-III) 1 C
Ap 21, '44:444-447 (-IV) 1 C*, S*
Je 1, '45:621 (-IV) C*
Air Pict, D '51:360-363 1
Can Avn, S '44:80 1
Comm Avn, D '43:53+ (-I) 4
Flight, Ap 22, '43:418a (L-57) 1
S 23, '43:331-335 (-I&-III) 1 C*, S
O 7, '43:393 (-III) 1
Ap 20, '44:428-430 (-IV) 1 S
N 8, '45:498-500 (-V) 4 C

2910. "Plus"
Flight, Ag 3, '39:114-116 1 C, S

TAYLORCRAFT USA

2911. L-2
Redesignation of O-57
Aero Digest, F 15, '45:71-75 (Pt 1) 1 S*
Mr 1, '45:76-79 (Pt 2) 1 S

2912. TG-6
Training glider version of L-2
Aero Digest, Ap '43:228-229 S*

TECHNICAL UNIVERSITY (Muegyetemi Sportrepulo Egyesulet) Hungary

2913. M. 24 "Nebulo"
RAF Fly Rev, Je '63:49 1

TEICHER-HUNT USA

2914. "Humming Bird"
Am Helicop, My '47:14-15+ 1

TEMCO USA
See also LING-TEMCO-VOUGHT
Nav Avn News, F '58:8-9 (US Naval aircraft)

TEMCO (cont.)

2915. T-35 "Buckaroo" or "Plebe"
Civil designation: 33
Aero Digest, D '51:20-27 1 C, S*
Air Prog, O '66:38-40+ 4 C
Avn Wk, Ag 20, '51:14 1
RAF Qtr, Ja '53:9+ 1

2916. TT-1 "Pinto"
US Navy designation
Civil designation: 51
Later development: AMERICAN JET INDUSTRIES T-610 "Super Pinto"
Avn Wk, O 15, '56:98-103 1 C, S*
D 30, '57:67-71 1
Ja 20, '58:95 1
S 21, '59:128-142 4
Flying, Jl '59:40-41+ 4
RAF Fly Rev, Ag '58:26-29 1 Pr, S*

2917. 1A and 1B "Buckaroo"
Developed into 33 "Plebe"
Avn Wk, Ja 24, '49:14 (-A) 1 C

2918. 33 "Plebe"
Military designation: T-35
Avn Age, D '53:68-71 1
Can Avn, Mr '54:36 1

2919. 51
US Navy version: TT-1 "Pinto"
Avn Age, F '57:28-29 1 C, S*
West Avn, My '56:24 1

TEXAS A & M USA

2920. AG-1
Avn Wk, Jl 16, '51:42-49 1
Ag 24, '53:15-16 1
Flying, Jl '50:34+ 1
West Fly, Ja '51:20-21 1

TEZIUTLAN Mexico

2921. Trainer (1943)
Skyways, F '43:72-73+ 1

THADEN USA

2922. T. 4
Aviation, Ja '31:60-61 1

THK (Turk Hava Kurumo) Turkey

2923. 11
Fly Rev Intl, Mr '69:71 1

THOMAS MORSE USA
AAHSJ, Win '67:235-253 (Military racing aircraft with photos)
Pop Avn, Jl '36:23-24+ (Scouts with photos)

2924. MB-3
AAHSJ, Fall '61:170 (-A) C

2925. MB-4
AAHSJ, Fall '67:188-189 1 Pr

2926. MB-6 and MB-7
Air Prog, O '67:46-47+ 1

2927. S-4 "Scout"
Air Prog, '56/'57 Annual:50-51 (-B&-C) 1 Pr
Ag/S '63:52-57+ (-B, -C&-E) 1
RAF Fly Rev, Jl '56:11 1

THURSTON USA

2928. HRV-1
Hydro-research aircraft
Air Prog, F '68:38-39+ 6 (Hydrofoil)

TIMM USA

2929. N2T-1 "Tutor"
West Fly, O '43:86-87+ 3

2930. PT-160-K
Comm Avn, D '39:34 1
US Air Svcs, D '39:36 1

2931. T-840
Aviation, Ap '38:40-41+ 1
Pop Avn, Ap '38:12 1
West Fly, Mr '38:24 1

TIMMONS Canada

2932.	"Royal Gull"		
	Can Avn, Ag '56:61-62+	4	C

TIPSY Belgium/GB

2933.	"B"		
	Flight, Ag 11, '38:122c-123	4	C
	O 6, '38:301	3	S
	Je 1, '39:565-566	4	C*
	Pop Fly, Jl '37:200	1	

TK (de Havilland Aeronautical Technical School) GB

2934.	2		
	Flight, Ag 22, '35:198-200	1	Pr, S*

TOKYO GAS & ELECTRICAL CO. (Tokyo Gasu Denki Kabushiki Kaisha) Japan

2935.	"Chidori-Go" (Plover)		
	Aeroplane, F 28, '34:356	1	

TOMASHEVICH Russia

2936.	"Pega"		
	Fly Rev Intl, Ag '69:69	1	

TOWLE USA

2937.	T-2		
	Flight, F 14, '30:215-216	1	
2938.	TA-3		
	Aero Digest, F '31:58-59	1	
	Aviation, Ja '31:59-60	1	
	Pop Avn, Ja '31:43	1	
	US Air Svcs, N '30:24	1	

TRANSALL Germany/France

2939.	C-160		
	Aeroplane, Ag 5, '65:18-20	1	
	Air Prog, D '68:25-27+	1	C*, S*
	Avn Wk, Je 10, '63:78-83	1	
	Flight, Ap 25, '68:614-620	1	C*, S*
	Fly Rev Intl, Ja '64:30-32	1	

TRANSALL (cont.)

2939.	C-160 (cont.)		
	Interavia, No. 9, '60:1156-1158	1	S*
	No. 1, '62:66	1	
	No. 2, '63:215-217	1	
	No. 12, '63:1908-1913	1	S
	No. 9, '66:1344-1346	1	S*
	No. 4, '68:442-443	2	
	RAFs Qtr, Vol. 3, No. 3:168		P
2940.	C-161		
	Aeroplane, O 4, '67:20-21	1	
	N 22, '67:29 (-J)	1	
	Avn Wk, Je 19, '67:34	1	
	Interavia, No. 4, '67:460-461	1	

TRANSCEDENTAL USA

2941.	1-G "Convertiplane"		
	Am Helicop, Ap '55:14+	1	

TRANSLAND USA

2942.	Ag-2 "Duster"		
	Air Pict, N '54:346	1	
	Avn Wk, S 13, '54:21-22	1	
	N 17, '58:98-118	4	C
	West Avn, Je '56:14	1	

TRAVEL AIR USA

Flying, N '64:90-93

2943.	Monoplane (1927)		
	Flying, Ja '64:65		D
2944.	"Mystery S"		
	Air Cl, D '69:44-53 (Hawks' "Texaco 13"-)	1	C, P
	:50 (Davis' #31-)		Pr*
	Aviation, Ag '30:81-85	1	
	Pop Avn, Ag '32:99+	1	
2945.	"R"		
	Air Prog, Ja '66:28-30+	1	
	AAHSJ, Fall '69:173-177 (Pt 1)	1	Pr
	Win '69:236-239 (Pt 2)	1	Pr
	Avn Wk, O 6, '69:36-37	1	D*

TRAVEL AIR (cont.)
2945. "R" (cont.)
Flight, Ap 24, '31:358-361 1 C*, Pr
US Air Svcs, Mr '31:12 ("Texaco 13") P
Ap '31:38 ("Texaco 13") C*

TUCKER USA

2946. XP-57
Fly Rev Intl, O '63:62 1

TUPOLEV Russia
See also ANT
Fly Rev Intl, Ap '69:58-61+ (Early jet designs)

2947. [Fiddler]
Air Pict, My '62:140-141 1
Fly Rev Intl, Ap '64:28-30 1 Pr

2948. SB-2 and SB-3
Civil designation: ANT 40
Fly Rev Intl, Mr '67:446, 448-451 (-2) 1 Pr*
RAF Fly Rev, Je '58:53 (-2&-3) 1
Je '63:37-38 (-2 in Spain) 1

2949. Tu-2 [Bat]
Fly Rev Intl, Mr '68:157-159 1 Pr*
Interavia, N '49:642 P
RAF Fly Rev, D '62:43 1

2950. Tu-10 [Frosty] and [Brawny]
Bomber version = [Brawny] Fighter version = [Frosty]
Avn Age, Jl '51:38-39 1 S*
Skyways, O '51:13+ [Brawny] 1

2951. Tu-12
First Soviet-designed jet bomber
Flying, F '52:12-13+ 1 S*

2952. Tu-14 [Bosun]
Fly Rev Intl, Ap '69:58-61 1 P, S*

2953. Tu-16 [Badger]
Aeroplane, Ja 23, '59:106-107 1
Air Pict, Ja '63:10-11 1 Pr
Avn Wk, Mr 23, '59:62-63 1

TUPOLEV (cont.)

2953.	Tu-16 (cont.)		
	Fly Rev Intl, My '64:21-25	1	Pr
	Ag '69:46-47		P*
2954.	Tu-20 [Bear]		
	Air Pict, Jl '63:212-213	1	Pr
	Avn Age, My '57:42-49	1	
	Avn Wk, Je 11, '56:27	1	
	Ap 8, '57:53-57	6 (Engine)	
2955.	Tu-22 [Blinder]		
	Earlier designation: [Beauty]		
	Air Pict, O '62:318-319	1	Pr
	Avn Wk, Ag 7, '61:31	1	
	S 30, '68:94	1	
	S 22, '69:30 (-B)	1	P
	D 22, '69:41	6 (Eng Nozzles)	
	Fly Rev Intl, Ag '64:28-31	1	Pr
2956.	Tu-70 [Cart]		
	Transport version of Tu-4 (BOEING B-29)		
	RAF Fly Rev, Ap '57:49	1	
	West Avn, S '51:12-13+	1	
2957.	Tu-81		
	Fly Rev Intl, Ap '69:58-61+	1	
2958.	Tu-89		
	Fly Rev Intl, Ap '69:58-61+	1	P
2959.	Tu-95 [Bear]		
	Cargo version: Tu-114 [Cleat]		
	Avn Wk, Je 10, '68:58-59	1	
	Interavia, No. 4, '57:337-338	6 (Engine)	
2960.	Tu-104 [Camel]		
	Aeronautics, Ap '56:38-39		P
	Aeroplane, Ap 6, '56:211-215	1	P
	Ap 4, '58:480-482	4	
	My 15, '59:566-567 (-A)		C
	Air BP, No. 7:28 (Aeroflot-)		Pr*
	No. 28:16 (-A:Ceskoslovenske)		Pr*
	Air Pict, My '56:151-153	1	
	Avn Age, My '56:72-77	1	
	Avn Wk, Ap 2, '56:28-31	1	
	Ap 16, '56:46-47, 126-127	1	

TUPOLEV (cont.)

2960.	Tu-104 (cont.)			
	Can Avn, My '56:45	1		
	Ag '58:28-29	4		
	Flight, Mr 30, '56:372+	1		
	Ap 6, '56:393	1		
	Ap 13, '56:427-428	1		
	Jl 6, '56:69	1		
	N 16, '56:784-785	1		C*
	Jl 25, '58:107 (-A)	1		
	Fly Rev Intl, My '66:550	1		
	Interavia, No. 5, '56:359	1		
	Sp/Aeron, Mr '60:107-110	6	(Anti-skid)	S*
2961.	Tu-110 [Cooker]			
	Air Pict, Mr '66:115	1		
	Avn Wk, S 30, '57:40-43	1		
	Flight, Jl 25, '58:107	1		
2962.	Tu-114 [Cleat]			
	Military version: Tu-95 [Bear]			
	Aeroplane, F 7, '63:20-21	1		
	Avn Wk, D 9, '57:38-39	1		
	F 17, '58:38-40	1		
	F 24, '58:38-39	1		
	Jl 6, '59:38-39	4		
	Jl 13, '59:114-115	1		
	Jl 20, '59:68-69	1		C
	Ag 17, '59:50-51	1		
	Flight, F 28, '58:286-287	1		S*
	Jl 25, '58:108-109	1		S*
	Ag 21, '59:43-45	1		
	Interavia, No. 1, '58:61	1		
	No. 8, '59:986-987	1		
	RAF Fly Rev, Je '58:20-21			S*
2963.	Tu-114 [Moss]			
	Early-warning variant			
	Fly Rev Intl, D '69:26-27	1		
2964.	Tu-124 [Cookpot]			
	Aeroplane, My 31, '62:28	1		
	D 24, '64:20-21, 25	1		
	F 10, '66:4-7	1		
	Air Prog, N/D '65:14-15	1		C
	Avn Wk, Ag 22, '60:43	1		
	Je 11, '62:44-45	1		

TUPOLEV (cont.)

2964.	Tu-124 (cont.)		
	Flight, Ag 16, '62:229-230+	1	
	D 24, '64:1071-1072	4	
	Fly Rev Intl, My '66:551	1	
	Interavia, No. 6, '64:817-819	1	
2965.	Tu-134 [Crusty]		
	Aeroplane, F 10, '66:4-7	1	C
	Ag 21, '68:27-28	1	
	Air Prog, N/D '65:20-21	1	
	Avn Wk, O 12, '64:43-46	1	
	N 9, '64:61	1	
	Je 14, '65:90-91	1	C
	N 4, '68:33-37	4	
	N 18, '68:43-45	4	
	Je 23, '69:57 (-A)	1	
	Flight, Ag 22, '68:288-290	1	
	N 21, '68:819-820	4	
	Fly Rev Intl, D '69:56-58	1	P*, Pr
2966.	Tu-144 [Charger]		
	Aeroplane, Jl 7, '66:4-7	1	
	Air Force, O '68:52-53	1	
	Air Pict, Mr '69:85+	1	
	Avn Wk, Je 28, '65:32	1	
	Je 26, '67:40-41	1	
	Ag 19, '68:18-19	1	
	Ag 26, '68:32-33	1	
	Ja 13, '69:27-29	2	C
	Ag 25, '69:26-28	1	S*
	Can Avn, F '69:43	1	
	Fly Rev Intl, My '66:551	1	
	Mr '67:425	1	
	Ap '69:25-27	1	
	Interavia, No. 2, '69:171-173	1	
2967.	Tu-154 [Careless]		
	Aeroplane, O 6, '66:16-19	1	S*
	Avn Wk, O 10, '66:56-59	1	
	N 18, '68:32-33	1	
	D 2, '68:23	2	
	D 16, '68:36	1	P
	Jl 7, '69:54-67	1	C
	Jl 21, '69:57+	6 (Engine)	
	Flight, S 29, '66:543-544 (Pt 1)	1	
	O 6, '66:582 (Pt 2)	1	

TUPOLEV (cont.)

2967. Tu-154 (cont.)
Fly Rev Intl, O '69:25-26 1
Interavia, No. 2, '67:224 1

UCC (United Consultant Corporation) USA

2968. "Twin Bee"
Original design: REPUBLIC "Seabee"
Flying, F '66:46-47+ 4

UDET Germany

2969. U 12 "Flamingo"
Air Pict, Ag '56:289 1
Fly Rev Intl, D '69:53 (-Austrian) Pr*

UFAG (Ungarische Flugzeugwerke, AG) Hungary

2970. C. I
Fly Rev Intl, S '69:50 Pr*

UMBAUGH USA

2971. 18
Flying, Ja '62:44-45+ 4 C
Interavia, No. 12, '60:1519-1520 1

UMBRA Italy

2972. A.U.T.-18
Fly Rev Intl, Ag '66:767 1
RAF Fly Rev, Ag '61:46+ 1

UNITED HELICOPTERS USA
Am Helicop, Ja '49:9-11+
West Fly, Mr '50:14-15+

2973. C-4 "Commuter"
Avn Wk, Jl 7, '47:29-30 1

UNITED STATES AIR SERVICE USA
See ENGINEERING DIVISION

URMV-3 Rumania

2974. IAR-821
Fly Rev Intl, Ja '69:26 1

URMV-3 (cont.)

2975. IS-23
Fly Rev Intl, Ja '69:26-27 1

UTVA Yugoslavia

2976. 56
Avn Wk, Ag 1, '60:87-90 1

2977. 65 "Privrednik" (Economist)
Avn Wk, Jl 12, '65:113 1

VANCE USA

2978. Flying Wing (1933)
Pop Avn, My '33:284 1
Ja '34:16 4

VANGUARD USA

2979. 2C "Omniplane"
Avn Wk, My 25, '59:129-135 1 S

VEB (Volkseigener Betriebe) East Germany

2980. 152
Earlier prototype: BAADE 152
Avn Wk, Mr 21, '60:76-85 1
Ag 21, '61:72-79 1
Flight, My 20, '60:699 1 S*
Fly Rev Intl, Mr '69:70-71 1

VEF (Valsts Elektrotechniska Fabrika) Latvia

2981. J. 12
Flight, Ag 17, '39:156g-h 1

VEGA USA
See also LOCKHEED

2982. "Starliner"
West Fly, O '38:34 1

2983. 35
Can Avn, Ap '41:64+ 1
Comm Avn, Ap '41:62 1

VERTIGYRO USA

2984. VG-1, VG-2 and VG-4
Flight, Ap 30, '64:732 1

VERTOL USA
See BOEING-VERTOL

VERVILLE USA

2985. "Air Coach"
Can Avn, F '30:19 1
Flight, Ag 29, '30:966 1

2986. "Racer"
Air Prog, Je '67:37-39+ 1
Jl '68:52-55+ (-R-3) 1
Flying, O '58:71 1 D

VFW (Vereinigte Flugtechnische Werke) Germany

2987. H3
Fly Rev Intl, D '68:29-30 1

2988. H5 and H7
Fly Rev Intl, D '68:30 1

2989. VAK-191B Hovering Rig
Avn Wk, Je 14, '65:305-309 1
My 13, '68:87 1
Je 2, '69:244-245 3
Fly Rev Intl, Ag '64:15+ 1
Ag '69:26-27 1
Interavia, No. 9, '67:1377-1381 1 S*

2990. VC 400
Aeroplane, Mr 27, '68:24 1
Air Force, Mr '69:114 1
Avn Wk, Jl 21, '69:67 2 S
Interavia, No. 10, '66:1556-1557 1
No. 7, '68:882-885 1 S*

2991. 614
Aeroplane, Ja 24, '68:26-27 1
Avn Wk, O 10, '66:47-49 1
F 5, '68:37-39 1
Je 2, '69:254-255 1 C

VFW (cont.)
2991. 614 (cont.)

Flight, My 26, '66:866-867	1	C
Ja 25, '68:119-120	1	
Fly Rev Intl, S '69:28-29	1	
Interavia, No. 7, '64:1017-1019	1	S
No. 5, '66:708-709	1	S*
No. 4, '68:438-441	1	C*, S*

VICKERS GB

Air Pict, My '52:136-139 (Pt 1) (Bombers: '17-'52)
Je '52:168-169 (Pt 2) (Bombers: '17-'52)
Flight, Ja 6, '61:18-30 (Photo history: 148 aircraft)
Pop Avn, Je '39:40-47+ (Many aircraft photos)

2992. "COW-Gun Fighter" F. 29/27
COW = Coventry Ordnance Works

Air Pict, Je '64:187	1	

2993. "Jockey"

Flight, F 5, '32:108-110	1	P

2994. "Valencia"

RAF Fly Rev, D '58:11	1	
Jl '63:45-46	4	

2995. "Valetta"

Aeronautics, S '48:102-106	1	S*
Aeroplane, O 17, '47:526-528	1	
Avn Wk, Ap 4, '49:18	1	
Flight, S 4, '47:253	1	C

2996. "Valiant"

Aeronautics, S '51:75		P
N '53:34-35 (-II)	1	P
Mr '54:54-57	1	P
Je '56:48-49		P
Aeroplane, O 2, '53:478-479 (-II)		P
Jl 19, '57:86-87	4	
O 25, '57:620-624	1	
My 16, '58:688-693	1	S*
Air Pict, Mr '65:80-81	1	
Air Power, Win '55:110		P
Avn Wk, Je 4, '51:15	1	
Flight, Mr 30, '56:360-361	4	
Jl 19, '57:86	4	
Jl 4, '58:13-20	1	S*

VICKERS (cont.)

2996. "Valiant" (cont.)

Fly Rev Intl, My '68:269 (-I:18 Sq) Pr*
Interavia, No. 2, '58:140-141 4
RAF Fly Rev, Mr '55:22-23 1
Mr '57:8-9 (-I) 1 P*
S '60:16+ 1

2997. "Vanguard"

See AVIATION TRADERS "Merchantman" for -953 conversion

Aeronautics, Mr '58:48-51 1
Mr '61:24-25 P, S*
Aeroplane, O 21, '55:645-646 (-900) 1
Je 1, '56:494-495 1
S 6, '57:388-389 (-951) S*
D 20, '57:929-931 6 (Air Cond)
D 19, '58:887-900 (-870) 1 S*
Ja 30, '59:125-127 1 P
F 24, '61:205-210 (-951, -953) 1 C
Air BP, No. 17:10 (BEA-) Pr*
Air Pict, N '56:377-378 1
Avn Age, Ap '56:38-43 1
Ap '57:16-21 4
F '58:36-37 1
My '58:40-45 1
Avn Wk, Je 18, '56:43 1
S 2, '57:76-81 3 C
D 29, '58:32-33 1
S 14, '59:54-65 (-I) 4 C, S
Can Avn, Mr '59:26-29 1 S*
Ag '60:16-17 4
Flight, O 21, '55:650-651 (-900) 1
D 9, '55:861-864+ (-900) 1
Je 1, '56:671-676 1 S*
Jl 6, '56:43-46 1 S*
My 10, '57:620-621 (-951) 3
N 1, '57:689-692 (-950) 1 S*
Jl 25, '58:129-131 (-950) 1 S*
Ag 8, '58:191-192 6 (Props)
D 26, '58:976 (-951) 1 Pr
Ja 9, '59:61-76 1 C, S*
Ja 30, '59:150 2
S 4, '59:134 P
O 2, '59:341 4
N 20, '59:616-617 S*
Mr 18, '60:361-363 4 C

VICKERS (cont.)

2997. "Vanguard" (cont.)
Flight, Mr 3, '61:279-286 4
Interavia, No. 5, '59:531-534 1 S*

2998. "Varsity"
Aeronautics, D '49:46-47 P
Aeroplane, F 3, '50:127-130 1 C, S
F 1, '52:126-129 4 C, P
Flight, Mr 30, '50:406-410 1 S
Ap 25, '52:484-485 4
O 19, '56:634-635 (Hussein's-) 1
RAF Qtr, O '52:310 P

2999. "Vellox"
Aeroplane, Je 27, '34:994 P

3000. "Venom"
Air Pict, Ag '67:278-279 1 Pr

3001. "Vernon"
Air Pict, My '67:168-169 1

3002. "Vespa"
Held 1932 altitude record
Aero Digest, N '32:34 1
Acft Eng, O '32:259-260 1
Flight, S 30, '32:912-914 1

3003. "Viastra"
Aeroplane, S 24, '30:713-718 1 S
Air Pict, Ap '63:96-99 1
Flight, Je 27, '30:673 1
S 26, '30:1059-1064 (-I) 1 Pr, S
My 25, '33:488-489 (Prince of Wales'-) 1

3004. "Victoria"
Air Pict, F '64:55 (-V) 1
RAF Qtr, O '32:452 P

3005. "Viking"
Aeronautics, D '45:50-53 1 P
Ap '47:26-35 (King's Flight-) 1 P, S*

VICKERS (cont.)
3005. "Viking" (cont.)
Aeroplane, N 30, '45:625-626 D*
Ag 16, '46:191-197 1 S*
Ja 10, '47:43-44 (King's Flight-) D*
Ja 14, '49:46-47 (Nene-) 4
Ja 2, '53:17-20 4 C
Can Avn, O '45:64-65 1
Flight, My 24, '45:556a-557 1 S*
O 4, '45:364-366 1
Ja 30, '47:117-119 (King's Flight-) 1
F 13, '47:166a-d (King's Flight-) 1
Mr 6, '47:192b-d (King's Flight-) 1
Ap 24, '47:361 (-IB) 4
D 2, '48:657-659 (King's Flight-) 4 C
Fly Rev Intl, Ag '67:813 (Nene-) 1
Interavia, No. 6, '47:27-32 (-IB) 1

3006. "Vildebeest"
Air Cl, F '69:22-37+ 1 P
:26-27, 34-35 (-III:36&100 Sq) Pr*
Flight, N 6, '31:1097-1098 1
Ap 8, '32:297-298 1 P
Ag 19, '32:viii P
Mr 9, '33:227 1
Pop Fly, S '33:293 P
Je '35:140-141 D
RAF Fly Rev, Mr '62:39 4
RAF Qtr, Ap '33:xxiii P
Ja '34:xxi D

3007. "Vimy"
Aeronautics, My '54:47 (Alcock & Brown-) P
Aeroplane, Je 16, '44:672-673 4 C
Avn Wk, N 14, '66:35 (Alcock & Brown-) P*
Can Avn, Jl '44:44-46+ (Alcock & Brown-) 4
Comm Avn, Jl '41:54 (Alcock & Brown-) 4
Flight, Je 15, '44:636-639 (Alcock & Brown-) 4
D 7, '44:607 (England to Australia-) 4
Je 12, '69:959-962 (Alcock & Brown-) 4 C
Flying, Je '58:44-45+ (Alcock & Brown-) 4
Fly Rev Intl, My '68:269 (-216 Sq) Pr*
RAF Fly Rev, Jl '57:22-24 (Alcock & Brown-) 4

VICKERS (cont.)

3007. "Vimy" (cont.)
RAF Qtr, Ap '38:175-195 ("Silver Queen"-) 4
Sentinel, Je '69:42-45 (Alcock & Brown-) 4 P

3008. "Vimy Commercial"
Air Pict, F '58:51-52 1
My '67:168-169 1

3009. "Vincent"
RAF Qtr, Ap '35:Supp 1 P

3010. "Virginia"
Air Pict, N '69:410-412 1
Fly Rev Intl, My '68:268 (-VII:7 Sq) Pr*
RAF Qtr, O '34:368 P

3011. "Viscount"
Aero Digest, My '50:36-37+ (-700) 1
Ap '55:40-46 (-700) 6 (Press) S*
Aeronautics, Ja '50:30-31 (-700) 1 P
My '53:40-48 (-700&-800) 1 P*
Aeroplane, S 10, '48:346-347 1
N 26, '48:673-677 1 S*
N 4, '49:618-620 4
F 17, '50:190 (-630) 6 (De-ice)
O 26, '51:558 S*
F 20, '53:226-227 1
Mr 27, '53:369-384, 396-
:396-399 (-700) 1 S*
:385-395 (-700) 6 (Engine) S*
:400-404 (-700) 2
:405-408 (-700) 4
S 28, '56:490-491 (-802) C, P
O 12, '56:560-561 (-802) 1 S*
Jl 18, '58:99-103 (-812) 1 C, S*
Air BP, No. 12:6 (-768:Indian Airlines) Pr*
No. 24:17 (-837:Austrian Airlines) Pr*
Air Pict, My '60:157-161 (Pt 1) 1
Je '60:201-207 (Pt 2) 1
Jl '60:240-247 (Pt 3) 1
S '60:325-327 (Supp) 1
Avn Age, Mr '53:27-29 4 S
Ap '55:168-171 1
Mr '57:32-33 (-810) 1 S*

VICKERS (cont.)

3011. "Viscount" (cont.)

Avn Wk,	N 6, '50:21-26 (-700)	1		
	My 12, '52:25-36 (-700)	1		S*
	My 4, '53:76 (-701)	4		
	My 11, '53:28-38 (-701)	4		
	S 13, '54:99-101 (-700)	4		
	F 21, '55:77-82	4		
	Je 6, '55:88-90	6	(Props)	
	F 4, '57:51-52 (-810D)	1		
Can Avn,	Mr '49:20-21+	4		
	Ap '53:36-40+	1		S*
	:41+ (-700)	4		
	Ag '54:18-19+	3		
	D '54:54+ (-724)	1		
	F '55:50	4		C
	Ap '55:30-31	3		
	Jl '58:27 (-810)	1		
Flight,	N 20, '47:568-571 (-I&-II)	1		C*, S
	S 9, '48:313-314	1		
	D 30, '48:788	4		
	Mr 24, '49:333 (-700)	1		
	Je 23, '49:725 (Tay-)	1		S
	Ag 25, '49:221-224 (Proto-)	4		C, P
	F 16, '50:236-237	6	(De-ice)	
	S 7, '50:281-283 (-701)	1		S*
	Ag 1, '52:125-128 (-700)	2		
	N 28, '52:682-683 (-724)	1		
	Ja 16, '53:69-72	1		C
	F 20, '53:216-217 (-700&-800)	1		
	Mr 20, '53:357-367, 377-381 (-To 800)	1		S*
	:368-371	6	(Eng & Prop)	S*
	:372-376	4		C*
	:382-386	4		
	N 13, '53:634-635 (-700)	1		
	D 18, '53:802-803	3		
	Jl 15, '55:83-93	1		C, S*
	O 14, '55:622-624	4		
	Mr 23, '56:322-324	3		
	Jl 6, '56:19-22	1		S*
	F 22, '57:241-245 (-802)	1		S*
	:246 (-802)	4		
	N 1, '57:674-676 (-810)	1		S*
	Ja 31, '58:132-133 (-790)	1		
	F 28, '58:271-275+ (-810&-812)	1		C, S*
	Jl 11, '58:42-51	1		P
	Jl 25, '58:124-126 (-810)	1		S*

VICKERS (cont.)

3011.	"Viscount" (cont.)		
	Interavia, No. 4, '50:199-203 (-700)	1	C, S*
	No. 8-9, '50:456		P
	Skyways, Je '53:10-11+ (-700&-800)	1	
	Ap '55:8-9+ (Americanized-)	4	
3012.	"Vulture"		
	Pop Fly, Mr '35:635	1	D
3013.	"Warwick"		
	Aeronautics, Ja '45:45	1	P*
	Aeroplane, Mr 31, '44:361		P
	Flight, Mr 30, '44:332-333	1	
	RAF Fly Rev, S '58:26+	1	
3014.	"Wellesley"		
	Aeroplane, D 7, '38:746-749	3	S*
	Air Cl, Mr/Ap '65:69 (RAF 14 Sq-)		Pr*
	Flight, Jl 14, '38:42	4	
	N 10, '38:426-428 (Distance Record-)	4	
	D 8, '38:517-523	3	S
	Pop Fly, Ja '39:502-503+ (Distance Record-)	4	
	RAF Fly Rev, Ap '56:15	1	
	RAF Qtr, Jl '37:xxxii-xxxiii		P
3015.	"Wellington" and "Wimpy"		
	Aeronautics, O '39:22-25	1	S*
	Aeroplane, Jl 5, '39:18-25	1	S*
	N 8, '40:510-521 (-I to -III)	1	P, S*
	:522-523	4	
	F 12, '43:181		P
	Air BP, No. 20:2-5	4	C
	No. 36:30-33	1	C, P
	Flight, Jl 6, '39:8a-j	1	C, S
	D 19, '40:522a-e	4	
	Ja 1, '42:8		C
	Je 11, '42:588a	1	
	Ja 3, '46:16-18	4	
	Fly Rev Intl, My '68:269 (-IC:150 Sq)		Pr*
	Roundel, D '53:40-44	4	
	My '64:33 (-II:RCAF)	1	P
	RAF Fly Rev, D '53:29-30	1	
	My '62:43 (-V&-VI)	1	
	RAF Qtr, S '43:242-243	4	
	Je '46:137		P

VICKERS (cont.)

3016.	"Windsor"		
	Air Pict, F '66:52-56	1	
	Fly Rev Intl, O '67:941	1	
	RAF Fly Rev, O '55:9	1	
	My '58:8+	1	
3017.	F.B. 5 "Gun-Bus"		
	Aerosp Hist, Spg '69:56		P*
	Flight, Ag 11, '66:231-232	1	
	Ap 4, '68:485	1	P*
	Fly Rev Intl, S '68:63 (Danish-)		Pr*
	Hist Avn, Ag '69:30-31	1	
	RAF Qtr, Ap '37:Supp	1	D
3018.	F.B. 7		
	Air Pict, S '61:269-273	1	
3019.	F.B. 12		
	Air Pict, Jl '59:274	1	
3020.	F.B. 14		
	Air Pict, Jl '58:256	1	
3021.	F.B. 16 "Pot Belly"		
	Air Pict, S '60:322-323+	1	
3022.	F.B. 19 "Bullet"		
	Air Pict, N '58:397	1	
3023.	F.B. 26 "Vampire"		
	Air Pict, Ja '60:19+	1	
3024.	G. 4/31 Type 253		
	Air Pict, Mr '67:94-95	1	Pr
3025.	M. 1/30 Type 207		
	Flight, Mr 23, '33:265		P
3026.	OR-330		
	Supersonic bomber project		
	Fly Rev Intl, Ja '68:52	1	

VICKERS (cont.)

3027. V. 1000

Cancelled while under construction

Aeroplane, Ja 13, '56:56-66	1		S*
Air Pict, N '68:425	1		

3028. VC 10

RAF designation: C. 1

Aeronautics, Ap '60:24-28	1		
Aeroplane, N 20, '59:505-510	1		S*
N 18, '60:685-690	3		S*
F 10, '61:141-143	4		
Ag 17, '61:181-186	6	(Flight Cntrl)	S*
Ja 11, '62:38-43	6	(Press)	S*
Jl 5, '62:31-39	1		
:40-42	6	(Auto Landing)	
N 28, '63:29-33	4		C
Ap 30, '64:Supp:3-5	1		
:6-9, 15-16, 29-35	4		
:10, 18-21	6	(Nav & Cabin Press)	
:25-27	2		
Je 2, '66:4-6 (-C.1)	1		
My 17, '67:27-28 (-C.1)	4		
Air Pict, My '64:127-131	1		S*
Acft Eng, D '65:372-378 (-C.1)	1		
Jl '66:26-27	6	(Interior)	
S '66:49-50+, 56+	6	(Auto Landing)	
Avn Wk, S 21, '59:38-41	1		C
My 2, '60:132-137	6	(Controls)	
Ja 22, '62:67-68	6	(Auto Landing)	
S 7, '64:53-57	4		
Flight, Ja 17, '58:66-67	1		
Jl 25, '58:119	1		
O 2, '59:326-327	1		
N 20, '59:618-619			S*
F 15, '62:240-241	3		
My 10, '62:728-742	1		C, S*
Jl 12, '62:53	2		
Mr 28, '63:425-426	3		
My 30, '63:784-787	1		C*
Jl 18, '63:86	4		
Ag 29, '63:335			P
My 7, '64:756-761, 768-772	4		
:762-765	6	(Engine)	S*

VICKERS (cont.)
3028. VC 10 (cont.)
Flight, O 28, '65:725-726 4
Ja 13, '66:59-61 (-C.1) 1 S*
Ja 5, '67:14-15 (-C.1) 4
F 16, '67:255-259 6 (Auto Lndg) C
Fly Rev Intl, Ag '64:42-49+ 1 Pr, S*
Ap '66:445-447 (-C.1)1 P*, Pr
Interavia, No. 8, '60:975-976 6 (Auto Landing)
No. 9, '61:1202 3
No. 9, '62:1108-1110 1 S*
No. 12, '64:1780-1783 1 S*
Sp/Aeron, F '61:59-61 6 (Hydraulic)

3029. VC 10, Super
Aeroplane, Ap 1, '65:Supp:3-4, 11-12 1
:5-6 4
:14 3
S 18, '68:18 1
Air Pict, My '64:127-131 1 S*
Acft Eng, My '64:137 6 (Fuel) S*
Ap '65:109-137 1 S*
Jl '66:19-20, 25 6 (Interior)
Avn Wk, D 30, '63:31 1
S 7, '64:56-57 3
Jl 12, '65:43-45 1
Flight, Jl 1, '60:23+ 1
N 18, '60:824-825 S*
Ap 1, '65:494-498 1
:499-503 4 C
My 18, '67:781-782 1
Interavia, No. 12, '64:1780-1783 1
No. 7, '65:1141 4

VICKERS-SUPERMARINE GB
See SUPERMARINE

VIKING USA

3030. V-2
Aviation, Ag '31:489 1

VOISIN France

3031. L
Flying, Jl '60:53 (Nungesser's-) 1 D

VOISIN (cont.)

3032. II (1909)
Pegasus, Ag '55:14-15 1

VOUGHT USA
Air Pict, D '57:419-421 (Pt 1) (Corsair story)
Ja '58:18-20 (Pt 2) (Corsair story)
AAHSJ, Win '60:275-284+ (FU, XF2U-1 and XF3U-1 fighters)
Bee-Hive, Spg '47:6-8 (30th Anniversary)
Flying, Mr '54:23-25 (Aircraft genealogical chart)
Nav Avn News, Je '47:23-25 (US Naval aircraft)
Mr '57:12-16 (US Naval aircraft)

3033. AU-1 "Corsair"
Aero Digest, D '52:26-48 1 S*
Avn Wk, Ap 7, '52:32-35 1
Nav Avn News, My '52:36 1

3034. F4U "Corsair"
Later development: GOODYEAR F2G
Aero Album, Sum '68:58-62 (-1D) 1 C, Pr†
Win '68:5, 7-8 (-17 Fighter Sq) P
Aero Digest, D '52:26-48 1 S*
Air News, Ja '43:52-53 1
AAHSJ, Win '64:294-295 4
Win '67:308 D
Aviation, Ap '45:189-193 (-1) S*
Avn Wk, Ja 26, '53:16 (-7) 1
Bee-Hive, Je '40:5 (X-/-1) 1
F/Mr '43:17-20 3
Ap/My '43:3-5 4
:19 6 (Wing)
Je/Ag '43:18-19 3
Ja '46:14-16 (-4) 1
Spg '50:27-29 (-5N) 1
Ja '53:29-31 1
Comm Avn, Jl '40:62 (X-) 1
Flight, Mr 18, '43:284a (-2) 1
My 13, '43:502-503 1
Flying, Jl '44:34-35 1
Jl '45:58+ 4
:59-62 (-1D) 1 S
Je '66:25 1 D

VOUGHT (cont.)

3034. F4U (cont.)
Fly Rev Intl, F '66:376 (X-&-1:British,
N.Z. & Pierce's) Pr*
:378 (-1 to -7:8 profiles) Pr*
Hist Avn, My '69:20-21 1
Nav Avn News, Jl 15, '44:30 4
O '48:18-19 (-5) 1 C
Ja '50:33 (-4) 2
Mr '53:1-5 1
F '69:27 (Kepford's-) P
RAF Fly Rev, F '61:18, 20-21 (-7) Pr, S*
:19-21+ 1
Skyways, F '44:18-23+ 1 P*, S*
US Air Svcs, Ag '45:28-30 (-4) 4
Ag '50:18-19 1 P
West Fly, O '43:88-89 (BREWSTER-) 3
O '44:100-102+, 142
(GOODYEAR-) 3
O '49:17-18+ (To -5) 1 S

3035. XF5U-1 "Skimmer" or "Flying Pancake"
See also V-173
Air Pict, F '59:51-54 1 Pr
Air Prog, Ap/My '64:52-53 1 Pr†
Bee-Hive, Fall '46:22-23 1
RAF Fly Rev, N '55:11 1
West Fly, S '46:52 1

3036. XF6U-1 "Pirate"
Avn Wk, My 24, '48:13 6 (After-burner) P
Bee-Hive, Ja '47:12-13 1
Can Avn, F '47:21 1
Nav Avn News, F '47:28 3
RAF Fly Rev, Ag '57:35-36 1
Skyways, S '48:21-23+ 1
US Air Svcs, Ja '48:18 P
My '48:10 6 (After-burner)

3037. F7U "Cutlass"
Aero Digest, Ja '52:26-32 1
Aeronautics, F '55:42-43 (-3) P
Aeroplane, O 5, '56:515 (-3) 1 S*
Am Helicop, D '48:3 (X-/-1) 1
Avn Wk, N 29, '48:12-13 (X-/-1) 1
Ja 28, '52:23-35 (-1) 6 (After-burner)
Mr 22, '54:26-27 (-3) 3

VOUGHT (cont.)

3037.	F7U (cont.)		
	Bee-Hive, Ja '49:3-6 (-1)	1	
	Spg '51:32		P
	Fall '51:24-26	2	
	Flight, O 12, '56:607	1	S*
	Nav Avn News, D '48:23 (-1)	1	
	Ja '49:5 (-1)	1	
	Je '54:1-5 (-3)	4	
	N '54:1 (-3P)		P
	RAF Fly Rev, Je '56:48-49 (-3)	1	
	D '56:38-41 (-3)	1	Pr, S*
	Skyways, My '49:16-17+ (X-/-1)	1	
	S '52:10-11+	4	
	US Air Svcs, D '48:26 (X-/-1)	1	
	West Avn, F '52:26 (-3)	1	
	O '56:54 (-3)	1	S*
3038.	F8U "Crusader"		
	Air Pict, S '57:312-313 (-1P)	1	Pr
	O '58:353-355	1	Pr
	Avn Age, F '57:26-27, 67 (-1)	6 (Wing Tank)	D*
	Ap '57:32-33 (-1)	3	
	Avn Wk, Je 27, '55:15 (X-/-1)	1	
	Ja 23, '56:29 (-1)	1	
	Ap 9, '56:42-43	1	D
	Ja 28, '57:101 (-1)		D*
	F 4, '57:60-63 (-1)	3	
	Jl 22, '57:30-32 (-1P)	2	P
	Je 9, '58:21 (-3)	1	
	Jl 14, '58:54-67	1	
	Je 27, '60:84-85 (-2N)	1	P
	N 19, '62:126-127 (-2NE)	1	
	Flight, My 24, '57:691-696 (-1)	1	S*
	Flying, Jl '57:39 (-1)	1	
	Fly Rev Intl, Jl '64:30-33	1	Pr, S*
	Nav Avn News, O '56:11-13 (-1)	4	
	D '58:22-24 (-1)	4	
	Ag '59:39 (X-/-1)	1	
	My '60:28 (-2N)	1	
	RAF Fly Rev, O '55:52-53 (X-/-1)	1	P*
	D '56:64-65 (-1)	1	P*
	Ag '58:16-18 (-1)	4	
	D '58:16-17 (-3)	1	P*
	Ag '60:22-23 (-2N)	1	P*
	Je '62:22-23 (-1T)	1	
	US Air Svcs, Jl '55:9 (X-/-1)	1	
	S '56:19-20 (-1)	4	

VOUGHT (cont.)

3038.	F8U (cont.)		
	West Avn, O '55:60 (X-/-1)	2	
	F '57:10-12 (-1)	1	
3039.	O2U "Corsair"		
	Aero Digest, F '30:73		C
	Avn Wk, O 28, '68:54		D*
	Nav Avn News, Ap '55:22 (-1)	4	
3040.	O3U "Corsair"		
	AAHSJ, Sum '64:102 (-3)		P
	Bee-Hive, Ag '35:1 (-3)		P
	Pop Avn, Mr '38:59 (-1)		P
	US Air Svcs, My '32:44+ (-4)	1	P
3041.	OS2U "Kingfisher"		
	Aviation, S '44:185		S*
	Bee-Hive, Je '39:4 (X-)	1	P
	Flight, O 22, '42:444b (-1)	1	
	RAF Fly Rev, My '59:42 (-1 to -3)	1	
	US Air Svcs, Jl '39:16 (X-)	1	P
3042.	SB2U "Vindicator"		
	Air Prog, Ja '43:19 (-3)		P*
	AAHSJ, Spg '69:21-25	1	P
	Bee-Hive, Ag '37:10 (-1)	1	P
	Flying, Ja '42:38		P*
	Pop Avn, S '36:20+ (X-/-1)	1	
	N '37:2 (-1)		D
	D '37:64 (-1)	1	
	RAF Fly Rev, N '59:9	1	
	US Air Svcs, S '37:23 (-1)	1	P
3043.	SBU-1 "Corsair"		
	Air Prog, Ap '65:18-19		Pr†
	Bee-Hive, S '35:5-6	1	P
	US Air Svcs, F '35:28	1	
3044.	XSO2U-1		
	Bee-Hive, Ja '40:7	1	
	RAF Fly Rev, O '62:57	1	
	US Air Svcs, Mr '40:13	1	P

VOUGHT (cont.)

3045.	SU "Corsair"		
	Air Pict, Ja '58:18-19 (-2)	5	Pr
	Pop Avn, My '33:283	1	Pr†, S*
	US Air Svcs, F '35:22 (-4)		P
3046.	XTBU-1 "Seawolf"		
	Later version: CONVAIR TBY-1 and TBY-2		
	RAF Fly Rev, O '57:11+	1	
3047.	V-50 "Corsair"		
	US Air Svcs, My '31:41	1	P
3048.	V-80 "Corsair"		
	Aero Digest, Jl '33:43	1	
	Bee-Hive, Ag '33:8-9	1	
	Pop Avn, S '33:154+	1	
	US Air Svcs, Jl '33:28	1	
3049.	V-90 "Corsair"		
	Bee-Hive, Mr '34:5	1	
	Flight, Ap 12, '34:364-365	1	
3050.	V-100 "Corsair-Junior"		
	Aero Digest, Mr '35:40-41	1	
	Bee-Hive, Mr '35:7+	1	
	Pop Avn, My '35:356-358	1	Pr, S*
	US Air Svcs, Mr '35:23	1	P
3051.	V-143		
	Aero Digest, S '36:16		P
	AAHSJ, Jl/S '56:106		P
	:167-168	1	
	Bee-Hive, S '36:9	1	
	Pop Avn, N '36:44	1	
	S '37:40	1	Pr
	O '37:45	1	
	RAF Fly Rev, Ap '59:48	1	
	Jl '60:11+	1	
3052.	V-173 "Flying Pancake"		
	US Navy designation: XF5U-1		
	Am Helicop, S '46:22+	1	
	Bee-Hive, Fall '46:22-23	1	
	Flying, Jl '47:33-34+	1	

VOUGHT (cont.)

3053. V-326
High altitude test aircraft
Aero Album, Win '69:42 P

3054. VE-7
AAHSJ, Fall '61:170 C

VOUGHT-HILLER-RYAN USA

3055. XC-142A
Later version: LING-TEMCO-VOUGHT XC-142

Air Prog, O/N '63:76-77	1	
Avn Wk, Je 11, '62:69	1	
S 3, '62:24-25	1	C
Ja 14, '63:60-69	1	S*
Can Avn, O '64:28-29	1	
Flight, O 29, '64:738	1	
Flying, Ja '63:48+	1	
Interavia, No. 2, '62:208-209	1	S*
No. 9, '64:1368-1369	1	S*

VOUGHT-SIKORSKY USA
See SIKORSKY and VOUGHT

VULTEE USA
See also CONSOLIDATED
West Fly, F '40:14-17+

3056. YA-19
Pop Avn, S '38:55 1 S

3057. A-31 "Vengeance"
See also V-72, Mks 1 to 3

Air News, Je '43:44-48	1	
Air Pilot, Je '43:42	6 (Wing)	
RAF Fly Rev, Je '57:11 (-A to -C)	1	

3058. A-35 "Vengeance"
See also V-72, Mk IV

Flight, Jl 12, '45:36-37	4	
RAF Fly Rev, Je '57:11 (-A&-B)	1	

3059. XA-41
RAF Fly Rev, O '62:59 1

VULTEE (cont.)

3060.	BT-13 "Valiant" or "Vibrator"		
	US Navy designation: SNV		
	Aero Digest, N '41:163-167+	2	
	Ja 15, '45:86-91	1	S*
	Air Cl, D '69:38-43+	4	P*
	Air Tech, Je '44:15-23+	1	C*, P*, S*
	Skyways, Jl '47:32-35+	4	C
3061.	P-48 and P-66 "Vanguard"		
	Aeroplane, Ap 19, '40:566	1	
	Flight, My 2, '40:398-399	1	
	F 18, '43:176a	1	
	Pop Avn, Mr '40:31	1	
	RAF Fly Rev, Ag '58:6	1	
	N '62:37	1	
3062.	XP-54		
	Air Force, F '45:37		P
	RAF Fly Rev, D '55:11	1	
	D '61:59+	1	
3063.	SNV "Valiant"		
	USAAF designation: BT-13		
	Air Tech, Je '44:15-23+	1	C*, P*, S*
3064.	V-1A		
	Aero Digest, Ap '33:46-48	1	C
	Ag '34:58-60	1	
	S '35:44, 46-47	6 (Instruments)	C
	Acft Eng, O '34:273-274	1	
	AAHSJ, Ja/Mr '58:25-30	1	
	Aviation, Ap '33:128-129	1	
	Flight, Mr 22, '34:276-277	1	
	O 18, '34:1101		C
	O 3, '35:356-358	4	
	Fly Rev Intl, My '68:298	1	
	West Fly, Ap '33:15+	1	
3065.	V-11		
	Aero Digest, Ja '36:74-77	1	
	Air Pict, F '69:55	1	S
	Air Prog, '39/'40 Annual:53		P
	Aviation, Ja '36:31-33	1	
	Ja '38:32-33+ (British-)	1	C, S

VULTEE (cont.)

3065. V-11 (cont.)

Flight, O 31, '35:462-463	1	C*
Ja 9, '36:42c-d	1	
Fly Rev Intl, N '67:1005 (Russian-)	1	
Pop Avn, Ja '36:21	1	
Ag '37:22	1	S
RAF Fly Rev, Ja '62:47+	1	
West Fly, Mr '36:25	1	

3066. V-12

Aviation, D '38:32-33	1	S

3067. V-72 "Vengeance"

RAF version of A-31 and A-35

Air Pict, D '68:465	4	
Fly Rev Intl, N '68:42 (-IV:India)		Pr*
RAF Fly Rev, Je '57:11 (-I to -IV)	1	
US Air Svcs, O '43:34-36	3	

WACO USA

Air News, N '42:31-33

3068. C

Pop Avn, My '38:59		P
West Fly, Jl '35:17-19+	3	

3069. CG-4 "Hadrian"

Aero Digest, F 1, '45:78-83	3	C, S
Air Pilot, D '43:15-17+	1	
Air Tech, Ag '44:35-37+	1	
Je '45:17-24+	1	S
Aviation, N '43:121-131	3	
Can Avn, Ag '43:90-92	1	
Comm Avn, Ja '43:46+ (FORD-)	1	
Flight, S 30, '43:364	1	
Fly Rev Intl, O '63:62 (Argentinian-)	1	
West Fly, Ag '43:48-49+	4	
F '44:46-47+ (GENERAL AIRCRAFT-)	3	

3070. YCG-13

Enlarged CG-4

Aero Digest, F 1, '44:71	1	
Air Force, F '44:54-55	1	
Can Avn, O '44:96	1	
Flying, Ap '44:34-35	1	

WACO (cont.)

3071.	D		
	Aero Digest, Je '35:34-36	1	C
	AAHSJ, Spg '64:40-41	1	P
	:76		Pr
	Sum '64:134-135 (Military-)	1	Pr
	Bee-Hive, N '34:7	1	
	Flight, F 27, '36:225-226 (Military-)	1	
	Fly Rev Intl, O '68:53 (-7: Uruguay)		Pr*
	Pop Avn, Ja '35:21	1	
3072.	F		
	Military designation: PT-14		
	Air Prog, My/Je '66:50-51+ (-7)	1	Pr†
	Je '68:84 (RN-)		P*
	Flying, Ja '69:38-43 (UP-/-7)	1	C, P*
	West Fly, Je '35:23-25	3	
3073.	N		
	Aviation, O '37:40	1	
	D '37:34+	1	
	Pop Avn, Ja '38:24	1	
	West Fly, O '37:26-27	1	
3074.	PT-14		
	Civil designation: F		
	Aero Digest, Ag '40:91-92 (Y-)	1	
	Air Prog, My/Je '66:50-51+	1	Pr†

WAGNER USA

3075.	W-18		
	Avn Wk, Jl 27, '64:66-72	1	
	Flight, My 7, '64:752-753	1	S*

WALDEN USA

3076.	I to IX		
	Flying, Ja '58:34-39+	1	

WALLIS GB

3077.	WA-116		
	Flight, Mr 31, '66:516-518	1	
	:518-520	4	S*

WATANABE Japan

3078. E9W1 [Slim]
RAF Fly Rev, Mr '62:43+ 1
Ap '62:47 1

3079. K3M-3 Type 90 [Pine 2]
RAF Fly Rev, Ja '60:40 1

WATERMAN USA

3080. W-5 "Arrowbile"
Aviation, My '37:32-34 1
Pop Avn, My '37:20 1

3081. "Arrowplane"
West Fly, Ag '35:13-14 1

WEDELL-WILLIAMS USA

3082. Racers
Air Cl, Ag '69:14-19 (Turner's-) 1 P
Air Pict, Mr '64:78-80 1 Pr
Bee-Hive, N '32:2 4 P

WEICK USA

3083. W-1A
Aviation, Ja '36:17-19 1
Flying, Je '45:46-47+ 1 S*

WEIR GB

3084. W. 5 and W. 6
RAF Fly Rev, D '55:45-46 1

WEISS MANFRED Hungary

3085. WM-21 "Solyom" (Falcon)
Fly Rev Intl, S '64:46 1
S '69:51 Pr

3086. WM-23 "Ezust Nyil" (Silver Arrow)
Air Pict, Ag '66:292-293 1
Fly Rev Intl, My '66:579 1

WESTLAND GB

Comm Avn, D '44:147-152
Flight, S 12, '40:204a-206
Ap 15, '55:477-496 (40th Anniversary, with aircraft types)
My 27, '65:823-831 (50th Anniversary, with aircraft types)
Fly Rev Intl, F '64:49-51 (Helicopter development)
RAFs Qtr, Vol. 2, No. 3:198-201 (Helicopters)

3087. "Belvedere"

Aeronautics, F '62:32		P
Aeroplane, Je 15, '61:697-701	1	S*
N 16, '61:645-647	4	
Flight, F 8, '62:211-214	4	C

3088. "Dragonfly"
Original design: SIKORSKY S-51

Aeroplane, Mr 5, '48:275-278	1	C, S*
O 29, '54:634-639	4	C
Air Pict, O '56:369	5	
Am Helicop, Je '50:8-9+	4	
Flight, Je 19, '47:567-569	1	C
Ap 1, '48:355-358	1	C, S*
Ja 23, '53:103	1	S
N 6, '53:617-620	4	C
Fly Rev Intl, My '66:582		P*

3089. "Limousine IV"

Acft Eng, My '29:90-93	1	Pr

3090. "Lysander" or "Lizzie"
See also P. 12

Aeroplane, Je 8, '38:717-721 (-I)	1	C*, S*
Jl 27, '38:113-116	2	
Air BP, No. 27:17-20	4	C, P
Air Pict, Ap '67:122-126 (Pt 1)	1	
My '67:162-165 (Pt 2)	1	Pr†
Je '67:202-204 (Pt 3)	1	
Air Prog, Ja '65:80	1	
My '68:44-47 (Restored-)	1	P
Can Avn, O/N '39:25-26+	3	S*
D '39:40	1	
My '68:18-19 (Canadian Museum's-)	1	P
Comm Avn, Jl '39:36	1	

WESTLAND (cont.)

3090. "Lysander" or "Lizzie" (cont.)

Flight, O 22, '36:414c-415	1	
Je 9, '38:569-576	1	S*
My 11, '39:470-472	4	
F 11, '43:148a (-II)	1	
Mr 21, '46:292a-293	4	
Fly Rev Intl, Jl '66:700, 702 (-II&-III)		Pr*
N '68:42-43 (-II:India)		Pr*
Pop Avn, Ap '37:22	1	
Ag '39:33	1	
Roundel, Ap '62:33 (Canadian-)	4	P
RAF Fly Rev, Ag '58:42	1	
Sentinel, Ap '68:28-29	3	
Skyways, Ap '43:61		P

3091. "Pterodactyl"

Flight, D 18, '31:1232-1233	1	P
S 6, '34:914-916 (-V)	1	
Pop Avn, Mr '32:156 (-IV)	1	
Pop Fly, F '35:560, 568-569+ (-V)	4	D, P
RAF Fly Rev, Je '60:11 (-I to -V)	1	

3092. "Rotodyne"

Aeronautics, F '62:33		P
Avn Wk, S 18, '61:41	1	

3093. "Scout"

Flight, F 14, '63:232-233	1	
Je 20, '63:963-964	4	C
Jl 30, '64:183-189	1	S*
Fly Rev Intl, F '65:27-30	1	

3094. "Wallace"

Prototype: P.V. 6

Air Pict, Jl '57:245	1	
Acft Eng, Mr '33:59-63	4	C
My '33:99-100+	4	
Flight, My 4, '33:411-416	1	Pr,
Pop Avn, D '33:373	1	
My '36:318-319	1	
RAF Fly Rev, Mr '63:20	4	
Je '63:47+	1	
US Air Svcs, My '33:28-29	4	

WESTLAND (cont.)

3095.	"Wapiti"		
	Flight, D 15, '32:1197	6 (Camera)	S*
	My 17, '34:487 (Altitude record-)	6 (Engine)	
	Fly Rev Intl, My '68:268 (30 Sq-)		Pr*
	N '68:42 (-IIA:India)		Pr*
	Roundel, S '64:33 (RCAF-)	1	P
	RAF Fly Rev, Je '63:47+	1	
3096.	"Wasp"		
	Flight, F 14, '63:232-233	1	S
	Jl 30, '64:183-189	1	S*
	Fly Rev Intl, F '65:27-30	1	S
3097.	"Welkin"		
	Aeronautics, Ag '45:48-49		P
	Jl '46:45	1	P*
	Air Pict, F '60:36-38+	1	
	Am Helicop, My '46:39	1	
	Comm Avn, Ag '45:66	1	
	Flying, My '46:36		P*
	RAF Fly Rev, Ja '58:38	1	
3098.	"Wessex" Helicopter		
	Aeroplane, Jl 19, '57:81-85	1	
	N 29, '57:807-814	1	S*
	Ag 12, '65:17-23 (-60)	1	S*
	S 18, '68:43 (-60)	1	
	Am Helicop, Ag '56:17	1	
	Je '57:10-13+	6 (Engine)	S
	Flight, Jl 19, '57:91+	1	
	N 29, '57:837-842	1	S*
	N 11, '60:749-752	4	
	F 14, '63:229-231+ (AS. 1)	1	S*
	Ag 1, '63:165-166 (HAS. 1)	4	C
	Ap 9, '64:581-584	4	
	Ag 28, '69:Cover (HAS. 3)		P*
	Fly Rev Intl, D '68:35 (-3)		P*
	Interavia, No. 11, '61:1514-1517	1	S*
099.	"Wessex" Transport		
	Flight, O 3, '30:1082-1087	1	Pr
	D 26, '30:1484	1	
	O 5, '33:1007 (8-seat-)	1	

WESTLAND (cont.)

3100.	"Westminster"			
	Aeronautics, N '58:43			P
	Je '59:34-39	1		
	Aeroplane, Jl 11, '58:69-73	1		S*
	Jl 22, '60:106-107	1		
	Am Helicop, F '59:9+	1		
	Avn Wk, Je 30, '58:25	1		
	Flight, Ag 15, '58:230-235	1		C, S*
	Interavia, No. 9, '58:925	1		
3101.	"Whirlwind" Fighter			
	Air Pict, Ja '60:2-4	1		
	Aviation, My '42:203+	1		
	Flight, Mr 5, '42:194-196	1		
	Je 18, '42:614b	1		
	Ag 1, '46:110-111	4		
	Fly Rev Intl, Ap '64:50-52 (253 Sq-)	1		Pr*
	RAF Fly Rev, Je '57:13+	1		
	Skyways, Mr '43:51			P
3102.	"Whirlwind" Helicopter			
	Original design: SIKORSKY S-55			
	Aeroplane, Je 29, '51:814-816	1		C
	Je 25, '54:819-825	1		S*
	D 17, '54:888-889			S*
	F 11, '55:176-177	6	(Floats)	
	Am Helicop, Ja '55:15	1		
	N '56:4 (-V)	6	(Engine)	
	Flight, N 23, '50:448-449	1		C
	Je 29, '51:758	1		
	Jl 6, '51:14-15	1		C
	Ja 23, '53:103	1		S
	D 30, '55:978 (-V)	1		
	N 6, '59:496+ (HCC. 8)	1		C, S
	F 14, '63:228	1		
	Ag 1, '63:167 (HAR. 10)	4		C
3103.	"Widgeon"			
	Aeroplane, N 11, '55:740-744	1		C, S
	O 12, '56:554-557	4		
	Am Helicop, N '55:10+	1		
	D '56:9	1		
	Flight, D 30, '55:976-978 (Proto-)	1		
	O 12, '56:610-613 (Series 2)	4		
	F 22, '57:238-239	1		S

WESTLAND (cont.)

3104.	"Wizard"			
	Fly Rev Intl, N '64:56	1		
3105.	"Wyvern"			
	Aeronautics, Jl '52:30-33			P
	Ja '54:32-35			P
	Aeroplane, Je 16, '50:688-689 (-2)	1		P
	Je 15, '56:549-556	1		C, S*
	Flight, Ap 24, '47:372-373	6	(Prop)	S*
	Jl 3, '47:5 (-I)	1		
	Jl 28, '49:113 (Python-)	1		
	Mr 27, '53:398-399 (Python-)	6	(Prop & Fuel)	S*
	Mr 9, '56:264-267	4		
	Je 15, '56:774-779	1		C*, S*
	RAF Fly Rev, S '56:44-47+	1		Pr, S*
3106.	F. 7/30			
	Pop Fly, F '36:595	1		P
	RAF Fly Rev, Ap '59:11	1		
3107.	P. 12 "Pregnant Perch" Experimental "Lysander"			
	Fly Rev Intl, Ap '64:61	1		
3108.	P.V. 3 "Houston-Westland" 1933 Mount Everest aircraft			
	Acft Eng, Mr '33:59-63	4		
	My '33:99-100+	4		
	Flight, F 2, '33:93-96	4		
	Pop Fly, Ap '33:38	4		
	US Air Svcs, My '33:28-29	4		
3109.	P.V. 6 "Wallace" prototype			
	Aeroplane, Mr 2, '32:374-375	1		
	Flight, Mr 4, '32:192-193	1		Pr
	RAF Qtr, Jl '32:Plate 5			P
3110.	SH-3D "Sea King" British-built SIKORSKY SH-3D			
	Flight, N 30, '67:911-916	1		S*
	N 27, '69:852-853	1		P
	Fly Rev Intl, N '69:25-26	1		P
	Interavia, No. 1, '69:70-73	1		C*, S*

WESTLAND (cont.)

3111.	WE. 01		
	Aeroplane, Jl 31, '68:30	1	S*
3112.	WG. 13		
	Avn Wk, My 13, '68:85	1	
	Flight, Ap 11, '68:561	1	
	Ag 15, '68:262-263	1	
	Fly Rev Intl, D '68:29	1	
	Interavia, No. 6, '68:771	1	D

WHITE USA

3113.	"New Standard" D-25		
	AAHSJ, Win '68:285-287	1	

WHITEHEAD AIRCRAFT CO. GB

Air Pict, N '65:386-389

WHITWORTH GLOSTER GB

See HAWKER SIDDELEY

WIBAULT France

3114.	281-T. 12		
	Flight, Jl 17, '31:699	1	
3115.	282-T. 12		
	Air Pict, Jl '68:257	1	
	Pop Avn, My '33:302	1	
3116.	360-T. 5		
	Flight, F 26, '32:175-176	1	

WINTER Germany

3117.	"Zaunkönig" (Wren)		
	Flying, S '48:38-39	1	
	Fly Rev Intl, N '65:189	1	

WNF (Wiener Neustadt Flugzeugwerke) Germany

3118.	342 (1943)		
	World's first jet helicopter		
	RAF Fly Rev, F '59:23-24	1	

WREN USA

3119. 460
Flying, Je '63:38-39+ 4 C
Ja '65:36-39 4
Jl '66:58-60 1

WRIGHT USA

3120. "Flyer I" (1903)
Aero Digest, Jl '53:28-37 1 S*
:65 D*
Aeronautics, D '53:28-35 1
Air Prog, '52 Annual:62-63 S*
Bee-Hive, Ja '53:3-6 1
:7-11 6 (Engine) P
Spg '53:20-24 6 (Prop)
Flight, D 11, '53:787-788 1 S
Flying, D '53:13 4
Hist Avn, D '68:10-11 1
Pegasus, Ag '55:14-15 1 S
Roundel, D '63:33 1 P
US Air Svcs, D '37:10-13+ 4
F '48:14-16 4
Ap '53:8-9 1 P
D '54:9-10 1

3121. "Flyer II" and "Flyer III"
Aeronautics, D '53:28-35 1

3122. "A" Type
Hist Avn, D '68:12-13 1

3123. "B" Type
Pop Avn, Ap '38:48+ 4

3124. "EX" Type
Hist Avn, O '69:11-26 (Rodger's "Vin Fiz"-) 1 P

WSK/MIL (Wytwornie Sprzetu Komunikacyjnego) Poland/Russia

3125. Mi-2
Original design: MIL Mi-2
Fly Rev Intl, D '68:31 1

YAK (Yakovlev) Russia
RAF Fly Rev, Je '58:11+ (Fighters 1 to 11)

3126.	[Fiddler]		
	See TUPOLEV [Fiddler]		
3127.	[Freehand]		
	Avn Wk, Ap 7, '69:22	1	P
	Ap 14, '69:33	1	P
	Flight, F 15, '68:243	1	
	Fly Rev Intl, Jl '69:26-27	1	
	Interavia, No. 5, '69:603	1	P
3128.	[Mandrake]		
	Development of YAK 25		
	Air Prog, Ag '67:56		Pr
	Fly Rev Intl, Mr '64:23	1	Pr
3129.	1 "Krasavec" (Beauty)		
	Name is for prototype		
	Flight, Jl 22, '43:97	1	
	RAF Fly Rev, N '55:11	1	
3130.	2		
	RAF Fly Rev, Je '63:47	1	
3131.	3		
	See also YAK 15		
	Aeroplane, Mr 22, '46:346-348	1	C*
	Flying, F '49:46	1	
	Fly Rev Intl, N '63:47 (Normandie-Niéman Sq-)	5	
	Je '64:60 (Luganskii's-)		Pr*
	:61-63	1	S*
3132.	4		
	Developed from YAK 2		
	RAF Fly Rev, Je '63:47	1	
3133.	9 [Frank]		
	Fly Rev Intl, N '63:47 (Normandie-Niéman Sq-)	5	
	RAF Fly Rev, Je '58:11	1	
	N '58:27-29	1	
	:26, 28-29 (-P)		Pr†, S*

YAK (cont.)

3134.	11 [Moose]		
	Fly Rev Intl, F '69:73	1	
	RAF Fly Rev, Je '58:24-27	1	Pr, S*
3135.	12 [Creek]		
	Avn Wk, S 3, '56:101 (-P)	1	P
	Ag 24, '59:99 (-A&-M)	1	C
	RAF Fly Rev, N '56:56-57	1	P*
	US Army Avn, My '59:16-17	1	
3136.	14		
	Flying, O '51:22-23+	1	
	Fly Rev Intl, Ap '66:500	1	
3137.	15		
	Development of YAK 3		
	RAF Fly Rev, Ap '61:20-21	1	Pr*
3138.	16 [Cork]		
	Avn Wk, Ag 21, '50:33	1	S*
3139.	17 [Feather] and [Magnet]		
	RAF Fly Rev, Je '58:50+	1	
3140.	18 [Max]		
	Aeroplane, Ag 31, '56:275-276	4	
	Avn Wk, Ap 4, '60:134 (-P)	1	S*
	Jl 17, '67:107-108 (-T&-PM)	1	
	Flight, Je 24, '48:696	1	
	Ja 16, '53:64	1	
	D 8, '66:992-993 (-PM)	4	C
	Flying, Ja '67:40-41+ (-PM)	4	
	Fly Rev Intl, Ja '67:295-296+	1	P*
	D '69:53 (Austrian-)		Pr*
3141.	23 [Flora]		
	Development of YAK 15 and YAK 17		
	RAF Fly Rev, Je '54:17	1	
	Je '57:11	1	
3142.	24 "Letayuchy Vagon" (Flying Boxcar) [Horse]		
	Aeroplane, Ag 26, '55:295		P
	Air Pict, O '59:364-365+	1	
	Avn Wk, D 28, '59:46-65	1	
	Ja 11, '60:43	1	
	RAF Fly Rev, S '57:66+	1	S*

YAK (cont.)

3143. 25 [Flashlight] and [Mangrove]
Trainer code name is [Mangrove]

Air Force, My '58:50-51	1	
Air Pict, Ap '58:128-129	1	Pr
Avn Wk, Ap 16, '56:33	1	
Ag 12, '57:27	1	
Ap 13, '59:72-73	1	
Fly Rev Intl, Mr '64:19-23 (-A to -D)	1	Pr
RAF Fly Rev, Ja '59:20-21		S*

3144. 28P [Firebar]
All-weather fighter
Trainer version code name: [Maestro]
Multi-purpose version code name: [Brewer]

Air Pict, Mr '63:76-77	1	Pr
Avn Wk, Jl 31, '67:62	1	
Fly Rev Intl, Ja '64:24-26	1	Pr

3145. 30 [Mantis]

Fly Rev Intl, Ap '68:217-218	1	P*

3146. 40 [Codling]

Aeroplane, My 3, '67:23-24	1	S*
Avn Wk, Ap 24, '67:62	1	
Je 19, '67:30-32	1	
Jl 21, '69:96-102	1	C
Flight, My 11, '67:763-764	1	
Fly Rev Intl, N '69:30-31	1	
Interavia, No. 5, '67:718	1	

3147. 42 [Backfin]

Air Force, My '60:44-45	1	
Avn Wk, Ap 6, '59:28-29	1	
RAF Fly Rev, N '60:13	1	
Sp/Aeron, O '58:40-46	1	

YOKOSUKA Japan

3148. E14Y1 [Glen]

Air Pict, Jl '65:242-243	1	
Nav Avn News, Mr '46:11	1	

YOKOSUKA (cont.)

3149.	MXY-7 "Ohka" (Cherry Blossom) [Baka] (Fool) Kamikaze-attack piloted bomb		
	Air Force, S '45:28-29	1	S*
	Air Prog, Ap/My '63:74-75		C, S*
	Nav Avn News, Ag 1, '45:37	1	
	RAF Fly Rev, My '57:52+	1	
	Skyways, S '45:80	1	
3150.	R2Y1 "Keiun" (Lucky Cloud)		
	RAF Fly Rev, Mr '59:27-29	1	

ZEPPELIN — Germany

3151.	"Rammer" (WW II)		
	Air Pict, D '65:443	1	
	Je '66:219	1	
3152.	C. I		
	Cr & Cockade, Aut '68:205-214	1	P
3153.	C. II		
	Cr & Cockade, Aut '68:205-214	1	P, Pr
	Fly Rev Intl, Mr '69:43 (Swiss-)		Pr*

ZEPPELIN-LINDAU — Germany
See DORNIER

ZEPPELIN-STAAKEN — Germany
See STAAKEN

ZEPPELIN/SUD-OUEST — Germany/France

3154.	Z. SO 523		
	RAF Fly Rev, N '57:9+	1	

ZLIN — Czech

3155.	26 "Trener" (Trainer)		
	Aeroplane, Ja 14, '49:48	1	
3156.	226 "Bohatyr" (Hero) and "Trener" (Trainer)		
	Aeroplane, Ag 3, '56:154-156	4	C
	Flight, Ag 3, '56:194-196 (-6)	4	
3157.	526 "Trener Master"		
	Air Prog, N '69:23-26+	4	C

Part One

Aircraft Addendum

Notes indicate restrictions in article content. Where notes do not appear, it may be assumed that the article is broadly historical.

ADVANCED MANNED STRATEGIC AIRCRAFT (AMSA)
Sp/Aeron, Je '66:62-75

AGRICULTURAL AIRCRAFT

AAHSJ, Sum '69:115-118	Early experiments
Flight, Ag 21, '69:277	In late 1960's
Flying, N '61:52-53+	
My '64:26-27+	
Fly Rev Intl, O '66:111-112+	
Interavia, No. 6, '60:704-706	
No. 10, '65:1574-1575	In Russia
No. 10, '68:1248-1249	New types
Pegasus, Jl '57:7-9	

AIRLINERS--Origins
Flight, Ap 11, '58:490-493 (Pt 1)
Ap 25, '58:579-581 (Pt 2)
My 9, '58:632 (Pt 3)
My 23, '58:712-714 (Pt 4)

ANTISUBMARINE AIRCRAFT

Fly Rev Intl, Ja '69:30-35	In 1969
Lock Hor, No. 7, Ja '69:35-55	1945-1969
Sp/Aeron, F '66:58-71	In 1966 and future

ATTACK AIRCRAFT
Nav Avn News, S '69:10-17

ARMAMENT (See also BOMBS and GUNS)
--British

Flight, Je 28, '34:638-644	
Ja 23, '36:91	Armstrong-Whitworth turret

ARMAMENT (cont.)

--General

Aeroplane, Mr 29, '39:402-404	In 1939
Air Force, Mr '69:64-68	New fire-control systems
Air Pict, S '55:258-261	In 1955 and future
Air Prog, Spg '59:80-88	In WW I
Air Tech, Ja '43:12-14 (Pt 1)	Shotguns to cannon
F '43:42-46+ (Pt 2)	" " "
Aviation, Ap '43:227-231 (Pt 1)	Turret development
My '43:247-253 (Pt 2)	" "
Je '43:227-231 (Pt 3)	" "
Flight, Ja 24, '35:88-90	In 1935
Mr 30, '39:323-325	In 1939
D 21, '39:508a-509	Single-seater in 1939
N 12, '42:518-522 (Pt 1)	WW II fighters
N 19, '42:553-557 (Pt 2)	" " "
N 26, '42:579-584 (Pt 3)	" " "
Ag 24, '50:215-219 (Pt 1)	Analysis and history of
S 21, '50:338-340 (Pt 2)	fighter armament
O 5, '50:379-381+ (Pt 3)	" "
N 23, '50:459-461 (Pt 4)	" "
D 7, '50:535-537 (Pt 5)	" "
Ja 4, '51:6-8 (Pt 6)	" "
Ja 25, '51:108-110 (Pt 7)	" "
Ja 28, '55:105-110 (Pt 1)	Fighter armament
:114-122+ (Pt 2)	Missiles and projectiles
Fly Rev Intl, Mr '68:133-139	
Interavia, Ja '48:38-42	
F '48:96-98 (Pt 1)	Fighter aircraft
Mr '48:151-154 (Pt 2)	" "
Ap '48:211-214 (Pt 3)	" "
Nav Avn News, Jl '53:1-5	Rockets
Pegasus, D '43:1-4	Turrets in 1930's and 1940's
Pop Avn, My '37:26-28+	WW I
Je '40:10-11+	In 1940
Pop Fly, Jl '35:195-197+	WW I

--German

Avn Age, N '52:45	Atypical aircraft

--Russian

Avn Age, My '52:34-35 (Pt 1)	
Je '52:40-41 (Pt 2)	

--Swedish

Interavia, No. 10, '51:541-547

--US Naval

Nav Avn News, F 15, '44:17-23	
Ap 15, '45:23-31	Rockets in 1945
S '63:19-27	In 1963

AUTOPILOTS

Aeroplane, Je 13, '34:930	P. B. Twin-Axes
Flying, My '44:48-49+	In WW II
Sp/Aeron, Ap '65:46-52	Self-adaptive

BOMBS

Flight, Je 22, '67:1007+	Israeli "Concrete Dibber"
D 14, '67:1013-1014	" " "
Fly Rev Intl, Ja '64:50-52	"Highball" and "Upkeep"
Roundel, Jl/Ag '52:33-37	Atomic

BOMBSIGHTS--Norden

Flying, Jl '45:28-30+	

CENTURY-SERIES (USA) AIRCRAFT

Air Pict, Ag '68:297	Non-production types
Flying, Ag '56:22-24+	
Interavia, No. 1, '57:37-41	

COMPARISONS

Air Pict, Ap '52:98-99+	MIG-15 vs F-86
Ja '54:7+	F-100 vs "Grognard"
Avn Wk, My 18, '53:13-15	F-86 vs MIG-15
My 31, '54:67-70	"Viscount" vs C-131C
My 23, '55:12-15	Russia vs USA
Comm Avn, S '43:90-100 (Pt 1)	WW II German fighter
N '43:90-100 (Pt 2)	structural comparisons
Flight, Ja 16, '64:82	DC-9 vs BAC 111
Flying, Ja '44:21-25+	WW II best
F '44:49+	B-17 vs B-24
N '50:22-23	P-51, P-51H, F-86
Mr '51:13-15+	F-80 vs MIG-15
D '51:22-23	F6F, F8F-2, F9F-5
Nav Avn News, Je '49:9	FJ-1 vs F8F
Pegasus, Ag '55:14-15	Wright "Flyer" vs 1908 Voisin
RAF Fly Rev, F '59:13-14	F-100D & J, F-104A & C, N-156F
RAF Qtr, Jl '53:251-252	F-86 vs MIG-15
Skyline, Vol. 9, No. 1, '51:14-16	" " "
Skyways, F '43:46-47	WW II best
Ja '44:38-39+	" " "
F '45:48-49+	" " "
Sp/Aeron, My '69:83	F-4 vs MIG-21
US Air Svcs, Je '53:10-12+	F-86 vs MIG-15

COUNTERINSURGENCY (COIN) AIRCRAFT

Fly Rev Intl, Ja '65:11-15+	
O '65:89+	
Interavia, No. 8, '64:1119-1123	

EJECTION SEATS

Avn Wk, F 17, '69:39	Parawing and rotor
Flight, Ag 19, '48:199-201	Martin-Baker w. sketches
N 14, '52:612-613	Martin-Baker development
N 19, '54:748-751	Martin-Baker, Mks 2, 3, 4
Ag 8, '63:207	G.Q. for SAAB "Draken"
S 17, '64:525	Martin-Baker, Mk 8VA
Interavia, No. 2, '62:191-193	In 1962
Nav Avn News, Je '47:17-19	New developments
S '49:20-23	US Navy procedure
N '59:13	North American Aviation
O '61:7-11	US Navy projects
Jl '65:23	Martin-Baker
RAF Fly Rev, Ja '58:35	Convair bobsled
RAF Qtr, O '52:340-345	Martin-Baker development
US Air Svcs, Ap '55:22	Republic Aviation
D '56:21	Convair RESCU Mk 1
West Avn, S '55:19	Douglas X-3
Mr '57:12	Lockheed

EXPERIMENTAL AIRCRAFT

Air Pict, My '56:155-156	Table of all RAF types
Je '56:212	Addendum to above
Interavia, No. 10, '56:804-807	USA to X-17
West Avn, O '59:16+	Eleven USA types

GUN SIGHTS--British

Flight, N 25, '43:590-591	WW II

GUNS (See also ARMAMENT)

Aeroplane, My 2, '34:753	Hispano-Suiza 12 Yers Moteur-Cannon
Air Prog, Ja '65:25	Browning AN-M2
Acft Eng, Ap '43:100-102	Axis cannon and machine guns: 1939-1943
Avn Age, Ag '52:44-45	Mauser MG-213 cannon
Avn Wk, S 18, '67:33	Flechette
Cr & Cockade, Spg '64:69-73	Lewis guns w. photo
Sum '68:134-141	Lewis guns

GUNS (cont.)

Cr & Cockade, Aut '68:263-268	German WW I
Flight, Ap 14, '38:369	Oerlikon
Jl 27, '39:86f-g	Vickers K
My 2, '40:406-409	1940 shell guns
Je 19, '41:417	Oerlikon shell
F 11, '43:149-151	WW II control systems
Mr 18, '43:285-286	German MG 131
Ja 26, '67:124-128	AC-47 minigun
Flying, Ja '44:66-67+	US .50 caliber
Pegasus, Jl '43:10-11+	1914 to 1943
Pop Avn, F '38:10-13+	Twenty-year history
Mr '39:62-64+	Danish Madsen .20 & .23 mm
Skyways, D '43:24-26+	Gun development
US Air Svcs, Ja '41:15	US .37 mm cannon
US Army Avn, F '64:17-19	Helicopter M-6 system
O '64:1-3	" M-6E3 system
O '66:18-19	" M-5 grenade launch
My '67:25-27	Helicopter M-22 system
D '68:2-5	" " "

HEATERS

Comm Avn, Ja '44:59-68	"Janitrol"
Ag '44:76-82+	"Southwind"
Flight, S 21, '44:319-321	"Janitrol"

HELICOPTERS

Fly Rev Intl, F '64:13-21
Interavia, No. 1, '54:16-18
Lock Hor, Spg '65:58-80
Nav Avn News, Ag '51:1-7
Pop Avn, Mr '39:12-14+
Pop Fly, My '39:72-73+
Roundel, Ap '56:26-30 (Pt 1)
My '56:6-15 (Pt 2)

INSIGNIA OF AIRCRAFT

Aero Album, Fall '68:3-11	USA: 1919-1941
Aerosp Hist, Spg '69:21-26	History of USA
Air Force, O '63:44-48	History of US wings
AAHSJ, O/D '56:120-127	USA: 1915-1956, illus.
Ja/Mr '57:48-49	Bibliography
Jl/S '57:143-175	US Navy and Coast Guard w. bibliography
Spg '59:56	3d US Pursuit Group

INSIGNIA OF AIRCRAFT (cont.)

AAHSJ, Spg '60:25-30	USAF
Win '60:342-344	US 15th Air Force
Win '63:256-266	WW II Pacific area fighter markings
Cr & Cockade, Sum '64:145-167	US WW I
Aut '66:217-222	French Escadrille, in color
Win '66:395-399	Comments on above
Flight, Je 27, '35:705	1935 RAF squadrons
O 21, '55:657-658	RAF serial numbers, 1912-1955
Roundel, Ja/F '65:10-13	Canadian: 1917-1918
RAF Fly Rev, Ap '55:10+	German WW II
S '60:38	German in Battle of Britain
N '60:11	Croatian cross: 1944-1945
F '61:22-24	Some WW II tail insignia (in color)

JET AIRCRAFT

--German

Air Pict, D '53:358-360 (Pt 1)	WW II fighters
Ja '54:24-26 (Pt 2)	WW II bombers and research
Air Prog, '54/'55 Annual:52-61	Strange WW II types; many photos
AAHSJ, Fall '63:155-184	
Interavia, No. 2, '46:25-29	With other WW II types
RAF Fly Rev, Jl '54:17-19+	
US Air Svcs, N '45:26-28	
West Fly, N '45:74-75	

--Japanese

Air Pict, Ag '54:234-238	At end of WW II

PRESIDENTIAL AIRCRAFT--USA

Aerosp Hist, Win '68:10-13+	Flying presidents
AAHSJ, Sum '63:79-96	
West Fly, Jl '45:48-49+	Roosevelt's C-54

TRIPLANES

Air Prog, Spg '62:66-83
Flight, My 17, '57:666-669
Pop Avn, My '31:21-22+

V/STOL AIRCRAFT

Air Force,	Mr '65:25-29	
AF/Sp D Intl,	Ap '65:12-22	
Acft Eng,	My '66:8-33 (Pt 1)	
	Je '66:8-19 (Pt 2)	
Am Helicop,	Mr '56:6-9+	
Avn Wk,	Je 24, '68:Entire issue	
Flying,	Ja '61:32-35+	Future of V/STOL
Fly Rev Intl,	N '64:9-13	Transition into V/STOL
	D '67:1033-1038+	
	My '69:46-53	
	D '69:34-43	
Interavia,	No. 7, '58:692-697	US types
	:698-699	European types
	No. 5, '60:553-568	
	No. 1, '61:47-97	
	No. 2, '63:168-200	In 1963
	No. 1, '64:48-66	European development
	No. 2, '69:144-158	Fifteen-year review
Nav Avn News,	F '66:26-27	In 1966
Sp/Aeron,	My '65:54-75	
West Avn,	Jl '61:12-14+	US Army test beds

Part Two

AIRCRAFT ENGINES

Part Two

AIRCRAFT ENGINES

This section provides information for the individual who wishes to learn more about the power plant of a specific aircraft, to verify performance, trace development, or confirm other details. No appreciable difference exists between the format of this and the preceding "Aircraft" section. The symbols used are indicated on pages vi and vii.

A very short addendum appears at the end of this section. It cites some better-than-average general articles concerning aircraft engines.

ACE (American Cirrus Engines) USA

1. "Cirrus"
   Pop Avn, Jl '32:29+ 1
   West Fly, My '31:46-49+ 3

2. "High Drive Ensign"
   US Air Svcs, Ap '30:30 1
   West Fly, My '31:46-49+ 3

AI (A. Ivchenko) Russia

3. 25
   Avn Wk, O 6, '69:63 1

AIRCOOLED FRANKLIN USA
See FRANKLIN

AIRESEARCH USA

4. ATF-3
   Avn Wk, O 7, '68:74-78 1
   Fly Rev Intl, Ja '69:27-28 1
   Interavia, No. 12, '68:1532 1 S

AIRESEARCH (cont.)

5.	T-76		
	Avn Wk, Ag 24, '64:92-99	1	S*
	Fly Rev Intl, N '68:35 (Y-)		S
	Interavia, No. 10, '64:1503	1	
6.	TFE-731		
	Avn Wk, Ap 28, '69:92-93+	1	
	Fly Rev Intl, Jl '69:32	1	
	Interavia, No. 5, '69:629	1	
7.	TPE-331		
	Avn Wk, D 18, '61:70-71	1	
	Flying, Je '65:73-75	1	
	Mr '66:80-83	1	S
	Interavia, No. 6, '68:774-775	1	S*

ALFA-ROMEO Italy

8.	D. 2		
	Pop Avn, Mr '34:167	1	

ALLISON USA

9.	J-33		
	Avn Wk, Ag 18, '47:21-22	1	
	S 27, '54:58-62 (-A-37)	1	
	Nav Avn News, Ap '47:6 (-A)		S*
10.	J-35		
	Aero Digest, Ap '51:21+ (-A-23)	1	
	Air Prog, '53/'54 Annual:76-77		S*
	Avn Age, My '51:34-35 (-A-23)	1	
	Avn Wk, Mr 19, '51:14-15 (-A-23)	1	
	Ap 2, '51:25-29 (-A-23)	1	
	Flight, Ag 5, '48:163	1	S*
11.	J-71		
	Flight, My 27, '55:733	1	
12.	T-40		
	Aero Digest, Ja '50:18+	1	
	Avn Wk, D 5, '49:11-13	1	
	N 10, '52:44-45	3	
	Nav Avn News, Ja '50:8 (X-)	1	

ALLISON (cont.)

13.	T-63		
	Commercial designation: 250		
	Avn Wk, Ap 6, '59:73-83 (-A)	1	S*
	D 25, '61:46-50 (-A)	1	S
	Flight, F 15, '62:251-252 (-5)	1	
	Flying, O '61:48-49+	1	S
14.	T-78		
	Avn Wk, Mr 2, '64:60-64 (-A-2)	1	S*
	Flight, N 12, '64:834-836	1	S*
	Interavia, No. 11, '64:1690-1691	1	S*
15.	TF-41		
	Avn Wk, Jl 15, '68:48-53	1	S
16.	V-1710		
	Aero Digest, Je '37:50+	1	
	Mr '41:248	1	
	Je '42:184	1	S*
	Aeronautics, My '41:58-59+	1	
	Aeroplane, Ja 9, '42:40-42 (Pt 1)	1	S
	Ja 16, '42:70-71 (Pt 2)	1	
	Ja 23, '42:98-99 (Pt 3)	1	
	Ja 30, '42:126-127 (Pt 4)	1	
	Aerosp Hist, Aut '69:48-51	1	P
	Air Prog, Mr '43:53 (-C-15)		S*
	Air Tech, Jl '43:19-32	1	S*
	Aviation, Jl '42:190 (-C-15)		S*
	O '45:187, 189		S*
	Flight, Je 19, '41:419-420 (-E-4)	1	
	Mr 26, '42:281-284 (-C-15)	1	S*
	Ja 29, '48:130 (-E-27)	1	
	Flying, My '43:28-31+	1	
	Pop Avn, Ag '37:39-40 (-C-6)	1	
	Skyways, My '43:56-57 (-C-15)		S*
	West Fly, Je '37:32	1	
17.	V-1760		
	Aviation, Je '37:34	1	
18.	V-3420		
	Aero Digest, O 1, '45:75-79+	1	
	West Fly, Jl '44:118	1	

ALLISON (cont.)

19. 250
Military designation: T-63
Avn Wk, N 27, '67:83-88 1 S*
Flight, S 19, '58:485 (-B2) 1 S
Interavia, No. 12, '68:1530-1531 1 S

20. 501
Military designation: T-56
Aeroplane, N 21, '58:744-745 1 S
Avn Age, Ja '56:32-33 1 S*
Avn Wk, Ja 24, '55:80-83 1 S*
N 7, '55:32-43 1 S
O 14, '57:93-99 (-D13) 2
F 19, '62:54-63 3
Interavia, No. 7, '55:532-533 (-D10) 1 S

21. 610
Avn Wk, Je 27, '66:50-54 1 S

ALVIS GB

22. "Leonides"
Aeroplane, F 22, '39:240-241 2
Je 14, '46:697-701 1 S*
F 11, '49:153-161 (-23 HM) S*
Acft Eng, Mr '49:76-84 3
Air Pict, Mr '59:92+ 1
Am Helicop, My '49:6-7+ (-21 to -24 HM) 1
Flight, N 7, '46:491 1
F 10, '49:166-170 (-23 HM) 1 S*

23. "Leonides Major"
Aeroplane, N 2, '56:654-659 1 S*
Air Pict, Mr '59:92+ 1
Flight, S 19, '52:381-382 (-1-1&-1-2) 1 S
S 30, '55:561-562 (-1-1&-1-6) 2

ARGUS Germany

24. As 8
Acft Eng, F '31:29-30 1 S

25. As 10
Flight, Ag 7, '31:788 1

ARMSTRONG SIDDELEY GB

26.	"Adder"		
	Aero Digest, O '50:43+	1	
	Aeroplane, Mr 7, '52:275-279	1	S*
	Flight, F 15, '51:187-190	1	S*
27.	"Cheetah"		
	Aeroplane, Ag 31, '45:252-253 (-25)	2	
	Aviation, D '43:201 (-10)		S*
	Bris Sid Jnl, Vol. 6, No. 2, '65:24-25	1	P
	Flight, Ag 29, '35:224f	1	
	Je 24, '43:656-657 (-10)	1	S*
	D 25, '47:732	1	
28.	"Cougar"		
	Aeroplane, Ja 4, '46:10-11	1	
	Flight, Ja 3, '46:11	1	
29.	"Double Mamba"		
	Aeroplane, Ap 1, '49:362-365	1	S*
	Avn Wk, Jl 18, '49:21-22	1	
	Flight, Mr 31, '49:373-377	1	S*
	Mr 4, '55:272-275	1	S
	N 22, '57:815-819 (-8)	1	S*
	RAF Qtr, Ap '53:126-131	1	P
30.	"Genet Major"		
	Aeroplane, My 21, '30:970-972	1	
31.	"Jaguar Major"		
	Acft Eng, Ap '30:98-99	1	S*
	Flight, Mr 7, '30:271-275	1	S*
32.	"Leopard"		
	Flight, Ap 29, '32:369 (-IIIA)	1	
33.	"Mamba"		
	Aero Digest, Je '48:44-46	1	S*
	Aeronautics, Ag '48:43-48	1	S*
	Aeroplane, Mr 19, '48:333-336	1	S*
	Avn Age, Je '52:36-37	1	
	Flight, Mr 18, '48:306b-h	1	S*
	D 9, '48:696-699	1	
	Mr 30, '51:369-371 (ASM-3)	1	S*
	Interavia, My '48:259-262	1	S

ARMSTRONG SIDDELEY (cont.)

34.	"Python"		
	Aeroplane, Ag 26, '49:256-261	1	S*
	Flight, Ag 11, '49:163-168	1	S*
	O 6, '49:469	2	
	Interavia, No. 3, '46:49-52	1	P
35.	"Sapphire"		
	Design development: WRIGHT J 65 and T 49		
	Aeroplane, O 7, '55:572-580	1	P, S*
	O 19, '56:587-599	1	P, S*
	Avn Age, D '50:30-31	1	
	My '54:22-23	1	S*
	Avn Wk, O 23, '50:34	1	
	Flight, Ja 6, '56:17-22	1	S*
	N 9, '56:756-760 (-7)	1	S*
36.	"Screamer"		
	Aeroplane, Ag 24, '56:255-260	1	S*
	S 7, '56:370-371		S*
	Avn Wk, N 19, '56:52-55	1	
	Flight, Jl 27, '56:160-164	1	S*
37.	"Tiger"		
	Aeroplane, Jl 20, '38:85-86 (-IX)	2	
	Acft Eng, S '37:247-248 (-VIII)	1	
	O '37:276 (-VIII)		P
	Flight, Je 20, '35:666a-d (-VI)	1	S*
	Ag 19, '37:185-186 (-VIII)	1	
	N 17, '38:437-438 (-VIII)	4 (Superchrgr)	S*
38.	"Viper"		
	Aeroplane, Jl 31, '53:139-144	1	S*
	F 22, '57:270-272	1	
	Flight, Ag 7, '53:170-174	1	S*
39.	P-181 and P-182		
	Avn Age, Je '58:66-71 (-181)	1	S*
	Flight, F 21, '58:234-235+	1	

AVRO CANADA Canada

40.	"Chinook"		
	Avn Wk, My 3, '48:17	1	
	Flight, Ja 22, '48:101	1	

AVRO CANADA (cont.)

41.	"Orenda"		
	Aero Digest, Ja '50:22-23+	1	
	Avn Wk, O 17, '49:21	1	
	Ag 6, '51:21-31	1	
	O 20, '52:40-42	3	S
	Can Avn, My '50:27-36+	1	
	Flight, D 1, '49:709	1	
	Interavia, No. 1, '52:33-36	1	

BENZ Germany

See DAIMLER-BENZ

BESLER USA

42.	Steam Plant (1933)		
	Flying, D '62:30+	1	
	Pop Avn, Jl '33:8-10+	1	S*
	West Fly, Je '33:20	2	

BLACKBURN GB

43.	A. 129		
	Aeroplane, D 26, '58:928-931	1	S*

BLACKBURN/TURBOMECA GB/France

44.	"Palouste"		
	Flight, Mr 21, '58:405-406	1	S

BMW (Bayerische Motoren Werke) Germany

45.	III		
	Cr & Cockade, Sum '61:97-99 (-a)	1	
46.	003		
	Acft Eng, F '46:43-49+	1	P, S*
	Aviation, Mr '46:55-68	1	S*
47.	116		
	Aviation, D '40:84+	1	
48.	132		
	Acft Eng, S '43:248-252 (-K&-N)	4 (Fuel)	S*
	Flight, D 3, '42:604-607 (-K&-N)	1	S

BMW (cont.)

49.	801		
	Aeroplane, Je 11, '43:672		S*
	Air Prog, F '43:60		S*
	Air Tech, Ap '43:26		S*
	Acft Eng, Ag '42:223-227	1	S*
	O '43:293-302 (-A/1)	4 (Fuel)	S*
	Aviation, D '42:256-259 (-A)	1	S*
	Flight, Ag 13, '42:169-173 (-A) (Pt 1)	1	S*
	Ag 20, '42:201-202 (-A) (Pt 2)	4 (Lube & Fuel)	S*

BOEING USA

50.	T-50		
	Commercial version: 502		
	Boeing Mag, F '62:10-11	1	
51.	T-60		
	Commercial version: 520		
	Boeing Mag, N '62:12-13	1	
52.	500		
	Acft Eng, Ag '49:242-243+	1	
	Avn Wk, Jl 28, '47:31	1	
53.	502		
	Military version: T-50		
	Air Pict, My '50:150+	1	
	Acft Eng, Ag '49:242-243+	1	
	Avn Age, Ap '54:22-23	1	
	Avn Wk, Jl 28, '47:31	1	
	Jl 20, '59:97-101	1	S
	Boeing Mag, D '49:3-5+	1	
	Ag '50:6-7	1	
	Ja '51:10-11	1	
	N '51:6-7+	1	
	Ag '56:12-13 (-10C)	1	
	Interavia, Je '49:353-356	1	P, S
	West Avn, Jl '54:16 (-10)	1	
54.	520		
	Military version: T-60		
	Boeing Mag, D '59:10-11	1	S

BOEING (cont.)

55.	551		
	Boeing Mag, F '65:6-7	1	
	Jl '65:11	2	

BRAMO Germany

56.	"Fafnir 323"		
	Acft Eng, S '43:252-253 (-P-1)	4 (Fuel)	
	Flight, D 3, '42:603-604 (-P-1)	1	

BRISTOL GB

Aeroplane, F 7, '34:230-231 (Developments in 1934)
F 15, '39:207-212+ (Sleeve-valve engs)
Pop Fly, F '33:624-626 (Developments in 1933)

57.	"Aquila"		
	Pop Avn, N '35:299-300	1	
58.	"Centaurus"		
	Aeroplane, Jl 6, '45:11-17	1	S*
	Acft Eng, My '51:124-133 (-20)	1	S
	Flight, Jl 5, '45:13-17	1	S*
	Ja 8, '48:32-33 (-630)	2	
	S 1, '49:250 (-660&-661)	2	
	N 16, '56:790-791 (-373)	1	
59.	"Coupled Proteus"		
	Aeroplane, D 9, '49:783-787	1	P
	Flight, Ag 5, '48:153-157	1	S
60.	"Hercules"		
	Aero Digest, Je '42:126		S*
	Aeroplane, O 6, '37:422	2	
	N 28, '41:599-601	3	
	Mr 13, '42:298-299		S*
	O 2, '42:394-397	1	S*
	N 6, '42:529	1	
	N 12, '43:551	1	
	N 8, '46:542-543 (-230)	1	
	Air Prog, O '42:53		S*
	My '43:53		S*
	Air Tech, Jl '43:49		S*
	Acft Eng, Ag '37:218-222	1	P
	Aviation, Je '42:129		S*
	Ap '44:195		S*
	S '44:187		S

BRISTOL (cont.)

60.	"Hercules" (cont.)		
	Can Avn, Ja '44:77-79	1	S*
	Flight, O 7, '37:359	1	
	N 27, '41:379-383 (-III)	1	S*
	O 1, '42:359		S*
	N 5, '42:504-507	1	S*
	N 11, '43:528-530	1	S*
	My 24, '45:558-560 (HE-10M)	1	
	N 7, '46:506-507 (-230)	1	
	Ja 8, '48:32-33 (-730, -750, -760)	2	
61.	"Jupiter"		
	Flight, Jl 25, '30:838 (-XI-F)	1	
	O 31, '30:1191-1192 (-X-F. AM & BM)	2	
62.	"Lucifer"		
	Aeroplane, S 18, '42:349	1	
63.	"Mercury"		
	Flight, Mr 11, '32:207-211 (IV-S2)	1	P
	O 4, '34:1032-1033 (VI-S)	2	
	N 5, '36:466-466c	1	S*
64.	"Orion"		
	Avn Age, Ap '57:34-41	1	S*
	Flight, O 26, '56:677-682	1	S*
65.	"Olympus"		
	Later development: BRISTOL SIDDELEY "Olympus"		
	Aeronautics, Ag '52:36-37	1	S
	Aeroplane, Jl 4, '52:8-9	1	P
	F 4, '55:135-136 (-101)	1	
	F 11, '55:170		P
	N 4, '55:700-713	1	P, S*
	Air Pict, Ag '52:228	1	
	Avn Age, Ag '52:45	1	
	Avn Wk, Jl 21, '52:44-45	1	
	Flight, Jl 4, '52:2-3	1	
	F 11, '55:164-165 (-101)	1	
	D 9, '55:869-876	1	S*
	F 15, '57:200-201 (-6)	1	
	Mr 7, '58:302-304	1	

BRISTOL (cont.)

66. "Orpheus"
Aeroplane, Je 3, '55:737 (-1) 1
N 22, '57:779-782 (-2) 1 S*
Avn Wk, N 25, '57:51-63 1
Flight, Je 3, '55:751 1
Jl 5, '57:21-22 (-3) 1
N 22, '57:809-812 (-3) 1
F 13, '59:219-223 1 S*
Interavia, No. 7, '57:721 (-3) 1
Sp/Aeron, D '59:59-62 (-803) 1 S*

67. "Pegasus"
Aeroplane, O 5, '32:634-636 1
N 4, '36:562-568 1 S*
N 10, '37:565-566+ (-18) 1 S*
Acft Eng, O '32:259-260 2
Aviation, Mr '42:91 (-22) S*
Bris Sid Jnl, Sum '61:112 1
Vol. 7, No. 1, '66:8-10 1 S*
Flight, Mr 4, '32:187-190 (Pt 1) 1 P
Mr 11, '32:207-211 (Pt 2) 1 P
O 4, '34:1032-1033 (-3&-4) 2
Ap 9, '36:384-386 (-10C) 1
N 5, '36:466-466c 1 S*
N 4, '37:440 (-18) 1

68. "Perseus"
Aeroplane, Ap 13, '38:453 (-XII) 1
Je 15, '38:751-752 (-X, -XI, -XII) 1
Acft Eng, Ag '37:218-222 1 P
Flight, N 23, '33:1177 1
Ap 14, '38:372 (-XII) 1
O 20, '38:332-333 (-XIIc) 1
Pop Avn, N '35:299-300 1
S '38:59 (-XII) 1

69. "Phoebus"
Avn Age, Je '52:38-39 1

70. "Phoenix"
Aeroplane, My 16, '34:802 2
Acft Eng, Jl '33:149 1
Flight, My 17, '34:487 1
My 24, '34:504-505 2

BRISTOL (cont.)

71.	"Proteus"		
	Aeronautics, S '52:83-87 (-3-700 Series)	1	S
	Aeroplane, Ag 6, '48:162-164	1	
	D 9, '49:783-787	1	S*
	Je 6, '52:685-695 (-705)	1	P, S*
	Ag 8, '52:192-193 (-III)	1	
	Air Pict, Ag '52:228 (-III)	1	
	Avn Age, Ag '52:40-41 (-3)	1	S
	Avn Wk, O 24, '49:21-27	1	
	Jl 14, '52:22-30 (-705)	1	S
	Ja 12, '59:77-85 (-705)	2	
	Flight, Ag 18, '49:191-196	1	S*
	Je 6, '52:687-692 (-705)	1	S*
	My 8, '53:581	1	
	Jl 11, '58:37-40	1	
	Interavia, No. 2, '53:72		S*
72.	"Theseus"		
	Aeronautics, My '49:43-48 (-501)	1	S*
	Aeroplane, S 13, '46:307-309 (-I)	1	S
	Acft Eng, N '46:366-371 (-I)	1	S
	Aviation, N '46:53-55 (-I)	1	S
	Can Avn, Mr '46:50+ (-I)	1	
	My '47:42-46	1	
	Flight, D 6, '45:597-599 (-I)	1	
	O 10, '46:386-387 (-I)	1	S
	Mr 27, '47:270-273	2	S
	Interavia, No. 3, '46:52-54	1	
73.	BE 25		
	Aeroplane, Ap 29, '55:563-565	2	S*
	Flight, Ap 29, '55:545-547	2	
	Interavia, No. 7, '55:533-534	1	
74.	BE 58		
	Bris Sid Jnl, Aut '59:3-5	1	

BRISTOL SIDDELEY GB

Bris Sid Jnl, Spg '64:58-62

75.	"Artouste"		
	Auxiliary power plant		
	Bris Sid Jnl, Sum '63:91-93	1	

BRISTOL SIDDELEY (cont.)

76. "Cumulus"
Bris Sid Jnl, Sum '63:91-93 1

77. "Gnome"
Earlier development: DE HAVILLAND "Gnome"
USA version: GENERAL ELECTRIC T-58
Bris Sid Jnl, Sum '62:95-97 1
Vol. 7, No. 4, '66:2-7 1 S*

78. "Olympus"
Earlier development: BRISTOL "Olympus"
Bris Sid Jnl, Sum '60:96-99 1 P, S
Flight, F 24, '61:233-235 (-104&-301) 1 S*

79. "Olympus 593"
Aeroplane, Ja 6, '66:24-25 1
Ap 12, '67:Supp:28-34 1
Mr 27, '68:22-23 1 S
Air Pict, F '66:64-65 1
Acft Eng, Ap '63:100-101 1
F '64:40-42 1 S*
Mr '66:16-17 (-B) 1
Ap '67:5-9 2
:10-16 4 (Control)
Ja '69:20-22 1 S
Avn Wk, My 29, '67:206-217 1
Bris Sid Jnl, Win '62/'63:36-38 1 S
Vol. 6, No. 3, '65:10-14 1
Vol. 7, No. 3, '66:16-23 1
Can Avn, O '66:22-23 1

80. "Palouste"
Auxiliary power plant
Bris Sid Jnl, Sum '63:91-93 1 S*

81. "Viper"
Acft Eng, F '63:39-43 (-520) 1
Avn Wk, My 29, '61:74-75 (-20) S
Bris Sid Jnl, Win '59/'60:48-52 1 P, S
Aut '61:8-9 1
Flight, D 20, '62:984-985 1

82. BS 53 "Pegasus"
Not to be confused with BRISTOL "Pegasus" (1938
Aeroplane, Ag 12, '60:177-179 1 I
Bris Sid Jnl, Aut '60:16-19 1 I

BRISTOL SIDDELEY (cont.)

83.	BS 75		
	Aeroplane, Ja 6, '61:13-14	1	
	Ap 5, '62:31-32	2	
	Acft Eng, Jl '62:205	1	
	Bris Sid Jnl, Spg '62:74-77	1	S*
	Interavia, No. 4, '62:433-435	1	
84.	BS 100		
	Avn Wk, S 7, '64:86-89	1	
85.	BS 347		
	Bris Sid Jnl, Vol. 7, No. 4, '66:15	1	

BRISTOL SIDDELEY/SNECMA GB/France

86.	"Mars" M. 45		
	Interavia, No. 9, '65:1398-1399	1	

BUDA-GUIBERSON USA

See GUIBERSON

CHRYSLER USA

87.	16-Cylinder Inverted "V"		
	Aviation, Ja '46:89-90	1	S

CIRRUS GB

88.	"Bombardier"		
	Aeroplane, O 22, '48:530-533	1	S
	Ap 22, '49:451	3	
	Flight, O 21, '48:487-491 (-702)	1	S*
89.	"Hermes"		
	Acft Eng, Je '29:129		S
	Flight, Jl 8, '32:637-638 (-IV)	1	
	West Fly, My '31:46-49+	3	
90.	"Major"		
	Aeroplane, Je 12, '35:689-690 (-I)	1	S*
	Ap 18, '52:457-461 (-III)	1	S*
	Aviation, Ap '41:71+	1	
	Flight, Je 13, '35:Supp:a-d	1	S
	Ap 21, '38:388c-d	2	
	My 25, '39:545-548 (-150)	1	S*

CIRRUS (cont.)

91.	"Mark III"		
	Acft Eng, Ap '29:61-63	3	P, S
	West Fly, My '31:46-49+	3	
92.	"Minor"		
	Aeroplane, F 27, '35:241-243	1	S*
	O 6, '37:421-422	1	
	O 15, '43:444	1	
	Aviation, Ap '41:71+	1	
	Can Avn, O '45:114 (-II)	1	
	Flight, F 28, '35:218-219	1	S
	F 20, '36:201	1	
	Ap 21, '38:388c-d	1	

CLERGET France

93.	Heavy-Oil Engine		
	Acft Eng, Mr '31:53-54	1	

CONTINENTAL USA

94.	A		
	Aero Digest, Ap '31:178 (-70)	1	
	Ap 15, '45:82-85+	1	S*
	Acft Eng, Ap '31:99 (-40)	1	
	US Air Svcs, Mr '30:50-51 (-70)	1	P
	West Fly, Ja '38:30+ (-50)	1	
	My '38:21-22 (-40)	3	S
95.	J-69		
	Avn Age, My '55:34-35 (Y-)	1	S*
	Can Avn, Ap '55:40	1	
96.	O-470		
	Avn Wk, Ja 26, '53:18 (-A)	1	
	Flying, D '61:78 (GI-)	1	
97.	R9-A		
	Aero Digest, Ap '46:84-87	1	
98.	R-670		
	Acft Eng, Je '32:149	1	
	Pop Avn, F '33:104	1	
	US Air Svcs, Ap '32:48, 50	1	

CONTINENTAL (cont.)

99. XT-51
Am Helicop, Jl '54:8-10+ (Pt 1) 2
Ag '54:6-8+ (Pt 2) 2

100. T-65
Avn Wk, F 18, '63:79-82 1
Flying, My '64:49+ 1

CURTISS USA
See also CURTISS-WRIGHT and WRIGHT

101. OX-5
Air Prog, Fall '60:30-40 1
Flying, D '42:57+ 1
Pop Avn, F '36:105-106 3
West Fly, Ja '32:34-36 (Pt 1) 3
F '32:30-32 (Pt 2) 3

102. OXX-6
West Fly, Ja '32:34-36 (Pt 1) 3
F '32:30-32 (Pt 2) 3

CURTISS-WRIGHT USA
See also CURTISS and WRIGHT

103. "Challenger" R-600
Aero Digest, Mr '31:70 1
Ap '31:178 (Correction) 1
West Fly, D '32:23-24 3

104. "Conqueror" V-1570
Aero Digest, S '32:66 2
US Air Svcs, S '32:27-28 (-F) 2

105. "Zephyr" TJ-38
West Avn, Je '57:26 1

DAIMLER-BENZ Germany

06. DB 600
Flight, S 2, '37:232 1

DAIMLER-BENZ (cont.)

107. DB 601
Aero Digest, Je '42:126 (-N) S
Aeroplane, N 1, '40:490-491 (-A) 1
Jl 16, '43:66-69 (-A) 1
Jl 23, '43:109 (-N) 1
Air Prog, F '43:53 (-N) S
Air Tech, Ap '43:24 S
Acft Eng, O '41:270-279 (-A) (Pt 1) 1
N '41:300-302+ (-A) (Pt 2) 2
Aviation, O '40:61+ 1
F '43:149 (-A) S
My '43:207 S
O '43:177-178 S
Ja '44:195 S
Flight, N 7, '40:390f-391 (-A) 1
Ap 16, '42:365-369 (-N) 1 S

108. DB 605
Acft Eng, Jl '44:190+ (-A1) 1

109. PTL-6
Civil version: DB 720
Interavia, No. 9, '64:1383-1386 1

DE HAVILLAND G

110. "Ghost"
Aero Digest, O '49:42-43+ 1
Aeronautics, Ja '49:43-46 1 S
Aeroplane, S 16, '49:390
My 2, '52:529-534 1 S
Aviation, Ap '47:42-43 1 S
Avn Opns, O '49:29-30+ 1
Flight, F 6, '47:142-145 (DGT/50) 1 S
Jl 22, '48:93 2
RAF Qtr, Jl '48:xviii D

111. "Gipsy"
Aeroplane, Ap 2, '30:582-588 (-I to -III) 1
Acft Eng, Mr '29:13-15 1 S
West Fly, Je '31:50-52 3

DE HAVILLAND (cont.)

112.	"Gipsy Major"			
	Aeroplane, Ap 25, '41:466-467	3		
	Acft Eng, N '32:277-278+	1		P, S
	Flight, Ja 14, '43:44-45 (-I)	2		
	Je 7, '45:611 (-I)	1		
	Interavia, No. 2, '47:49 (-10)	1		
113.	"Gipsy Minor"			
	Aeroplane, N 2, '38:534-536	1		S*
	Can Avn, D '38:20	1		
	Flight, N 10, '38:424f-425	1		S*
114.	"Gipsy Queen"			
	Aeronautics, N '46:143-144 (-Series 70)	1		
	Aeroplane, D 28, '45:747-750 (-71)	1		S*
	Ap 5, '46:405 (-71)	1		
	Flight, O 8, '42:395-396 (-IV)	1		
115.	"Gipsy Six"			
	Aeroplane, Ja 24, '34:143-146	1		S*
	O 31, '34:525-528 (Racing-)	1		S*
	Ja 6, '37:15-18 (-II)	1		S*
	My 2, '41:502-503	3		
	Acft Eng, F '34:47-48	1		P
	Can Avn, Mr '34:10-11	1		
	D '36:20 (-II)	1		
	Flight, Ja 25, '34:84-86	1		
116.	"Gipsy Twelve"			
	Aeroplane, Je 22, '38:776-778	1		S*
	Acft Eng, Ag '38:248-249	2		P, S*
	F '39:57-60	3		
	Can Avn, Ag '38:24	1		
	Flight, Mr 31, '38:331	4	(Cooling)	S
	Je 23, '38:612e-615	1		P, S*
117.	"Gnome"			
	Later development: BRISTOL SIDDELEY "Gnome"			
	Aeroplane, Ag 15, '58:225-228 (-1)	1		S*
	Am Helicop, O '58:8-9			P
	Flight, Ag 15, '58:236-238	1		S*
	Ap 29, '60:587-588	1		

DE HAVILLAND (cont.)

118.	"Goblin"		
	Aero Digest, My '46:98-99+ (-II)	1	
	Aeroplane, F 22, '46:223-230 (-I&-II)	1	S*
	Acft Eng, Ap '46:125-128 (-I&-II)	1	S*
	Aviation, O '46:91 (-II)		S*
	Avn Wk, F 7, '49:21-22 (-II)	2	
	Can Avn, Je '47:80-82+	1	
	Jl '47:46+	3	
	F '51:16-19	3	
	Comm Avn, Ja '46:51	1	
	Flight, N 1, '45:472-474	1	
	F 21, '46:185-190 (-II)	1	S*
	My 15, '47:442-443, 450	1	
	Roundel, N '48:48	2	
119.	"Gyron"		
	Aeroplane, S 9, '55:416-419	1	
	Mr 1, '57:300-305	1	S*
	Ag 2, '57:154-158 (-2)	1	S*
	Air Prog, Spg '58:90-91		S*
	Avn Age, O '57:26-31	1	S
	Avn Wk, Ag 19, '57:31		S
	Ag 26, '57:62-65	1	
	Flight, Ag 2, '57:163-168 (-1&-2)	1	S*
120.	"Gyron Junior"		
	Aeroplane, S 7, '56:369	1	
	Mr 1, '57:300-305	1	S*
	Flight, S 7, '56:425-426	1	
	Ag 8, '58:187 (-10)	1	

DELAGE France

121.	12		
	Acft Eng, F '34:51	1	

DESCHAMPS USA

122.	Diesel		
	Aero Digest, My '34:38-40	1	
	Aeroplane, Ag 1, '34:145-146	1	S
	Flight, My 24, '34:512-514	1	S

DEVANEY			GB
123.	"Aeolus"		
	Aeroplane, Ap 18, '34:698	1	
EGAO (Egyptian General Aero Organization)			Egypt
124.	E-300		
	Avn Wk, Je 24, '68:42-43 (-A&-C)	1	P
	Fly Rev Intl, O '68:32	1	
FAIRCHILD			USA
125.	J-44		
	Avn Age, D '53:28-31	1	S*
	Avn Wk, N 30, '53:30-39	1	S*
	F 6, '56:53-61	1	S
126.	6-390		
	Aero Digest, F '30:134-136	1	
	Can Avn, F '30:17-18	1	
FAIRCHILD CAMINEZ			USA
127.	447 Cam Engine		
	Air Prog, O/N '64:78-80+	1	P
FARINA			Italy
128.	"Algol"		
	Acft Eng, Jl '29:164-165	1	S, P
129.	"Alioth"		
	Acft Eng, Jl '29:164	1	S
FARMAN			France
130.	12		
	Acft Eng, F '34:51-52 (-BRS)	1	
FEDDEN			GB
131.	"Cotswold"		
	Flight, Je 26, '47:600a-601	1	S
	Interavia, O '47:31-35	1	S*

FIAT			Italy
	Interavia, No. 6, '59:706-708		
132.	A.N. 1 Diesel		
	Flight, Ja 16, '31:48	1	
133.	A.S. 6		
	Aero Digest, My '33:46	1	
	Aeroplane, My 29, '35:627-630	1	
134.	4002		
	Aeroplane, Je 21, '57:882-884	1	S*
	Avn Age, My '57:32-33	1	P, S
FLYGMOTOR			Sweden
	Bee-Hive, Spg '64:22-25		
135.	RM 8		
	Swedish military version of PRATT & WHITNEY JT8D-22		
	Avn Wk, My 29, '67:278-281	1	
	Flight, Jl 11, '68:61	1	
	Interavia, No. 4, '65:524-525	1	
	No. 11, '65:1742-1743	2	
FRANKLIN			USA
	Flying, Ag '63:54-55+		
136.	O-425		
	Civil version: 6V6-245		
	Am Helicop, Ag '54:9+ (-1)	1	P
	N '54:13-14 (-13)	1	
	Avn Age, S '54:74-75 (-2)	1	
137.	6V6-245		
	Military version: O-425		
	Avn Age, Jl '53:30-31 (-B16F)	1	S*
138.	6VS-335		
	Am Helicop, Ap '59:11	1	
139.	8AC-398		
	Aviation, Jl '41:57+	1	
140.	12AC-596		
	Aviation, Jl '41:57+	1	

GARRETT-AIRESEARCH USA
See AIRESEARCH

GENERAL ELECTRIC USA
Flying, F '63:50-51+ (Jet engines)

141.	CF6		
	Acft Eng, F '69:20-24	1	
	Avn Wk, S 25, '67:28-30 (-34)	1	S
	O 28, '68:50	1	P
	N 4, '68:76-77	3	
	O 20, '69:34-35	2	
	O 27, '69:52-58+	1	S*
	Flight, O 31, '68:714-717	1	S
	Fly Rev Intl, N '69:31-32	1	
	Interavia, No. 12, '67:1880-1881 (-34)	1	S*
	No. 3, '69:273-275	1	
	No. 10, '69:1710 (-50A)	1	
142.	CF 700		
	Aeroplane, F 20, '64:24	1	
	Avn Wk, Ap 6, '64:98-106	1	
	Flight, D 17, '64:1050-1052	1	
	Interavia, No. 5, '67:769		S
143.	CJ 610		
	Military version: J 85		
	Interavia, No. 6, '63:857	1	
144.	CJ 805		
	Avn Wk, Mr 2, '59:60-66 (-3)	2	
	D 21, '59:29 (-3)	2	
	Jl 17, '61:82-86 (-3&-23)	1	
	Flight, D 26, '58:982 (-21)	1	
	Interavia, No. 12, '58:1266-1268 (-21)	1	
	Sp/Aeron, Ag '60:63-66 (-23)	1	S*
	West Avn, Je '57:25	1	
145.	CTF 39		
	Avn Wk, O 31, '66:121-123	3	
146.	GE 4/J5		
	Avn Wk, Je 7, '65:24-27	1	
	S 26, '66:42-45	1	
	O 31, '66:106-121	3	
	Fly Rev Intl, D '69:31-32	1	

GENERAL ELECTRIC (cont.)

147.	I 16		
	Aviation, Ja '46:43-50	1	S*
148.	I 40		
	Aero Digest, Ja '46:61+	1	
	Aviation, Ja '46:51-59	1	S*
149.	J 47		
	Civil version: TG 190		
	Aero Digest, F '54:28-32	1	S*
	Avn Age, Ag '51:30-31 (-GE-21)	1	
	D '53:56-67	1	S*
	S '56:86-91 (Pt 1)	1	
	O '56:40-45 (Pt 2)	1	
	Avn Opns, N '49:39-40	1	
	Avn Wk, D 4, '50:21-27 (-GE-17& -GE-23)	1	
	Jl 2, '51:21-22 (-GE-21)	1	
	Ap 14, '52:21-35	1	
	O 26, '53:27-44	1	S*
150.	J 53		
	Flight, Ap 29, '60:594-595 (X-)	1	
151.	J 79		
	See also Canadian-built ORENDA J-79-OEL-7		
	Avn Wk, My 27, '57:29-30	1	
	Ag 11, '58:69-74	1	
	Ap 19, '65:52-58 (-J1B, -10, -17)	1	S
	Can Avn, O '59:28-29	1	
	Sp/Aeron, F '60:59-61	1	S
152.	J 85		
	Civil version: CJ 610		
	Avn Wk, Ap 11, '60:56-70	1	S
	Ag 18, '69:58 (-13&-21)	1	
	Flight, D 11, '59:723	1	
	Interavia, No. 11, '59:1371-1373	1	
	No. 7, '61:966-969	1	
153.	J 93		
	Avn Wk, S 16, '63:32-33	1	
	O 18, '65:64-67 (Y-)	2	
	Flight, S 26, '63:543-544 (Y-)	1	

GENERAL ELECTRIC (cont.)

154. J 97
Avn Wk, Ag 11, '69:72 1

155. T 58
See also BRISTOL SIDDELEY and DE HAVILLAND "Gnome"
Aeroplane, Ja 10, '58:46-48 1 S*
Am Helicop, F '56:9 1
Avn Wk, D 16, '57:69+ 2 S*
Ap 29, '63:48-51 1
Flight, Ja 17, '58:79-83 1 S*
Interavia, No. 7, '61:968 1 S*
Nav Avn News, Jl '56:36-37 1
US Army Avn, Jl '59:23-26 1

156. T 64
Avn Wk, Je 8, '59:61-67 1
O 30, '61:60-64 1 S
N 8, '65:52-55 (-12) 1
Flight, Ja 13, '61:63-64 1 S
Je 18, '64:1019-1023 1 S*
Interavia, No. 7, '61:969 1 S*
No. 1, '63:57-59 1 S*

157. TG 100
Aero Digest, N '48:37-40+ 1 S*

158. TG 180
Avn Wk, Jl 7, '47:36-43 1 S*
Jl 14, '47:29-33 3 S*

159. TG 190
Military version: J 47
Avn Wk, Ja 16, '50:40 (-B) 1

GNÔME GB/France

160. "Monosoupape" Rotary
Pop Fly, Ap '35:33-34 1 S

GUIBERSON USA

161. A-980
Aero Digest, Ap '32:74-75 1
Aviation, Ap '32:195 1
Pop Avn, My '32:299-300+ 1 P, S

GUIBERSON (cont.)			
161.	A-980 (cont.)		
	US Air Svcs, Ap '32:39-40	1	P, S
	West Fly, Ap '32:56-58	1	
162.	A-1020		
	Aero Digest, Je '40:53	1	
	Aviation, Jl '40:59+	1	
	Can Avn, O '40:35+	1	
	Flight, Ja 2, '41:9	1	S*
163.	A-1090		
	Aeroplane, S 27, '40:354	1	
HEINKEL-HIRTH			Germany
164.	He S 011		
	Aviation, S '46:63-64	1	
HERMANN			USA
165.	Cam Engine		
	Am Helicop, O '57:6-7+	1	
HILLER			USA
166.	8RJ2B		
	Am Helicop, N '54:11+	2	
HIRTH			Germany
167.	H-M 60		
	Acft Eng, S '31:227	1	
HISPANO-SUIZA			France
168.	"Nene"		
	French-built ROLLS ROYCE "Nene"		
	Avn Age, Mr '52:34-35	1	
	Interavia, No. 6, '52:320-324	3	S*
169.	A, E, and I		
	Cr & Cockade, Win '69:329-344 (-A)	1	P, S
	West Fly, My '35:26-28+	3	
170.	12B		
	Avn Age, F '52:42-43 (-40)	1	
	Flight, Ag 11, '49:161-162	1	

HISPANO-SUIZA (cont.)

171. 12X and 12Y
Pop Avn, N '36:33-34 (-Crs) 1

ISOTOV Russia

172. GTD-350
Fly Rev Intl, N '68:32 1

JACOBS USA

173. L-4
West Fly, Jl '35:17-19+ 3

174. L-6
Aero Digest, Ap '38:60+ 1
Comm Avn, Mr '39:38+ 1

JUNKERS Germany

175. "Jumo"
RUSSIA M-012 derived from -004
Aeroplane, S 5, '34:283-285 (-004) 2
O 19, '38:473-474 (-211) 1
Ag 6, '43:165-167 (-211A) 1
Ag 13, '43:181-182 (-211B &-211D) 1
Acft Eng, Je '38:180-181 (-210) 1 S*
Ja '43:8-18 (-211B) 1
D '45:342-347 (-004) (Pt 1) 1 S*
Ja '46:10-14+ (-004) (Pt 2) 1 S*
Aviation, S '40:52-53+ (-211) 1
Ja '41:63+ (-205&-207) 1
N '45:115-130 (-004) 1 S*
F '46:167 (-222) S*
Flight, My 1, '31:375 (-004) 2
Jl 14, '38:47-48 (-204&-205) 1 S*
Ja 18, '40:45-50 (-211A) 4 (Fuel Injec) S*
Ag 5, '43:147 (-211D) 2
Ag 26, '43:228-231 (-211) 1
S 13, '45:282-284 (-004B) 1 S*

176. SL-1
Acft Eng, Mr '30:62-63 1

KEN-ROYCE USA

See LE BLOND

KINNER USA

177. R-53
Can Avn, S '41:37+ 1

178. -5
Pop Avn, F '36:107 (R-/-Series 2) 1
West Fly, Ja '31:44-46 (K-) 3
F '33:19-20 (B-, C-, K-, R-) (Pt 1) 3
Ap '33:16-17 (B-&K-) (Pt 2) 3
My '33:19 (B-, C-, K-, R-) (Pt 3) 3
Ag '35:23-26 (B-&K-) 3

LE BLOND USA

179. 7F
Aviation, Jl '39:58+ 1

180. 60, 70, and 85
West Fly, Jl '31:40-42+ (-60&-85) 3
Jl '36:19-20+ 3

181. 90
West Fly, Jl '31:40-42+ 3

LEVER GB

182. Redrup-Lever "Fury"
Aeroplane, F 12, '30:262-264 1 S

LIBERTY USA

183. 12
Air Prog, Ap '66:33-35+ 1
Cr & Cockade, Win '69:350-354 1
Pop Avn, Ap '34:239+ 1
Jl '35:19+ 1
US Air Svcs, Mr '31:38 1

LORRAINE France

184.	"Petrel"		
	Aeroplane, S 13, '33:471-472	1	S*
	Flight, My 25, '33:497-498	1	S*

LYCOMING USA

Flying, D '61:30-31+ (Piston engines)
Interavia, No. 12, '58:1274 (50th Anniversary)

185.	O-145	
	West Fly, F '38:34	1
186.	O-235	
	Aero Digest, F '41:141-145	1
	Can Avn, F '41:90	1
187.	O-290	
	Aero Digest, F '41:141-145	1
	Can Avn, F '41:90	1
188.	O-350	
	Aviation, F '41:139	1
	Can Avn, F '41:90	1
189.	O-435	
	Aviation, F '41:139	1
	Avn Age, Ag '55:40-43	2
	Can Avn, F '41:90	1
	Flying, Je '43:110 (GO-)	1
190.	R-530-D	
	Aero Digest, D '37:76	1
	Aviation, Ja '38:38	1
	West Fly, D '37:36	1
191.	R-680	
	Aero Digest, Mr '30:130-132	1
	My '32:62-63 (-BA)	1
	Can Avn, Mr '30:32, 34	1
	US Air Svcs, My '30:52-53	1
	West Fly, Mr '32:30-31 (Pt 1)	3
	My '32:31-32 (Pt 2)	3
192.	XR-7755	
	Aero Digest, D '46:82-83	1
	Aviation, D '46:86-87	1

LYCOMING (cont.)

193.	T-53		
	Aeroplane, O 10, '58:559-561	1	S
	Am Helicop, Ap '56:14 (X-)	1	
	Mr '57:6-9+	1	
	Avn Age, O '56:26-27 (-L-1)	1	
	Avn Wk, Ag 27, '56:62-64 (-L-1)	1	
	D 16, '57:73+ (-L-1)	2	
	Je 9, '58:46-51	1	S*
	Can Avn, My '57:32 (X-)	1	P
	Flight, Ap 27, '56:509 (X-)	1	
	Ja 17, '58:79-83 (-L-1)	1	S*
	US Army Avn, F '60:29-32	3	S
194.	T-55		
	Am Helicop, Ap '57:16	1	
	Avn Wk, Ag 10, '59:87-97	1	
	Flight, Jl 18, '63:96-97	1	S*
	Sp/Aeron, N '58:30-31	1	

MANNERSTEDT Sweden

195.	M		
	Flight, O 28, '43:469-471	1	S*

MATHIS France

196.	42-E-00		
	Flight, S 6, '45:257	1	

MENASCO USA

197.	"Buccaneer" B6S		
	Aero Digest, Jl '36:46-50	1	
	Flight, S 10, '36:280-281	1	S
	Pop Avn, Ag '36:34-35	1	
198.	"Pirate"		
	Aero Digest, Jl '36:46-50 (C4S-)	1	
	Flight, S 10, '36:280-281 (C4S-)	1	S
	Pop Avn, Ag '36:34-35 (C4S-)	1	
	West Fly, F '38:22-24 (B4-&C4S-)	3	
199.	"Super-Buccaneer" C6S		
	Aviation, O '37:42+ (-4)	1	

MENASCO (cont.)

200.	"Unitwin"		
	Aero Digest, F '39:55-59	1	
	Aviation, Ag '38:32-35	1	
	My '39:30-31+	2	
	Can Avn, Mr '39:20	1	
	US Air Svcs, Mr '39:36	1	

MERCEDES-BENZ Germany

See DAIMLER-BENZ

METEOR GB

201.	"Mark I"		
	Flight, S 2, '32:833-834	1	
202.	"Mark II"		
	Flight, D 6, '34:1300	1	S*

METROPOLITAN-VICKERS GB

203.	F. 2		
	Aeroplane, Mr 8, '46:281-287	1	S*
	Flight, Je 12, '47:552	2	
204.	F. 3		
	Aviation, Jé '46:66-67	1	S*
	Flight, Ap 25, '46:420-423	1	S*
205.	F. 5		
	Flight, Ja 2, '47:18	1	

MICHEL France

206.	AM 14		
	Flight, Ap 11, '30:413-414 (-II)	1	

MITSUBISHI Japan

207.	"Kinsei"		
	Aeroplane, O 9, '42:420-421	1	
	Aviation, Ag '42:110-117+	1	P, S

NAPIER GB

Flight, Je 27, '58:893-900

Interavia, No. 7, '58:738-739

NAPIER (cont.)

208.	"Coupled Naiad"		
	Aero Digest, Je '50:42-43+	1	
	Aeroplane, S 30, '49:482-483	1	
	Flight, S 15, '49:342	1	S
209.	"Dagger"		
	Aeroplane, Jl 10, '35:57-59 (-II&-III)	1	
	Ja 11, '39:46-50 (-VIII)	1	S*
	Air Prog, Ja '43:53 (-H)		S*
	Flight, Jl 11, '35:48-50 (-III)	1	P, S*
	Ag 4, '38:112 (-VIII)	1	
	Ja 12, '39:38e-39 (-VIII)	1	S*
	D 7, '39:464c (-VIII)	1	P
210.	"Eland"		
	Aeroplane, Jl 23, '54:110-116 (-1)	1	S*
	D 13, '57:881-883	1	S*
	Avn Age, O '54:136-138	1	
	My '56:26-27	1	S
	O '56:28-33	3	S
	Avn Wk, D 30, '57:64-66	1	
	Flight, Jl 23, '54:113-118	1	S*
	N 27, '59:637-639	1	S*
211.	"Gazelle"		
	Aeroplane, Ap 19, '57:535-540 (-1)	1	S*
	S 6, '57:391 (-1)		S*
	Am Helicop, Je '57:10-13+	1	
	Avn Age, Jl '57:32-33	1	S*
	Flight, My 10, '57:630-635	1	S*
212.	"Gazelle Junior"		
	Am Helicop, F '58:12	1	
	Flight, F 7, '58:166	1	
213.	"Lion"		
	Flight, Je 27, '58:893, 896	1	S
214.	"Naiad"		
	Aeronautics, S '49:83-87	1	S*
	Aeroplane, Ag 13, '48:191-195	1	S*
	Flight, S 4, '47:254-255	1	S*
	Ag 12, '48:180-185	1	S*

NAPIER (cont.)

215.	"Nomad"		
	Aeroplane, Mr 20, '53:334-335	2	
	Ap 30, '54:528-540	1	S*
	Air Pict, Je '54:165-166	1	S*
	Avn Age, Ja '52:40-41	1	
	Avn Wk, Ap 6, '53:30-41 (-6)	1	
	My 17, '54:30-34 (-6)	1	S
	Flight, Mr 20, '53:348 (-6)	1	
	Ap 30, '54:543-552	1	S*
	Interavia, No. 6, '54:357-360 (-6)	1	S
216.	"Oryx"		
	Aeroplane, Ag 5, '55:207-213 (-1)	1	S*
	Am Helicop, O '55:9-11+ (-1)	1	
	Avn Age, N '55:156-159	1	
	D '55:28-29	1	S
	Avn Wk, N 28, '55:37	1	
	Flight, Ag 5, '55:189-193	1	S*
217.	"Rapier"		
	Aeroplane, Ja 3, '34:33-34 (-II&-IV)	1	
	Acft Eng, Jl '30:166	1	
	F '34:44-46	1	
	Flight, Ja 11, '34:38-39 (-II&-IV)	1	
	Mr 14, '35:266-269 (-II&-IV)	1	S
218.	"Sabre"		
	Aero Digest, O 15, '44:122-124	1	
	Aeroplane, Ja 14, '44:42-43	1	
	Mr 24, '44:323-329 (-IIA)	1	S*
	N 23, '45:586-589 (-VA&-VII)	1	
	N 8, '46:535-536	3	S
	Aviation, Ag '44:201		S*
	Flight, Ja 13, '44:34-36	1	
	Mr 23, '44:309-315 (-II)	1	S*
	N 22, '45:550-553 (-VII)	1	
	Jl 25, '46:99 (-VII)	1	
219.	"E. 97"		
	Aeroplane, Je 22, '32:1148-1150	1	
	Acft Eng, Ag '32:210-212	1	S
	Flight, Je 24, '32:576-578	1	

NORTHROP USA

220.	XT-37 "Turbodyne"		
	Avn Wk, Je 5, '50:14	1	

ORENDA Canada

221.	"Iroquois"		
	Avn Wk, Jl 29, '57:26-27	1	
	Can Avn, Ag '57:75	1	
	Flight, S 6, '57:412-413	1	
222.	"Orenda"		
	Aeroplane, Je 3, '55:741-747 (To -14)	1	S*
	Can Avn, Ap '56:33+ (PS-13)	2	
223.	J-79-OEL-7		
	Development of GENERAL ELECTRIC J-79		
	Can Avn, My '61:12-15+	1	
	Ag '61:22-23	3	

PACKARD USA

224.	Diesel		
	Aero Digest, Ap '30:93-97	1	
	Aeron (US), My '30:398+	1	
	Aeroplane, Ap 30, '30:794-798	1	S
	Acft Eng, Jl '30:163-165+	1	S
	Aviation, Ap 5, '30:684-691	1	P
	Flight, My 16, '30:527-531	1	
	Hist Avn, Ap '69:10-11	1	
	West Fly, Ap '30:94-96+	1	

PIRNA East Germany

225.	014		
	Avn Wk, Ap 4, '60:91-95	1	

PLYMOUTH USA

226.	Aero-Engine (1935)		
	Pop Avn, O '35:237	1	

POBJOY GB

227. "Niagara"

	Aeroplane, Ja 31, '34:183-185	1	
	Ap 3, '35:384-386 (-II)	1	S*
	Flight, Ap 4, '35:361-362 (-II)	1	S*

228. "R"

	Acft Eng, Jl '32:180	1	
	Flight, Ap 3, '31:292	1	
	Je 5, '31:496-498	1	S*
	Ap 22, '32:349-350	1	
	O 13, '32:955-957	1	S*

POTEZ France

229. 12 As

	Aeroplane, Ag 14, '35:208-209	1	S*

PRATT & WHITNEY USA

Bee-Hive, Ag '45:1-19 (20th Anniversary)
Comm Avn, S '40:36 (15th Anniversary)
Nav Avn News, Ag '50:27-29 (US Naval enterprises)
US Air Svcs, Jl '50:9-11 (25th Anniversary)
West Fly, S '40:22-23 (15th Anniversary)

230. "Double Wasp" R-2800

	Aviation, D '41:120		S*
	Bee-Hive, Fall '46:10-12 (-CA)	1	
	Spg '52:16-17	1	
	Interavia, No. 6, '46:48-49 (-CA)	1	
	US Air Svcs, Ap '50:11 (-CB-16)	1	P

231. "Hornet" R-1690 and R-1860

	Aero Digest, Ja '33:49 (T1C)	1	
	Ja '35:42-43 (S1E-G)	1	
	Aviation, Ja '35:30 (S1E-G)	1	
	Bee-Hive, F '31:3-5 (Series B)	1	
	My '31:8-10+	3	
	Ja '32:6+	4 (Fuel Injection)	
	Ja '33:3-4 (T1C)	1	
	Ag '34:3-4 (S3D1G)	2	
	Ja '35:3-5 (S1E-G)	1	
	Flight, F 26, '32:171	4 (Fuel Injection)	
	Pop Avn, Mr '35:159-160 (S1E-G)	1	S

PRATT & WHITNEY (cont.)

231. "Hornet" (cont.)
US Air Svcs, Ag '31:48 (Series B) 1
O '31:48+ (-B-1) 1
F '32:35-36 4 (Fuel Injection)
Ja '33:25-26 (T1C) 1
Ja '35:15-16 (S1E-G) 1
West Fly, Mr '31:40-41 (-A) 3
F '32:56-57 4 (Fuel Injection)
Ja '35:16+ (S1E-G) 1

232. "Twin Hornet"
Bee-Hive, Jl '37:6 1
US Air Svcs, Jl '37:19 1

233. "Twin Wasp" R-1830 and R-2000
Aero Digest, Ap '33:52-54 1
Air Tech, My '43:50-54+ (-C3) 1
Acft Eng, Je '33:134-135 1
Bee-Hive, Ap '33:3-6+ 1
Pop Avn, O '33:243-244+ 1
Ag '36:29-30+ 1
US Air Svcs, Ap '33:28-31 1
My '36:19-21 1
West Fly, My '33:21 1

234. "Twin Wasp E" R-2180
Aero Digest, Mr '49:114 1 S*
Avn Wk, N 22, '48:20-22 1 S*

235. "Twin Wasp Junior" R-1535
Aero Digest, My '32:68+ 1
Ap '33:52-54 1
Aeroplane, My 11, '32:848-850 1
Bee-Hive, Ap '33:3-6+ 1
Pop Avn, O '33:243-244+ 1
Ag '36:29-30+ 1
US Air Svcs, My '32:29-30 1
Ap '33:28-31 1
My '36:19-21 1
West Fly, My '33:21 1

236. "Wasp" R-1340
Aero Digest, O '34:36-38 (S1H1-G) 1
Acft Eng, Mr '35:70-72 (S1H1-G) 1 S
Aviation, O '34:334-336 (S1H1-G) 1
Bee-Hive, O '34:3-4+ (S1H1-G) 1

PRATT & WHITNEY (cont.)

236.	"Wasp" (cont.)		
	Can Avn S '44:117-122	3	
	Pop Avn, D '33:391 (S1D1)	1	
	West Fly, Mr '31:40-41+ (-C)	3	
237.	"Wasp Junior" R-985		
	Bee-Hive, Ja '30:1-4	1	
	F '30:14-16+	2	
	Ap '30:8-9	1	P
	Ja '54:23-25	1	
	Flight, Ja 10, '30:101-102	1	
	US Air Svcs, F '30:47	2	
	West Fly, Mr '31:40-41+	3	
238.	"Wasp Major" R-4360		
	Aero Digest, Ja '46:65	1	
	Bee-Hive, Ja '46:8-9	1	
	Boeing Mag, Ap '46:5	1	
	Can Avn, Ja '46:42	1	
	Flight, Jl 3, '47:12a-d	1	S*
	Interavia, No. 1, '46:52-54	1	
	US Air Svcs, D '45:18	1	
	D '48:20 (-VDT)	1	
	West Fly, Ja '46:46	1	
239.	J 42 "Turbo Wasp"		
	Civil designation: JT 6		
	Aero Digest, My '49:44-46	1	
	Avn Age, My '50:24-25	1	S*
	Bee-Hive, Sum '50:4-6	1	
240.	J 48 "Turbo Wasp"		
	Civil designation: JT 7		
	Aero Digest, Ap '50:17-18	1	
	Avn Opns, My '50:23-25	1	
	Avn Wk, Mr 6, '50:15	1	
	Jl 24, '50:22-25	3	
	Bee-Hive, Spg '50:18-21	1	
	Nav Avn News, Ap '50:11	1	
241.	J 57		
	Civil designation: JT 3		
	Aeroplane, N 27, '53:716-717	1	S*
	My 21, '54:635		P
	Avn Age, Ag '57:34-35	1	S
	Avn Wk, N 16, '53:17-18	1	
	My 3, '54:44-48	1	

PRATT & WHITNEY (cont.)

241.	J 57 (cont.)		
	Bee-Hive, Ja '54:3-6	1	
	Flight, N 27, '53:697-699	1	S
	My 14, '54:619-620	1	
242.	J 75		
	Civil designation: JT 4		
	Avn Wk, My 4, '59:60-65	1	
243.	JT 3		
	Military designation: J 57		
	Interavia, No. 11, '59:1374-1375	2	
244.	JT 3D		
	Avn Wk, Ja 26, '59:48-51 (-1)	1	S
	Bee-Hive, Sum '60:25-26	1	
	Flight, D 19, '58:935 (-1)	2	
245.	JT 4		
	Military designation: J 75		
	Avn Wk, My 4, '59:60-65	1	
246.	JT 8D		
	See also FLYGMOTOR RM-8		
	Avn Wk, O 14, '63:52-59	1	S*
	Ap 5, '65:37-39 (-1)	2	
	Bee-Hive, Sum '68:2-8	1	
	Flight, Je 29, '61:892 (-1)	1	S
	My 9, '63:681-682, 688 (-1)	1	S*
247.	JT 9D		
	Avn Wk, Ap 18, '66:42-43 (-1)	1	
	O 31, '66:125-132	3	
	O 9, '67:43-49	1	
	O 20, '69:33	3	
	D 1, '69:54-59	3	
	Bee-Hive, Spg '69:12-15	3	
	Fly Rev Intl, N '69:35, 37		P*
	Interavia, No. 7, '66:1033-1035 (-1)	1	
	No. 6, '67:882-883	1	
	No. 5, '69:625-627	1	
248.	JT 12		
	Avn Wk, O 20, '58:56-62	1	
	Interavia, No. 7, '61:974-976	1	

PRATT & WHITNEY (cont.)

249.	JT 18D			
	Avn Wk,	O 9, '67:43-49	1	S*
	Interavia,	No. 12, '67:1880-1881	1	
250.	JTF			
	Avn Wk,	D 6, '65:67-71 (-14E)	1	
		F 28, '66:44-47 (-17A)	1	
		Ag 29, '66:50-57 (-17A)	1	S
		O 31, '66:123-125 (-17A)	3	
251.	PT 6T "Turbo Twin Pac"			
	Military version: T 400			
	Fly Rev Intl,	O '69:31	1	
252.	T 34 "Turbo Wasp"			
	Avn Age,	N '54:36-37 (Y-)	1	S*
	Avn Wk,	S 11, '50:12-14	1	
		S 20, '54:19	1	S
		O 4, '54:26-43	1	S
	Bee-Hive,	Fall '50:30-31	1	
		Sum '53:22-26	1	
	Nav Avn News,	O '50:10	1	
	US Air Svcs,	O '50:8-9	1	P
253.	T 400			
	Civil version: PT 6T			
	Fly Rev Intl,	O '69:31-32	1	
254.	TF 30			
	Avn Wk,	Mr 28, '66:68-75 (-P-1)	1	

PRATT & WHITNEY - CANADA Canada

255.	JT-15-D			
	Avn Wk,	O 9, '67:96-97	1	
		S 9, '68:33	2	
	Can Avn,	N '67:51	1	
	Flight,	O 26, '67:679 (-5)	1	
256.	PT 6			
	Military designation: T 74			
	Avn Wk,	S 21, '59:31	1	S
		N 6, '61:74	2	
	Can Avn,	O '60:41-42	1	
		My '61:28	1	

PRATT & WHITNEY - CANADA (cont.)

256. PT 6 (cont.)
Flying, Ap '61:28-29+ 1
Fly Rev Intl, Je '66:619 (-A) 1
Interavia, No. 7, '61:974-976 1
No. 8, '69:1002-1004 1

257. T 74
Civil version: PT 6
Interavia, No. 10, '64:1503 1

RANGER USA
Pegasus, S '44:1-16

258. SGV-770
Aero Digest, Ap '38:46-48+ (-B-5) 1
Je '40:72-73+ (-B-7) 1
Je 1, '45:72-76+ (-C-2) 3 S*
Aviation, Je '40:48-49+ (-B-7) 1 S
Can Avn, N '44:98 1
Comm Avn, N '44:102+ (-D-5) 1
Pegasus, Mr '44:1-5 4 (Cylinders)
West Fly, D '44:138 (-D-5) 1

259. V-770
Aero Digest, Ap '32:83 1
Ap '38:46-48+ (-B-4) 1
Flight, Je 6, '35:622 1

260. 6-390B
Flight, Je 6, '35:622 1

261. 6-410B
Aviation, Mr '38:41-42 1

262. 6-440C
Aero Digest, S '39:64 (-2) 1
Je '40:76-77+ 1
Aviation, S '39:38-42 (-2) 1
Can Avn, Mr '42:49-60+ 3 S*
Comm Avn, Ja '44:82-84 3

RATEAU France

263. A. 65
Aviation, Ap '47:42 1

REDRUP-LEVER GB
See LEVER

REGNIER France

264. 4-EO
Aviation, Je '39:36 (USA "Monsoon") 1

RHÔNE France/GB

265. Rotary (80 H.P.)
Cr & Cockade, Win '60:59-66 1 S*

ROLLS ROYCE GB
Aeroplane, Mr 7, '34:402-405 (Operations in 1934)
Acft Eng, D '31:307-312 (Engine types to 1931)
Flight, Jl 27, '39:86i-91
My 7, '54:571-584 (Engine types to 1954)

266. "Avon"
Aeronautics, Ap '53:43 1
Aeroplane, My 7, '54:570 (-RA 14) P*
D 16, '55:944-957 (Series) 1 P, S*
Ja 25, '57:118-121 (-100 & -200) 3
F 15, '57:238-242 (-100 & -200) 3 S*
O 11, '57:552-557 (-200) 1 S*
O 24, '58:622-624 (-RA 29) 1 S
Air Pict, O '54:296 P
Avn Wk, Ap 13, '53:39-42 1
Flight, D 16, '55:900-906+ 1 S*
O 11, '57:569-574 (-200) 1 S*
Ag 1, '58:159 (-RA 29) 3
Interavia, No. 10, '57:1024-1025 3
No. 12, '58:1269-1272 (-RA 29) 1 S

267. "Buzzard" or "H"
Acft Eng, D '29:327-336 1 P

268. "Condor"
Aeroplane, Ap 30, '30:791B-794 (-III-B) 1 S
Acft Eng, D '32:308-309 1
Flight, N 17, '32:1094 (-C.I) 1 P

ROLLS ROYCE (cont.)

269.	"Conway"			
	Aeroplane,	N 20, '59:511-513	1	S
		Ja 15, '60:78-84	1	S*
		Jl 5, '62:47+ (-42 Mk 540)	1	
		D 27, '62:8-9 (-42)	2	
	Acft Eng,	Je '62:179-181 (-42)	1	S*
		S '65:277+ (-12&-42)	4 (Bearings)	
	Avn Age,	Ag '57:192-195	2	
		My 2, '60:138-147 (-12 Mk 509)	1	S*
		N 6, '61:42 (-42)	1	S*
	Boeing Mag,	S '57:10-11	1	
	Flight,	F 8, '57:183-186	2	
		Ja 15, '60:77-82	1	S*
	Interavia,	No. 11, '59:1376-1377	3	
		No. 3, '60:341-342	1	S*
270.	"Dart"			
	Aero Digest,	My '55:36-40	3	S*
		D '55:52-56	2	
	Aeroplane,	D 10, '48:733		P
		S 9, '49:361-364	1	S*
		Mr 27, '53:385-388 (-505)	1	S*
		F 28, '58:289-293 (-7)	1	S*
	Air Pict,	D '50:354-356 (Pt 1)	1	
		Ja '51:12-14 (Pt 2)	1	S
	Acft Eng,	Mr '65:86-92 (-531)	1	S*
	Avn Age,	O '51:34-35	1	S*
		Ja '56:34-45	1	S*
	Avn Wk,	My 11, '53:28-38	2	
		D 13, '54:47-54 (-506)	1	
		N 14, '55:68-70 (-7)	1	
		N 21, '60:95-103	3	
	Can Avn,	My '54:37 (-502)	1	
	Flight,	D 2, '48:651 (-I)	1	
		Mr 31, '49:362-363+	1	
		S 8, '49:279-281	1	S
		Ag 31, '51:249-250	2	
		Mr 20, '53:368-371	1	S*
		Ja 14, '55:45-48 (-506&-510)	1	S*
		F 28, '58:276-280 (-525)	1	S*
	Flying,	Mr '62:40-41+	1	S
	Interavia,	No. 1, '63:60-62	1	S
		:69-71	4 (Props)	

ROLLS ROYCE (cont.)

271.	"Derwent"		
	Aeroplane, O 26, '45:477-480 (-B. 37)	1	
	F 8, '46:167-173 (-V)	1	S*
	Acft Eng, Jl '46:245-248+ (-I)	3	
	Aviation, Ja '46:69-72	1	
	Can Avn, D '45:64-66+	1	S*
	Comm Avn, O '45:26-48	1	S*
	Flight, O 25, '45:447-450	1	
	Ja 3, '46:12a-b		S*
	RAF Qtr, S '46:239	2	P
272.	"Eagle"		
	Aeroplane, Je 16, '44:674-675 (-8)	1	S*
	Ap 25, '47:420-426 (-22)	1	S*
	My 7, '54:568 (-8)		P*
	Flight, Ja 30, '47:116	1	
	Ap 24, '47:368a-371 (-22)	1	S*
	Interavia, No. 5, '47:32-35 (-22)	1	P, S*
273.	"Griffon"		
	Aeronautics, Jl '46:48-51 (-130)	1	
	Aeroplane, S 21, '45:328-335 (-65)	1	S*
	Ja 11, '46:45-47 (-130)	1	
	Can Avn, Jl '44:70	1	
	Flight, Mr 9, '44:254-256	1	
	S 20, '45:309-316 (-65)	1	S*
	Ja 10, '46:34-36+ (-130)	1	
274.	"Kestrel" or "F"		
	Acft Eng, D '29:327-336	1	P, S
	Jl '35:164-172	1	P, S*
	Flight, Mr 28, '35:326-328	4 (Superchrgr)	S*
275.	"Merlin"		
	Aeroplane, Ap 26, '39:519-528 (-II)	1	P, S*
	F 27, '42:248-249 (-20)	1	
	D 18, '42:705, 708 (-61)	1	S*
	:706-707 (-20)		S*
	Ja 25, '46:105-106 (-130)	1	
	My 7, '54:569 (-II)		P*
	Air Prog, O '42:60		S*
	Ap '43:53 (-61)		S*
	My '43:58-59 (-20)		S*
	Air Tech, F '43:16-19+ (-20)	1	S*

ROLLS ROYCE (cont.)

275.	"Merlin" (cont.)			
	Acft Eng, Ag '36:223			P
	Jl '46:218-226	1		S*
	Aviation, Ag '40:67-68+	1		
	Je '42:207+ (-20)	1		
	Can Avn, Ja '42:22-24+ (Pt 1)	1		
	F '42:39-40+ (Pt 2)	3		
	S '44:60-63+	1		S
	Jl '50:27+	1		
	Comm Avn, S '42:72-78 (-20)	1		
	Jl '43:110-114 (-Packard)	1		
	D '44:86-93+	1		
	Flight, S 23, '37:308-311 (-I&-II)	1		S*
	D 7, '39:464a	1		P
	F 26, '42:176a-g (-20)	1		S*
	D 17, '42:655-659 (-61)	1		S*
	S 23, '43:346 (-61)	4	(Superchrgr)	S*
	Ja 24, '46:92-94+ (-130)	1		S
	Jl 25, '46:99 (-620)	1		
	Flying, Mr '46:52-53+ (-Packard)	3		
	Skyways, My '43:59 (-20)	1		S*
276.	"Nene"			
	See also HISPANO-SUIZA "Nene"			
	Aeronautics, Je '46:75-76	1		
	S '48:95-100	1		S*
	Aeroplane, Ap 19, '46:457-464 (-I)	1		S*
	Mr 12, '48:305-306			P*
	S 16, '49:391			P
	Aviation, My '46:73-76	1		S*
	Bee-Hive, Sum '47:28-31	1		S*
	Can Avn, D '54:60+	1		
	Flight, Ap 18, '46:389-394 (-I)	1		S*
	Mr 13, '47:216a-217	1		S*
	Interavia, No. 2, '46:37-42	1		P, S*
277.	"Spey"			
	Aeroplane, Ja 4, '62:18	1		
	Acft Eng, Ja '62:7-11	1		S*
	:11	4	(Fuel Control)	
	My '63:148	1		S
	Je '64:182-183	1		S
	Ap '69:27-28 (-25 Mk 512)	1		S
	Avn Wk, Je 5, '61:59-60	1		S
	Ja 8, '62:78-80	1		
	Je 3, '63:64 (-25)	1		

ROLLS ROYCE (cont.)

277.	"Spey" (cont.)		
	Avn Wk, O 4, '65:64-69	1	S
	Flight, N 16, '67:805-811	1	
	:808-809 (-25R)		S
	Interavia, No. 4, '62:432-433	1	S*
	No. 5, '63:625-627	1	S
278.	"Spey Junior"		
	Acft Eng, My '67:54-55	1	S
279.	"Trent"		
	Aeroplane, Je 14, '67:23-24	1	
	Avn Wk, S 12, '66:30-31	1	
	F 6, '67:39	1	
	F 13, '67:50		S*
	My 29, '67:273-278	1	S
280.	"Tyne"		
	Aeroplane, O 21, '55:630	1	
	D 19, '58:900-901	1	P
	Ap 22, '60:482-488	1	S*
	Ag 5, '60:159-163	3	S*
	Air Prog, Win '61/'62:38-39		S*
	Acft Eng, S '63:260-264 (-12)	1	S*
	Avn Wk, N 14, '55:68-70 (-109)	1	
	N 9, '59:98-104 (-1, -11, -12)	1	S
	Flight, Ap 22, '60:563-567	1	S*
	Interavia, No. 3, '63:372-374	4 (Props)	
	Roundel, Ja/F '60:8 (-12)		S*
	Sp/Aeron, Je '60:121-123	1	S*
281.	"Viper"		
	Interavia, No. 11, '69:1775 (-600)	1	S
282.	"Vulture"		
	Flight, Je 4, '42:557	1	
283.	O-240-A		
	Built under CONTINENTAL license		
	Flight, Jl 24, '69:125-128	1	P, S*
284.	O-300		
	Built under CONTINENTAL license		
	Aeroplane, Mr 8, '62:250-251		S*

ROLLS ROYCE (cont.)

285.	R		
	Aeroplane, Mr 5, '30:388-392	1	S
	O 7, '31:864-872	1	
	Acft Eng, O '31:244-246	2	
	Ap '32:91-95	1	S*
286.	RB 108		
	Aeroplane, Ja 25, '62:93-96	1	S*
	Flight, Ag 26, '65:336-340	1	
	Interavia, No. 8, '62:969-971	1	
287.	RB 109		
	Flight, D 17, '54:859	1	
288.	RB 145		
	Flight, Ag 26, '65:336-340	1	
289.	RB 162		
	Acft Eng, D '62:350-352	1	
	Ap '69:27-28	1	
	Avn Wk, S 24, '62:91	1	P
	Ap 18, '66:54-61	1	
	Flight, Je 13, '63:922-925	1	S*
	Ap 14, '66:586-587	1	
	Interavia, No. 2, '63:180-182	1	
290.	RB 163		
	Flight, Mr 11, '60:332	1	S*
291.	RB 172		
	Avn Wk, Ag 2, '65:70-75	1	
	Interavia, No. 9, '65:1397-1398	1	
292.	RB 178		
	Avn Wk, Jl 18, '66:40-41	1	
	Flight, My 12, '66:796	1	S*
	Interavia, No. 7, '66:1031-1033	1	S*
293.	RB 207		
	Developed from "Trent" and RB 178		
	Air Prog, O '68:14		S
	Avn Wk, My 29, '67:273-278	1	
	Interavia, No. 12, '67:1874-1875	1	S*

ROLLS ROYCE (cont.)

294.	RB 211		
	Aeroplane, N 8, '67:19	1	S
	S 11, '68:14	1	
	Air Force, Je '69:22	1	S
	Air Prog, O '68:18-19		S*
	Acft Eng, Ap '68:22-24+	1	S
	S '69:25-26	1	S*
	Avn Wk, My 29, '67:273-278	1	
	N 6, '67:38-50	1	S
	Jl 1, '68:33	2	
	Ap 14, '69:115	2	
	Flight, N 9, '67:754-756	1	S
	Fly Rev Intl, Je '69:39-40	1	
	N '69:34		P*
	Interavia, No. 12, '67:1881	1	S*
	No. 5, '69:549-552	1	S*
295.	RB 360		
	Avn Wk, S 23, '68:20	1	
	Fly Rev Intl, N '68:34 (-07)		S

ROLLS ROYCE/SNECMA — GB/France

296.	M-45H		
	Avn Wk, Je 2, '69:254-255	1	S
	Fly Rev Intl, Mr '69:27-28	1	

ROLLS ROYCE/TURBOMECA — GB/France

297.	"Adour"		
	Flight, N 13, '69:734	1	S

ROVER — GB

298.	T.P. 90		
	Flying, Ja '62:48-49+	1	

RUSSIA — Russia

299.	AM-34		
	Acft Eng, Je '45:159-165	1	S
300.	K		
	Avn Age, Jl '57:50-55	1	
	Avn Wk, Ap 8, '57:53-57	1	
	Interavia, No. 4, '57:337-338	1	

RUSSIA (cont.)

301. M-012
Developed from JUNKERS "Jumo" 004
Interavia, No. 11, '50:586-588 1 S*

302. M-40F
Avn Age, Ap '52:44-45 1

SALMSON GB

303. A.D. 9
Acft Eng, Jl '30:167-170 1 S
Mr '32:65-68+ 3 S
Flight, N 6, '31:1100-1104 1 S
Je 15, '33:569-570 (-R) 1

SECURITY USA

304. S-5-125
Aviation, N '37:42 1
West Fly, N '37:31 1

SIEMENS Germany

305. S.H. 20
Acft Eng, Jl '30:171-172+ 2 S

SNECMA (Société Nationale d'Etude et de Construction de Moteurs d'Aviation) France
Bris Sid Jnl, Spg '63:72-75

306. "Atar"
Acft Eng, Jl '49:206 (-100) S
Sp/Aeron, S '59:97-98+ (-8&-9) 1

307. "Atar 101"
Aeroplane, N 28, '52:718-721 1
Avn Age, O '45:60-61 1 P,
My '56:112-119 (Pt 1) 3
Jl '56:78-83 (Pt 2) 3
Ag '56:98-103 (Pt 3) 3
Interavia, Je '49:345-349 (-B) 1 S
No. 4, '56:287-290 1

SNECMA (cont.)

308. "Super Atar"
Interavia, No. 12, '58:1291-1292 1

309. TF 106
Development of PRATT & WHITNEY JTF 10A
Interavia, No. 5, '63:632 1

SNECMA/TURBOMECA France

310. "Larzac"
Interavia, No. 5, '69:629 1 S

SOCEMA (Société de Constructions et d'Equipements Méchaniques pour l'Aviation) France

311. TGA-1 bis
Aeronautics, Mr '49:65-66 1
Acft Eng, Jl '49:207 S

SOLAR USA

312. "Jupiter" T 400
Avn Age, N '52:40-41 1 S

STAL (Svenska Turbinfabriks ab Ljungström) Sweden

313. "Dovern"
Avn Age, D '52:33 1
Interavia, No. 5, '53:258-260 1 S*

SUTOR USA

314. Diesel
Pop Avn, Mr '32:159 1

SZEKELY USA

315. SR-3
West Fly, Je '32:28-29+ (Pt 1) 3
Ag '32:26-28 (Pt 2) 3 S*

TOROIDAL USA

316. Aircraft Engine
Pop Avn, Mr '36:175-176 1 S, P

TURBOMECA			France
317.	"Aspin"		
	Aeroplane, F 9, '51:170-171	1	S*
	Flight, F 1, '51:135-136 (-I)	1	S
318.	"Astafan"		
	Fly Rev Intl, Ag '69:31	1	
319.	"Astazou"		
	Acft Eng, Ja '65:13-16 (-10)	1	S*
	S '67:33-37+ (-14)	1	S
	Interavia, No. 7, '61:970-973	1	S*
320.	"Aubisque"		
	Flight, F 15, '62:252	1	
	Interavia, No. 5, '63:631-632	1	
321.	"Bastan"		
	Interavia, No. 8, '61:1115 (-IV)		S*
322.	"Marbore"		
	Aeroplane, F 27, '53:258-261 (-2)	1	S*
323.	"Piméné"		
	Flight, Ja 12, '50:53	2	
324.	"Turmastazou"		
	Fly Rev Intl, Ag '69:32	1	
325.	"Turmo"		
	Interavia, No. 7, '61:973 (-IIIC&-IIID)	1	S*

UNITED AIRCRAFT OF CANADA — Canada
See PRATT & WHITNEY - CANADA

VELIE			USA
326.	M-5		
	West Fly, O '32:28-30	3	

VILLIERS			GB
327.	"Maya"		
	Flight, Ap 23, '36:427-428	1	
	F 25, '37:187	2	

WALTER			Czech
328.	N. Z. 60		
	Acft Eng, Je '29:134-135	1	S
WARNER			USA
329.	"Scarab"		
	West Fly, Ap '31:44-46	3	
330.	"Scarab Junior"		
	West Fly, Ap '31:44-46	3	
331.	"Super Scarab"		
	Aero Digest, Jl '39:110	1	
	Aviation, Ag '39:41	1	
	West Fly, Mr '35:19-21+	3	
WESTINGHOUSE			USA
332.	J 34		
	Aero Digest, D '49:40-41+	1	S*
	Avn Opns, D '49:42-43+	2	
	Nav Avn News, Ap '47:6	1	S*
333.	19B "Yankee"		
	Aeronautics, O '46:32-34	1	
	Acft Eng, Ag '46:254-258	1	
	Aviation, Ja '46:60-68	1	S*
	Interavia, No. 7, '46:51-57	1	P, S
WHITTLE			GB
334.	Jet Engines		
	Aeroplane, O 19, '45:445-452	1	
	N 2, '45:503-507	1	
	N 9, '45:543-546	1	
	Flight, O 11, '45:399-401	1	
	N 3, '49:580	1	
WOLSELEY			GB
335.	"Aquarius"		
	Aeroplane, My 8, '35:542-544	1	S*
	Flight, My 16, '35:526-529	1	S

WOLSELEY (cont.)

336. "Scorpio"
Flight, Jl 18, '35:Supp:b-d 1 S*

337. A.R. 9
Aeroplane, Je 21, '33:1139-1140 (-1) 1 S
Acft Eng, Jl '33:157-159 1 S*
Flight, Je 22, '33:612-614 1 S*

WRIGHT USA
See also CURTISS and CURTISS-WRIGHT

338. "Cyclone 7" R-1300
Avn Age, My '52:36-37 (-3) 1
Jl '52:38-39 (-3) 1 S*
US Air Svcs, Mr '45:48 1

339. "Cyclone 9" GR-1820
Aero Digest, Je '38:67-68 (-G-100) 1
Aviation, Je '38:34-35 (-G-100) 1
Flight, O 15, '36:388d (-G) 1
US Air Svcs, D '36:32+ (-G-100) 1

340. "Cyclone 9" R-1820
Aero Digest, O '32:40-41 1
Je '35:30-32+ (-F-50) 1
S '36:32-35 (-G) 1
Jl '39:113-114 (-G-200) 1
Acft Eng, Ap '35:91-94 (-F-1) 1 S
O '36:284-287 (-G) 1 P
Aviation, Ag '36:37-41 (-G) 1
Jl '39:56+ (-G-200) 1
Pop Avn, Je '36:420-422+ (-G) 1
Ja '37:26-28+ (-G) 1
US Air Svcs, O '32:21-23 (-F) 1
Jl '35:31 (-F-50) 1
Ap '36:23 (-G) 1
Ag '36:34 (-G) 1
My '40:21 (-G-200) 1
F '46:34 (-HD) 1
Mr '47:26 (-HE) 1
West Fly, Ag '36:28 (-G) 1

WRIGHT (cont.)

341.	"Cyclone 14" R-2600		
	Aero Digest, O '39:117-118	1	
	Air Tech, Mr '43:7-13	1	S*
	Aviation, O '39:36-37	1	
	Flight, O 26, '39:334f	1	
	F 11, '43:156-157	1	
	Skyways, My '43:43-44	1	S*
	US Air Svcs, Ap '38:30	1	
	O '39:22	1	P
342.	"Cyclone 18" R-3350		
	Aero Digest, Je '47:58-61+ (-18BD)	1	S*
	Avn Wk, My 24, '48:26-31	1	
	Comm Avn, Ja '44:98+	1	
	West Fly, D '43:138	1	
343.	"Flier"		
	Aeroplane, D 18, '53:827-828	1	S*
	Bee-Hive, Ja '53:7-11	1	P
344.	"Gipsy" L-320		
	West Fly, Je '31:50-52	1	
345.	"Turbo Compound" 972TC18DA		
	Avn Age, F '54:30-31	1	S*
	Avn Wk, D 14, '53:48-54	3	
346.	"Turbo Cyclone 18" R-3350		
	Avn Opns, Jl '50:20-23	1	S*
	Avn Wk, My 29, '50:27-32	1	S*
	F 18, '52:35-40	1	S*
	US Air Svcs, Ap '48:20	1	
	O '48:17	1	
347.	"Typhoon" XJ-37		
	US Air Svcs, Ap '48:14-15	1	P
348.	"Whirlwind" E-760		
	US Air Svcs, S '35:36	1	
349.	"Whirlwind" J-5		
	Pop Avn, F '29:40+	2	
350.	"Whirlwind" R-540		
	West Fly, F '31:46-48+	3	

WRIGHT (cont.)

351.	"Whirlwind" R-760		
	Aero Digest, S '33:50-51 (-E)	2	
	Ag '35:48-49+	1	
	Aviation, Ag '35:31-32 (-E)	1	
	West Fly, F '31:46-48+	3	
352.	"Whirlwind" R-975		
	Aero Digest, Ag '35:48-49+	1	
	Aviation, Ag '35:31-32 (-E)	1	
	US Air Svcs, S '35:36 (-E)	1	
	West Fly, F '31:46-48+	3	
353.	"Whirlwind" R-1510		
	Aero Digest, Ja '33:42-43	1	
	US Air Svcs, F '33:12-14	1	
354.	J65		
	Military version: T49		
	Original design: ARMSTRONG SIDDELEY "Sapphire"		
	Aero Digest, D '54:31-35	1	S*
	Avn Wk, D 22, '52:21-22	1	
	Je 14, '54:28-42	1	S*
	Flight, Je 18, '54:787-788	1	S*
355.	T49		
	Civil designation: J65		
	Avn Wk, D 20, '54:28-29	1	

Part Two

Aircraft Engine Addendum

Aviation,	Ap '46:78-81	British turbojet development
Flying,	My '67:48-67	Development to 1967
Fly Rev Intl,	Jl '66:686-688	Russian engines
Interavia,	No. 11, '59:1362-1363	Jet engines: 1939-1959
Pegasus,	N '56:10-15	Early development
Pop Avn,	D '28:64-66+	Development to 1928
	Ap '29:84+	Engine evolution
	Mr '32:161-162+	Short history
	Ag '32:101-102+	Diesel advantages
	Mr '37:36-38+	Diesel history

Part Three

BIOGRAPHY

Part Three

BIOGRAPHY

The elimination of those personalities who have provided impetus and direction to the growth of aviation, whose exploits have thrilled both young and old, and whose genius for organization and design has produced the aircraft, the air forces, and corporations with which we are familiar, would have materially lessened the value of this book. One would wish that far more could be included here, but practicality demands that only the major figures be cited. These are primarily designers, heads of corporations, fighter aces, pioneers, military figures, and recipients of outstanding awards and honors. Occasionally, though, an article provides such detailed treatment of a lesser-known figure that inclusion is desirable.

This section can be of great value to the modeler, too. Aircraft insignia and other details are often shown in the photographs which accompany biographical sketches of famous fliers and fighter aces. Further, a model has more meaning for its builder when the life and exploits of its original creator or those who flew it are known.

Format

The name of the individual is followed by a descriptive note indicating position or accomplishment. This is brief for reasons of space and serves primarily as verification.

Notes accompanying individual citations define any limitations on the article or indicate special features such as portraits, photographs, drawings of aircraft used, obituaries, and so on. If a note does not follow the citation, it should be assumed that the article is a general biography or describes a number of exploits. Obviously, particular attention should be given the date on which the biography appeared in print.

Addendum

Interest in military aces as a group seems to justify a separate listing. Such articles are usually well researched and provide an excellent supplement to the individual biographies indexed in the main body of this section.

ADAMS, Donald E. (US ace, Korea)
US Air Svcs, S '52:16+

ADER, Clément (French aero theorist and teacher)
Interavia, No. 12, '47:19-21 — w por

ALER, Isaac A. (President, KLM, 1954)
Interavia, No. 4, '54:249

ALEXANDER, William M. (RCAF ace, WW I)
AAHSJ, Fall '66:228 — Drawing of his triplane
CAHS, Fall '65:62-65 — w pors

ALISON, John R. (US ace, WW II)
RAF Fly Rev, F '59:33-35 — CBI exploits

ALLEN, Cecil (US long-distance pilot)
Bee-Hive, Ja '66:28-32 — 1931 "Clasina Madge" exploit

ALLEN, Edmund T. "Eddie" (Boeing test pilot)
US Air Svcs, D '39:1, 23-24 — w por
Mr '43:9-10, 38+ — On death, w por
O '51:20-22+ — Character, por (fp)

ALLEN, James (US Army Chief Signal Officer)
US Air Svcs, Mr '33:11-12 — Obit

ALLMENRÖDER, Karl (German ace, WW I)
Cr & Cockade, Win '61:344-347 — Victory log, photos, aircraft

ALTHAUS, Ernst--Freiherr von (German ace, WW I)
Cr & Cockade, Sum '61:115-119

ANDREWS, Frank M. (USAAF General)
Aerosp Hist, Sum '69:6-7 — w por
US Air Svcs, Ap '35:16
Mr '36:5 — Por (fp) only
Ag '36:16-18+ — w por (fp)
Je '43:11 — Obit

ANTONOV, Oleg K. (Russian designer)
RAF Fly Rev, N '56:17-18 w designs

ARNOLD, Henry H. "Hap" (Commander-in-Chief, USAAF)
Bee-Hive, My/Je '41:2-3 Promotion, w por
Roundel, Ja '52:38-43 w por
Skyways, O '43:28-29+
US Air Svcs, Ag '34:13-14
N '38:11-12 w por
F '46:10-11 Retirement, w por
Ja '50:5-6 Obit

ARNOLD, Leslie P. (Around-the-world flyer)
Pop Avn, Mr '40:49-50+ w por

AUDET, Richard J. (Canadian ace, WW II)
CAHS, Fall '64:73-74

AURIOL, Jacqueline (French aviatrix)
RAF Fly Rev, N '55:19-20 Exploits

BADER, Douglas R. (RAF ace, WW II)
Air Prog, Ja '64:36-37+
RAF Fly Rev, Mr '54:12-14 Exploits

BAER, Paul F. (US ace, WW I)
US Air Svcs, Ag '38:20-22+ w por

BALBO, Italio (Italian Air Minister)
US Air Svcs, F '31:35-38 US flight, w por (fp)
Mr '33:22 por only
Je '33:23

BALCHEN, Bernt (Polar flyer)
US Air Svcs, Ap '51:4-5 w por (fp)
Mr '54:10

BALDWIN, Frederick Walker "Casey" (First Canadian to fly)
Roundel, Jl/Ag '60:24-25 w por
US Air Svcs, Ag '48:22 Obit

BALDWIN, Thomas S. (US pioneer airman)
Pop Avn, S '28:46-50+

BALL, Albert (British ace, WW I)
Cr & Cockade, Aut '63:243-244 Last flight
Fall '69:218-234+ Exploits

BALL (cont.)
Pop Fly, F '33:632-633 w por
Ja '34:548 w por
D '34:469 w por and grave
RAF Fly Rev, F '55:20-21+ Exploits
RAF Qtr, Ja '32:1, 88-91 Exploits, diary, por

BANE, Thurman H. (Commander, McCook Field, Ohio)
US Air Svcs, Ap '32:23 Obit, w por

BANFIELD, Gottfried (Austrian "Ace of Trieste," WW I)
Pop Fly, Ap '39:32-33 Exploits, w por

BARACCA, Francesco (Italian "Ace of Aces," WW I)
Pop Fly, O '32:394-395 w por

BARKER, William G. (Canadian ace, WW I, Victoria Cross)
Avn Wk, Je 1, '64:36-37 w aircraft in color
CAHS, Win '65:108-111 Aircraft story and drawing
Flying, Ja '60:19 Aircraft drawing
Pop Fly, Je '35:146-148 w por
RAF Qtr, Ap '30:233, 386-387 w por

BARKHORN, Gerhard (German ace, WW II)
Air Prog, Jl '69:30-33+ Exploits, por in 1969

BARLING, Walter H. (Designer, Barling Bomber)
Air Prog, F '43:32+
US Air Svcs, Ap '55:7 On retirement

BARNWELL, Frank (Chief designer, Bristol, 1929+)
Flight, Mr 19, '54:338-342

BARTON, Cyril Joe (British "Halifax" pilot, Victoria Cross)
RAF Qtr, S '44:xv Victoria Cross citation

BASHKO, Jezup S. (Russian bomber leader, WW I)
Cr & Cockade, Win '69:325-326 w por

BATESON, Robert "Bob" (British Group Captain, WW II)
RAF Fly Rev, Ap '55:28-29+ Exploit

BÄUMER Arno Paul (German ace, WW I)
Cr & Cockade, Win '64:301-320 Exploits, aircraft, pors

BAUR, Hans (Hitler's pilot)
Pop Fly, Mr '39:590-592+ WW I experiences, w por

BAYLES, Lowell (US Gee-Bee racing pilot)
Pop Avn, My '32:291-292+

BAYLIES, Frank (US ace, WW I)
US Air Svcs, Ja '39:10-12+

BEACHEY, Lincoln (US pioneer airman)
AAHSJ, Sum '61:105-111+ Last flight
Spg '64:56 Dirigible exploits
Bee-Hive, Sum '57:18 Aircraft vs. auto race
Flying, D '53:16-17+
Pop Avn, N '31:16+
Ag '36:36 w photo in Curtiss "Tripod"
F '37:19-20+ His mechanic's reminiscences

BEAIRD, Henry Grady, Jr. (Republic test pilot)
Bee-Hive, Sum '59:26-29 w por in F-105

BEAMONT, Roland P. (British test pilot)
Roundel, My '56:30-31

BEAUCHAMP-PROCTOR, Andrew W. (British ace, WW I, Victoria Cross)
Pop Fly, Jl '33:184 VC exploit, w por (fp)
S '34:293 w por

BECKER, Richard S. (US ace, Korea)
US Air Svcs, Ap '52:13-14

BEECH, Walter H. (Founder of Beech Aircraft)
Skyways, F '51:8 Obit, w por

BELL, Hilliard Brooke (Canadian ace, WW I)
CAHS, Sum '63:43-46
Fall '63:83-88 (Pt 1) WW I exploits, w pors
Win '64:15-17 (Pt 2) WW I exploits

BELL, Lawrence D. (US manufacturer and designer)
Am Helicop, S '46:24+ w por
US Air Svcs, Je '44:28-30+

BELLANCA, Giuseppe Mario (US manufacturer and designer)
Flying, F '53:36-37+
US Air Svcs, N '30:1, 58 w por

BENDIX, Vincent (Founder of Bendix Aviation)
US Air Svcs, Ap '45:40 Obit

BENNETT, Floyd (US airman)
Pop Avn, Je '28:33-34 Obit, w por

BERA, Frances ("Powder Puff Derby" winner)
Flying, O '62:32-33+

BERRY, Matt (Canadian Arctic pilot)
Pop Fly, F '39:564-565+

BERTHOLD, Rudolf (German ace, WW I)
Air Prog, Ap '67:38-40+ w aircraft profile

BEURLING, George F. "Screwball," "Buzz," or "Lone Eagle of Malta" (Canadian ace, WW II)
RAF Fly Rev, My '55:34-36+ w por
Sentinel, Ja '68:44-45

BIRKIGT, Marc (Designer of Hispano-Suiza engine)
Fly Rev Intl, Ap '64:48-49 w por

BISHOP, William "Billy" (Canadian ace, WW I, Victoria Cross)
Air Force, N '56:107-115
Avn Wk, F 16, '59:149+ w plane and por
CAHS, Fall '64:71
Win '67:103-106
Pop Avn, Ag '35:79-80+ Exploits
Pop Fly, Jl '33:186 w por
N '33:410 w por in mufti
Ag '34:240 w por
Mr '36:646-650+
RAF Fly Rev, D '57:31-33+ WW I exploits
US Air Svcs, O '56:7 Obit

BISSELL, Clayton L. (US ace, WW I)
Cr & Cockade, Spg '66:40 Por only

BLACKBURN, Robert (British manufacturer and designer)
RAF Fly Rev, Ap '59:16-18+

BLAIR, Charles F., Jr. (Harmon International Trophy winner, 1952)
Flying, Ag '59:24-30 "Excalibur III" arctic flight

BLAND, Lilian (British pioneer aviatrix)
Flight, Ja 23, '64:140-141 Her "Mayfly"

BLECKLEY, Erwin R. (WW I, US Medal of Honor)
Avn Wk, D 7, '59:96-97 — w aircraft

BLERIOT, Louis (French pioneer airman)
Airpwr Hist, Ap '59:94-102 — Channel crossing
RAF Fly Rev, Jl/Ag '59:21-23 — Channel crossing, w por

BOEING, William E. (US aircraft manufacturer)
Boeing Mag, Ap '34:34
N '56:1, 3-5 — Obit, w por (color)
N '69:10-11 — Early life
US Air Svcs, Ap '41:21-22
My '42:11 — Rejoins Boeing for WW II
O '56:7 — On his death

BÖLCKE, Oswald (German ace, WW I)
Air Prog, O '68:24-25+ — w por and aircraft
Flying, Ja '59:67 — w Fokker E-1
Pop Avn, O '32:237-238+ — Who shot him?
Pop Fly, N '33:432-433 — w por
RAFs Qtr, Aut '68:179-184 — w por

BOLLING, Raynal C. (US air officer)
Flying, Ag '59:44-45+

BOLT, George B. (New Zealand pioneer airman)
Air Power, Aut '55:27-31

BONG, Richard I. (US ace, WW II)
Aero Album, Sum '68:70-71 — His P-38 aircraft
US Air Svcs, Ja '45:16 — Exploits, w por
S '45:8 — Tribute on his death

BOYINGTON, Gregory "Pappy" (US Marine ace, WW II)
Skyways, Jl '44:40-41+ — Exploits, w por

BOYLE, Dermot A.--Sir (Air Chief Marshal, RAF)
Air Power, Win '55:80 — w por

BRABAZON OF TARA--Lord (Britain's first pilot)
Flight, My 28, '64:877 — On his death
RAF Fly Rev, Mr '60:38-40

BRACK-PAPA, Francesco (Italian racing pilot, 1920's)
Aero Album, Win '69:26-30 — Exploits, w photos

BREGUET, Louis (French designer and pioneer)
Am Helicop, S '47:16+ w por

BREITHAUPT, William R. (Canadian ace, WW II)
CAHS, Fall '67:81

BRETT, George H. (US Assistant to Chief of Air Corps)
Bee-Hive, My/Je '41:2-3 w por
US Air Svcs, Mr '39:16 w por
N '40:10-11 w por

BREWER, Griffith (First Englishman to fly)
US Air Svcs, O '36:1, 34+ w por

BRIDGEMAN, William B. (US test pilot, Douglas "Skyrocket")
Flying, O '51:13-15+

BROWN, Arthur R. "Roy" (Canadian ace, WW I)
Avn Wk, My 11, '59:155+ w por and aircraft
CAHS, Sum '63:41-42 Exploit against Richthofen
Spg '67:10-11
Flying, Ap '59:19 Drawing of downing of Richthofen
Pop Fly, O '34:350 w por
RAF Fly Rev, Ap '55:22-24+ Exploit against Richthofen

BROWN, Donald Lamont (President, United Aircraft)
Bee-Hive, F '40:1-11 On his death, w por

BROWN, Frederick E. (Canadian ace, WW I)
CAHS, Sum '64:43+

BROWN, Ralph M. (US pioneer airman)
AAHSJ, Win '68:309 w por

BRUNER, Donald L. (US night-flying pioneer)
Aerosp Hist, Aut '69:16-17+ Accomplishments

BÜCKER Carl C. (German designer)
Air Prog, Ag/S '63:60-69+

BULLOCH, Terence "The Bull" (Britain's best destroyer of U-boats, WW II)
RAF Fly Rev, F '58:20-22+ Exploits

BURDEN, Henry J. "Hank" (Canadian ace, WW I)
CAHS, Ja '63:8-9 (Pt 1)
Spg '63:20-24 (Pt 2) w pors

BURKE, Arleigh A. "31-Knot" (Chief of Naval Operations, US Navy, 1956)
Flying, Ja '56:22-23+

BURNELLI, Vincent (US manufacturer and designer)
AAHSJ, Fall '69:163-165

BYRD, Richard E. (US polar flight pioneer)
Nav Avn News, Ja '69:14-15 — w por
Pop Avn, O '31:13-14+ — Controversy w Fokker

CAIN, E. J. "Cobber" (New Zealand ace, WW II)
RAF Fly Rev, O '57:39-40+ — Exploits

CALDERARA, Mario (Italian pioneer airman and naval commander)
US Air Svcs, S '30:37-38 — Early experiences, w por

CALDWELL, C. R. (Australian ace, WW II)
Air Prog, S '66:45-48+ — w Spitfire profile

CALDWELL, Keith L. "Grid" (New Zealand ace, WW I)
Cr & Cockade, Fall '69:193-212 — Exploits

CALLAHAN, Lawrence K. (US ace, WW I)
Cr & Cockade, Spg '66:42 — Por only

CAMM, Sidney--Sir (Director, Hawker Siddeley)
Interavia, No. 5, '66:667 — On his death

CAMPBELL, Andrew C. (US Sergeant pilot, WW I)
Cr & Cockade, Sum '63:198-201

CAMPBELL, Douglas "Doug" (US ace, WW I)
Avn Wk, Je 8, '59:123+ — w aircraft
Cr & Cockade, Spg '65:30-42 — Interview

CAMPBELL, Hugh (Chief of Air Staff, RCAF)
Roundel, S '62:16A-16D — On retirement

CANIFF, Milton (US cartoonist)
Pegasus, Jl '55:1-5

CAPRONI DI TALIEDO, Gianni (Italian manufacturer and designer)
Pegasus, D '56:Supp:1-20 — w por and papers
N '57:17 — On his death, w por

CARBERRY, J. E. (US JN-4 pilot, Mexican Punitive Expedition)
Avn Wk, Ag 21, '61:52-53 Drawing of aircraft incident

CAREY, Frank R. (RAF ace, WW II)
RAF Fly Rev, Ap '63:18-20+

CARR, Bruce W. (US ace, WW II)
Aero Album, Sum '69:20-21 Exploits and P-51 aircraft

CASTLE, Frederick W. (US Commanding General, Third Air Division, WW II)
US Air Svcs, F '45:11-12 On his death
Ap '46:14-15 Medal of Honor, w por
My '48:13 Posthumous Legion of Honor
Jl '52:12-14

CAYLEY, George--Sir (British pioneer designer)
Flight, N 12, '54:701-703

CHADWICK, Roy (Early Avro designer)
Flight, Mr 30, '56:362-364

CHAMBERLAIN, Cyrus Foss (Famous US pilot)
Cr & Cockade, Sum '65:187-193 w pors

CHANUTE, Octave (US pioneer airman)
Aerosp Hist, Aut '68:30-31, 44-47

CHAPMAN, Victor E. (Lafayette Escadrille pilot, WW I)
Pop Avn, Ap '33:215 w por

CHENNAULT, Claire L. (Commander, American Volunteer Group, China, WW II)
Airpwr Hist, Ap '59:88-93
US Air Svcs, Mr '42:32

CHESHIRE, Geoffrey Leonard (Brit airman, Victoria Cross)
Roundel, N '55:44-47 w por
RAF Fly Rev, My '54:18-20 Exploits, w por (fp)
RAF Qtr, Jl '50:214-219 w por (fp)

CHETVERIKOV, Igor (Russian designer in 1930's and 1940's)
Fly Rev Intl, Ag '67:803-805 On aircraft he designed

CHETWOOD, Kenneth (8th Air Force "trainbuster," WW II)
Skyways, Ja '45:50-51+ Exploits

CHEVALIER, Godfrey (US pioneer airman)
Avn Wk, Ag 19, '68:38-39 USS Langley landing, w drawing

CHRISTOFFERSON, Harry P. and Silas (US pioneer airmen)
AAHSJ, Spg '64:21-22 w pors

CHURCH, Ellen (First airline stewardess)
US Air Svcs, Je '55:11+ w por

CIERVA, Juan de la (Autogiro inventor)
US Air Svcs, Mr '32:10-12
Ja '37:12-13 On his death, w por (fp)

CLOSTERMANN, Pierre (Top Free French ace, WW II)
Roundel, F '52:27-32 w por

COANDA, Henri (Early French aircraft designer)
Air Prog, Ag/S '65:6-12 w many photos of aircraft
Flying, My '56:32-33+
Mr '67:64-68
RAF Fly Rev, S '56:40+

COBBY, A. H. (Top Australian ace, WW I)
Pop Fly, F '35:573-574+ Exploit w aircraft photo

COCHRAN, Jacqueline (US aviatrix)
Pop Avn, D '38:61-62+
RAF Fly Rev, Ja '54:31-32 Exploits
US Air Svcs, Ja '38:1, 24-28 w por

COCHRANE, Ralph A. (Air Chief Marshal, RAF)
RAF Qtr, O '49:96-99 w por

COCKING, Robert (Britain's first parachutist)
Pop Fly, D '37:497+

CODY, S. F. "Sam" (American pioneer in GB)
Air Force, My '55:113-118
Flight, My 16, '58:690-691
My 23, '58:699
Flying, N '57:24-29
Pop Fly, Je '34:123

COHEN, Lionel F. W. "SOS" (Wing Commander, RCAF)
Roundel, My '53:9-12

COLLETT, Clive F. (New Zealand ace, WW I)
Cr & Cockade, Win '65:335-340 w pors

COLLISHAW, Raymond (Canadian ace, WW I)
Air Cl, D '68:38-46+ In Russia after WW I
Avn Wk, O 21, '63:36-37 w aircraft (in color)
Cr & Cockade, Win '61:357-358 Victory log
Sum '64:179-182 Victories
Sum '67:138-157 His art, aircraft, w pors
Pop Fly, F '37:580 Por only
Roundel, My '64:4-10 (Pt 1)
Je '64:18-23 (Pt 2)
Jl/Ag '64:20-24 (Pt 3) w por
S '64:18-22 (Pt 4)
O '64:18-22 (Pt 5)
RAF Fly Rev, O '62:45+ w por
Jl '63:45-46 1940 incident

COOK, Arthur Byron (Chief, BurAir, US Navy)
US Air Svcs, Mr '37:1, 27 w por

COPPENS, Willy (Belgian ace, WW I)
Avn Wk, Mr 5, '62:54-55 w aircraft (in color)
Flying, F '60:79 Aircraft
Pop Avn, Mr '34:147-148 Exploits

CORRIGAN, Douglas "Wrong-Way" (US Trans-Atlantic pilot)
Pop Avn, D '38:10-13 w por
US Air Svcs, Ag '38:17 w por
S '38:13+ Atlantic flight
Je '53:9 " "
Ap '56:22 " "

CRABB, Earl Frederick (Canadian-born ace, WW I)
CAHS, Spg '69:26 Exploits

CREIGHTON, Richard D. (US ace, Korea)
US Air Svcs, Ap '52:14

CROIL, George M. (RCAF's first Chief of Air Staff)
Pop Avn, Jl '40:41+
Roundel, Je '59:30-31 On his death, w por

CULLEY, Stuart (Destroyed Zeppelin L. 53, WW I)
RAF Fly Rev, N '56:47-48 Exploit

CUMMINGS, Robert "Bob" (Hollywood actor, arcrft enthusiast)
Flying, Mr '60:44-46+

CUNNINGHAM, Alfred A. ("Father of US Marine Corps Aviation")
Nav Avn News, Ja '66:6-7 w por
US Air Svcs, O '44:11

CUNNINGHAM, John (RAF night-fighter ace, WW II)
RAF Fly Rev, Je '54:18-20 Exploit, w por (fp)

CURTISS, Glenn (US pioneer airman)
Pegasus, D '55:10-15 w early aircraft
Mr '56:6-15 " " "
Pop Avn, N '33:293-294 Early days
US Air Svcs, Ag '30:40 On his death

CUSTER, Willard R. (US designer of Channel Wing)
Flying, S '47:45-46+

CUTTER, William P. (Southwest US pioneer pilot)
Pegasus, Mr '45:10-13+

DAHL, Walter (German ace, WW II, Leader "Suicide Brigade")
RAF Fly Rev, D '58:27+ Göring exploit, w por

DAHLBERG, Kenneth L. (US ace, WW II)
Skyline, Vol. 6, No. 1, '45:8 Exploit

DALLAS, Roderic (RNAS ace, WW I, Leader 40th Squadron)
RAF Fly Rev, Ja '59:24-25+

D'ANNUNZIO, Gabriele (Italian poet and airman, WW I)
Avn Wk, N 27, '61:34-35 w aircraft (in color)

D'ASCANIO, Corradino (Italian rotary-wing pioneer)
Am Helicop, O '48:15-16

DARGUE, Herbert A. (General, USAAF and JN-4 pilot)
AF News Ltr, F '42:35+ Obit w por
Avn Wk, Ag 21, '61:52-53 Incident w aircraft in Mexico
Flying, Ap '49:38-39+ " " " " "
US Air Svcs, N '38:30

DAUGHERTY, Earl S. (Barnstormer; first air-to-air refueling)
Avn Wk, Je 12, '67:52-53 w aircraft (color drawing)

DAVIDS, Rhys (British ace, WW I)
Cr & Cockade, Win '67:382-399 w pors, victory log

DAVIDSON, R. T. P. (Canadian ace, WW II)
CAHS, Sum '64:34

DAVIS, David R. (US designer of "Davis" wing)
Pop Avn, Je '40:36-37+ Birth of "Davis" wing

DAVIS, George A., Jr. (US ace, Korea)
US Air Svcs, Ap '52:13

DAVIS, Walter (US aircraft designer in 1920's)
Air Prog, Jl '65:16-27 (Pt 1)
Ag/S '65:44-48+ (Pt 2)

DAVIS, William V., Jr. (Vice-Admiral, US Navy)
Nav Avn News, Je '60:29 On retirement

DAY, William R. (Top Canadian Far East ace, WW II)
CAHS, Fall '65:66

DE BOTHEZAT, George (Pioneer helicopter designer)
Am Helicop, Jl '57:6-11+

DE HAVILLAND, Geoffrey--Sir (British manufacturer and designer)
Air Pict, Jl '65:220-223 On his death
Flight, My 27, '65:812 " " "
Je 3, '65:859 " " "
RAF Fly Rev, Jl/Ag '59:29-30+

DEBRÖDY, György (Hungarian ace, WW II)
RAF Fly Rev, D '61:33-34+

DEPEW, Richard H., Jr. (US pioneer and Fairchild associate)
Air Pilot, O '44:8
Pegasus, N '43:8-12+
US Air Svcs, F '48:27+ On his death

DERRY, John D. (British test pilot and Squadron Leader)
Interavia, No. 11, '52:631 On his death

DICKENS, Clennel H. "Punch" (Canadian arctic and bush pilot)
Pop Fly, My '35:74-75 w por

EAGLESTON, Glenn T. (US ace, WW II)
Aero Album, Fall '68:26-27 Exploits, w P-51 profile

EAKER, Ira C. (Air Commander-in-Chief, Mediterranean Allied Air Forces WW II)
US Air Svcs, Ap '39:1, 25+ w por
Mr '43:13 On becoming 8th AF CG
My '47:8, 10-13+ On retirement, w por

EARHART (Putnam), Amelia (US aviatrix)
Airpwr Hist, Ja '61:39-44 Last flight
AAHSJ, Sum '63:114-120 " "
Pop Avn, D '39:10-13+ (Pt 1) Is she alive?
Ja '40:29+ (Pt 2) " " "
Skyways, N '42:44-47+ How did she die?
US Air Svcs, Je '32:10-14 Her Atlantic crossing
F '35:19-20 Her Honolulu-Oakland flight

EDWARDS, Hughie Idwal (Australian RAF Wing Commander, Victoria Cross)
RAF Fly Rev, D '62:37-39
RAF Qtr, Ja '51:51-54 w por

EGTVEDT, Claire (President, Boeing Aircraft Co.)
Boeing Mag, F '57:6-7 Accomplishments, w por
US Air Svcs, Jl '35:1, 32 w por
Ap '47:11 Por only

EIELSON, Carl Ben (Arctic pioneer pilot)
CAHS, Sum '68:36-41 (Pt 1) Search for him
Fall '68:78-82 (Pt 2) " " "

ELLEHAMMER, Jacob C. (Early Danish designer)
Air BP, No. 44:14-15
Flying, Je '57:42-43+

ELLYSON, Theodore Gordon "Spuds" (First US Naval aviator)
Nav Avn News, Ja '65:10-11

ELWORTHY, Charles--Sir (Air Chief Marshal, RAF)
RAFs Qtr, 1963, No. 3:174-176 w por
1967, No. 1:71 w por

ELY, Eugene (First to land on ship, 1911)
Nav Avn News, Ja '51:6 Exploit, w photo
Ja '66:6 w por
US Air Svcs, Ja '45:12-15+ Exploit, w photos

EMBRY, Basil--Sir (British Air Vice Marshal)
RAF Fly Rev, Jl '57:19-20

EMMONS, Delos C. (Commanding General, GHQ Air Force)
Bee-Hive, F/Mr '39:8-9 w por
US Air Svcs, Mr '39:16 w por
N '40:1, 10-11 w por

ERWIN, William "Bill" (US ace, WW I)
Pop Avn, D '35:355-358+

ESCHWEGE, Lt. Rudolph von (German ace, WW I)
Avn Wk, D 9, '63:62-63 w aircraft drawing (in color)

FAHRNEY, Delmar S. (US Naval officer, "Father of the Guided Missile")
Am Helicop, O '50:9-12+ w por

FAIRCHILD, Muir S. (Vice Chief of Staff, USAAF)
US Air Svcs, My '48:10 w por

FAIRCHILD, Sherman L. (US manufacturer and aerial surveyor)
Pop Avn, Ja '36:15-16+ w por
Skyways, S '51:18-19+

FAIREY, Richard--Sir (British manufacturer and designer)
Flight, O 5, '56:573 On his death
Mr 13, '59:347

FARMAN, Henri (French pioneer airman)
Aeroplane, D 8, '37:699 Exploits
Interavia, No. 5, '47:12-15 (Pt 1) w diary excerpts
No. 6, '47:14-16 (Pt 2) " " "

FARMAN Brothers (French pioneer airmen)
Aeroplane, D 8, '37:699 Exploits

FAULKNER, William (US author and aviator)
Cr & Cockade, Sum '68:159-162 w bibliography and pors

FECHET, James E. (Chief, USAAF, 1927-1931)
US Air Svcs, F '48:27 On his death

FIERRO, Roberto (Mexican Chief of Army Air Corps)
Air Pilot, D '44:16-17+
Pegasus, O '47:9-11

FINDLEY, Earl N. (US editor and newsman)
Pegasus, Ap '53:6-12

FINUCANE, Paddy (RAF Irish ace, WW II)
RAF Fly Rev, O '55:27-29+ w por

FIRESTONE, Roger S. (President, Firestone Aircraft Co.)
Am Helicop, My '47:20+ w por

FISCHER, Harold E., Jr. (US ace, Korea)
AAHSJ, Fall '69:189-195 Korean exploits

FISHER, Bernard F. "Bernie" (USAF pilot, Vietnam, Medal of Honor)
Air Force, Mr '67:42-44
Air Prog, My '67:42-45+ Medal of Honor exploit

FITCH, Willis (US pilot with Italian Air Force, WW I)
Avn Wk, S 24, '62:36-37 Bombing Pola, drawing (in color)
Flying, N '61:1 Bombing Pola, drawing

FLEET, Reuben H. (Founder, Consolidated Aircraft Co.)
Aeron (US), O '29:28+ w por
West Fly, D '40:29 w por

FLEISCHER, Alfred (German ace, WW I)
Cr & Cockade, Win '65:358-361 w pors

FLETTNER, Anton (German rotary-wing designer)
Avn Wk, N 29, '54:26-33

FLOREZ, Luis de (Rear Admiral, US Navy)
Nav Avn News, F '63:28-29
S '68:7-11 w por
US Air Svcs, O '54:17 w por

FOKKER, Anthony H. "Tony" (Dutch manufacturer and designer)
Flight, Ja 4, '40:7-8 On his death
Pop Avn, O '31:13-14+ Controversy w Byrd
Pop Fly, D '36:464-465+
US Air Svcs, Ja '40:9 On his death

FONCK, Rene (French ace, WW I)
Flying, My '61:71 Caudron G. IV exploit
Pop Avn, S '32:148+ Exploits, w por
Pop Fly, S '32:338-339 w por

FOSS, Joseph J. (US Marine Corps ace, WW II)
Air Prog, Ap/My '63:25+
US Air Svcs, My '43:11-13 w por (fp)

FOULOIS, Benjamin D. "Bennie" (US Chief of Air Corps)
Air Force, Je '67:85-91
Bee-Hive, Ja '32:1, 3-5 w por
Pop Avn, My '34:291-292+ w por
US Air Svcs, Ja '36:11 Last flight on active duty

FOWLER, Frederick Bernard (British aviation pioneer)
RAF Fly Rev, S '59:44-46+

FOWLER, Harlan D. "Flaps" (Inventor of Fowler flap)
US Air Svcs, N '52:18

FOWLER, Hedley (Test pilot and POW escapee, WW II)
RAF Fly Rev, D '57:23-24+ POW escape

FRAKES, F. F. "Bowser" (US stuntman)
Flying, Jl '51:26-27+

GABRESKI, Francis S. (US ace, WW II and Korea)
Aerosp Hist, Spg '68:4-5
US Air Svcs, S '52:12-13 w por (fp)

GABRIEL, Walter and Willi (German aces, brothers, WW I)
Cr & Cockade, Win '62:333-350 w pors and aircraft photos

GALLAND, Adolf (Luftwaffe General and ace, WW II)
Interavia, N '49:650
Roundel, D '55:30-36

GANN, Ernest (US author and pilot)
Flying, Je '67:58-63

GARRETT, John C. "Cliff" (Founder, Garrett Corp. and AiResearch)
Flying, Jl '57:42-43+

GARROS, Roland (French ace, WW I)
Airpwr Hist, Jl '64:87-88
Flying, Jl '61:75 Morane N exploit
Pop Avn, F '35:79-80+ Exploits

GATTY, Harold (US long-distance pilot)
US Air Svcs, Ag '31:15-18 — Around-the-world flight, w por (fp)

GEHLBACH, Lee (US racing and test pilot in 1930's)
US Air Svcs, O '35:1, 33-34 — w por

GENTILE, Don S. (US ace, WW II)
RAF Fly Rev, Ag '58:22+ — Teamwork w John Godfrey
US Air Svcs, F '51:17 — On his death, w por

GENTRY, Viola (US aviatrix, late 1920's)
AAHSJ, Sum '68:125-126

GIBSON, Guy (RAF "dam-buster," Victoria Cross)
RAF Fly Rev, My '53:23-25+ — Exploit
Mr '55:24-26+ — Exploit, w por

GIBSON, Ralph D. (US ace, Korea)
US Air Svcs, Ap '52:14

GIBSON, William Wallace (Canadian aviation pioneer)
Can Avn, N '43:48-50+ — w aircraft

GILBERT, Walter E. (Canadian McKee Trophy winner)
Pop Fly, My '35:74-75 — w por

GILMORE, Lyman, Jr. (California aviation pioneer)
AAHSJ, Sum '67:125-129
Pop Avn, Ap '36:247-248+ — w por

GLENN, John H. (US pilot and astronaut)
Bee-Hive, Fall '57:19-22 — Vought F8U speed record

GODDARD, Robert H. (Rocket pioneer)
Airpwr Hist, O '60:216-221 — w bibliography
Bee-Hive, Spg '58:2-7
Pop Avn, Mr '37:31-32+ — Experiments
RAF Fly Rev, O '58:35-36+

GODFREY, Arthur (US radio/TV personality, aero enthusiast)
Flying, O '58:40-41+
Skyways, My '48:18-21+

GODFREY, John T. (US ace, WW II)
RAF Fly Rev, Ag '58:22+ — Teamwork w Don Gentile

GOETTLER, Harold E. (US WW I pilot, Medal of Honor)
Avn Wk, D 7, '59:96-97

GOLDWATER, Barry M. (US Senator and air enthusiast)
Air Prog, Je '67:10-14+
AAHSJ, Fall '68:157-168 WW II experiences
Flying, S '67:56-61 Flying technique and experience

GONTERMANN, Heinrich (German ace, WW I)
Pop Fly, N '36:414-415 Exploits, w por

GÖRING, Hermann (German Air Minister, WW II)
Interavia, No. 4, '46:12-18 (Pt 1) w pors
No. 5, '46:12-18 (Pt 2) WW I, his 8th victim, w por
Pop Fly, Jl '33:190 Most thrilling combat, w por
S '34:283-285
Je '38:128-129+ WW I experience and rise
Skyways, Ja '44:28-29+ As a "muddler"

GORRELL, Edgar S. (Air historian; President, Air Transport Assoc.)
Pop Avn, F '40:27-28+
US Air Svcs, Je '32:16-20 w por (fp)
O '32:24-30 In Mexican Punitive Expedition, 1916
Mr '36:1, 38 w por
Ap '45:38

GÖTTSCH, Walter (German ace, WW I)
Cr & Cockade, Aut '65:237-241 Last flight, w por

GRAF, Hermann (German ace, WW II)
Air Pict, S '68:348-349

GRAHAME-WHITE, Claude (British air pioneer)
Flight, Ap 20, '50:491-495 London-Manchester flight, 1910
RAF Fly Rev, My '55:5, 41-42+ Early exploits

GRANDY, John (Chief of Air Staff, RAF)
RAFs Qtr, Spg '67:70 w por

GRANVILLE, Edward H. (Gee-Bee designer)
Bee-Hive, Spg '57:12-15

GRAY, Robert Hampton (Canadian; last WW II Victoria Cross award)
RAF Fly Rev, O '62:30-31+ Victoria Cross exploit

GREEN, Murlis (British ace, WW I)
Pop Fly, Jl '37:208-209+

GREENAMYER, Darryl (US racing pilot)
Air Prog, Jl '67:26-27+

GREENE, Carl F. (US aeronautical engineer)
Pop Avn, F '40:37-38+

GROSS, Robert E. (President, Lockheed Aircraft)
Interavia, No. 1, '46:79-80 w por
US Air Svcs, F '39:1, 24+ w por
West Fly, O '40:26 w por

GUTHRIE, K. M. (Air Vice Marshal, RCAF)
Roundel, Jl/Ag '55:35-36 w por

GUYNEMER, Georges (French ace, WW I)
Avn Wk, Ag 31, '64:54-55 w drawing of aircraft
Cr & Cockade, Aut '66:262-263 Aircraft profile
:307-308 Aircraft drawing
Sum '67:168-172
Pop Avn, Je '33:351-352+ Exploits
Pop Fly, Mr '33:678-680+ "

HALE, Frank Lucien "Bud" (US ace, WW I)
Cr & Cockade, Win '62:328-332 w por

HALFORD, Frank B. (British engine designer)
Flight, F 6, '59:183-184+

HALL, Bert (US pilot, Lafayette Escadrille, WW I)
Avn Wk, O 9, '61:86-87 Aircraft incident
Cr & Cockade, Aut '63:274-277
Pop Avn, My '36:321-322+ w por

HALL, Donald A. (Ryan "Spirit of St. Louis" designer)
Nav Avn News, S '60:12-15

HALL, Frederick Vincent "Vin" (Royal Naval Air Service ace, WW I)
Cr & Cockade, Aut '65:228-236

HALL, James Norman (US author and ace, WW I)
Aerosp Hist, Spg '68:44-47, 53
Airpwr Hist, Ap '64:34-38 Last flight

HALLETT, George E. (US pioneer mechanic)
Flying, Ap '58:40-46+

HAMILTON, Charles K. (US pioneer airman)
AAHSJ, Fall '69:159-162 w por
Pop Avn, Ag '28:38-44
US Air Svcs, Ag '34:23-25

HAMMANN, Charles H. (US Navy, WW I, Medal of Honor)
Nav Avn News, Jl '68:16-17 Exploit

HAMMOND, A. W. (British gunner, WW I)
Avn Wk, Ja 7, '63:36-37 w aircraft drawing (in color)

HAND, Earl McNabb (Canadian ace, WW I)
CAHS, Fall '66:73-74

HANDLEY PAGE, Frederick--Sir (British pioneer and designer)
Flight, My 3, '62:715 On his death
RAF Fly Rev, Je '62:37-38

HANKS, Eugene R. (US Navy ace, WW II)
Flying, O '44:90 w por (color)

HANKS, Peter Prosser (RAF ace, WW II)
Aero Album, Spg '69:40-41 Exploits w Spitfire profile

HARGRAVE, Lawrence (Australian pioneer airman)
Pop Avn, Ja '32:78+
RAF Fly Rev, S '58:69-70+

HARMON, Hubert R. (Superintendent, US Air Force Academy)
US Air Svcs, Ag '54:22

HARRIS, Arthur "Bomber" (Air Chief Marshal, RAF)
RAFs Qtr, Aut '67:207-208 Leadership, w por

HARRIS, Harold L. (First military emergency bailout)
Avn Wk, S 11, '67:92-93 Color drawing of incident

HARTMANN, Erich (German "Ace of Aces," WW II)
Air Cl, D '68:20-36+ Exploits, w pors and aircraft

HARTNEY, Harold E. (US airpower advocate)
US Air Svcs, N '45:22-23 On his death, w por (fp)

HAWKER, Harry George (British pioneer aviator)
Aero Album, Sum '69:12-17 Early racing exploits
RAF Fly Rev, S '61:24-26

HAWKER, Lanoe G. (British pilot, WW I, Victoria Cross)
Avn Wk, My 27, '63:58-59 w color drawing of incident
Flying, Ag '60:73 Drawing of battle w Richthofen
Pop Fly, F '35:575 w por
Ja '36:550-552+ Exploits

HAWKS, Frank Monroe (US racing pilot)
Air Cl, D '69:44-53+ Exploits w aircraft photos
Bee-Hive, Fall '64:33-36 Non-stop N. Y. -Los Angeles & return
Pop Avn, Je '31:23+
O '38:56 On his death, w por
US Air Svcs, S '30:41-42 Exploits
Je '31:19 1931 records w por
Ja '33:21-22
Mr '35:11-12

HEALY, James A. (US ace, WW I)
Pop Avn, Ap '34:217-218+ w por

HEATH, Ed B. (US pioneer designer)
Flying, Jl '45:56-57+
D '53:28-29+

HEDLEY, J. H. (British flyer, WW I; Richthofen's 72nd victim)
Pop Avn, O '30:12-14 Duel w Richthofen

HEFTY, Frigyes (First parachute descent from burning arcrft)
RAF Fly Rev, Ja '60:34-35 Exploit

HEGGENBERGER, Albert Francis (General, USAF)
US Air Svcs, Ap '35:12-15 w por (fp)
Je '37:18-21 First Hawaiian flight

HEINKEL, Ernst (German designer)
Flight, F 14, '58:204 On his death

HENN, Peter (German ace, WW II)
RAF Fly Rev, F '55:28-31 Exploit leading to capture, w por

HEPPES VON BELENYES, Aladar de (Hungarian ace, WW II)
Air Prog, Je '67:26-29+ w Me-109G profile

HICKAM, Horace M. (US officer; Hawaiian airfield named in his honor)
US Air Svcs, F '34:15 Por only (fp)

HIDDESSEN, Ferdinand von (German pilot, WW I; dropped message over Paris)
Avn Wk, Ap 23, '62:36-37 Drawing of incident (in color)

HILLER, Stanley (US helicopter designer and manufacturer)
Am Helicop, Jl '47:15+ w por
Interavia, No. 3, '51:149-151
Skyways, Ag '45:46-47+ w por

HODGKINSON, Colin "Hoppy" (Legless British military pilot, WW II)
RAF Fly Rev, O '54:26-28 Exploits, w por (fp)
My '57:23-25

HOFER, Ralph K. "Kid" (Canadian ace, WW II)
AAHSJ, Sum '64:116-128 Exploits, w pors

HOIDGE, Reginald T. C. (RAF ace, WW I)
Cr & Cockade, Win '66:381-388 w por

HOLMAN, Charles W. "Speed" (US racing pilot)
Pop Avn, Jl '31:21-22 On his death, w por

HOPKINS, John J. (Founder, General Dynamics)
Interavia, No. 6, '57:584-585 On his death, w por

HORNELL, David E. (Canadian pilot, WW II, Victoria Cross)
Roundel, Mr '61:27+ Exploit

HOWARD, Beverly "Bevo" (US acrobatic pilot)
Pegasus, Ap '57:7-9

HOWELL, Cedric E. (Australian ace, WW I)
Cr & Cockade, Win '61:324-329

HOXSEY, Arch (US air pioneer)
Aerosp Hist, Win '67:211-217
Pop Avn, O '28:78-82+

HUGHES, Howard (US manufacturer and airman)
Interavia, No. 7, '47:50-53
US Air Svcs, Ap '37:1, 24-26+ w por
Ag '38:15-16+, 34+ Around-the-world flight, 1938

HURT, Robert M. (Commander, US Navy; pilot for three Chiefs of Naval Operations)
Nav Avn News, Ag '66:10 w por

HYDE, Russell N. (Commander, US Navy and Aviator No. 181)
Nav Avn News, Jl '53:33

IMMELMANN, Max (German ace, WW I)
Avn Wk, N 28, '60:100-101 w aircraft
Flying, My '59:65 " "
Pop Avn, Jl '30:12-14+ Account of his death
Pop Fly, Mr '35:618-622+ Who killed him? w por
RAFs Qtr, Win '68:309-314 Exploits, w por

INCE, Arthur S. (First Canadian aerial victor, WW I)
CAHS, Ja '63:4-5

INGALLS, David S. (First US Naval ace, WW I)
Nav Avn News, S '68:22-23 w pors

INGALLS, Laura (US aviatrix)
US Air Svcs, Je '34:1, 32-33 w por

INSALL, Gilbert S. M. (British pilot, WW I, Victoria Cross)
Pop Fly, Jl '33:183 VC exploit, w por (fp)

JABARA, James (US ace, WW II and Korea)
Air Prog, F '67:23+ On his death, w por in cockpit
US Air Svcs, S '51:20 On becoming first jet ace
Ap '52:13

JABS, Hans-Joachim (German ace, WW II)
Air Prog, O '66:34-37+ w aircraft profile

JACKSON, Joe M. (Lt.Col. pilot, USAF, Medal of Honor, Vietnam)
Air Force, Mr '69:98-101 Exploit, w por

JACOBS, Joseph Peter (German ace, WW I)
Cr & Cockade, Win '65:307-334 w pors

JAMIESON, George E. "Jamie" (New Zealand ace, WW II)
RAF Fly Rev, Ag '57:31-32+

JANE, John Frederick T. "Fred" (Founder, "Jane's All the World's Aircraft")
RAF Fly Rev, Mr '60:45

JANNUS, Rodger W. (US air pioneer)
AAHSJ, Spg '69:57-58 — w por

JEPPESON, E. B. (Aero manual publisher)
Flying, O '58:52-53+

JOHNSON, James E. "Johnnie" (RAF ace, WW II)
RAF Fly Rev, D '54:32-34+ — Exploits, w por (fp)
D '56:23-26 — Exploits
RAFs Qtr, 1965, No. 3:168 — w por

JOHNSON, Philip G. (President, Boeing Corp.)
Bee-Hive, S/O '44:10 — On his death, w por
Boeing Mag, O '44:1-4+ — " " " " "
US Air Svcs, Ja '30:35 — w por
Ag '33:34
O '44:10-11, 22-23+ — On his death, w por
Ja '53:10-15 — Anecdotes, w por
West Fly, S '40:21 — w por

JOHNSTON, Alvin Melvin "Tex" (Boeing test pilot)
Boeign Mag, Ag '52:7-9+
US Air Svcs, My '52:18

JONES, Charles Sherman "Casey" (Founder, US aeronautical schools)
US Air Svcs, Ja '30:49 — w por
F '32:30
My '50:24

JOYCE, Temple N. (Vice-President, Berliner-Joyce)
Cr & Cockade, Sum '62:95-112 — WW I experiences, w por
US Air Svcs, N '32:1, 37 — w por

KAIN, Edgar James "Cobber" (First WW II Allied ace, RAF)
RAFs Qtr, Win '69:287-291 — w por

KAMAN, Charles H. (US helicopter designer)
Am Helicop, N '47:22+ — w por

KAMMHUBER, Josef (German Commander-in-Chief, Luftflotte 5, WW II)
RAF Fly Rev, O '59:15+

KARMAN, Theodore von (Founder, USAF Scientific Advisory Board)
Air Force, O '57:41-48
West Aerosp, Je '63:Cover, 10-11 — On his death, w por (color)

KASLER, James H. (US ace, Korea)
US Air Svcs, S '52:18

KATAJAINEN, Nils E. (Finnish ace, WW II)
RAF Fly Rev, Ap '59:19-21

KAY, C. E. (Air Commodore, RNZAF)
RAF Qtr, Ap '51:157-158 — w por

KAZAKOV, Alexander A. (Russian ace, WW I)
Avn Wk, Jl 11, '60:90-91 — w aircraft and por
RAF Fly Rev, Je '58:17-19 — Exploit, w por

KELLETT, W. Wallace (US designer)
US Air Svcs, S '40:1, 24-26+ — w por

KELLY, Colin P., Jr. (First US air hero, WW II)
US Air Svcs, Ja '42:13-14 — w por
Je '43:12 — Por only

KELLY, Dennis P. W. "John" (British Naval aerobatic pilot)
RAF Fly Rev, Mr '58:21-23+ — w por

KELLY, Fred W. (US pilot, WW I, and barnstormer)
AAHSJ, Fall '68:193-203 (Pt 1)
Win '68:258-268 (Pt 2)

KENNEY, George C. (General, USAF)
US Air Svcs, O '42:38
Je '47:11 — Por only (fp)

KEPFORD, Ira (US Navy ace, WW II)
Air Prog, Ja '64:34-35+
Nav Avn News, F '69:27 — Aircraft in flight

KINCHELOE, Iven C., Jr. (US ace, Korea, and test pilot)
Aerosp Hist, Win '69:7-8
US Air Svcs, S '52:13+

KINDELBERGER, James H. "Dutch" (Designer, North American Aviation)
Interavia, No. 12, '50:634-636
Skyline, Vol. 5, No. 1, '44:6-7 — w pors
US Air Svcs, Ag '39:1, 33-34+ — w por
West Fly, Mr '40:26-27 — w por (fp)

KINDLEY, Field E. (US ace, WW I)
Aerosp Hist, Win '69:25-27 — Exploits
AAHSJ, Win '59:233-252 — w pors

KING, Bernadine Lewis (US stunt pilot and skywriter)
Pop Avn, Mr '39:57-58+

KING, Ernest J. (Chief, BurAir, US Navy)
US Air Svcs, My '33:27 — w por
N '33:1, 28 — w por
Jl '42:12 — Por only (fp)

KINGSFORD SMITH, Charles--Sir (Australian long-distance pilot)
Flight, N 21, '35:525 — On his death

KIRSCHSTEIN, Hans (German pilot, WW I)
Avn Wk, Ja 7, '63:36-37 — Incident w aircraft (in color)

KLEIN, Martin (German pilot, WW I)
Cr & Cockade, Win '65:391-395 — His death by McCudden

KNIGHT, Clayton (RAF pilot, WW I)
Cr & Cockade, Spg '67:29-42 — w pors

KNOKE, Heinz (German ace, WW II)
Roundel, Mr '54:3-9 — w por

KNOWLES, James (US ace, WW I)
Cr & Cockade, Win '69:355-365 — Recollections of 95th Aero Sq

KOJEDUB, Ivan N. (Russian "Ace of Aces," WW II)
RAF Fly Rev, Ja '58:23-24

KOLLSMAN, Paul (Aircraft instrument designer)
Pop Avn, Ja '40:26-28+

KRUTEN, Evgraph N. (Russian ace, WW I)
Cr & Cockade, Win '69:319-320 — w por

KUROE, Y. (Japanese ace, WW II)
RAF Fly Rev, S '59:39

KUTTELWASCHER, Karel (Czech ace flying w RAF, WW II)
RAF Fly Rev, N '59:31-32

LACEY, James Harry "Ginger" (Top "Battle of Britain" ace)
RAF Fly Rev, Je '62:25-26+

LA GUARDIA, Fiorello H. (Pilot, WW I and Mayor, New York City)
US Air Svcs, D '37:1, 19-20 — w por

LAHM, Frank Purdy (US pioneer military aviator)
Air Force, Ag '63:74-75
AF News Ltr, O '41:19+ — w por
Flying, Ag '57:41-51
US Air Svcs, S '31:21-24
O '41:9
Ja '42:10, 17 — Retirement, w por (fp)

LAIRD, Emil Matthew "Matty" (US aircraft designer)
Air Prog, My '67:26-29+ (Pt 1) — Many photos of aircraft designed
Je '67:45-47+ (Pt 2) — " " " " "
Jl '67:37+ (Pt 3) — " " " " "
Ag '67:52-54 (Pt 4) — " " " " "
S '67:44-45 (Pt 5) — " " " " "

LANSDOWNE, Zachary (US Navy lighter-than-air advocate)
Avn Wk, D 18, '67:36-37 — "Shenandoah" tragedy, drawing in color

LARICHELIERE, Joseph E. P. (Canadian ace, WW II)
CAHS, Sum '63:36

LAROCHE, Raymonde de--Baroness (French pioneer aviatrix)
Air Prog, F/Mr '64:35-37+
Flying, Mr '57:34+

LASSEN, Clyde E. (US Navy, Medal of Honor)
Nav Avn News, Ap '69:19 — Exploit

LATHAM, Hubert (British air pioneer)
Flight, Mr 30, '51:365-366

LATSHAW, Robert T. (US ace, Korea)
US Air Svcs, S '52:18

LAW (Oliver), Ruth (US woman daredevil)
Bee-Hive, Sum '58:25-27

LAWRENCE of Arabia (British author, soldier-of-fortune)
RAF Fly Rev, Ja '58:30-32

LE GLOAN, Pierre (French ace, WW II)
Air Prog, Je '68:48-49+ w aircraft profile

LEAR, William P. (US aircraft manufacturer)
Air Prog, Ja '66:31

LECKIE, Robert (Air Marshal, RCAF and "Zeppelin killer," WW I)
Roundel, S '58:28-29 WW I exploit

LEDUC, Rene (French designer of ramjets)
Interavia, No. 7, '53:401

LEIGH-MALLORY, Trafford--Sir (Air Chief Marshal, RAF)
RAFs Qtr, Aut '67:206-207 Leadership, w por

LEMAY, Curtis E. (US Commanding General, SAC)
US Air Svcs, S '44:15 Becomes Chief, 20th Bomber Command, w por

LE PAGE, Wynn L. (US helicopter designer)
Am Helicop, O '47:19+ w por

LESSUPS, Jacques de--Count (French pioneer airman)
Can Avn, N '44:64-66+

LEWIS, David G. (Richthofen's last victim)
Flying, Je '68:53-55 Account of air battle w Richthofen
Pop Fly, My '34:74-75 " " " " " "

LIDDELL, John Aidan (Holder of Victoria Cross)
Pop Fly, Mr '35:623

LILIENTHAL, Otto (German aero pioneer)
Air Prog, '54/'55 Annual:14-15
Flying, Jl '58:26-27+

LINDBERGH, Charles A. (Most famous US airman)
Bee-Hive, Spg '67:15-21 N.Y.-Paris flight preparations
Flying, O '62:34-35+
Pop Avn, O '39:10-11+ On return from Germany, 1939
US Air Svcs, My '37:12-13 w por
My '39:16-17 On return from Germany, w por
N '45:44 Role in WW II
My '47:24-27+ Reprint of Paris flight
N '49:11-12 Role in WW II
West Fly, My '30:50-53+ As glider pilot
My '37:18-19 As "Jenny" instructor
:22-23+ His flights

LINK, Edwin A. (US designer, ground trainers)
Nav Avn News, D '68:6-11 w por
Pop Avn, Ag '38:46-48+
US Air Svcs, Je '49:19+ w por

LINKE-CRAWFORD, Frank (Austro-Hungarian ace, WW I)
Cr & Cockade, Win '60:21-27
Win '61:350-354 His death

LITTLE, Robert Alexander (British Royal Navy ace, WW I)
Pop Fly, Ap '37:45 w por

LLOYD, Hugh P.--Sir (Air Marshal, RAF)
RAF Qtr, Ap '51:124-127 w por

LOCKHEED, Allan H. (US designer and manufacturer)
Air Pilot, Je '44:11-13 w aircraft designs
AAHSJ, Spg '69:59-60 w por

LOENING, Grover (US designer and manufacturer)
Air News, Ap '42:14-17+
Air Prog, D '68:54-55+ Interview
Pop Avn, Jl '38:21+

LOHNER, George (Canadian pioneer aeronaut, Austrian-born)
CAHS, Spg '69:17-20

LORD, David S. A. (Awarded Victoria Cross)
RAF Fly Rev, N '54:19-20 VC exploit
RAF Qtr, O '50:309-314 w por

LOVE, Robert J. (US ace, Korea)
US Air Svcs, S '52:16

LOW, James F. (US ace, Korea)
US Air Svcs, S '52:20

LÖWENHARDT, Erich (German ace, WW I)
Cr & Cockade, Win '60:56-58 His death in air accident
Pop Fly, N '37:450-451+ w por
Mr '38:680-684+ Account by his 7th victim

LUFBERY, Raoul (French-American ace, WW I)
Air Force, Ja '57:94-101
Airpwr Hist, Ja '61:13-22
Skyline, Vol. 10, No. 1, '52:2 w por
US Air Svcs, Je '38:17-20 w por (fp)

LUKE, Frank (US "Balloon Buster," WW I, Medal of Honor)
Air Force, Mr '55:84-90
Air Prog, F '68:53-54+ Exploits
Avn Wk, Ag 3, '59:131+ w por and aircraft
Pop Avn, N '32:280-282+ Exploits
Pop Fly, O '38:348-351+ w por and aircraft

LUNARDI, Vincent (First balloon ascent in England, 1784)
Pop Fly, D '34:476

LUUKKANEN, Eino (Finnish ace, WW II)
Fly Rev Intl, O '63:39-40 Exploit

LYM, Arthur (Chief of Staff, Cantonese Air Force)
US Air Svcs, F '34:23+ w por

MCCAIN, John S. (Chief, Bur. of Aeronautics, US Navy, WW II)
Flying, F '43:51 Por (in color)
Nav Avn News, O 15, '45:11 On his death

MCCONNELL, John P. (Sixth USAF Chief-of-Staff)
Air Force, S '69:56-57 Retirement ceremony

MCCRUM, R. T. "Bob" (Engineer, Bird-Wing Co.)
Hist Avn, S '69:4-6+ (Pt 1) w por
O '69:4-7 (Pt 2) w por

MCCUBBIN, George R. (Credited with shooting down Immelmann, WW I)
Avn Wk, N 28, '60:100-101 w aircraft

MCCUDDEN, James T. B. (British ace, WW I, Victoria Cross)
Cr & Cockade, Win '61:356-357 Victory log
Win '65:391-395 His 53rd & 54th victories, w por
Pop Fly, My '33:67-70, 84-85 w drawing of his flight
Ja '35:529 w por
Mr '37:620-622+ In 56 Sq
S '37:333-334+ Views on air tactics
RAF Fly Rev, Je '62:33-36 Exploits

MCCURDY, J. A. D. (Canadian air pioneer)
Bee-Hive, Sum '53:27-31
Flying, Ag '54:19+

MACDILL, Leslie (US military pilot; Florida airbase named in his honor)
US Air Svcs, D '38:8 On his death

MCDONNELL, James S., Jr. (Founder, McDonnell Aircraft)
Am Helicop, Ap '47:27+ w por
Interavia, No. 9, '51:496 w por
US Air Svcs, Jl '55:7-8 w por

MCDONNELL, Sanford N. "Sandy" (President, McDonnell Aircraft)
Fly Rev Intl, Je '69:17

MCELROY, George E. H. (British ace, WW I)
Pop Fly, Ap '35:24-27+ 40th Sq exploits, w por
Je '36:132-133+

MCGUIRE, Thomas B., Jr. (US ace, WW II, Medal of Honor)
US Air Svcs, Ap '46:31 Medal of Honor award

MCKEE, James Dalzell (Founder, Canada's McKee Trophy)
CAHS, Win '64:107-109 First Trans-Canada seaplane flight
Flying, F '59:45+
Roundel, Ja/F '62:24-26

MCKEEVER, Andrew E. (Canadian ace, WW I)
Avn Wk, Ja 25, '60:84-85 w aircraft and por
CAHS, Win '66:92-94+ " " " "
Cr & Cockade, Sum '66:155-162 Exploits, w pors
Flying, Mr '61:53 Exploit, w drawing of aircraft
Pop Avn, S '37:29-30+ Exploits
Pop Fly, Ja '38:580-581 Exploits, w por
RAF Fly Rev, Mr '58:27-28+ Exploit, w por

MCKELLAR, Archibald A. (Scottish RAF ace, WW II)
RAF Fly Rev, My '60:19-20 w por

MCKENNON, Pierce (US ace, WW II)
AAHSJ, Win '66:244-252

MACLACHLAN, J. A. F. (RAF one-armed ace, WW II)
Skyways, Jl '43:60-61 How he lost arm, w por

MCLEOD, Alan (Canadian pilot, WW I, Victoria Cross)
Avn Wk, Ja 7, '63:36-37 w drawing of aircraft (color)
Roundel, Mr '58:24-25 VC incident, w por

MCNAMARA, Frank H. (Australian pilot, Victoria Cross)
Pop Fly, Jl '35:204 VC incident, w por
RAF Fly Rev, F '62:34-35 VC incident

MACREADY, John (Famous US military pilot)
Air Force, Ag '56:290-299

MCSWEYN, Allan F. (Australian pilot, WW II; POW escapee)
RAF Fly Rev, Je '55:34-36+ (Pt 1) POW escapes
Jl '55:24-26+ (Pt 2) " "
Ag '55:28-30+ (Pt 3) " "

MACH, Ernst (German physicist)
Bee-Hive, Ja '51:8 w por
Flying, Ja '52:15+ w por
Roundel, D '51:45-46 w por

MAHLAPUU, Jaan (Estonian ace, WW I)
Cr & Cockade, Win '69:314-319 Exploits, w por

MAHURIN, Walker M. "Bud" (US ace, WW II)
Air Prog, Mr '68:55-57+ w P-47 profile

MALAN, Adolph G. "Sailor" (So. African RAF ace, WW II)
Fly Rev Intl, D '63:47 w por

MALCOLM, Hugh G. (Wing Commander, RAF, Victoria Cross)
RAF Qtr, Je '43:xi — Citation for VC
Ap '50:122-126 — w por

MALLORY, William L. "Slim" (Founder, Mexican Aviation Co.)
Pegasus, F '51:13-15

MANNOCK, Edward "Mick" (British ace, WW I, Victoria Cross)
Avn Wk, Mr 16, '59:123+ — w aircraft and por
Cr & Cockade, Sum '64:179-182 — Victories
Flying, Jl '59:19 — w aircraft
Pop Fly, Jl '32:210-211 — w por
Jl '34:178 — w por
Ja '35:526-528
Jl '35:205-207+
Mr '36:662-665+ — Exploits
Jl '38:188-189 — Last flight
RAF Fly Rev, D '58:38-40+
RAF Qtr, Ja '32:79-87 — Exploits, w por

MANTZ, Paul (US stunt flyer, aircraft collector)
Flying, Jl '62:42-43+
Skyways, Je '48:14-16+

MARCHETTI, Alessandro (Italian designer)
Pop Avn, Ap '38:44-46+ — w por

MARMIER, Lionel de (DeGaulle's pilot, WW II)
RAF Fly Rev, S '59:55-56+

MARRIOTT, Frederick (British designer of "Aerial Steam Carriage")
Pop Fly, Je '34:115-117+ — w por

MARSEILLE, Hans-Joachim (German ace, WW II)
Air Prog, Ja '69:46-47+ — Exploits and aircraft
RAF Fly Rev, Mr '61:17-18

MARSHALL, Winton W. (US ace, Korea)
US Air Svcs, Ap '52:16

MARTIN, Glenn L. (US designer and manufacturer)
Air News, Je '41:24-26 — w pors
Flying, Ja '50:21-26+
Pegasus, My '56:10-15 — Activities to 1916
Pop Avn, Jl '37:23-24+

MARTIN (cont.)
US Air Svcs, Je '30:18, 33-35 w por
D '48:11-12 Flight w mother, w por
D '55:7-8 On his death

MASSON, Didier (Lafayette Escadrille pilot, WW I)
Flying, D '53:19+ Exploits in Mexican Punitive Expedition

MATHY, Heinrich (German Zeppelin Commander, WW I)
Pop Fly, O '35:370-371+ w por

MATTERN, James "Jimmie" (US racing pilot)
Pop Avn, O '35:235+ w por
US Air Svcs, Jl '33:40

MAXIM, Hiram Stevens (Inventor of steam-powered aircraft)
Bee-Hive, Sum '58:7-11

MAXWELL, William Calvin (US military pilot; Alabama air-base named in his honor)
Aerosp Hist, Spg '69:6-9

MAY, Wes (US barnstormer; first air-to-air refueling)
Avn Wk, Je 12, '67:52-53 w drawing of aircraft (in color)

MAY, Wilfred Reid "Wop" (Canadian pilot, WW I)
CAHS, Sum '63:41-42 Against Richthofen
Pop Fly, My '35:74-75 w por

MELVILLE, S. A. (Air Chief of Staff, So. African Air Force)
Air Power, Win '56/'57:139-140 On retirement, w por

MENCKOFF, Karl (German ace, WW I)
Pop Fly, O '36:351-352 Escape from French prison camp, w por

MEOS, Edgar (French "Les Cicognes" pilot, WW I)
Cr & Cockade, Aut '66:260-265

MEYER, Cord (US pioneer airman)
AAHSJ, Sum '66:106-107 w por

MEYER, John C. (Vice Chief of Staff, USAF, WW II ace)
Air Force, Ag '69:44-45 w por
Air Prog, Je '69:28-29+ Exploits, w pors and aircraft

MIDDLETON, Rawdon H. (RAAF pilot, Victoria Cross)
RAF Qtr, Mr '43:xi — Citation for VC

MILLER, F. R. (Air Chief Marshal, RCAF)
Roundel, Jl/Ag '64:5 — w por

MILLS, Frank (US pioneer airman)
AAHSJ, Sum '66:107-108 — w por

MITCHELL, R. J. (British Spitfire designer)
Flight, Ja 29, '54:116-118

MITCHELL, William "Billy" (US airpower advocate)
Aeroplane, F 27, '42:228 — Prophecies
Air Cl, Mr/Ap '65:41-43+ — In WWI
Flying, Ag '42:18-20+ (Pt 1) — His court-martial
O '42:43+ (Pt 2) — " " "
Pop Avn, Je '36:407-408+ — w por
US Air Svcs, O '33:11 — Por only (fp)

MITSCHER, Marc A. (Admiral, US Navy)
Nav Avn News, Mr '47:17 — On his death, w por

MOFFETT, William A. (Admiral, US Navy)
Bee-Hive, My '33:2-4 — On his death, w por
US Air Svcs, My '33:9 — On his death

MOISANT, John Bevins (US pioneer pilot)
Air Force, Ag '60:118-120
AAHSJ, Spg '63:3-12
Pop Avn, Jl '28:40-44+
US Air Svcs, Mr '32:24-28

MOISANT, Matilde (US pioneer aviatrix)
Pop Avn, F '29:46-52+

MÖLDERS, Werner (German ace, WW II)
RAF Fly Rev, Ag '61:27-28

MOMYER, William Wallace (Commanding General, TAC)
Air Force, Jl '68:29-33 — As TAC Commander

MONTEE, James W. "Dad" (Designer and aero safety exponent)
Nav Avn News, Je '52:12

MONTGOMERY, John Joseph (California air pioneer)
Flying, S '46:52-53+
Skyways, Ag '46:40-41+
US Air Svcs, Ap '46:7-8 On motion picture about his life
West Fly, Ag '46:20-21+ " " " " " "

MOORE, Guy Borthwick (Richthofen's Canadian victim, number 77)
CAHS, Fall '63:90-93

MOORE, Robert H. (US ace, Korea)
US Air Svcs, S '52:13

MORATO, Garcia (Facist ace for Spain)
RAF Fly Rev, My '57:16-19 Exploits and death in 1939 w aircraft profile

MORRIS, Douglas--Sir (Air Marshal, RAF)
RAFs Qtr, 1965, No. 2:87-88 w por

MOSLEY, Zack (US cartoonist of "Smilin' Jack")
Pop Avn, F '39:16-18+
Skyways, Ag '48:32-34+

MOTTERSHEAD, Sgt. Thomas (British pilot, WW I, Victoria Cross)
Avn Wk, Mr 9, '64:46-47 w aircraft drawing (in color)
Pop Fly, Jl '33:188 Exploit, w por

MOYLE, Donald (US long-distance pilot)
Bee-Hive, Ja '66:28-32 "Clasina Madge" flight, 1931

MYER, Nicholas (Managing Director, Beagle Aircraft, Ltd.)
Fly Rev Intl, D '68:19 w por

NAVARRE, Jean (French ace, WW I)
Pop Avn, Ag '37:17-18+

NEUMANN, Gerhard (Engineer, General Electric)
Air Prog, S '67:36-37+

NICHOLS, Ruth (US aviatrix)
AAHSJ, Spg '66:34-37 Flying Lockheed Vegas
US Air Svcs, Jl '34:1, 36 w por

NICOLSON, James Brindley (British pilot, Victoria Cross)
RAF Fly Rev, Jl '58:39+

NOLTENIUS, Friedrich (German ace, WW I)
Cr & Cockade, Win '66:307-342, 409 War diary, w pors, photos, & acft profile

NORDHOFF, Charles (US author and WW I airman)
Aerosp Hist, Spg '68:45-53

NORSTAD, Lauris (US Commanding General, SHAPE)
Pegasus, Je '56:4-7 w por

NORTH, John D. (British designer)
RAF Fly Rev, F '63:37-38 w por

NORTH, Russell B. (Michigan air pioneer)
AAHSJ, Spg '64:24-25 w por

NORTHROP, John K. (US manufacturer and designer)
US Air Svcs, Ap '35:1, 36 w por
West Fly, Je '40:26+ w por

NUNGESSER, Charles (French ace, WW I)
Air Prog, Ja '65:52-57+ w Nieuport 24 profile
Cr & Cockade, Sum '66:181-184 Decorations, w pors
Flying, Jl '60:53 w aircraft drawing
Interavia, No. 5, '52:255-256 Atlantic crossing w Coli
West Fly, F '31:41-43 His character

NUR KHAN, M. (Air Marshal, Pakistan)
Fly Rev Intl, D '68:19 w por

OCKER, William C. (US pioneer airman)
US Air Svcs, O '42:9-10, 27-28 On his death

OFFENBERG, Jean "Pyker" (Belgian ace, WW II)
RAF Fly Rev, Ja '57:21-24 Exploits

O'HARE, Edward H. "Butch" (US Navy ace, WW II)
Skyways, O '44:22-23+ Exploits
US Air Svcs, My '42:9 Medal of Honor award

OHRSTROM, George L. (Shot down last enemy aircraft for US, WW I)
Avn Wk, N 7, '66:62-63 Drawing of incident (in color)

OLDS, Robert (US General; organized Air Corps Ferry Command)
Aerosp Hist, Sum '68:4-7
Pop Avn, O '38:50 — DFC awarded
US Air Svcs, F '38:1, 32 — w por
My '43:18 — On his death

OMLIE, Phoebe F. (First US woman transport pilot)
AAHSJ, Fall '68:186-188

ORLEBAR, A. H. (Wing Commander, RFC; established world speed record)
Pop Fly, O '33:352 — w por

ORLOV, Ivan A. (Russian ace, WW I)
Cr & Cockade, Win '69:321-323 — w por

ORTMANS, Victor M. M. (Belgian RAF ace, WW II)
Aero Album, Fall '69:42-43 — Exploits and profile of Hurricane

OVINGTON, Earle (First US airmail pilot)
Flying, Mr '61:38-39+ — Winner 1st US cross-country race, 1911

PAINTER, John F. "Jack" (Early acft newsreel cameraman)
Bee-Hive, Win '63:7-10

PATRICK, Mason M. (First Chief, US Army Air Corps)
AF News Ltr, F '42:3-4+ — On his death
US Air Svcs, Ag '38:11-14
F '42:7-8 — On his death

PATTLE, Marmaduke Thomas St. John "Pat" (So. African RAF ace, WW II)
RAF Fly Rev, Je '61:20-21+ — w por
RAFs Qtr, Sum '69:119-126 — Exploits, w por

PEEL, Robert W. (Pilot, RNAS, WW I)
Cr & Cockade, Win '68:351-365 — w pors and photos

PEGOUD, Adolphe (French pioneer and ace, WW I)
Pop Avn, Je '28:27-28+ — w por
RAF Qtr, Ja '39:86-88 — Flying ability

PIASECKI, F. N. (US helicopter designer and manufacturer)
Am Helicop, Mr '47:18+ — w por

PINEDO, Marchese de (Italian naval aviator)
Flight, S 7, '33:886 On his death

PLATZ, Reinhold (Fokker designer)
RAF Fly Rev, My '59:30-34

PLESMAN, Albert (Founder and President of KLM)
Interavia, No. 2, '54:104-105 On his death

PLESS, Stephen W. (US Marine Corps, Medal of Honor)
Nav Avn News, Ap '69:19 Exploit, w por

PLÜSCHOW, Gunther (German WW I pilot in China)
Avn Wk, Ja 25, '65:58-59 Color drawing of aircraft
Cr & Cockade, Sum '62:153-161 China exploits, w por

POKRYSHKIN, Alexander I. (Russian ace, WW II)
RAF Fly Rev, Ja '58:24-25
US Air Svcs, Ag '44:40

PONIATOWSKI, Jan (Polish ace, WW II)
Air Pict, Ap '66:151 Exploits

PORTAL, Charles F. A. (Air Chief Marshal, RAF, WW II)
Flying, S '42:111 Por (color)
RAF Qtr, S '40:408 w por
RAFs Qtr, Aut '67:202-203 Personality, w por

POST, Wiley (US long-distance aviator)
Airpwr Hist, Jl '64:66-70
Pop Avn, O '33:225
US Air Svcs, Ag '31:15-19 Around-the-world flight, por (fp)
Ag '33:20-23 Solo around-the-world flight, w por

POUYADE, Pierre (French Commander, Normandie Regiment in Russia, WW II)
RAF Fly Rev, N '57:21-22+

POWELL, L. F. (Canadian observer to McKeever)
Flying, Mr '61:53 Flight exploit WW I, drawing of aircraft

PREDDY, George E., Jr. (US ace, WW II)
Aero Album, Win '68:34-35 Exploits, w P-51D profile
AAHSJ, Win '69:242-245 w aircraft

PRIDE, Alfred M. (Chief, BurAir, US Navy)
US Air Svcs, Ag '47:16-17+ w por

PUTNAM, David (US ace, WW I)
US Air Svcs, Jl '39:12-14

RADFORD, Arthur W. (US Navy pilot; Chairman, Joint Chiefs of Staff)
US Air Svcs, S '53:4-6 w por (fp)

RAESCH, Josef (German pilot, WW I)
Cr & Cockade, Win '67:307-335 From diary, w photos & pors

RAICHE, Bessica (America's first aviatrix)
Pop Avn, Jl '31:44+

RASCHE, Thea (Germany's "Flying Fräulein")
AAHSJ, Win '66:280-282

READ, Albert C. (US Navy Commander; first aircraft to cross Atlantic)
Nav Avn News, Ja '66:6-7 w por

REES, Lionel W. B. (British pilot, WW I, Victoria Cross)
Avn Wk, Ag 5, '63:64-65 Drawing of VC incident (color)
Pop Fly, Ag '33:256-257 Exploit of 1915
N '35:429 Exploit, w por

REID, William (British Lancaster pilot, Victoria Cross)
RAF Qtr, Mr '44:xi VC citation

REITSCH, Hanna (German aviatrix)
Flying, Jl '50:22-23+
RAF Fly Rev, O '53:21-22 Last mission for Nazis

RENTSCHLER, Frederick B. (President, United Aircraft)
Bee-Hive, Sum '56:2-5 On his death

RHODES-MOORHOUSE, William B. (British 2nd Lt., Victoria Cross)
Pop Fly, Ap '35:23 w por

RHYS DAVIDS, Arthur P. F. (British ace, WW I)
Pop Avn, Ja '37:17-18+ w por
Pop Fly, My '33:69-70 Fight w Voss, w por
Jl '33:191-192+ Shooting down of Schaefer

RICHARDSON, Holden C. (US Navy test pilot)
Nav Avn News, O '60:6 — On his death

RICHTHOFEN, Manfred von (German "Ace of Aces," WW I)
Avn Wk, Ap 13, '59:171+ — w por and aircraft
CAHS, Sum '63:41-42 — Canadians in his last battle
Cr & Cockade, Sum '63:153-167 — Last flight. Who killed him?
Aut '64:257-263 — His honors, w photos
Spg '67:43-51 (Pt 1) — Who killed him?
Sum '67:189-194 (Pt 2) — " " "
Sum '68:170-187 (Pt 3) — " " " w por and bibliography
Aut '68:287-289
Sum '69:97-152 — Photo history, w pors
Win '69:289-313 — His death
Flight, Mr 14, '30:294-295 — w por
Flying, My '53:32-33+ — Mother's account of life
Ap '59:19 — Drawing of death by Brown
Ag '60:73 — Drawing of aircraft battling Hawker
Je '68:50 — w por (in color)
:53-55 — Battle w David Lewis, w por (fp)
Pop Fly, Ag '32:269-271 — w por
D '32:497-499+ — Last flight and burial
S '33:307-309 — Mother's account of death
Ja '34:534-535+ — Museum at his home
F '34:576-579+ — His "Circus"
My '34:74-75 — Victory over D. G. Lewis
Ja '35:527 — Drawing of last flight
My '35:71-72+
Ag '38:244-245 — Mother's diary of WW I
RAF Fly Rev, Ap '55:22-24+ — Last flight

RICKENBACKER, Edward Vernon "Eddie" (US ace, WW I, Medal of Honor)
Air Force, Jl '58:78-82
Airpwr Hist, Ap '60:117-124
Am Helicop, F '49:6-8 — w por
Avn Wk, Ja 15, '62:36-37 — w aircraft drawing (color)
Flying, F '61:77 — First victim, w aircraft drawing
Pop Avn, F '31:39-40+ — Exploits
Jl '32:37-38 — w por
Ap '36:257-258+ — w por
Pop Fly, Ja '33:582-583 — w por
Skyways, Ja '43:10-11+ — w por

RICKENBACKER (cont.)
US Air Svcs, Je '30:64
D '30:32-33 Medal of Honor award
Je '39:1, 8, 26-28+ w por
D '42:7-8 Pacific rescue

RIDDER, Ernst de (German pilot, WW I)
Cr & Cockade, Spg '65:57-70 w pors and photos

ROBERTS, Daniel T. (US ace, WW II)
Aerosp Hist, Aut '69:24-27+

ROBINS, Augustine Warner (Asst. Chief of Staff, US Air Corps)
US Air Svcs, F '35:27 w por

ROBINSON, W. Leefe (British pilot, WW I; destroyed Zeppelin S. L. 11, Victoria Cross)
Flight, Ag 25, '66:302-303
RAF Fly Rev, Ap '57:21-22 Exploit

ROCKWELL, Robert L. (Commanding Officer, Lafayette Escadrille)
Pop Avn, Ag '33:87-88+

RODGERS, Calbraith "Cal" (First to fly across the US)
Bee-Hive, Ja '61:25-29
Flying, N '49:26-27+
Hist Avn, O '69:11-26 "Vin Fiz" flight w pors & photos

ROGERS, William W. (Canadian fighter ace, WW I)
CAHS, Sum '67:41-43 w por

ROLLS, William T. E. (RAF ace, WW II)
Aero Album, Sum '69:38-40 Exploits

ROOSEVELT, Quentin (Pilot, WW I; son of US President)
Pop Avn, F '31:10-12+ His death in air combat

ROPER, Albert (Secretary-General, International Civil Aviation Organization, ICAO)
Interavia, No. 5, '50:237

ROYCE, Henry--Sir (Rolls Royce partner)
Flight, Ap 27, '33:396 On his death
O 5, '56:575-577

RÜDEL, Ulrich (German ace, WW II)
Air Pict, S '69:349 Exploits

RUTHERFORD, W. J. (Canadian ace, WW I)
Can Avn, My '32:14-15+

RYAN, John D. (Seventh USAF Chief of Staff)
Air Force, Ag '69:42-45

RYAN, T. Claude (US manufacturer and designer)
Pop Avn, My '39:10-15+
US Air Svcs, S '42:32-34+ w por
West Fly, Ag '41:25+ w por
S '42:47 w por

SAKAI, Saburo (Japanese ace, WW II)
RAF Fly Rev, D '59:19-21+ Exploits

SALMOND, John M.--Sir (Chief of Air Staff, RAF)
Flight, Ap 25, '68:599-600 Obit w por
Pop Fly, Jl '33:176 w por
RAFs Qtr, No. 3, '64:192-198 w por
Sum '68:114 Obit w por

SALMOND, William Geoffrey H.--Sir (Air Chief Marshall, RAF)
Flight, Jl 29, '32:704 RAF resignation
My 4, '33:418-419 Obit

SANDER, Hans (Focke-Wulf test pilot)
RAF Fly Rev, S '57:41-42+ Experiences

SANTOS-DUMONT, Alberto (Brazilian-born dirigible and aircraft designer)
Air Pilot, Ja '45:16-18 His flights
Flying, Je '59:38-40+
Hist Avn, Ag '69:11-26 w pors and aircraft

SAUNDBY, Robert H. M. S.--Sir (Air Marshal, RAF)
RAFs Qtr, Sum '67:152 w por

SAUNDERS, Hugh W. L.--Sir (Air Marshal, RAF)
RAF Qtr, Ja '50:4-7 w por (fp)

SAUNDERS, K. F. "Ken" (Canadian bush pilot and pioneer)
CAHS, Win '66:95

SAUVAGE, Roger (French ace, WW II)
Air Prog, Ja '66:26-27+

SCHÄFER, Emil (German ace, WW I)
Cr & Cockade, Win '61:344, 346, 348-349 Victory log, aircraft (in color)

SCHÄFER, Karl (German ace, WW I)
Pop Fly, Jl '33:191-192+ Who shot him down?

SCHILT, Christian F. (US Marine pilot, Navy Medal of Honor)
Avn Wk, O 28, '68:54-55 Exploit and color drawing

SCHRAM, Richard A. (US Navy "Flying Professor," barnstormer)
Nav Avn News, Ag '69:15 On his death, w por

SCOTT, Blanche S. (US pioneer aviatrix)
Bee-Hive, Sum '68:18-22 w pors

SEAMANS, Robert C., Jr. (Ninth Secretary, USAF)
Air Force, Mr '69:Cover, 40-42 Por (color)

SEIBEL, Charles M. (US helicopter designer and manufacturer)
Am Helicop, Ap '50:12-13+ w por

SEVERSKY, Alexander P. de (US designer and manufacturer)
Pop Avn, Ag '35:87-88+ w por
US Air Svcs, Ag '37:1, 18-20 w por

SEWALL, Sumner (US ace, WW I)
Cr & Cockade, Win '63:322-323

SHANK, Robert "Bob" (Early US airmail pilot)
Bee-Hive, Spg '60:24-27

SHANNON, Sam (Convair chief test pilot)
Flying, Ja '54:20-21+

SHOMO, William A. (US pilot, WW II, Medal of Honor)
Air Cl, D '68:6-11 Exploits, w aircraft profile

SIKORSKY, Igor Ivan (Russian and US designer and manufacturer)
Aeron (US), Mr '30:31-32 w por
Am Helicop, Mr '46:34-38 w photos of helicopters
O '46:13+ w por

SIKORSKY (cont.)
Skyways, Ag '52:20-21+
US Air Svcs, Ja '30:25-26+
N '52:5-6

SINCLAIR, Laurence F. (Air Commodore, RAF)
RAF Qtr, Ja '52:43-46

SLEMON, Charles Roy (Air Marshal, RCAF)
Roundel, Mr '53:24-25 w por
Jl/Ag '64:3
S '64:23 Retirement ceremony

SLESSOR, John--Sir (Air Chief Marshal, RAF)
RAF Qtr, Jl '49:48-51 w por

SMITH, Art (US inventor of skywriting)
Pop Avn, Ap '37:28+

SMITH, Ernie (First to fly California-Hawaii, in "City of Oakland")
Pop Avn, O '39:42-43+

SMITH, James H., Jr (US Asst. Sec. of the Navy for Air)
Nav Avn News, S '53:12

SMITH, Joan Merriam (US around-the-world solo aviatrix)
Flying, Ag '65:80-84 On her death

SMITH, John L. (US Marine Corps ace, WW II)
US Air Svcs, N '42:17 Exploits while top US ace, w por

SMITH-BARRY, Robert R. (RFC instructor)
RAF Fly Rev, Mr '63:33+

SMUTS, Jan Christiaan ("Father of the RAF")
Roundel, Ja '51:26-30 w por

SOUCEK, Apollo (Chief, BurAir, US Navy)
Bee-Hive, Jl '30:3-4 Altitude record
Nav Avn News, Je '57:25-27
US Air Svcs, Jl '53:5 On becoming Chief
Ag '55:5 On his death

SPAATZ, Carl A. "Tooey" (Chief of Staff, USAAF)
Flying, My '46:32-33+
Pop Avn, Je '37:17-18+ In WW I
US Air Svcs, F '46:6-7 w por (fp)

SPARROW, Isaac E. (Early British balloonist)
Pop Fly, My '35:82-83 Exploits

SPERRY, Elmer A. (Founder, Sperry Co.)
Pop Avn, O '32:230 w por

SPOTSWOOD, Denis--Sir (Air Chief Marshal, RAF)
Fly Rev Intl, F '69:19

SPRINGS, Elliott White (US ace, WW I and author)
Air Prog, N '67:49-51+ w S.E. 5 profile
US Air Svcs, Ap '50:26 Role in textile industry
Ag '51:16

STENSETH, Martinus (US naval ace, WW I)
Pop Avn, My '33:281-282+ Exploits, w por

STEWART, James (US film actor)
Pop Avn, My '40:53-54+

STINSON, Edward A. (US aircraft designer)
Air Prog, O/N '63:16-31
US Air Svcs, Mr '32:49 On his death

STOUT, William "Bill" (US aircraft designer)
Air Prog, Ap '68:54-57+ w por and aircraft photos

STRANGE, Louis A. (US pilot, WW I)
Avn Wk, O 17, '60:36-37 w aircraft

STREIB, Werner (German ace, WW II)
Air Prog, Ja '65:12-16+

SWAAB, Jacques M. (US ace, WW I)
Cr & Cockade, Aut '63:313 On his death

SWETT, James E. (US Marine ace, WW II, Medal of Honor)
Skyways, N '44:24-25+ Exploits

SYMONS, Harry Lutz (Canadian ace, WW I and author)
CAHS, Win '64:113

SZNYCER, Bernard (US designer and helicopter manufacturer)
Am Helicop, S '49:14-15+ w por

TALBOT, Ralph (US Marine pilot, WW I, Medal of Honor)
Cr & Cockade, Sum '67:162-167 w por

TANK, Kurt (Focke-Wulf designer)
RAF Fly Rev, Jl '60:26-29

TANSKI, Czeslaw ("Father of Polish Aviation")
Flight, Jl 27, '56:174

TAYLOR, C. Gilbert (US Taylorcraft designer and founder)
Flying, S '47:44+

TAYLOR, Charles Edwards (Mechanic for Wright Brothers)
US Air Svcs, Ag '40:1, 20-22 — w por

TAYLOR, H. Lumsdon (British pilot, WW I)
Cr & Cockade, Aut '65:237-241 — w por

TEDDAR, Arthur William--Lord (Air Chief Marshal, RAF)
Roundel, Ap '53:4-14 — w por
RAF Qtr, Ap '49:71-72 — w por
RAFs Qtr, Aut '67:193-195 — w por
:204-205 — Leadership
Skyways, Je '45:42-43+ — In WW II

THIEFFRY, Edmond (Belgian ace, WW I)
Pop Fly, N '32:450-451 — w por

THOMPSON, Stephen W. (First US pilot to shoot down enemy aircraft, WW I)
Cr & Cockade, Sum '68:103-113 — w pors, photos
US Air Svcs, D '36:11-14 — Exploit, w por

THYNG, Harrison R. (US ace, Korea)
US Air Svcs, S '52:18

TIMM, Otto (US designer)
West Fly, Ag '34:19-21

TINKER, Clarence L. (Commanding General, Air Force in Hawaii, WW II)
US Air Svcs, Jl '42:7 — On his death

TKACHEV, Vyatcheslav M. (Commander-in-Chief, Russian Air Force, 1917)
Cr & Cockade, Win '69:323-325 — w pors

TOWERS, John H. (US Navy Chief, BurAir, WW II, Navy Aviator #3)
Bee-Hive, Je '39:2-3 — w por
Flying, Ja '42:35 — w por (color)

TOWERS (cont.)
Nav Avn News, Ja '67:6-7 w por
US Air Svcs, Ag '35:1, 16-17 w por
O '42:8-9 On promotion to Admiral
Ap '48:25 On Bd. of Directors, PAA

TOWNSEND, Peter (RAF ace, WW II)
RAF Fly Rev, D '55:23-26

TRAVIS, Robert F. (Commanding Officer, Fairfield-Suisun AFB)
US Air Svcs, Ag '50:22 On his death

TRENCHARD, Hugh M. "Boom"--Lord (Chief of Air Staff, RAF, 1918-1929)
Air Force, Ag '68:34-37
Flight, F 24, '56:214-215 On his death
F 28, '58:292-293
Roundel, Ap '56:3 On his death, w por

TRENT, Leonard (British pilot, Victoria Cross)
RAF Fly Rev, My '55:20-22+ VC exploit

TROLLOPE, John (RAF ace, WW I; six in one day)
RAF Fly Rev, My '63:20-21+ Exploit

TRULUCK, John H., Jr. (US ace, WW II)
Aero Album, Win '69:40-41 Exploits, w P-47D profile

TRUSCOTT, Keith W. "Bluey" (Australian ace, WW II)
Air Prog, Mr '69:38-39+ Exploits, w por and aircraft

TUCK, R. R. Stanford (RAF ace, WW II)
Air Prog, N '66:27-29+ w por and Hurricane profile
RAF Fly Rev, Ja '55:26-28 Exploits, w por
F '57:23-25 " " "

TUPOLEV, Andrei N. (Russian aircraft designer)
Flying, D '56:24-27+

TURNER, Roscoe (US racing pilot)
Air Cl, Ag '69:8-19+ w pors and aircraft
AAHSJ, Win '68:300 w por
Pop Avn, F '34:81-82
Ag '39:10-12+ Racing exploits
US Air Svcs, O '34:1, 36 w por

TWINING, Nathan F. (Chief of Staff, USAF)
US Air Svcs, F '46:28 On becoming Commanding General, Air Technical Service Command
Je '53:6-7 On becoming Chief of Staff, w por

UDET, Ernst (German ace, WW I)
Avn Wk, Mr 13, '61:224-225 w por and aircraft
Cr & Cockade, Win '63:409 His SSW DIV aircraft
Pop Fly, Jl '33:187 w por
My '36:76-78 Exploits

VALENTIN, Leo (French parachutist)
RAF Fly Rev, Je '55:19-21+

VANDENBERG, Hoyt S. (First Chief of Staff, USAF)
Flying, Jl '48:27+
US Air Svcs, Ap '48:9-10 On becoming Chief of Staff, w por
Ap '54:6 On his death

VAUGHN, George A. (US ace, WW I)
Cr & Cockade, Sum '61:162-169

VAUGHN, Stanley (US air pioneer and designer)
Pop Avn, O '38:49-50+

VERDON-ROE, Alliott--Sir (British designer and manufacturer)
Pop Fly, Ag '35:244-248 (Pt 1) w por
S '35:314-317 (Pt 2)
O '35:366-369+ (Pt 3)

VERVILLE, Alfred V. (US designer)
Nav Avn News, Ja '62:6 w por

VIALLET, Marcel (French ace, WW I)
Pop Avn, Je '36:423-424+ Exploits

VILJOEN, B. G. (Air Chief of Staff, South African Air Force)
Air Power, Win '56/'57:140-141 w por

VON GROENHOFF, Hans H. (Aero photographer)
Flying, Ja '67:64-69 w pors

VOSS, Werner (German ace, WW I)
Flying, Je '60:73 Drawing of aircraft
Pop Fly, My '33:69-70 Account of last flight by Rhys Davids, w por

VOUGHT, Chance M. (US designer and manufacturer)
US Air Svcs, S '30:26-27 On his death, w por

VUIA, Traian (Rumanian pioneer airman)
Flight, Mr 30, '56:365-366

WACKETT, L. J. (Australian Wing Commander, WW II; General Manager, Commonwealth Aircraft)
Pegasus, Ag '43:4-5+

WADE, Horace M. (General, USAF)
Fly Rev Intl, N '68:19

WAGNER, Boyd "Buzz" (First US ace after Pearl Harbor)
Air Cl, Mr/Ap '65:55

WALD, Charles (US pioneer airman)
AAHSJ, Sum '66:105-106 w por

WALDHAUSEN, Hans (German pilot, WW I)
Cr & Cockade, Sum '67:103-116 w pors and bibliography
Aut '68:300 w por

WALSH, Raycroft (Vice-Chairman, United Aircraft)
Bee-Hive, Fall '52:28-30 w por

WARNEFORD, R. A. J. (British pilot, WW I, Victoria Cross)
Avn Wk, N 12, '62:36-37 w drawing of aircraft (color)
Flying, N '60:75 Drawing of aircraft
Pop Fly, F '35:570-572+ On his death, w por

WATERMAN, Waldo Dean (US designer)
AAHSJ, Win '66:231-243 w history of his aircraft

WEEMS, Philip Van Horn (US Naval Commander and navigator)
Pop Avn, Ag '38:31+

WENDEL, Fritz (Messerschmitt test pilot)
RAF Fly Rev, N '55:15-18+ Test experiences

WESCOTT, William H. (US ace, Korea)
US Air Svcs, S '52:16

WESTIN, Brian E. (US Navy Cross recipient)
Nav Avn News, Ja '68:27 Navy Cross exploit

WESTOVER, Oscar (Chief, USAAC)
Bee-Hive, Mr '32:11 As Asst. Chief of Air Corps
F '36:1, 3 As Chief of Air Corps, w por
US Air Svcs, Je '33:1, 40 w por
Ja '36:4, 18 w por (fp)
O '38:9 On his death

WHEADON, E. Melvin (US ace-in-one-day, WW II)
Skyways, Mr '44:24-25+

WHISNER, William T., Jr. (US ace, Korea)
US Air Svcs, Ap '52:16

WHITE, Thomas Dresser (Fourth Chief of Staff, USAF)
Air Force, Jl '61:105-108
US Air Svcs, O '34:27 w por

WHITEHEAD, Gustave (Early US pioneer airman)
Pop Avn, Ja '35:22-34+ w por
US Air Svcs, Ag '45:9 His flights termed myth

WHITEHOUSE, Arch (US author and pilot, WW I)
Cr & Cockade, Win '63:313-321 Exploits

WHITTLE, Frank--Sir (Designer of first jet engine)
Flying, Ja '49:29+
Roundel, S '54:20-21 w por

WILDMAN, Frank "Doc" (US pioneer aviator)
Pop Avn, Je '37:25-26+

WILLARD, Charles F. (Early US barnstormer)
AAHSJ, Sum '69:125-127 w por
Flying, F '56:30-31+
Pop Avn, F '32:95-96+ Exploits

WILLIAMS, Alford J. "Al" (US racing pilot)
Nav Avn News, Jl '55:16-18
Pop Avn, Ap '31:31+ w por
My '36:337-338
US Air Svcs, Jl '33:1, 38 w por
Ja '36:19-20

WILSON, Charles "Engine Charley" (Fifth US Secretary of Defense)
Pegasus, Ag '57:7-9

WILSON, Eugene Edward "Gene" (President, United Aircraft)
US Air Svcs, Je '40:1, 26+ w por
O '49:26

WILSON, Gil Robb (US air enthusiast, writer, editor)
Flying, D '66:54-57 On his death, w por

WINSLOW, Alan (First US pilot to down German aircraft in WW I)
Pop Avn, D '33:363-364+ Exploits, w por

WISEMAN, Fred J. (California air pioneer)
Flying, O '47:34+

WITMER, Charles C. (US pioneer airman)
AAHSJ, Spg '63:42-43 w por

WOODS, Robert J. (Bell "Aircuda" designer)
Pop Avn, Jl '39:53-54+ w por

WRIGHT, Orville (First heavier-than-air flight, 1903)
Interavia, Mr '48:129-130
Pop Avn, Ap '34:223-224+ Interview, w por
US Air Svcs, My '39:12-15+ Visit to Kitty Hawk
F '48:5-8 On his death, w por
Mr '48:7-9 Press coverage on his death

WRIGHT, Wilbur (US pioneer airman)
Flight, D 11, '53:796-797
US Air Svcs, N '34:22-23 His last flight
Mr '35:13-15 Memories
O '39:10-12+ At Hudson-Fulton Celebration, 1909
D '55:4-6 His fame, w por (fp)

WRIGHT BROTHERS
Flying, D '59:36-39+ Primarily early life
Hist Avn, D '68:4-15+
Interavia, No. 12, '53:692-693
Pegasus, Ap '55:6-11
Pop Avn, D '28:11-16+
D '36:17-18+ Bicycle shop "factory"
RAF Fly Rev, D '53:17-18 First flight
D '54:29-30+ How they learned to fly
US Air Svcs, Jl '32:20-24 w Wilbur Wright por in 1908
D '37:10-13+ First flight
My '38:13-15 Their contributions

WRIGHT BROTHERS (cont.)
US Air Svcs, D '43:9-30 — Entire issue on their contributions, w pors
Ag '50:13-17 — What they really invented
Ap '55:16-19 — Early flights

WÜSTHOFF, Karl (German ace, WW I)
Pop Fly, O '37:401-403 — w por

WYCKOFF, Clarence (US air pioneer)
Flying, Ag '52:14-15+

YEAGER, Charles E. "Chuck" (Bell X-1 test pilot)
RAF Fly Rev, S '55:57-58+

YOUNG, Gerald O. (US pilot, Medal of Honor)
Bee-Hive, Sum '68:30-32 — Exploit, w por

YOUNT, Barton Kyle (US military pilot)
US Air Svcs, D '38:1, 12 — w por
Jl '49:14 — On his death

ZELEZNY, Walter (Austrian pilot, WW I; sank submarine)
Avn Wk, F 18, '63:36-37 — Drawing of aircraft (color)
Cr & Cockade, Spg '63:23-25 — Sinking of "Foucault"

ZEPPELIN, Ferdinand von--Graf (German lighter-than-air designer)
Pop Avn, Ja '32:21-22+ — Early adventures

ZURAKOWSKI, Janusz "Zura" (British test pilot)
RAF Fly Rev, Ap '59:24-25+

Part Three

Biographical Addendum

Fighter Aces
(Five or more enemy aircraft destroyed in the air)

Air Pict, S '67:328-329	Top-scoring aces
AAHSJ, Spg '69:42-46	Men who became aces in one day

CANADIAN

CAHS, Spg '63:17-18	Obscurity of Canadian aces
Fall '64:61-66	WW I Naval aces w pors
Win '64:110-112	Against WW II German jets

FRENCH

Cr & Cockade, Aut '66:254-259	A list w photos

GERMAN

Air Prog, Jl '69:30-33+	Present occupations and past exploits
Cr & Cockade, Aut '62:216-228	List of 458 WW I aces w pors
Fly Rev Intl, D '65:242-246	Color pors and enemy numbers destroyed
RAF Fly Rev, S '60:23	Top-scoring WW II aces

ITALIAN

Air Pict, Ja '67:25	Top aces 1936-1945

RUSSIAN

Air Pict, Mr '68:101	Leading WW II aces
RAF Fly Rev, Ja '58:23-25	" " " "

UNITED STATES

Flying, Ag '44:26-27+	To 1944
Pop Avn, O '38:29-30+	To 1938
US Air Svcs, Ap '51:15-16	10th and 14th AF WW II list
My '51:18-19	Marine Corps " " "
Je '51:16-18	Navy " " "

UNITED STATES (cont.)

US Air Svcs,	Jl '51:26	15th AF WW II list
	Ag '51:26	12th " " " "
	S '51:23-24	5th " " " "
	O '51:25 (Pt 1)	
	N '51:22, 24 (Pt 2)	
	D '51:24 (Pt 3)	
	Ja '52:24 (Pt 4)	
	F '52:23 (Pt 5)	
	Je '52:24 (Pt 6)	8th AF WW II list
	F '53:22	9th and 13th " " " "

Part Four

AIR FORCES

Part Four

AIR FORCES

The interest and importance accorded the air forces of the world is evidenced by the numerous articles written about them. In fact, one must be highly selective in choosing those to be cited, since writings concerning the United States military air effort alone could fill a substantial reference work.

Part Four, "Air Forces," is divided into four major sections as follows:

1. HISTORY AND ORGANIZATION: Arranged alphabetically by country, the subjects, when there are two or more, are subdivided into air force components; e.g., Army Aviation, Naval Aviation, Air National Guard, Royal Flying Corps, Royal Navy, etc. Notes specify those articles which are limited in time or scope and the absence of a topical note indicates a general history.

2. EQUIPMENT: These are articles concerning types of aircraft used by the various countries, as distinguished from the history and organization of the air forces. When an article includes both subject areas, it is cited in HISTORY AND ORGANIZATION. Format is similar in both cases.

3. AIR BASES: The military periodicals of many nations often provide historical articles about their air bases. These are frequently well researched and worthy of inclusion. Unfortunately, because there are so many bases, only one article has usually been written about any given station. This, however, makes that article all the more valuable. The air bases are geographicall arranged and format is similar to the preceding sections

4. UNIT HISTORIES: This section is arranged first by those units having a number as part of their designatio Following the numbered units are those organizations

that are better known (or only known) by name. Examples of the latter group are aerobatic teams, schools, institutes, and famous named squadrons or flights.

Air Forces: History and Organization

ARGENTINA

AAHSJ, Spg '61:23-30
Avn Wk, My 16, '55:45-52
Flight, Mr 6, '69:385-386 — In 1969

AUSTRALIA

Royal Australian Air Force (RAAF)

Air Force, Ja '65:42-48
Air Pict, Ja '69:17-20 (Pt 1)
F '69:50-54 (Pt 2)
Mr '69:90-97 (Pt 3)
Ap '69:140-144 (Pt 4)
My '69:174-178 (Pt 5)
Je '69:212-216 (Pt 6)
Jl '69:253-258 (Pt 7)
Ag '69:294-300 (Pt 8)
S '69:336-341 (Pt 9)
O '69:378-384 (Pt 10)
N '69:418-423 (Pt 11)
D '69:460-466 (Pt 12)
Flight, Ap 9, '36:388-389 — Air defense, 1936
S 22, '38:254a-255
N 16, '39:387-390
Mr 21, '40:269
Ap 13, '51:426-427
Ag 24, '56:296-297 — In 1956
Flying, N '42:22-24+ — In WW II
RAF Fly Rev, Mr '61:22-26+
RAF Qtr, Ap '31:292-297
O '33:385-391 — 1932-1933
Jl '35:301-311
Ja '38:2-8 — In 1937
Je '44:128-151 — WW II operations
RAFs Qtr, Vol. 1, No. 3:215-222
Roundel, Ja/F '60:10-14

Royal Navy

Flight, D 19, '68:1045

AUSTRIA
Flight, F 1, '68:156-158
Fly Rev Intl, D '69:49-55
RAF Fly Rev, N '60:9+ — 1938-1960

AUSTRO-HUNGARY
Air Prog, Sum '61:75-81
Cr & Cockade, Win '62:281-295 — 1910-1918

BELGIUM
Cr & Cockade, Win '64:335-347, Back Cover — To 1918
Sum '66:103-111 — 1914-1918
RAF Fly Rev, Jl '57:40+ — In 1957
O '60:35-36+ — In Congo, 1914-1916

BRAZIL
US Air Svcs, Mr '45:31 — WW II fighter sq

CANADA

Army Aviation

US Army Avn, F '63:46-48

Aviation Corps

Roundel, O '59:17-21

Naval Aviation

CAHS, Fall '63:80-81 — Beginnings
Flight, Ap 6, '56:390-392 — In 1956
Nav Avn News, My '51:13-15 — To 1951

Royal Canadian Air Force (RCAF)

Air Power, Win '53:203-212 — In 1953
Can Avn, Ja '59:37-44
CAHS, Fall '64:84-95 — Numerical register
Spg '68:4-8 (Pt 1) — To 1944
Sum '68:42-45 (Pt 2) — 1944-1967
Fall '68:68-71 (Pt 3) — Squadron continuity
Win '68:99-103 (Pt 4) — " "
Comm Avn, D '41:27-33 — 1920-1941
Flight, Ap 25, '40:378e-h
Flying, Ap '42:42-44+ — 1939-1942
Nav Avn News, O '50:11-13 — In 1950
Pegasus, My '53:1-15 — Including civil aviation
Pop Avn, D '38:42-45+ — In 1938
Pop Fly, O '34:354-355

CANADA Royal Canadian Air Force (cont.)

Roundel,	O '53:20-28	Organization in 1953
	Ja '55:1-7	" " 1954
	Ja/F '59:4-31	On 35th Anniversary
	Jl/Ag '61:2-31	In 1961
RAF Qtr,	Jl '35:296-300	WW I
	S '46:198-199	In 1945

CEYLON

RAF Qtr, Jl '52:261

CHINA -- Before 1946

Cr & Cockade, Win '62:361-365	In WW I
Flight, Mr 4, '43:230-233	
Pop Avn, Mr '40:24-26+	
Pop Fly, Ap '38:11-13+	

CHINA, COMMUNIST

Air Prog,	Mr '67:12-17+	
Nav Avn News,	F '53:16-17	1937-1953
RAF Fly Rev,	Jl '55:17	In 1955
	O '58:13-15	In 1958
	Ja '59:11-15	In 1959

CHINA, NATIONALIST

Nav Avn News,	Je '53:20-21	1949-1953
RAF Fly Rev,	O '58:13-15	In 1958
	Ja '59:11-15	In 1959

COCHIN CHINA

RAF Fly Rev, N '57:13, 17	In WW II; aircraft insignia in color

CZECHOSLOVAKIA

Fly Rev Intl, Je '64:56-57	1939-1944
Nav Avn News, Jl '52:12-13	1923-1952
Pop Fly, My '38:76-77+	
RAF Fly Rev, N '54:24-25	In 1954

DENMARK

Avn Wk,	Mr 31, '69:55-59+	Operations in late 1960's
Fly Rev Intl,	S '68:62-67	w list of aircraft and profiles
Interavia,	No. 7, '63:1074-1075	
RAF Fly Rev,	Ja '62:26-29 (Pt 1)	
	F '62:37-38 (Pt 2)	

ETHIOPIA

Pop Fly, Je '36:122-123+	1928-1936
RAF Fly Rev, Ap '56:41-42	

FINLAND

Air Pict, Je '68:192-196 (Pt 1)	"Winter War"
Jl '68:231-234 (Pt 2)	1941-1945
Flight, D 14, '51:749-750	
Fly Rev Intl, D '68:48-53 (Pt 1)	
Ja '69:42-45+ (Pt 2)	
Interavia, No. 3, '68:349-350	50th Anniversary
Nav Avn News, N '52:22-23	1939-1952
RAF Fly Rev, Mr '62:20-21	In 1962
RAFs Qtr, No. 4, '65:279-284	1939-1965
Skyways, Ag '52:13+	

FRANCE

Air Force

Air Force, O '68:80-83	Operations in 1968
Air Pict, My '63:128-132 (Pt 1)	1939-1940
Je '63:169-172 (Pt 2)	" "
Jl '63:216-217 (Pt 3)	" "
Cr & Cockade, Aut '66:205-231	Inventory of WW I Escadrilles w insignia in color
Flight, Jl 24, '41:48a-c	Free French, WW II
Flying, Ap '44:21-23+	Defeat by Germany, WW II
Fly Rev Intl, My '65:17-22	In 1965
Nav Avn News, Ja '53:20-21	1940-1953
Pop Avn, O '39:12-15+	In 1939
Roundel, My '51:42-45	1940-1951
RAF Fly Rev, Jl '58:42	1940 fighter strength
S '58:31-32	" " "
Je '60:21-24	1936-1940
Mr '63:17-19+	In 1963
RAF Qtr, Ja '51:33-38	1940-1950
RAFs Qtr, Win '68:269-275	In 1967
Skyways, Ag '43:26-28+	Free French, WW II

Naval Aviation

Interavia, No. 10, '60:1236-1240	In 1960, primarily aircraft types
No. 2, '61:231-233	In 1961, primarily aircraft types

GERMANY

Air Force

Aeroplane, F 15, '39:202 — Reorganization
My 29, '42:606-609 — 1941-1942
Air Force, My '47:29-39+ — In WW II
AF News Ltr, Ag/S '42:20-21 — 1941-1942
Air Pict, Mr '68:82-88 — Against RFC, Mr/Ap 1918
Acft Eng, Ap '48:102-104 — Maintenance and organization, 1948
Airpwr Hist, Jl '58:139-151 — 1919-1932
Jl '60:165-177 — 1914-1916
Flight, O 13, '38:313-315
O 12, '39:300h-301 (Pt 1)
O 19, '39:324-325 (Pt 2)
Flying, Mr '44:50-53+ — In WW II
Mr '55:14-15+ — Organization, 1955
Ja '57:28-30+ — In 1957
Fly Rev Intl, My '66:533-536 — In 1966
Pegasus, D '57:1-9
Pop Fly, Ap '38:28-29 — Pors of leaders, 1938
RAF Fly Rev, Je '55:17-18+ — New Luftwaffe in 1955
D '62:15-17+ — In 1962
RAFs Qtr, Win '69:269-275 — In 1969

Naval Aviation

Cr & Cockade, Aut '62:185-195 — In WW I
Flight, Jl 22, '43:87-91

GHANA

Air Pict, N '64:362-363

GREAT BRITAIN

Air Forces Far East

Air Pict, Ap '66:122-125 — In 1966
Flight, My 29, '53:651-658, 667-669 — In 1953
My 8, '59:629-632 (Pt 1) — In 1959
My 22, '59:718-720 (Pt 2) — " "
Je 5, '59:772 (Pt 3) — " "

Air Forces Gulf

RAFs Qtr, Sum '67:87-92

Air Forces Middle East

Aeroplane, Je 6, '34:893-897 (Pt 1)
Je 20, '34:950-954 (Pt 2) — In Egypt

GREAT BRITAIN Air Forces Middle East (cont.)

Aeroplane, Je 27, '34:975-982 (Pt 3) In Iraq
Air Pict, My '65:144-148 In 1965
Flight, My 29, '53:651-658, 667-669 In 1953
F 26, '60:274-276 In 1960
Interavia, No. 1, '51:24-27 In 1951
RAF Qtr, D '46:26-46 Ja/My 1941
Mr '47:94-108 In Greece, 1941
RAFs Qtr, No. 3, '65:169-175 In 1965

Army Aviation

Interavia, No. 11, '67:1639-1641 In 1967
US Army Avn, Mr '62:13-17

Fleet Air Arm and Royal Navy

Air Pict, O '67:354-358 (Pt 1) WW II fighter operations
N '67:412-414 (Pt 2) " " " "
Bris Sid Jnl, Vol. 6, No. 1, '65:16-22 50th Anniversary
Can Avn, Ap '34:8+ (Pt 1) Beginnings to 1918
My '34:12-13+ (Pt 2) " " "
Flight, O 10, '40:290-296
Ap 20, '51:450-486 Types of aircraft used
Mr 22, '57:362-374 Organization and equipment, 1957
Flying, Je '44:28-31+
Interavia, No. 5, '66:705-707 In 1966
Nav Avn News, Ja 15, '44:14-25 In 1944
RAFs Qtr, Win '66:276-279

Royal Air Force (RAF)

Aeronautics, Ag '61:26-30 Evolution of uniform
Aeroplane, Mr 29, '39:392-399
Je 12, '68:18-20 Air Support Command history
Air Force, S '66:331-334 Strength (in detail), S 3, '39
Ag '68:30-33 On 50th Anniversary
Air Pict, O '62:322-326 (Pt 1) 1934-1939
N '62:338-342 (Pt 2) " "
D '62:381-383 (Pt 3) " "
Ap '68:115-124 On 50th Anniversary
Air Power, Spg '58:183-189 On 40th Anniversary
Flight, Je 27, '30:681-696
Je 26, '31:581-598 Many aircraft photos
Je 28, '34:622-637 Organization in 1934 w aircraft used, units in being and insignia shown
Je 27, '35:684-691 Methods of operation in 1935

GREAT BRITAIN Royal Air Force (cont.)

Flight, Je 24, '37:612-618 Organization and training in 1937
My 26, '38:506-516 In 1938
Ag 31, '39:198-200 RAF High Command at start, WW II
My 9, '40:422a-426
Ap 1, '43:343-348 1918-1943
Ja 31, '46:123-124 Coastal Command, WW II
F 7, '46:149-150 " " " "
Jl 6, '50:11-64
Je 28, '57:861-902 Work of each command, 1957
My 17, '62:776-797
Mr 28, '68:443-460 On 50th Anniversary
Flying, S '42:Entire issue
Jl '55:40-41+ New jet aircraft in 1955
Fly Rev Intl, Jl '64:58-59 Operations in 1934
My '68:239-240, 245-272 On 50th Anniversary
w aircraft lists, profiles, commands, etc.
Interavia, No. 5, '66:705-707
No. 11, '69:1803-1805 Reorganization in 1969
Nav Avn News, S 15, '43:14-23 In 1943
Pop Fly, Jl '38:174-179 In 1938
Roundel, S '59:17-23 On 41st Anniversary
RAF Fly Rev, Ap '58:19-20+ Inception, 1918
RAF Qtr, D '44:43-54 1944 statistical review
Mr '45:115-120 In 1944
Je '45:126-149 1939-1945
Mr '46:108-111 In 1945
RAFs Qtr, Fall '66:173-181 In Germany, 1966

Royal Flying Corps (RFC)

Air Pict, My '62:127-134
Mr '68:82-88 Against Germany, Mr/Ap 1918
Flight, Ap 13, '33:335-346 w aircraft photos

Royal Navy

See Fleet Air Arm and Royal Navy

Royal Observer Corps

Flying, Ap '53:26-27+

GREECE

Fly Rev Intl, Je '69:70-75+ (Pt 1) w aircraft profiles in color
Jl '69:46-47+ (Pt 2)

HAITI
Pop Avn, Ja '40:49+

HONDURAS
Air Pict, N '63:344-346
Pegasus, Ap '47:14-16

HONG KONG
Flight, Ap 25, '63:620a
S 11, '69:434-435 — Helicopter force

HUNGARY
Air Pict, My '64:144-148
Fly Rev Intl, S '69:50-57 (Pt 1)
O '69:56-63 (Pt 2)
RAF Fly Rev, My '57:15+

INDIA
Air Pict, Ja '66:12-16 — War w Pakistan, S 1965
Flight, S 17, '42:317-319
Ap 17, '53:483
My 29, '59:743-747 (Pt 1)
Je 5, '59:773-775 (Pt 2)
Ag 28, '59:89-90 (Pt 3)
Flying, Ja '45:26-28+
Fly Rev Intl, N '68:42-49 — List of all aircraft used, profiles, etc.
Roundel, Je '60:12-15
RAF Fly Rev, Mr '60:13-16 — In 1960
My '63:23-26 — In 1963

INDONESIA
Air Pict, Mr '67:72-76
RAF Fly Rev, My '59:13-16

IRAN
Interavia, No. 7, '50:361-364 — In 1950

IRAQ
Flying, D '43:47-48+ — In WW II

IRELAND
Flight, S 26, '63:544-546
Flying, Ag '50:23+
Mr '65:56-58

ISRAEL

Air Force, Ja '68:90-94
Air Pict, Ag '63:230-233
Air Prog, Je '68:18-23+ 1966-1968, w many photos
Fly Rev Intl, D '67:1015-1019+ Many aircraft photos
Interavia, No. 11, '67:1637-1638 Arab-Israeli conflict
RAF Fly Rev, Jl '55:21-23

ITALY

Air Pict, Ap '64:94-96 In Albania
Ag '68:276 Reorganization
S '69:332-335 Raids on Britain, WW II
Airpwr Hist, Jl '62:184-188 Pre-WW I
Flight, Ag 19, '43:193-196 In WW II
Fly Rev Intl, Ja '66:267-272+ In 1966
Interavia, No. 6, '50:290-294 In 1950
Pop Avn, Jl '40:10-13+ In WW II
RAF Fly Rev, S '62:19-21+ 1912-1962

JAPAN

Aeroplane, O 16, '42:451-454
O 23, '42:476-478
O 30, '42:512-513
N 6, '42:535-537
Aerosp Intl, Ja/F '69:48-50 Self Defense Force in 1969
Air Force, D '67:56-63 Self Defense Force
Air Prog, Jl '66:18-23+ Fighter regiments of WW II, w aircraft photos
S '66:62 Fighter regiments of WW II, addendum
Aviation, S '41:48-49+ To 1941
S '42:94-97+ In WW II
Flight, F 27, '41:172e-174
Ap 23, '42:392-397 (Pt 1)
Ap 30, '42:431-434 (Pt 2)
Flying, S '45:30-32+ In WW II
Ag '46:44-45+ Why it failed in WW II
Fly Rev Intl, My '54:13-15 Growth in 1954
D '66:245-254 Air defense 1952-1966
Pegasus, Jl '57:10-15 Self Defense Force to 1957
Pop Avn, O '36:19+
Pop Fly, Ag '38:234-235+
RAF Fly Rev, Ja '58:13-15 Self Defense Force to 1957
US Air Svcs, F '40:13-14+ Pre-WW II
West Fly, Je '44:46-47+ (Pt 1) In WW II
Jl '44:50-51+ (Pt 2) " " "

MALAYSIA

Fly Rev Intl, Je '66:599+	
Roundel, Jl/Ag '62:14	In 1960's
RAFs Qtr, Spg '66:37-42	

MANCHURIA

RAF Fly Rev, N '57:13-17	In WW II, w aircraft insignia in color

MEXICO

Air Pict, F '65:44-47	
West Fly, Je '44:52-53+	

NATO

See North Atlantic Treaty Organization

NETHERLANDS

Air Force

Air Pict, Jl '63:198-201	
My '68:152-157	
Flight, Ag 5, '48:158-159	
Pop Avn, Ap '38:28-30+	In 1938
RAF Fly Rev, My '57:31-33+	Against Germany, WW II
Ag '63:19-23	50th Anniversary
RAF Qtr, Ja '50:51-53+	After WW II
RAFs Qtr, Vol. 1, No. 2:98-105	In 1961
Skyways, N '44:26-27+	In WW II

Dutch East Indies Air Force

Air Pict, D '62:370-374	Operations against Japanese
Flying, D '41:18-20+	In 1941
RAF Fly Rev, N '58:6+	1940-1941
Skyways, N '43:16-17+	"Dutch Squadron," WW II

Naval Aviation

Air Pict, F '62:44-48	
Mr '62:75-78	
Fly Rev Intl, N '67:951-954	50th Anniversary
Interavia, No. 8, '56:628	

NEW ZEALAND

Air Pict, Ja '68:6-11	
:19-22	WW II, Europe
:23	Insignia
Air Power, O '56:55-56	In 1956
Aut '57:48-53	Territorial Air Force

NEW ZEALAND (cont.)
Flight, Ja 11, '40:30f-h
S 6, '45:256-257 In WW II
Mr 28, '58:416 21st Anniversary
Roundel, Ap '60:18-21
RAF Qtr, S '44:ix-xii, 181-190 In WW II
Jl '48:155-159 Pacific operations, 1942-1945
RAFs Qtr, Vol. 1, No. 4:325-328 1936-1962

NIGERIA
Fly Rev Intl, Mr '68:119-121

NORTH ATLANTIC TREATY ORGANIZATION (NATO)
Flying, Mr '52:11-13+
Interavia, No. 8, '52:419-439 In 1952
Roundel, Ja '54:20-22
RAF Fly Rev, D '61:22-27 In 1961

NORTH KOREA
Air Prog, My '68:38-40+ w list of aircraft types
S '68:48-51+ In Korean War, w list of aircraft types

NORWAY
Air Power, O '56:46-51 In 1956
Avn Wk, Mr 17, '69:40-45 In 1969
Flight, Mr 11, '43:255-256
Fly Rev Intl, Ap '69:50-55+ (Pt 1) w color profiles
My '69:54-56 (Pt 2)
Pegasus, N '54:1-9
RAF Fly Rev, Jl '62:17-20+ 50th Anniversary
RAF Qtr, Jl '50:238-243 1944-1950
Jl '52:224-229 1944-1952
RAFs Qtr, Sum '68:97-100 1950's and 1960's

OMAN
Air Pict, Ja '67:24

PAKISTAN
Avn Wk, D 2, '68:42-52 In 1968
Flight, F 21, '58:246-248 (Pt 1)
F 28, '58:282-284 (Pt 2)
Ja 16, '69:89-94 In 1969
Fly Rev Intl, F '69:42-45+ w color aircraft profiles
Roundel, N '59:20-24

POLAND

Air Pict, O '59:378-382+ (Pt 1)	Against Luftwaffe, WW II
N '59:394-400 (Pt 2)	" " " "
Je '65:194-198	1940-1945
Air Prog, Fall '59:63-72	w aircraft photos
Flight, Ag 6, '42:149-153	w RAF, WW II
Flying, Ja '43:30-31+	" " " "
Ag '53:30+	In 1953
Nav Avn News, My '52:16-17	
Pop Avn, N '39:12-13+	
Pop Fly, Ap '37:20-21	1918-1937
RAF Fly Rev, S '55:54-55+	In 1939
RAF Qtr, Je '45:173-177	At RAF Swinderby, 1940-1941

RHODESIA

Air Pict, D '65:416-419
Roundel, Jl/Ag '62:12-14

RUMANIA

Flight, Jl 1, '43:9-11

RUSSIA

Air Force

Airpwr Hist, Jl '63:90-95	1910 to WW I
AAHSJ, Fall '59:188-191	In WW I
Aviation, Ag '42:90-92+	In WW II
Flight, Jl 3, '41:6f-h	" " "
O 23, '41:274-276	" " "
Ja 29, '42:92e-93	" " "
Flying, O '41:14-17+	" " "
Ja '43:27-28+	Fighter tactics, WW II
My '45:21-23+	" " " "
Interavia, N '49:641-649	
No. 10, '50:510-516	
Nav Avn News, S '48:13-15	In 1948
Pop Fly, D '38:445-447+	In 1938
Roundel, N '48:22-26 (Pt 1)	
D '48:35-39 (Pt 2)	
Ja '49:14-17 (Pt 3)	
F '49:30-33 (Pt 4)	
RAF Fly Rev, Je '58:13-15+	Organization in 1958
:Supp:1-8	Sketches of 1958 aircraft
:28-30+ (Pt 1)	Bombing force
Ja '59:20-23+ (Pt 2)	Fighter force
RAFs Qtr, Vol. 1, No. 4:267-276	In 1961
Skyways, Ja '51:10-12+	

RUSSIA (cont.)

Naval Aviation

Nav Avn News, Ap '51:16-19

SAUDI ARABIA

Fly Rev Intl, Ap '66:472-473+ Air defense in 1966

SCANDINAVIA

Aeroplane, Ap 19, '40:546 In 1940
Interavia, No. 8/9, '50:438-442 In 1950

SERBIA

Cr & Cockade, Sum '67:127-137 (Pt 1) w bibliography and maps
Aut '67:238-251 (Pt 2)
Spg '68:73-84 (Pt 3)
Sum '68:163-169 (Pt 4)
Aut '68:291-299 (Pt 5) Pilot biographies

SOUTH AFRICA

Flight, F 22, '40:171-174
Roundel, Je '49:10-14 (Pt 1)
Jl '49:33-37 (Pt 2)
O '59:22-26

SOUTHERN RHODESIA

Flight, Ja 2, '54:7-8

SOVIET UNION

See RUSSIA

SPAIN

Air Pict, N '66:408-412
Nav Avn News, Mr '53:16-17 1921-1953
O '58:35

SWEDEN

Air BP, No. 34:18-24
Air Force, F '56:38-46
N '64:37-41
Air Pict, F '68:40-41+
Air Power, Sum '55:259-264
Avn Wk, Ap 27, '59:70-83 (Pt 1) Arctic air defense
My 4, '59:70-76 (Pt 2) " " "
My 11, '59:97-101 (Pt 3) " " "
Mr 24, '69:40-46+ In late 1960's

SWEDEN (cont.)

Flight, Je 11, '36:626a-d	In 1936
Ag 27, '42:234-237	In WW II
Jl 22, '48:97-100	In 1948
Flying, Jl '51:16-18+	In 1951
Interavia, No. 2, '58:116-119	
No. 3, '59:281	In 1959
No. 7, '60:836-839	In 1960
No. 7, '62:861-865	Defense in 1962
No. 2, '64:211-213	Roads as runways
Nav Avn News, Ag '50:30-31	In 1950
Pop Fly, F '38:614-617+	
RAFs Qtr, No. 2, '65:113-123	
Skyways, D '51:18-19+	

SWITZERLAND

Air Force, O '66:70-78	
Air Pict, Ag '62:226-227	In 1962
Jl '64:196-203	
Flight, N 27, '47:612-613	
D 21, '67:1029-1031 (Pt 1)	
D 28, '67:1074-1075 (Pt 2)	
Fly Rev Intl, Mr '69:42-49+	w color aircraft profiles
Pop Fly, Ag '37:274-277	In 1937

SYRIA

RAF Fly Rev, O '57:13-15	In 1957

THAILAND

Air Power, Win '53:159-164	
Flying, Ag '52:16-17+	Beginnings in 1952
Fly Rev Intl, My '67:527-530	
RAF Fly Rev, N '57:13-16	w aircraft insignia, in color

TURKEY

Air Power, Aut '54:30-32	
Flight, Ap 9, '42:341-343	
Flying, Je '42:26-27+	In WW II
RAF Fly Rev, F '58:13-15	In 1957
RAF Qtr, O '52:328-334	1911-1952

UNITED ARAB REPUBLIC

RAF Fly Rev, Mr '58:13-15	In 1958

UNITED STATES

Air Force

Air Force,	Ag '57:Entire issue	
	O '63:44-48	History of wing insignia
	S '67:Entire issue	
Air Pict,	D '64:378-387	In 1964
	D '66:436-440	Update to 1966
Air Power,	Win '56/'57:89-99	In 1956
Flying,	My '51:Entire issue	In 1951
Interavia,	No. 11, '57:Entire issue	50th Anniversary
Pegasus,	Ag '47:Entire issue	
RAF Qtr,	O '49:124-130	In Europe

Air National Guard

AAHSJ,	Ap/Je '56:27-32 (Pt 1)	Pre-WW II
	Jl/S '56:68-71+ (Pt 2)	WW II and after
Flying,	O '61:40-41+	

Air Service

Cr & Cockade,	Spg '62:1-5	WW I combat organization, Order of Battle
	Sum '65:135-147	WW I: Development in US
	Aut '65:258-271	WW I: Development in Europe
	Win '65:362-372	WW I: Training in US
	Spg '66:57-72	WW I: Training in Europe
	Sum '66:163-174	WW I: Operations in France
	Aut '66:289-297	WW I: Operations in France
	Win '66:365-376	WW I: Operations in France
	Spg '67:74-87	WW I: St. Mihiel
	Sum '67:173-188	WW I: Meuse-Argonne
Pop Fly,	Mr '38:702-704	In WW I

Army Air Force and Army Air Corps

Aero Album,	Spg '68:3-11	WW II combat groups, w list
Aviation,	Ag '41:54-55+	
Flying,	S '41:Entire issue	
	O '43:Entire issue	
Pop Avn,	D '28:17-27+	
	Ag '39:28-31+	
Pop Fly,	Ag '38:228-231+	In 1938
US Air Svcs,	Ag '39:11-14	30th Anniversary
	O '41:25-26+	Organization in 1941

UNITED STATES (cont.)

Army Aviation

Flying, F '52:9-11+	
Ja '56:24-25+	
Mr '61:22-23+	Changing tactics
Ap '61:40-42+	Jet growth
My '62:22-61	In 1962
Interavia, No. 7, '58:687-691	Requirements
Pegasus, O '53:4-7	
US Army Avn, Je '57:7-20	
Mr '58:2-60	1955-1957
Je '61:1-7	
Je '62:2-25 (Pt 1)	
Jl '62:10-31 (Pt 2)	
Ag '62:9-19 (Pt 3)	
S '62:28-37 (Pt 4)	
O '62:28-34 (Pt 5)	
N '62:34-47 (Pt 6)	
D '62:26-38 (Pt 7)	
Ja '63:34-39 (Pt 8)	
F '63:33-41 (Pt 9)	

Balloon Service

Aerosp Hist, Sum '68:26, 39-51	In WW I

Coast Guard

Flying, D '62:46-47+	
Nav Avn News, Je '47:3-7	
Pop Avn, F '28:10-12+	In 1928

"Confederate Air Force"

Air Prog, O '68:34-37+	At air show
S '69:36-38+	Their aircraft
Flight, My 29, '69:908-909	

Marine Corps Aviation

Flying, D '43:24-25+	In WW II
D '49:22-23+	In 1949
Nav Avn News, O 1, '43:14-19	
Ja '47:1-5	In 1946
My '52:15	40th Anniversary
N '54:1-5	Some exploits
Ja '61:30-35	1912-1961
Ag '63:12-15	
Pegasus, Jl '45:12-16	
Jl '53:1-5	Marine Air Transport

UNITED STATES Marine Corps Aviation (cont.)
Pop Avn, Mr '39:20-23+

Naval Aviation

AAHSJ,	Spg '61:3-16+	Birth
Aviation,	Ag '41:56-58+	
Flying,	Ja '42:Entire issue	In WW II
	F '43:Entire issue	" " "
	O '44:Entire issue	" " "
	N '51:Entire issue	In 1951
	Je '61:24-30+	50th Anniversary
Interavia,	S '49:544-546	Air navy of future
	No. 5, '52:260-264	In 1952
Nav Avn News,	Ja 1, '44:10-11	In 1943
	Ja 1, '45:1-13	In 1944
	Ja '48:1-9	In 1947
	Ja '49:1-3	In 1948
	Ja '61:2-23	50th Anniversary
	N '61:7-13	Naval Reserve
	Ap '67:12-13	WW I: Beginnings
	My '67:22-23	" " Ap/Je 1917
	Je '67:22-25	WW I: First unit in France
	Jl '67:16-19	WW I: Ground School at MIT
	Ag '67:24-27	WW I: Training by Royal Flying Corps
	S '67:24-25	WW I: Jl/S 1917
	O '67:22-24	WW I: Naval Aircraft Factory
	N '67:26-29	WW I: Lighter-than-air training
	D '67:22-23	WW I: O/D 1917
	Ja '68:14-17	WW I: 2nd Yale Unit
	F '68:18	WW I: Origin of Navy Wings
	Mr '68:18-19	WW I: Ja/Mr 1918
	Ap '68:22-23	WW I: Ap/Je 1918
	My '68:22-25	WW I: Flying bomb development
	Je '68:24-25	WW I: Against U Boat
	Jl '68:16-17	WW I: Medal of Honor exploit
	Ag '68:26-27	WW I: Jl/S 1918
	S '68:22-23	WW I: Navy's first Ace
	O '68:28-31	WW I: Northern Bombing Group
	N '68:12-14	WW I: At the end
Pegasus,	Ja '48:1-16	
	S '48:1-5+	Training
Pop Avn,	D '28:28-34+	
	Mr '37:14-16+	Beginnings

UNITED STATES Naval Aviation (cont.)

West Avn, Ja '61:14-23 — 50th Anniversary
West Fly, S '37:12-14 — In 1937

URUGUAY

Fly Rev Intl, O '68:50-53 — Aircraft profiles and list of types

YEMAN, SOUTH

Fly Rev Intl, F '68:98-99 — w color photos

YUGOSLAVIA

Air Pict, Ap '52:118-119
Air Power, Spg '54:279-284
Flying, Je '52:16-17+
Nav Avn News, Ja '52:20-21
Pegasus, O '53:1-3
RAF Fly Rev, S '53:21-24 — In 1953

Air Forces: Equipment

[See also Air Forces: History and Organization]

AFRICA

Fly Rev Intl, Jl '66:654-660 — w photos and insignia

AUSTRALIA

Avn Age, O '52:42-43

AUSTRIA

Cr & Cockade, Sum '66:112-131 — WW I naval aircraft

CZECHOSLOVAKIA

Aviation, S '34:278-281

FRANCE

Air Power, Aut '55:10-20 — In 1955
Aviation, Ap '33:101-105 — Pursuit and observation aircraft
Je '33:173-176 — Bombers, fighters, naval aircraft
RAF Fly Rev, S '54:22-23 — In 1954
Ag '57:16-20+ — 1957 fighters

GERMANY

Air Pict, D '67:430-436 — In 1967
Aviation, S '42:98-99+
N '42:200-201 — Firepower
N '45:180-183 — Unusual WW II aircraft
Cr & Cockade, Aut '60:1-11 — WW I aircraft
Flight, S 14, '39:242a-d — Early WW II aircraft
S 28, '39:270a-f — Dive bombers
O 5, '39:286a-287 — Fighters
O 12, '39:300a-e — Junkers bombers
O 19, '39:316a-d — Seaplanes
O 26, '39:334a-d — Cargo types
N 9, '39:372a-d — Dornier flying boats
Flying, Ja '46:76-77+ — WW II aircraft list
Mr '49:30-32+ — WW II light aircraft
Pegasus, D '45:1-4+ — WW II development
Pop Fly, Jl '33:196-197 — WW I types

GERMANY (cont.)

Pop Fly, Je '35:121-123	In 1935
RAF Fly Rev, F '59:Supp:1-8	1939-1945, w side profile and tables
Skyways, Je '43:34-47	1943 inventory w photos
Mr '44:38-49	1944 " " "

GREAT BRITAIN

Air Power, Jl '56:245-258	In 1956
Aviation, D '32:472-475 (Pt 1)	
Ja '33:46-50 (Pt 2)	
Avn Wk, S 16, '68:72-76	Latest designs
Flight, Je 26, '31:581-598	
My 18, '39:504-509	
O 10, '40:297-300	Fleet Air Arm aircraft
Ap 1, '43:329-336+	1918-1943
Nav Avn News, Ja '47:19-21	Fleet Air Arm aircraft
Pop Fly, D '32:514-516	In 1932
Jl '33:193-196	WW I types
:198-205	In 1933
Jl '34:172-175+	In 1934
Jl '35:187-194+	In 1935
RAF Fly Rev, Ap '58:Supp:1-16	1918-1958
Skyways, Mr '43:48-63	Photos of fighters & bombers in WW II
Ap '43:56-63	Photos of other types in WW II
Ap '44:42-54	Photos of types in 1944
Je '45:44-57	In 1944-1945 w photos

GREECE

Aviation, D '34:381-383	

HOLLAND

Air Pict, D '67:454-455	Interned aircraft, 1914-1918

HUNGARY

Air Pict, D '67:456-457	Trainer aircraft

ITALY

Air Pict, Jl '55:194-197 (Pt 1)	1939-1945
Ag '55:238-242 (Pt 2)	" "
S '55:266-268 (Pt 3)	" "
O '55:319-322 (Pt 4)	" "
N '55:352-353 (Pt 5)	w table of 1939-1945 types
Acft Eng, Ag '40:224-232+	w profiles, sketches, & photos
Aviation, O '33:313-317	Bombers

ITALY (cont.)

Aviation,	N '33:340-343	Observation & reconnaissance aircraft
	D '33:372-375	Pursuit and training aircraft
Cr & Cockade,	Aut '64:264-269	WW I naval aircraft
Flight,	Je 13, '40:527-528d	w photos
	Je 27, '40:562a-563	Single-seat fighters
Pop Avn,	O '37:24-26+	In 1937

JAPAN

Air Pict,	Je '56:189-192 (Pt 1)	Army aircraft: WW II and before
	Jl '56:225-228 (Pt 2)	Army aircraft: WW II and before
	Ag '56:270-273 (Pt 3)	Army aircraft: WW II and before
	S '56:305-307 (Pt 4)	Army aircraft: WW II and before
	D '58:410-413+ (Pt 1)	Naval aircraft: 1940-1945
	Ja '59:4-8 (Pt 2)	" " " "
	F '59:62-66+ (Pt 3)	" " " "
	Mr '59:74-77+ (Pt 4)	" " " "
	Jl '59:267-270 (Pt 5)	" " " "
Air Prog,	Je/Jl '64:35-57	List of WW II types w photo and code names
Aviation,	Jl '33:203-207	
Can Avn,	Ja '42:14-17+	
Flying,	Mr '46:39-42+	How good were Japanese aircraft?
RAF Fly Rev,	F '61:25-31	WW II types

MIDDLE EAST

Fly Rev Intl,	Ap '67:465-469+	In 1967

POLAND

Air Pict,	D '68:440-445+	1918-1968
Aviation,	D '34:381-383	
Flight,	S 7, '39:211-213	
RAF Fly Rev,	N '59:25-30+	Profiles and tables of types

RUMANIA

Aviation,	S '34:278-281	

RUSSIA

Air Pict,	Ap '50:114-116 (Pt 1)	
	My '50:138-140 (Pt 2)	
	F '51:34-38+	Jet fighters
Air Power,	Ap '56:171-181	In 1956

RUSSIA (cont.)

Acft Eng,	Je '42:152-159	
	Ag '61:218-226	Turbine aircraft
Aviation,	Ag '33:236-240	
	Ag '42:90-92+	
Avn Age,	Je '51:31-57	
Avn Wk,	Mr 15, '54:90-96	
	Je 24, '69:208-210	V/STOL aircraft
	Ag 5, '68:32-33	
Can Avn,	My '42:23-25+	
	Je '54:44-45+	
Flight,	Jl 20, '67:113-116	w many photos
	Ag 10, '67:229-234	In 1967
Flying,	Ag '50:15-17+	Sonic fighters
Fly Rev Intl,	Je '66:626-628	Cannon fighters of 1930's
Interavia,	No. 5, '55:327-330	In 1955
	No. 8, '56:584-590	At Tushino Air Show
	No. 10, '61:1434-1435 (Pt 1)	In 1961
	No. 11, '61:1520-1521 (Pt 2)	" "
	No. 11, '62:1453-1459	In 1962
	No. 9, '67:1372-1376	In 1967
Nav Avn News,	N '50:10-13	
	Ag '53:16-17	Sport planes, 1953
	O '54:13-15	
	My '56:20-23	In 1956
Pegasus,	Ap '51:4-7	Transports in 1951
Roundel,	Mr '49:6-9	Fighters in 1949
RAF Fly Rev,	Mr '55:17-21	Bombers in 1955
	O '57:25-27+	Transports in 1957
	F '63:21-25+ (Pt 1)	In 1963
	Mr '63:21-25+ (Pt 2)	" "
	Ap '63:21-25+ (Pt 3)	" "
RAFs Qtr,	No. 4, '63:256-265	" "
Skyways,	Mr '51:18-19+	In 1951
	S '52:12-15+	Fighter evolution, 1922-1952
Sp/Aeron,	Jl '68:64-72	Fighters in 1968
	Ap '69:52-59	Bombers in 1969

SOVIET UNION

See RUSSIA

SPAIN

RAF Fly Rev,	My '54:24-25	Types for Franco
	O '60:9	Fokkers in Civil War

SWEDEN

Fly Rev Intl,	Ag '65:5-8	Helicopters

SWITZERLAND

Fly Rev Intl,	O '63:49-51	Me-109s in WW II

TURKEY

Aviation,	D '34:381-383	

UNITED STATES

Aviation,	Je '32:247-253	General aircraft
	Jl '32:290-294	Pursuit "
	S '32:369-373	Bombardment "
	O '32:411-415	Naval "
	N '32:440-443	Trainer "
	O '42:93-95+	
Fly Rev Intl,	F '68:67-72+	Principal types in 1968, many in color
RAF Fly Rev,	My '60:29-36	In 1960, w profiles
Skyways,	D '43:36-51	Fighters and bombers in 1943
	Ja '44:40-50	Transports and trainers in 1943
	D '44:38-52	All combat types in 1944

Air Force

Air Pict,	Je '56:193-195	B-1 to B-66
Flying,	D '48:17-27+	Jet aircraft in 1948
Nav Avn News,	Mr '48:18-21	In 1948
Pop Avn,	N '36:14-16+	Bombers: 1918-1936
	Ap '37:14-18	1935-1937
	N '37:10-13+	Attack aircraft to 1937
Skyways,	Mr '49:54-64	In 1949
	Je '51:26-41	In 1951
	Je '52:18-35	In 1952

Army Aviation

US Army Avn,	Ag '62:9-19	Fixed-wing aircraft
	S '62:28-37	Rotary-wing aircraft
	O '62:28-34	V/STOL aircraft

Naval Aviation

Air News,	Je '41:4-13	In 1941
Air Pict,	F '55:40-42	Fighters in 1955
Nav Avn News,	F '61:7-11	50th Anniversary, w photos
	Mr '61:14-19	Biplane fighters
	Ap '61:22-28	Monoplanes and jet fighters
	My '61:6-11	Heavy attack & anti-sub aircraft
	Je '61:24-29	Dive bombers & attack aircraft
	Jl '61:25-29	Flying boats

UNITED STATES Naval Aviation (cont.)

Nav Avn News,	Ag '61:22-25	Patrol aircraft
	S '61:12-17	Observation & scouting aircraft
	O '61:25-29	Trainer aircraft
	N '61:33-38	Utility/transport "
	D '61:27-31	Rotary-wing "
Pop Avn,	Ap '37:14-18	1935-1937
	Je '37:14-16+	In 1937
	Ag '37:14-16+	Torpedo bombers to 1937
Skyways,	My '49:34-41	In 1949
	Jl '51:26-39	In 1951
	S '52:20-29	In 1952

YUGOSLAVIA

Aviation,	S '34:278-281	
Avn Wk,	Ap 8, '68:64-65	1968 military aircraft
RAF Fly Rev,	F '54:18-19	Post-WW II aircraft

CANADA

Bogotville, Quebec
Roundel, O '60:10-13
Camp Borden, Ontario
Roundel, Mr '52:14-18
Ja/F '60:24-27
Sentinel, S '66:22-23
Chatham, New Brunswick
Roundel, D '63:11-13
Comox, British Columbia
Roundel, S '59:27-32
Ja/F '65:4-8
Fort Churchill, Manitoba
Roundel, O '64:23-25
Foymont, Ontario
Roundel, Jl/Ag '60:4-7
Goose Bay, Labrador
Roundel, My '50:17-27
D '60:8-11
Greenwood, Nova Scotia
Roundel, My '62:11-14
Holberg, British Columbia
Roundel, Ap '65:11-13
Long Branch, Ontario
CAHS, Win '69:123-127
Namao, Alberta
Roundel, N '60:24-27
North Bay, Ontario
Roundel, Je '62:25-28
Portage la Praire, Manitoba
Roundel, Mr '60:12-15
Rivers, Manitoba
Roundel, O '61:10-15
Rockcliffe, Ontario
Roundel, O '49:3-9
S '61:10-14
St. John's, Newfoundland
Roundel, O '59:12-16
Sea Island, British Columbia
Roundel, Ja/F '61:7-10

CANADA (cont.)
Summerside, Prince Edward Island
Roundel, Je '60:5-9
Trenton, Ontario
Roundel, Ap '61:12-16 (Pt 1)
My '61:5-9 (Pt 2)
Uplands, Ontario
Roundel, Jl/Ag '63:12-16
Victoria Island, Ottawa
Roundel, O '62:11-15
Whitehorse, Yukon
Roundel, Ag '49:28-33
Winnipeg, Manitoba
Roundel, N '59:14-19

GERMANY
Bitburg
Bee-Hive, Ja '57:36-40

GREAT BRITAIN
Abingdon
Air Pict, Je '68:216
Brawdy
Air Pict, O '68:378-379
Brize Norton
Air Pict, Jl '67:244-246
Calshot
Flight, Mr 15, '34:237-242
RAF Fly Rev, Ap '61:43
Duxford
Air Pict, D '64:398-399
Exeter
Air Pict, Mr '68:95-96
Farnborough
Flying, S '55:39
Leuchars
Air Pict, F '68:50-52
Woodbridge
Air Pict, Ja '67:22-24

GREENLAND
Thule
Boeing Mag, O '52:3-5+

HAWAII
Ford Island NAS
Nav Avn News, O '65:22-25

HAWAII (cont.)
Hickam
Aerosp Hist, Aut '68:16-21
AAHSJ, Win '65:288-290
Pearl Harbor NAS
Nav Avn News, D 1, '44:12-17

ICELAND
Keflavik
Nav Avn News, Ag '62:34-35

ITALY
Foggia
Flying, Jl '44:31+
Naples NAS
Nav Avn News, D '64:22-24

JAPAN
Atsugi NAS
Nav Avn News, F '53:22
Oppama NAS
Nav Avn News, Ap '57:12-14

KWAJALEIN
Kwajalein
Nav Avn News, Ag '54:22

MOROCCO
Khouribga Base
Nav Avn News, D '54:8-9

PHILLIPINE ISLANDS
Cubi Point NAS
Nav Avn News, Ap '60:33-35
Ja '69:16-18
Sangley Point NAS
Nav Avn News, F '54:14-15

PUERTO RICO
Roosevelt Roads NAS
Nav Avn News, Mr '59:6-9
D '63:6-8

UNITED STATES
Abilene AFB, Texas
Boeing Mag, Je '56:14-15

UNITED STATES (cont.)
Alameda NAS, California
Nav Avn News, My '54:18-19
Beaufort, Marine Corps Aux. Air Station, South Carolina
Nav Avn News, Ja '60:32
Bolling AFB, Maryland
Airpwr Hist, O '62:231-239
Bradley Field, Connecticut
US Air Svcs, Jl '50:21-22
Brunswick NAS, Maine
Nav Avn News, D '59:15-17
Cecil Field NAS, Florida
Nav Avn News, Ag '62:37
Cherry Point, Marine Corps Air Station, North Carolina
Nav Avn News, Ja 1, '44:14-21
My '62:30-31
Corpus Christi NAS, Texas
Nav Avn News, Je 1, '44:27-35
Mr '56:1-7
Edwards AFB, California
Bee-Hive, Sum '58:2-5
Eglin Field, Florida
Bee-Hive, Ja '48:28-31
Pegasus, N '47:1-5
El Toro, Marine Corps Air Station, California
Nav Avn News, Mr '58:8-9
Elizabeth City, Coast Guard Station, North Carolina
Nav Avn News, D '60:23-25
Ellington Field, Texas
Skyline, Vol. 2, No. 4, '41:26-27
Fallon NAAS, Nevada
Nav Avn News, S '54:26-27
Fort Myer, Virginia
Air Pilot, O '43:6-8+
Hicks Field, Texas
Pegasus, N '43:14-16
Jacksonville NAS, Florida
Nav Avn News, S 15, '44:14-25
Kelly AFB, Texas
Air Prog, Ag '67:48-49+
Bee-Hive, Fall '51:20-22
Flying, My '43:96+
Langley Field, Virginia
AAHSJ, Spg '66:3-12
Lincoln NAS, Nebraska
Nav Avn News, F '50:25-27

Air Forces: Unit Histories

Notes indicate restrictions in article content. Where notes do not appear, it may be assumed that the article is broadly historical.

Where known, official or common names assigned numbered units are placed in parentheses.

1 AERO SQUADRON -- US
Airpwr Hist, O '57:207-212 — 1913-1917
AAHSJ, Fall '65:190-197 — In Mexico
Fall '68:228-229 (Pt 1)
Win '68:296-297 (Pt 2)
Spg '69:51-54 (Pt 3)
Fall '69:211-217 (Pt 4)
Pop Avn, F '36:89-90+ — In WW I

1 AIR COMMANDO GROUP -- USAAF
AAHSJ, Fall '68:178-185

1 AIR DEFENSE GROUP -- RAF
Flight, Ag 12, '32:752-755

1 AIR NAVIGATION SCHOOL -- RAF
Air Pict, F '65:54-56

1 BOMBARDMENT SQUADRON -- US
Airpwr Hist, O '63:111-117

1 DAY BOMBARDMENT GROUP -- US
Cr & Cockade, Spg '62:6-13

1 FIGHTER REGIMENT -- Czechoslovakia
Air Pict, Mr '66:92-94 — In WW II

1 FIGHTER SQUADRON -- US Navy
Nav Avn News, D '47:8-9 (Pt 1) — In WW II
Ja '48:24-25 (Pt 2) — " " "

1 FIGHTER WING -- USAF
Air Pict, Ap '62:103-105 Representative aircraft
Airpwr Hist, O '62:213-224 (Pt 1)
Ja '63:24-31 (Pt 2)

1 FLEET AIR WING -- US Navy
Nav Avn News, S 1, '45:20-25 In WW II

1 FLYING TRAINING SCHOOL -- RAF
Flight, Jl 21, '38:62c-f At Leuchars

1 HELICOPTER (Flying Angels) SQUADRON -- US Navy
Nav Avn News, Jl '53:16-17 1948-1953

1 JAGDSTAFFEL -- German AF
Cr & Cockade, Win '65:341-346 Origins

1 MARINE AVIATION FORCE -- US Marines
Cr & Cockade, Sum '65:173-186 Development
Aut '65:272-292 Employment, w bibliography

1 MARINE HELICOPTER EXPERIMENTAL SQUADRON -- US Marines
Nav Avn News, O '59:6-10

1 PURSUIT GROUP -- US
AAHSJ, Sum '67:142-144 1919-1939, w aircraft photos

1 RADAR AND COMMUNICATIONS WING -- RCAF
Roundel, Mr '51:40-46

1 SQUADRON -- Australian Flying Corps
Air Pict, F '69:50-54

1 SQUADRON -- RAAF
Air Pict, Jl '65:238-240

1 SQUADRON -- RAF
Air Pict, Jl '68:240-243
Cr & Cockade, Sum '61:104-110 1912-1919
Flight, S 7, '33:887-891
D 23, '48:747-754 70-year history

1 UTILITY SQUADRON -- US Navy
Nav Avn News, Ja '50:30-31 1925-1949

2 BOMB WING -- USAF
Air Pict, Jl '63:218-220

2 FIGHTER SQUADRON -- US Navy
AAHSJ, Fall '69:145-152 Aircraft and crews
:153-158 Brewster aircraft
Pop Avn, F '40:30-31+

2 GROUP -- RAF
Air Pict, Jl '68:256

2 SQUADRON -- RAF
Flight, D 19, '30:1453-1456 Army cooperation
Je 9, '49:675-682
RAF Fly Rev, D '53:23-25 VC awards
RAF Qtr, Ap '52:163-167 Korea: N '50-Ag '51

2 TRANSPORT SQUADRON -- US Navy
Nav Avn News, Ap '52:14-15 Flying the Martin "Mars"

3 AEROSPACE RESCUE AND RECOVERY GROUP -- USAF
Aerosp Hist, Aut '69:13-15+

3 BOMB WING -- USAF
Air Pict, N '62:358-360
Airpwr Hist, Jl '64:71-77

3 FLYING TRAINING SCHOOL -- RAF
Flight, Ag 17, '33:827-832
Jl 27, '39:86b-d

3 NAVAL WING -- RNAS
Cr & Cockade, Spg '63:88-91 In WW I
Roundel, Jl/Ag '63:4-10 " " "

3 SQUADRON -- RAF
Cr & Cockade, Spg '64:30-38 In WW I
Flight, Mr 17, '49:313-320

3 TRANSPORT SQUADRON -- US Navy
Nav Avn News, Ag '65:18

4 EARLY WARNING SQUADRON -- US Navy
Nav Avn News, Ap '63:14-17

4 FIGHTER WING -- USAF
Airpwr Hist, Ja '62:21-34

4 SQUADRON -- RAF
Flight, Ag 3, '39:112e-h
F 23, '50:255-262

4 TRANSPORT SQUADRON -- US Navy
Nav Avn News, F '63:36-38

5 AIR DEVELOPMENT SQUADRON -- US Navy
Nav Avn News, F '63:38-39

5 JAGDSTAFFEL -- German AF
Cr & Cockade, Aut '60:63-76 w photos of aircraft

5 SQUADRON -- RAF
Air Pict, My '67:158-160
Cr & Cockade, Spg '69:1-33 Pilot's diary in WW I
Flight, O 18, '57:618-623 (Pt 1)
O 25, '57:642-646 (Pt 2)
N 8, '57:745-746 (Pt 3)

6 AERO SQUADRON -- US
Airpwr Hist, Jl '60:178-182 1917-1947

6 AIR GROUP -- US Navy
Nav Avn News, S '49:14-15

6 BOMBER GROUP -- RCAF
Air Pict, Ja '67:12-15
Mr '67:93 Art work on Halifaxes
Roundel, D '50:20-24
Ap '63:16-23
RAF Qtr, S '46:199-200

6 SQUADRON -- RAF
Air Pict, Ap '64:103-105
Flight, D 28, '50:617-623 (Pt 1)
Ja 4, '51:11-16 (Pt 2)

6 TRANSPORT SQUADRON -- US Navy
Nav Avn News, Ap '63:23

7 FIGHTER SQUADRON -- US Navy
Nav Avn News, Ag '50:14-15 In WW II

7 JASTA -- German AF
Cr & Cockade, Win '65:307-334 Includes roster

7 SQUADRON -- RAF
Flight, Je 10, '32:503-509+
My 18, '51:589-593 (Pt 1)
My 25, '51:620-624 (Pt 2)

8 SQUADRON -- RNAS
Became 208 Squadron -- RAF
Cr & Cockade, Sum '61:157-161 In WW I
Flight, Ap 4, '58:463-464
Pop Fly, Ja '37:509-513 (Pt 1)
F '37:580 (Pt 2)
O '37:386-388+

8 TORPEDO SQUADRON -- US Navy
Air Prog, Ag '67:16-19+ In WW II
Skyways, Jl '44:19+ " " "

8 TRANSPORT SQUADRON -- US Navy
Nav Avn News, D '53:20-21

9 AIR FORCE -- USAAF
AAHSJ, Win '64:274-279 In Africa, WW II

9 FIGHTER SQUADRON -- US Navy
Nav Avn News, Jl '49:24-25 In WW II

9 FLYING TRAINING SCHOOL -- RAF
Flight, Jl 14, '38:40e-h At Hullavington

9 SQUADRON -- RAF
Cr & Cockade, Spg '69:34-54 WW I observer's account
Flight, O 23, '53:561-564

9 TORPEDO SQUADRON -- US Navy
Nav Avn News, F '48:8-9 In WW II

10 JASTA -- German AF
Cr & Cockade, Aut '62:204-207

10 SQUADRON -- RAF
Air Pict, Mr '62:70-72 w representative aircraft
Flight, S 12, '35:277-280

10 SQUADRON -- RNAS
See "Black Flight" for WW I "B-Flight" of this unit

10 TACTICAL RECONNAISSANCE WING -- USAF
Air Pict, O '63:330-331

11 AERO SQUADRON -- US
Cr & Cockade, Aut '61:179-197 In WW I

11 BOMB SQUADRON -- USAF
Aerosp Hist, Aut '68:14 "Jiggs" insignia

11 FIGHTER (Red Rippers) SQUADRON -- US Navy
Nav Avn News, Ag '56:14
Mr '67:16-17

11 FLEET AIR WING -- US Navy
Nav Avn News, S '64:19-22

11 JASTA -- German AF
Cr & Cockade, Aut '60:60-62 In WW I
Spg '64:74-79 Last days in WW I

11 PATROL SQUADRON -- US Navy
Nav Avn News, F '50:16-17 In WW II

11 SQUADRON -- RAF
Air Pict, D '63:398-400
RAF Fly Rev, O '54:24

11 SQUADRON -- RFC
Pop Fly, D '33:477-478 In 1917
F '35:578-581

11 TORPEDO SQUADRON -- US Navy
Nav Avn News, O '48:26-27 In WW II
N '48:24-25 " " "

12 AERO OBSERVATION SQUADRON -- US
AAHSJ, Fall '64:179-188 (Pt 1) In WW I
Spg '65:47-52 (Pt 2) " " "

12 GROUPE DE COMBAT -- French AF
See "Cigognes, Les"

12 SQUADRON -- RAF
Air Pict, F '63:48-50

14 FIGHTER (Top Hatters or High Hat) SQUADRON -- US Navy
Nav Avn News, S '65:18-19 46th Anniversary

14 SQUADRON -- RAF
Air Pict, Ja '62:12-14 Representative aircraft
Flight, Ja 18, '34:49-53

14 SQUADRON -- RNZAF
Air Pict, Ja '68:16-18

14 TORPEDO SQUADRON -- US Navy
Nav Avn News, Ag '49:16-17 In WW II

15 CARRIER AIR GROUP -- US Navy
Nav Avn News, Ja '57:24

15 FIGHTER SQUADRON -- US Navy
Nav Avn News, Ap '48:28-29 In WW II
My '48:16-17 " " "

15 RESERVE FLYING SCHOOL -- RAF
Flight, Mr 19, '54:331-334

16 JASTA -- German AF
Cr & Cockade, Win '67:359-381 WW I: Photos, rosters, etc.

16 SQUADRON -- RAF
Air Pict, Je '61:178-179 w representative aircraft
Flight, Ja 17, '35:71-73
Ap 27, '56:503-505

17 CARRIER AIR GROUP -- US Navy
Aero Album, Win '68:3-8 WW II, w aircraft photos
Nav Avn News, F '69:22-27 In WW II

17 COMBAT AVIATION GROUP -- US
US Army Avn, D '67:14-17

17 JASTA -- German AF
Cr & Cockade, Win '65:358-361

17 PHOTO RECONNAISSANCE SQUADRON -- USAF
AAHSJ, Fall '61:180-182+

17 SQUADRON -- RAF
Flight, Je 27, '30:691-694

18 AIR FORCE -- USAAF
Pegasus, F '53:1-10

18 SQUADRON -- Netherlands East Indies AF
Air Pict, Ap '65:124-126

18 SQUADRON -- RAF
Air Pict, S '64:288-290
Flight, Ja 27, '56:108-111 (Pt 1)
F 10, '56:164-167 (Pt 2)
F 17, '56:190-193 (Pt 3)

19 AUXILIARY WING -- RCAF
Roundel, Jl/Ag '54:8-15

19 AVIATION SQUADRON -- Russia
Air Pict, D '67:461 In WW I

19 SQUADRON -- RAF
Flight, Ja 5, '33:4-8
F 22, '52:211-214

20 AIR FORCE -- USAAF
AAHSJ, Fall '62:165-179+ (Pt 1) w bomber markings
Win '62:257-273 (Pt 2) " " "

20 FIGHTER WING -- USAF
Air Pict, Ja '63:12-14

20 SQUADRON -- RAF
Air Pict, Ja '65:24-26

20 SQUADRON -- RFC
Pop Fly, O '33:361-362+ On Jl 27, '17

21 BOMBER COMMAND -- USAF
Airpwr Hist, Ap '64:49-53 In defeat of Japan

21 ANTISUBMARINE SQUADRON -- US Navy
Nav Avn News, Ja '67:8-12

21 SQUADRON -- RAAF
Flight, N 9, '50:401-406

21 SQUADRON -- RFC
CAHS, Win '65:102-103

22 AERO SQUADRON -- US
Cr & Cockade, Sum '63:109-136 WW I: Photos, rosters, etc.
Aut '68:269-282 In WW I

22 SQUADRON -- RAF
Air Pict, F '61:36-37+
Comm Avn, Ja '41:38-39+ In WW I

22 TRANSPORT SQUADRON -- US Navy
Nav Avn News, Ap '67:17

23 FIGHTER SQUADRON -- US Navy
Nav Avn News, D '50:18-19 In WW II

23 SQUADRON -- RAF
Air Pict, D '65:434-436
Flight, Mr 2, '33:193-196

24 AERO SQUADRON -- US
AAHSJ, Sum '67:112-113 WW I, w photos
Cr & Cockade, Spg '64:53-68 " " " "

25 AERO SQUADRON -- US
AAHSJ, Spg '62:17-35 (Pt 1) Many group photos
Sum '62:126-130+ (Pt 2)
Spg '63:31-38 (Pt 3)

25 FLIEGERABTEILUNG -- German AF
Cr & Cockade, Spg '68:22-32 In WW I

25 SQUADRON -- RAF
Air Pict, D '61:374-376 w representative aircraft
Flight, D 8, '32:1163-1167

26 SQUADRON -- RAF
Flight, Ag 10, '33:807-811

27 AERO SQUADRON -- US
Cr & Cockade, Sum '60:11-42
Aut '62:229-238 On Ag 1, '18

27 COMPOSITE SQUADRON -- US Navy
Nav Avn News, Je '50:20-21

27 FIGHTER INTERCEPTOR SQUADRON -- USAF
Bee-Hive, Sum '60:14-17

27 SQUADRON -- RAF
Air Pict, O '69:370-372
Flight, Ja 21, '55:93-95

28 SQUADRON -- RAF
Cr & Cockade, Aut '67:273-303 In WW I

28 TORPEDO SQUADRON -- US Navy
Nav Avn News, Jl '50:18-19 In WW II

29 TORPEDO SQUADRON -- US Navy
Nav Avn News, Ag '48:20-21 In WW II
S '48:18-19 " " "

30 SQUADRON -- RAF
Air Pict, O '62:315-317 w representative aircraft
Cr & Cockade, Win '66:349-359 WW I: Rosters, reports, etc.
RAF Fly Rev, Jl '54:9

31 FIGHTER GROUP -- USAF
AAHSJ, O/D '58:242-243

31 FIGHTER (Tomcats) SQUADRON -- US Navy
Nav Avn News, S '50:18-19 In WW II

31 TACTICAL FIGHTER WING -- USAF
Air Pict, O '64:322-324

33 PHOTO RECONNAISSANCE SQUADRON -- USAAF
Aero Album, Sum '69:32-37 w aircraft photos

33 SQUADRON -- RAF
Flight, Je 27, '30:687-691
My 29, '59:741-742

34 BOMB GROUP -- USAAF
Aeroplane, Ag 24, '45:215 In WW II

35 AIR RESCUE SERVICE SQUADRON -- USAF
Air Cl, Mr/Ap '65:36-39+ In T-6's

35 ATTACK (Black Panthers) SQUADRON -- US Navy
Air Cl, Mr/Ap '65:32-35+

35 SQUADRON -- South African AF
RAFs Qtr, Vol. 1, No. 1:42-44

36 TROOP CARRIER SQUADRON -- USAAF
Pegasus, Mr '47:14-16

38 GROUP -- RAF
Flight, Je 7, '45:608-609 In WW II

39 SQUADRON -- RAF
Flight, My 20, '32:454

40 SQUADRON -- RAF
Flight, My 13, '32:411-416

40 SQUADRON -- RFC
Pop Fly, Ap '35:24-27+ In WW I

42 ATTACK SQUADRON -- US Navy
Bee-Hive, Spg '65:2-7 w Grumman A-6A aircraft

42 COMPOSITE SQUADRON -- US Navy
Nav Avn News, O '49:14-15 In WW II
D '49:26-27 " " "

42 SQUADRON -- RAF
Flight, S 8, '38:206a-d
Ja 29, '54:125-128

43 AIR REFUELING SQUADRON -- USAF
Boeing Mag, S '50:8-9

43 JASTA -- German AF
Cr & Cockade, Win '67:307-335 WW I pilot's diary

43 SQUADRON -- RAF
Air Pict, Mr '61:78-79 w representative aircraft
Flight, S 19, '30:1042-1047
F 17, '49:205-206 Squadron reformed
S 5, '52:318-322

43 SQUADRON -- RAAF
Air Pict, O '65:346-347

46 SQUADRON -- RAF
Air Pict, Je '67:196-198
Cr & Cockade, Aut '61:242-249 In WW I

47 BOMB WING -- USAF
Air Pict, F '62:36-38
Flight, Mr 25, '55:388-392

47 FIGHTER WING -- USAF
Skyline, Vol. 7, No. 3, '49:20-21 1948-1949

47 PATROL SQUADRON -- US Navy
Nav Avn News, Mr '56:14-16

47 SQUADRON -- RAF
Flight, Ap 8, '55:454-457+
Pop Fly, Je '37:144-146+ In South Russia, 1919

48 SQUADRON -- RAF
Flight, D 3, '36:604a-d

49 FIGHTER GROUP -- USAF
Airpwr Hist, O '61:220-226

52 FIGHTER GROUP -- USAF
Bee-Hive, Sum '57:19-23

52 PATROL-BOMBING SQUADRON -- US Navy
Nav Avn News, D '48:26-27 In WW II

53 FIGHTER (Blue Knights) SQUADRON -- US Navy
Nav Avn News, Ag '53:26-27 1948-1953

53 SQUADRON -- RAF
Air Pict, My '61:146-147 w representative aircraft

54 SQUADRON -- RFC
Pop Fly, D '37:506-508+ In WW I

55 ATTACK SQUADRON -- US Navy
Nav Avn News, Je '68:37

55 SQUADRON -- RAF
Aeroplane, N 10, '37:562-564

55 SQUADRON -- RFC
Pop Fly, My '35:79-81+
Ag '35:272-273+ On Jl 20, 1918

56 JAGDSTAFFEL -- German AF
Cr & Cockade, Win '60:69-74 In WW I

56 SQUADRON -- RAF
Air Pict, Ap '63:104-106
My '63:133 Aircraft insignia and colors

56 SQUADRON -- RAF (cont.)
Cr & Cockade, Win '60:1-15 In WW I
Flight, D 24, '36:676a-678 Mostly WW I
My 18, '56:611-613

57 FIGHTER INTERCEPTOR SQUADRON -- USAF
Flight, F 24, '56:216-220 In Iceland w F-89Cs

58 SQUADRON -- RAF
Flight, Je 10, '32:503-509+

60 SQUADRON -- RFC
Cr & Cockade, Aut '64:233-243 In WW I
Pop Fly, D '34:464-467+ (Pt 1) " " "
Ja '35:540+ (Pt 2) " " "
Ja '36:536-540+

63 SQUADRON -- RFC
Pop Fly, Mr '37:617-618+

65 SQUADRON -- RAF
Flight, Ja 1, '54:13-16

68 FIGHTER SQUADRON -- USAF
Airpwr Hist, O '62:249-252 In Korea

70 SQUADRON -- RAF
Air Pict, Je '65:206-207

71 SQUADRON -- RAF
See "Eagle Squadron"

72 SQUADRON -- RAF
Air Pict, O '66:380-382

74 FIGHTER (Bedevilers) SQUADRON -- US Navy
Nav Avn News, Je '62:20-23

74 PATROL SQUADRON -- US Navy
Nav Avn News, O '50:16-17 1940-1945

74 SQUADRON -- RAF
Nav Avn News, Je '62:20-23
RAF Qtr, Jl '33:290-295 In WW I

76 SQUADRON -- RAF
Flight, Mr 17, '38:256e-g

77 SQUADRON -- RAF
RAF Fly Rev, D '59:27-29+ In Korea

77 SQUADRON -- RAAF
Air Power, Ap '55:196 1950-1955: Korea
RAF Qtr, Jl '52:266-272 1950-1952: Korea

78 SQUADRON -- RAF
Air Pict, Ap '61:104-105 w representative aircraft

79 SQUADRON -- RAF
Air Pict, D '60:425-426 w four representative aircraft
Cr & Cockade, Sum '64:168-175 w pors

80 SQUADRON -- RAF
Air Pict, D '62:392-394

81 SQUADRON -- RAF
Air Pict, O '61:322-323 w representative aircraft
RAF Fly Rev, Mr '61:28-30+ w Russians in WW II

83 PATROL SQUADRON -- US Navy
Nav Avn News, F '49:22 In WW II
Mr '49:16 " " "

83 SQUADRON -- RAF
Air Pict, Ag '61:236-237+ w representative aircraft
Bris Sid Jnl, Sum '61:86-91 w Vulcans

86 FIGHTER GROUP -- USAF
AAHSJ, Spg '69:1-7

88 AERO SQUADRON -- US
AAHSJ, Fall '60:187-201 w pors
Pop Avn, Ja '34:11-12+ WW I exploit

89 AERO SQUADRON -- US
Cr & Cockade, Sum '68:124-133 WW I: w photos, pors

90 SQUADRON -- RAF
Flight, Ja 19, '39:62a-c

92 SQUADRON -- RAF
Air Pict, Jl '61:207-208 w representative aircraft
Cr & Cockade, Spg '66:1-15 WW I: w pilot roster and enemy aircraft destroyed

93 BOMB WING -- USAF
Air Pict, Jl '62:210-212 w representative aircraft

94 BOMB GROUP -- USAAF
Aeroplane, Jl 13, '45:42

95 BOMB WING -- USAF
Airpwr Hist, Jl '60:183-186

95 PURSUIT SQUADRON -- USAAF
Pop Avn, N '31:8-10+ In WW I
US Air Svcs, O '30:62
West Fly, Je '30:44-47

98 SQUADRON -- RAF
Air Pict, F '64:52-54

99 SQUADRON -- RAF
Air Pict, N '61:339-340 w representative aircraft
Cr & Cockade, Spg '67:1-18 In WW I
Flight, Ag 3, '33:771-773

100 SQUADRON -- RAF
Flight, Ag 19, '32:774-779
O 28, '55:673-676+

101 SQUADRON -- RAF
Flight, Ap 24, '31:353-357

103 SQUADRON -- RAF
Air Pict, D '67:448-449

104 PATROL SQUADRON -- US Navy
Nav Avn News, Ap '50:16-17 In WW II

106 AEROMEDICAL TRANSPORT GROUP (Air National Guard) -- USAF
Air Pict, My '63:138-139

107 BOMBING SQUADRON -- US Navy
Nav Avn News, F '49:22-23 In WW II
Mr '49:16-17 " " "

109 BOMBING SQUADRON -- US Navy
Nav Avn News, Ja '49:12-13 In WW II

111 SQUADRON -- RAF
See "Treble One Squadron"

112 MARINE FIGHTER SQUADRON -- US Marines
See "Wolf Pack"

112 SQUADRON -- RAF
Flight, Ja 16, '59:100-101

114 SQUADRON -- RAF
Air Pict, Je '66:220-221

115 SQUADRON -- RAF
Air Pict, Ja '66:22-23

119 PATROL-BOMBING SQUADRON -- US Navy
Nav Avn News, Ap '49:22-23 In WW II
Je '49:26-27 " " "

119 SQUADRON -- RAF
RAF Fly Rev, O '53:9

123 PATROL-BOMBING SQUADRON -- US Navy
Nav Avn News, Mr '48:16-17 In WW II

124 SQUADRON -- RAF
Air Pict, Ja '68:29

125 ATTACK (Skylancers) SQUADRON -- US Navy
Bee-Hive, Spg '67:22-25 Using A-4 aircraft

128 PATROL-BOMBING SQUADRON -- US Navy
Nav Avn News, Je '48:26-27 In WW II
Jl '48:26-27 " " "

134 SQUADRON -- RAF
RAF Fly Rev, Mr '61:28-30+ w Russians in WW II

135 AERO SQUADRON -- US
Cr & Cockade, Sum '65:152-165

138 AERO SQUADRON -- US
Cr & Cockade, Spg '60:16-18

147 SQUADRON -- RAF
Air Pict, S '69:349

148 AERO SQUADRON -- US
Airpwr Hist, Jl '62:157-165 In WW I

148 PURSUIT SQUADRON -- US
AAHSJ, Win '59:253-264 w pors and bibliography

148 SQUADRON -- RAF
Air Pict, My '69:181

149 SQUADRON -- RAF
Air Pict, F '68:65
RAF Fly Rev, D '54:47

160 SQUADRON -- RAF
Air Pict, Ja '69:25

162 SQUADRON -- RCAF
Roundel, Mr '61:26-27+

168 SQUADRON -- RAF
Air Pict, S '68:349

169 SQUADRON -- RAF
Air Pict, D '65:443

185 SQUADRON -- RAF
Air Pict, Ja '67:25

186 SQUADRON -- RAF
Air Pict, Mr '65:99

200 KAMPFGESCHWADER -- German AF
RAF Fly Rev, O '60:17-18+ WW II spy delivery

201 SQUADRON -- RAF
Flight, Ap 12, '34:347-351
F 3, '38:108a-d
O 26, '44:440-443
Ap 11, '52:439-444 (Pt 1) 1914-1919
Ap 18, '52:469-474 (Pt 2) 1929-1952
D 30, '55:961-963
F 8, '57:168-169 Disbanded

204 SQUADRON -- RAF
Air Pict, S '61:291-292 w representative aircraft
Flight, Je 12, '31:527-532

205 SQUADRON -- RAF
Air Pict, F '66:60-62
Mr '66:110
My '66:180

206 SQUADRON -- RAF
Air Pict, Je '62:172-173 w representative aircraft

207 SQUADRON -- RAF
Air Pict, Ja '61:19-20 w representative aircraft
Flight, O 12, '33:1022-1024
Ap 17, '53:486-490 (Pt 1)
Ap 24, '53:514-518 (Pt 2)

208 SQUADRON -- RAF
Flight, Ap 4, '58:463-464

209 (City of Hong Kong) SQUADRON -- RAF
Air Pict, My '69:161 Disbanded
Flight, Je 12, '31:527-535
Sentinel, Ap '66:40-41

213 AERO SQUADRON -- US
Cr & Cockade, Sum '65:121-134

213 SQUADRON -- RAF
Fly Rev Intl, D '64:44-45 In Libya, WW II

214 MARINE FIGHTER SQUADRON -- US Marines
See "Blacksheep Squadron"

214 SQUADRON -- RAF
Flight, Mr 3, '38:204a-b, g-h

216 SQUADRON -- RAF
Flight, N 18, '55:783

225 SQUADRON -- RAF
Flight, N 1, '62:724-725 Army close support in 1962

230 SQUADRON -- RAF
Air Pict, Jl '69:242-244
Flight, F 8, '57:168-169 Disbanded
D 5, '58:852-853

232 MARINE (Red Devils) FIGHTER SQUADRON -- US Marines
Nav Avn News, S '58:14-15

242 SQUADRON -- RAF
Air Pict, F '67:61

257 SQUADRON -- Royal Navy
Aeronautics, Jl '55:44-47

263 SQUADRON -- RAF
CAHS, Win '67:96-99 Operations against Norway, WW II

282 SQUADRON -- RAF
Air Pict, N '69:429

295 SQUADRON -- RAF
Air Pict, D '67:461

313 BOMB WING -- USAF
AAHSJ, Fall '64:189-197

315 AIR DIVISION -- USAF
Pegasus, S '56:7-9

325 FIGHTER (Checkertails) GROUP -- USAAF
AAHSJ, Fall '65:155-171 (Pt 1) 1942-1944
Win '65:254-265 (Pt 2) 1944-1945

342 SQUADRON -- RAF
See "Lorraine Squadron"

350 FIGHTER GROUP -- USAF
AAHSJ, O/D '58:242-243

350 SQUADRON -- Belgian AF
Air Pict, Ag '64:236-238

352 FIGHTER GROUP -- USAAF
AAHSJ, Spg '65:18-23

354 FIGHTER WING -- USAF
Air Pict, S '62:284-286

355 FIGHTER GROUP -- USAAF
Aero Album, Win '68:12-15 w aircraft photos

357 FIGHTER GROUP -- USAAF
AAHSJ, Ap/Je '56:43-46 (Pt 1)
Jl/S '56:107-111 (Pt 2)

370 FIGHTER GROUP -- USAF
Aero Album, Spg '68:15-17 WW II aircraft photos

376 BOMB WING -- USAF
Airpwr Hist, Jl '61:161-171

397 BOMB GROUP -- USAF
Aero Album, Sum '68:52-57 w aircraft photos

400 (City of Toronto) SQUADRON -- RCAF
CAHS, Win '64:33 Profiles of aircraft, 1932-1962
Roundel, Ja '50:3-6

401 (Ram) SQUADRON -- RCAF
Roundel, Je '50:10-15

402 (City of Winnipeg) SQUADRON -- RCAF
Roundel, Mr '50:31-35

403 (Wolf) SQUADRON -- RCAF
Roundel, N '50:3-8

404 SQUADRON -- RCAF
Air Pict, Mr '64:72-73

405 SQUADRON -- RCAF
Roundel, Ap '50:27-35

406 (Lynx) SQUADRON -- RCAF
Roundel, My '51:31-35

407 (Demon) SQUADRON -- RCAF
Roundel, Ja '54:2-10

409 (Nighthawks) SQUADRON -- RCAF
Roundel, Jl/Ag '60:8-12 (Pt 1)
S '60:21-25 (Pt 2)

410 BOMB GROUP -- USAAF
US Air Svcs, Ag '45:25-26+ In WW II

410 (Cougar) SQUADRON -- RCAF
Roundel, Jl/Ag '50:36-42

411 (Grizzly Bear) SQUADRON -- RCAF
Roundel, Ja/F '60:18-23 (Pt 1)
Mr '60:20-25 (Pt 2)

412 SQUADRON -- RCAF
Air Power, Sum '54:440-441
Roundel, N '53:8-9
Ag '58:4-9 (Pt 1)
S '58:18-22 (Pt 2)

413 SQUADRON -- RCAF
Roundel, S '63:22-28

414 SQUADRON -- RCAF
Roundel, Jl/Ag '54:38-46 In WW II

416 SQUADRON -- RCAF
Roundel, My '61:2-9

417 (City of Windsor) SQUADRON -- RCAF
CAHS, Spg '66:10
Roundel, N '54:11-21

418 (City of Edmonton) SQUADRON -- RCAF
Roundel, N '60:15-21 (Pt 1)
D '60:18-25 (Pt 2)
Ja/F '61:18-26 (Pt 3)
Mr '61:18-25 (Pt 4)
Ap '61:19-25 (Pt 5)

419 (Moose) SQUADRON -- RCAF
Roundel, N '58:20-25 (Pt 1)
D '58:20-25 (Pt 2)
Mr '59:10-17 (Pt 3)
Ap '59:19-24 (Pt 4)
My '59:12-18 (Pt 5)

420 (Snowy Owl) SQUADRON -- RCAF
Roundel, D '50:3-10

421 (Red Indian) SQUADRON -- RCAF
Roundel, Ja '51:3-10

422 (Tomahawk) SQUADRON -- RCAF
Roundel, N '53:35-43

423 SQUADRON -- RCAF
Roundel, D '57:20-26 (Pt 1)
Ja/F '58:10-16 (Pt 2)
Mr '58:10-16 (Pt 3)
Ap '58:8-13 (Pt 4)

425 (Alouettes) SQUADRON -- RCAF
Roundel, Ap '57:18-22 (Pt 1)
My '57:22-27 (Pt 2)
Je '57:15-20 (Pt 3)
Jl/Ag '57:5-11 (Pt 4)

426 (Thunderbird) SQUADRON -- RCAF
Air Power, Sum '54:439
Roundel, F '50:22-28
N '53:6-7
D '54:32-36 — Korean airlift
RAF Qtr, Jl '52:254-259 — Korea: Jl 1950-Jl 1952

430 SQUADRON -- RCAF
Roundel, Je '53:22-27

432 SQUADRON -- RCAF
Roundel, S '56:24-32 (Pt 1)
O '56:27-31 (Pt 2)
N '56:3-8 (Pt 3)

433 (Porcupine) SQUADRON -- RCAF
Roundel, Mr '55:35-44 (Pt 1)
Ap '55:5-15 (Pt 2)

434 (Bluenose) SQUADRON -- RCAF
Roundel, Mr '56:1-7

435 (Chinthe) SQUADRON -- RCAF
Air Power, Sum '54:439-440
Roundel, F '53:2-14 — In WW II
N '53:7

436 SQUADRON -- RCAF
Air Power, Sum '54:441
Roundel, N '53:9
Mr '62:17-22 (Pt 1)
Ap '62:18-24 (Pt 2)
My '62:24-28 (Pt 3)
Jl '62:21-24 (Pt 4)

438 SQUADRON -- RCAF
Roundel, S '50:5-9 — In WW II
Je '57:4-6

439 SQUADRON -- RCAF
Roundel, D '52:34-45

441 (Silver Fox) SQUADRON -- RCAF
Roundel, Ja '52:4-12

442 (Caribou) SQUADRON -- RCAF
Roundel, Jl/Ag '51:45-50

443 (Hornet) SQUADRON -- RCAF
Roundel, Je '52:4-12

466 SQUADRON -- RAAF
Air Pict, Ap '67:129-131

488 SQUADRON -- RNZAF
Air Pict, Mr '63:84-85

491 BOMB GROUP (Heavy) -- USAF
AAHSJ, Sum '64:79-95 (Pt 1) In WW II
Fall '64:207-218 (Pt 2) " " "

497 TACTICAL FIGHTER (Wolf Pack) SQUADRON -- USAF
AAHSJ, Fall '69:196-198 In Vietnam

501 (City of Bristol) SQUADRON -- RAF
Air Pict, O '69:389
Flight, S 13, '34:940-943

502 (Ulster) SQUADRON -- RAF
Flight, N 23, '33:1165-1169

504 (County of Nottingham) SQUADRON -- RAF
Flight, Je 13, '35:628-630

524 SQUADRON -- RAF
Air Pict, Je '67:213

533 NIGHT FIGHTER SQUADRON -- US Marines
Aero Digest, N 15, '45:40+

600 (City of London) SQUADRON -- RAF
Flight, Ag 21, '31:834-836
N 3, '49:585-592

601 (County of London) SQUADRON -- RAF
Flight, Ag 15, '30:922-925

602 (City of Glascow) SQUADRON -- RAF
Flight, S 16, '32:865-868

603 (City of Edinburgh) SQUADRON -- RAF
Flight, S 16, '32:868-871
Ap 23, '54:514-518 (Pt 1)
Ap 30, '54:554-557 (Pt 2)
My 7, '54:586-589 (Pt 3)

604 (County of Middlesex) SQUADRON -- RAF
Flight, Jl 2, '36:20a-21

609 (West Riding) SQUADRON -- RAF
Air Pict, Jl '66:263
RAF Fly Rev, Mr '57:17-19+ (Pt 1) Typhoon exploits, WW II
Ap '57:35-38 (Pt 2) Typhoon exploits, WW II

615 (County of Surrey, "Churchill's Own") SQUADRON -- RAF
RAF Fly Rev, N '54:35

616 (South Yorkshire) SQUADRON -- RAF
RAF Fly Rev, N '59:16-17+ In WW II

625 SQUADRON -- RAF
Air Pict, N '69:429

766 SQUADRON -- Royal Navy Air School
Air Pict, D '65:424

800 SQUADRON -- Royal Navy
Air Pict, N '64:364-365
Flight, O 19, '51:498-500

804 SQUADRON -- RAF
Flight, My 6, '48:496-498 In 1948

807 SQUADRON -- Royal Navy
Air Pict, Ag '62:244-245 w representative aircraft

809 SQUADRON -- Royal Navy
Air Pict, Ja '69:14-16
Flight, N 24, '49:675-679

810 SQUADRON -- Royal Navy
Flight, S 26, '35:326-329

812 SQUADRON -- RAF
Flight, My 6, '48:496-498 Equipment and organization in 1948

813 SQUADRON -- Royal Navy
Flight, S 25, '53:437-440

824 SQUADRON -- Royal Navy
Aeronautics, S '55:66-69
Air Pict, Ag '63:242-243

825 SQUADRON -- Royal Navy
Air Pict, N '55:354
Flight, O 28, '55:686

826 SQUADRON -- RAF
Air Cl, Mr/Ap '65:50-53+ Early "Pathfinding"

830 SQUADRON -- Royal Navy
Flight, Mr 9, '56:264-267

831 SQUADRON -- Royal Navy
Flight, Mr 9, '56:264-267

890 SQUADRON -- Royal Navy
Aeronautics, S '54:66-69

897 SQUADRON -- Royal Navy
Flight, Mr 23, '56:325-328

899 SQUADRON -- Royal Navy
Flight, Mr 23, '56:325-328

"AEROJETS" -- USAF Aerobatic Team
Skyways, Ag '50:22-24+ In 1950

AERONAUTICAL ENGINE LABORATORY -- US Navy
Nav Avn News, Mr '64:6-11

AIR CORPS TACTICAL SCHOOL -- USAAF
Pop Avn, F '40:34-36+

AIR FORCE ACADEMY -- USAF
Bee-Hive, Ja '57:28-32
Boeing Mag, My '55:8-10
Flying, O '53:14-15+ University of Mexico as inspiration--early thoughts
Ja '55:32-33+
Mr '56:20-22+ Cadet life in 1956
D '58:40-41 Soaring program
Mr '63:30-31+

"BLUE ANGELS" -- US Navy Aerobatic Team (cont.)
Nav Avn News, Ag '55:1-7
Ja '57:19-21
D '61:17-23 1946-1961
O '65:8-12 Maintenance, 1965
Je '66:12-15 20th Anniversary
Ag '67:11-13 Overseas tour
O '69:10-19 Photo story of maintenance

"BLUE DIAMONDS" -- RAF Aerobatic Team
RAF Fly Rev, O '61:27

"BOELCKE JASTA" -- German AF
Cr & Cockade, Win '68:313-350 WW I pilot losses, victory list, pors, photos

BOMBER COMMAND -- RAF
RAFs Qtr, Sum '66:88-96

CAMBRIDGE UNIVERSITY AIR SQUADRON -- RAF
Air Power, Ap '56:188-196
Flight, Jl 10, '31:674-676
Ag 8, '32:727-728
Jl 26, '34:758-760
RAF Qtr, Jl '52:234-236 In 1952

CENTRAL EXPERIMENTAL AND PROVING ESTABLISHMENT -- RCAF
Roundel, My '63:4-8

CENTRAL FLYING SCHOOL -- RAF
Air Pict, Jl '52:196-198
Je '69:208-210
Bris Sid Jnl, Spg '62:68-73
Flight, Jl 17, '31:681-686
Ag 18, '38:146a-d
Pop Fly, Jl '34:176-177+
RAF Fly Rev, S '58:50+ Aerobatic team

CENTRAL FLYING SCHOOL -- RCAF
Roundel, Je '54:4-13
Je '65:4-8

"CIGOGNES, LES" -- French AF
Cr & Cockade, Aut '60:40-54

"GOLDEN HAWKS" -- RCAF Aerobatic Team (cont.)
Roundel, My '61:2-4 In 1961
My '62:2-3 " 1962
My '63:2-3 " 1963
Ap '64:14 Disbandment
RAFs Qtr, Vol. 2, No. 2:134-135

IMPERIAL DEFENSE COLLEGE -- GB
Roundel, Jl/Ag '54:20-21

INSTITUTE OF TECHNOLOGY -- USAAF
Pegasus, Ja '47:1-5+

"KOSCIUSZKO SQUADRON" -- Polish AF
Air Pict, S '65:316-319

"LAFAYETTE ESCADRILLE" -- French AF
US pilots in French AF, WW I
Airpwr Hist, Jl '57:150-161
Cr & Cockade, Spg '61:1-64
Pop Fly, Ap '34:24-25

LIGHT BOMBING WING -- US Navy
AAHSJ, Jl/S '58:148-151

"LORRAINE SQUADRON" -- RAF
Free French airmen in RAF, WW II
RAF Fly Rev, Je '57:21-23 Bombing of Paris, WW II

"MEN ON THE FLYING TRAPEZE" -- US Aerobatic Team
US Air Svcs, O '34:32-33

NATIONAL WAR COLLEGE -- US
Nav Avn News, Je '60:16-18

NATO DEFENSE COLLEGE
See North Atlantic Treaty Organization Defense College

NAVAL AEROSPACE MEDICAL INSTITUTE -- US Navy
Nav Avn News, Ag '69:6-11

NAVAL AIR DEVELOPMENT CENTER -- US Navy
Nav Avn News, Ap '69:22-25

NAVAL AIR ENGINEERING CENTER -- US Navy
Nav Avn News, Jl '66:22-25

NAVAL AIR RESERVE -- US Navy
Nav Avn News, Jl '66:6-10 50th Anniversary

NAVAL AIR SYSTEMS COMMAND -- US Navy
Nav Avn News, N '66:7-13

NAVAL AIR TRANSPORT SERVICE (NATS) -- US Navy
Nav Avn News, Jl '48:13 Disbanded
Pegasus, F '45:1-4

NAVAL AVIONICS FACILITY INDIANAPOLIS -- US Navy
Nav Avn News, D '68:14-17

NAVAL ORDNANCE TEST STATION -- US Navy
Nav Avn News, Mr '49:11-13

NAVAL PHOTOGRAPHIC CENTER -- US Navy
Nav Avn News, Ap '63:20-22 1943-1963

NAVAL WAR COLLEGE -- US Navy
Nav Avn News, Mr '60:13-17

"NORMANDIE REGIMENT" -- Free French AF
Air Prog, Ja '66:26-27+

NORTH ATLANTIC TREATY ORGANIZATION (NATO) DEFENSE COLLEGE
Roundel, F '53:35
RAFs Qtr, No. 2, '64:113-118

NORTHERN BOMBING GROUP -- US Navy and Marines, WW I
Nav Avn News, O '68:28-31

OXFORD UNIVERSITY AIR SQUADRON -- RAF
Flight, Ag 1, '30:862-863
Ag 7, '31:784-787
Jl 22, '32:684-686
N 10, '32:1047-1048
Ag 1, '35:118-119

"RED ARROWS" -- RAF Aerobatic Team
Fly Rev Intl, O '66:102-105
S '68:50-53

"RICHTHOFEN GESCHWADER" -- German AF
Flight, Mr 11, '37:226-229
Pop Fly, D '35:486-489+
S '36:292-296+ After Richthofen's death
RAF Fly Rev, Ap '60:11+ Insignia
D '60:34-35

ROYAL AIR FORCE COLLEGE -- RAF
Aeroplane, My 28, '30:1028 Armorial bearings
Air Power, O '56:18-29 In 1956
Flight, Je 14, '34:581-586
F 5, '60:173-174
RAF Fly Rev, Ap '60:21 40th Anniversary
RAF Qtr, O '50:340-347
O '51:310-317 Selection policy
RAFs Qtr, Sum '66:105-111

ROYAL AIR FORCE FLYING COLLEGE -- RAF
Air Power, Aut '53:24-32
Roundel, D '51:47-48

ROYAL AIR FORCE STAFF COLLEGE -- RAF
Flight, Jl 20, '33:723-726
RAFs Qtr, No. 3, '62:194-197
No. 4, '63:249-255 Its activity
No. 2, '66:129-133

ROYAL AIR FORCE TECHNICAL COLLEGE -- RAF
Air Power, Spg '54:271-277

ROYAL AUSTRALIAN AIR FORCE COLLEGE -- RAAF
RAF Qtr, Ja '53:80-85 Aims and program

ROYAL FLIGHT -- RAF
RAF Fly Rev, N '61:28

ROYAL MILITARY COLLEGE -- RCAF
Roundel, S '51:4-14
Je '55:10-14

ROYAL ROADS -- Canada
Roundel, My '55:20-24

"SISKENS" -- RCAF Aerobatic Team
Roundel, S '58:4-6

TECHNICAL TRAINING AIR FORCE -- US
Pegasus, Ag '52:11-15
D '54:6-10

TECHNICAL TRAINING COMMAND -- RAF
RAFs Qtr, Spg '66:18-25 In 1966

TEST PILOT SCHOOL -- US Navy
Nav Avn News, Jl '63:8-10

"THUNDERBIRDS" -- USAF Aerobatic Team
Aerosp Hist, Spg '68:16-27 1953-1968, w rosters
Avn Wk, S 1, '69:36-39+ In F-4E aircraft
RAF Fly Rev, F '58:16 F-100C-10-NH color photo
Skyline, Vol. 14, No. 3, '56:42-44 In 1956; pilots and aircraft

"TREBLE ONE SQUADRON" -- RAF Aerobatic Team
RAF Fly Rev, S '56:49-50+ In 1956
S '58:13
D '60:40

UNIVERSITY OF LONDON SQUADRON -- RAF
Flight, Jl 30, '36:134-136

"WOLF PACK" -- US Marines
Flying, Mr '44:34+ (Pt 1) In WW II
Ap '44:63+ (Pt 2) " " "

YALE NAVAL AVIATION UNITS -- US Navy
Nav Avn News, N '66:14-16 First Unit
Ja '68:14-17 Second Unit in WW I

Part Five

MISCELLANY

Part Five

MISCELLANY

This final part contains a variety of topics, arranged alphabetically, none of which seems to fall naturally into the preceding sections. These articles were chosen because they impressed the author with their comprehensiveness or contained information often sought after by the researcher.

Notes will once again provide direction to the user and the citations follow a format similar to the earlier parts.

AEROBATIC TEAMS

Avn Wk, Ag 21, '67:50-73	Techniques of different countries, in color

AERONAUTICS, HISTORY OF

Pegasus, Ap '55:6-11 (Pt 1)	Wright brothers
Jl '55:6-11 (Pt 2)	Dunne, Cody, etc.
Ag '55:4-15 (Pt 3)	Farman, Voisin, etc.
D '55:10-15 (Pt 4)	Curtiss, AEA, etc.
Mr '56:6-15 (Pt 5)	Curtiss to 1914
My '56:10-15 (Pt 6)	Martin to 1916
Ag '56:10-15 (Pt 7)	Pfitzner, Benoist, etc.
S '56:10-15 (Pt 8)	Early testing, NACA
N '56:10-15 (Pt 9)	Early engines
F '57:1-10 (Pt 10)	European development to WW I
O '57:6-12 (Pt 11)	WW I

AIR FORCE CROSS

Aerosp Hist, Win '67:246-248

AIRBORNE UNITS

Pegasus, Je '54:1-15	History and development
Jl '54:1-11	" " "
Ag '54:1-15	" " "

AIRCRAFT CARRIERS

European

Nav Avn News,	My '63:30-33	In WW II

Japan

Nav Avn News,	O '62:23-27	Development
	Ap '63:27-29	Destruction in WW II

United States

Nav Avn News,	F '62:22-28 (Pt 1)	History of operations
	Mr '62:25-31 (Pt 2)	" " "
	My '62:24-29 (Pt 3)	" " "
	Je '62:6-11	Battle of Midway, WW II
	:32-37 (Pt 4)	History of operations
	Ag '62:22-27 (Pt 5)	" " "
	S '62:22-26 (Pt 6)	" " "
	N '62:22-26 (Pt 7)	" " "
	D '62:15-19 (Pt 8)	" " "
	Ja '63:26-28 (Pt 9)	" " "
	O '63:22-26 (Pt 10)	" " "
	N '63:23-27 (Pt 11)	" " "
	D '64:6-9 (Pt 1)	Modern aircraft carrier
	Ja '65:16-19 (Pt 2)	" " "
	F '65:14-16 (Pt 3)	" " "
	Ap '65:15-18 (Pt 4)	" " "
	Jl '65:13-18 (Pt 5)	" " "
	S '65:23-26 (Pt 6)	" " "
	N '65:15-19 (Pt 7)	" " "
	Ja '66:24-27 (Pt 8)	" " "
	Mr '66:24-29 (Pt 9)	" " "
	My '66:24-28 (Pt 10)	" " "
	D '67:15-19 (Pt 11)	" " "
Pegasus,	My '54:Supp:1-20	Development

AIRLIFT

Berlin

Air Force,	Jl '68:68-79	20th Anniversary history
Avn Opns,	Ap '49:Entire issue	
Bee-Hive,	Fall '48:1-32	
Interavia,	Ja '49:21-25	Many photos
Pegasus,	Je '49:1-7	Lessons and C-82 role
Skyways,	Jl '49:26-27+	Personal accounts
US Air Svcs,	S '48:8-10	
	F '49:7-11	

AIRLIFT (cont.)

Congo

Bee-Hive, Fall '60:24-27

Hungarian

Bee-Hive, Ja '57:2-16

ATLANTIC FLIGHTS

Aero Digest, D '31:33-37+	Attempts, 1919-1931
Can Avn, Spg '69:22-24	1919 race
Flight, Je 12, '69:959-962	Alcock and Brown
Flying, D '68:S-1 to S-16	
Interavia, O '49:599-602	"Oiseau Canari"
No. 5, '52:243-247	
:255-256	Nungesser and Coli
Nav Avn News, My '69:6-36, 45	US Navy, NC-4
Jl '69:12-17	US Navy, NC-4, 50th Anniversary
Pop Avn, Ag '27:5-9	Lindbergh
:10-12	Chamberlin's "Columbia"
S '27:5-8	Byrd's "America"
D '27:27-31+	US Navy, NC-4
Je '28:13-16	"Bremen"
O '33:231-232+	Balbo's Squadron
RAF Fly Rev, Jl '57:22-24	Alcock and Brown
Sentinel, Je '69:42-45	" " "
West Fly, Ag '30:42-44	Chances of success

BARNSTORMING

Flying, Je '66:72-88

Pop Avn, Je '39:60-62+

BATTLE OF BRITAIN

Air Pict, S '65:294-298	Order of Battle, S 7, 1940
Fly Rev Intl, O '63:24-26	Fighter losses
Roundel, Ap '50:14-21	Lessons
S '60:6-11	Canadian effort
RAF Fly Rev, Ag '56:26-27	Ag 15, 1940
S '60:31-35	
RAF Qtr, Je '47:180-203	Dowding's 1941 report

CAMOFLAGE

Flight, Ap 29, '37:408-410	British pre-WW II method
Fly Rev Intl, Mr '66:428-431	

MEDAL OF HONOR

Air Force,	Mr '67:45-47	USAF
Nav Avn News,	D '62:6-11	US Navy

MEXICAN PUNITIVE EXPEDITION, 1916

Aerosp Hist,	Aut '68:12-15, 34-39 (Pt 1)	Aviation role
	Win '68:30-36+ (Pt 2)	" "
	Spg '69:28-33 (Pt 3)	" "
Flying,	Ap '49:38-39+	" "
	Mr '56:30-31+	" "
US Air Svcs,	O '32:24-30	Gorrell's experiences

MUSEUMS, AIR

Aerosp Hist,	Spg '69:10-13+	RAF Museum, Hendon
Air BP,	No. 11:32-36	USAF Museum, Ohio
	No. 30:10-14	National Aero Collection, London
	No. 45:15-17	Fleet Air Arm, England
Air Cl,	D '69:26-35	Bangkok Air Museum, Thailand
Air Force,	Mr '69:102-109	USAF Museum, Ohio
Air Pict,	O '57:369-370	" " "
	D '59:452+	" " "
	Mr '67:89	Fleet Air Arm, England
Air Prog,	Ja '65:60-77	Musée de l'Air, France
	Ap '65:52-59	" " " "
	My/Je '65:50-61	" " " "
	My '68:52-53+	National Technical Museum, Czech
	Jl '68:56-57+	RAF Museum, Hendon
AAHSJ,	Sum '69:81-82	National Technical Museum, Czech
Can Avn,	F '69:14-15	National Aero Collection, Canada
CAHS,	Sum '63:49-54	Canadian War Museum, Ottawa
Cr & Cockade,	Sum '62:111-123	Imperial War Museum, London
	Aut '62:248-255	Science Museum, London
Flying,	N '56:42-43+	Henry Ford Museum, Michigan
	Mr '59:42-43+	Smithsonian Institute, Washington, D. C.
Fly Rev Intl,	F '68:110	Willow Grove Naval Air Museum, Pennsylvania
	Ag '68:464-466	Polish National Aircraft Museum, Krakow
Interavia,	No. 5, '67:750-752	Musée de l'Air, France
Nav Avn News,	Ap '49:14-15	Smithsonian Institute, Washington, D. C.
Pegasus,	N '48:13-15	Smithsonian Institute, Washington, D. C.

MUSEUMS, AIR (cont.)

Pegasus, Ap '55:12-15	Smithsonian Institute, Washington, D. C.
Pop Avn, My '35:297	Jarrett Museum, New Jersey
O '35:221-222+	Smithsonian Institute, Washington, D. C.
My '38:43	USAF Museum beginnings
Pop Fly, Ja '34:534-535+	Richthofen Museum, Schweidnitz
N '37:442-444+	Jarrett Museum, New Jersey
Roundel, Ja/F '61:27-29	National Aviation Museum, Canada
My '65:18-22	National Aviation Museum, Canada
RAF Fly Rev, D '56:45-46	Musée de l'Air, France
D '57:16-19	Shuttleworth Collection, England
RAFs Qtr, Spg '68:37-39	RAF Museum, Hendon

PACIFIC FLIGHTS

Bee-Hive, N '31:3-5	Pangborn and Herndon
Ja '66:28-32	"Clasina Madge"
Pop Avn, Ag '27:14-15	Hagenberger and Maitland
S '27:9-10	"City of Oakland"

PARACHUTES

Air Tech, Mr '43:35-42+	Care
Flight, Ag 8, '35:148-151 (Pt 1)	
Ag 15, '35:b-e (Pt 2)	
Ag 25, '38:168a-d	
Interavia, No. 8/9, '50:434-437	Early pioneers
No. 9, '53:522-523	
No. 2, '54:102-103	Development in US
No. 2, '58:146-149	Modern uses
Nav Avn News, N 1, '43:10-17	
Ap 1, '45:14-15	Care and preservation
My '54:14-15	
Pegasus, F '49:6-9	Evolution
Ja '56:6-9	Bailout incidents
Roundel, O '51:28-35	
US Air Svcs, S '45:32-34	

ROYAL AERONAUTICAL SOCIETY

Flight, Ja 15, '60:69-71 (Pt 1)	
Ja 29, '60:154-156 (Pt 2)	
F 26, '60:284-285 (Pt 3)	

TACTICS

Cr & Cockade,	Spg '62:60-62	WW I formation
	Sum '62:168-172	In WW I
Flight,	Ag 15, '35:168-172	British in 1935
	Je 25, '36:686a-b	" " 1936
	Ja 25, '40:78a-79	Dive bombing
	Je 10, '43:601-605 (Pt 1)	" "
	Je 17, '43:629-633 (Pt 2)	" "
	Je 24, '43:661-665 (Pt 3)	" "
Flying,	Ja '43:27-28+	Fighter tactics, WW II
	My '45:21-23+	" " " "
Interavia,	No. 2, '51:108-110	Interception
Pop Avn,	Ag '28:48-54	In 1928
	Je '32:348-350+	WW I tactics of aces
	Mr '39:28-30+	Dive bombing
Pop Fly,	Ap '33:26-27+ (Pt 1)	Fighter tactics, WW I, w illus
	Je '33:148+ (Pt 2)	" " " " " "
	Jl '33:207-208+ (Pt 3)	" " " " " "
	O '33:374-375+ (Pt 4)	" " " " " "
	N '33:438+ (Pt 5)	" " " " " "
	F '34:596 (Pt 6)	Bomber " " " " "
	Je '34:135 (Pt 7)	" " " " " "
	Jl '36:192-194+	German formation flying, WW I
	My '37:78-82	In WW I
	S '37:333-334+	McCudden's views on WW I tactics
	Jl '38:180-184+	Fighter vs. bomber, 1938
RAF Qtr,	Ja '31:70-75	Day bomber formations, 1918
	Ap '31:253-257	" " " "
	Jl '31:397-411	Reconnaissance, 1918
	:426-438	Battle aircraft, 1918
	O '33:400-406	Bomber defense
	Ja '36:1-15	Bomber formation in next war
	Ap '36:214-220	Reply to bomber formation in next war
	O '36:441-446	Bomber formations
Skyways,	Ag '43:20-24+	Dive bombing evolution
West Fly,	Ja '38:10-13+	

V-1

Air Tech,	F '45:28-29	w schematic diagram
Aviation,	N '44:190-191+	" " "
Flight,	Je 29, '44:690-693	" " "
	S 14, '44:292	
	S 21, '44:308-311	Operational use
	O 5, '44:364-367+	w schematic diagram

V-1 (cont.)

Flying, S '44:24-25+	w schematic diagram
US Air Svcs, Ag '44:16-17	" " "

V-2

Am Helicop, Mr '46:13-15+	w schematic diagram
Aviation, My '46:63-67	Engine
Avn Wk, N 26, '51:23-35	What we have learned from V-2
Flight, D 21, '44:670-673	
N 8, '45:508-509	Fuel and control systems
Interavia, No. 5, '47:27-32 (Pt 1)	Development
No. 6, '47:33-37 (Pt 2)	"

WORLD FLIGHTS

Aerosp Hist, Ap '64:45-48	US Air Service, 1924
Win '69:9-11+	"Lucky Lady II," 1949
Bee-Hive, Spg '49:2-7	"Lucky Lady II"
Boeing Mag, Mr '49:3-5+	" " "
Pop Avn, Ag '27:31-35	US Air Service, 1924
US Air Svcs, Ag '31:15-19	Wiley Post, 1931
Ag '33:20-23	Wiley Post, solo, 1933

INDEXES

INDEX OF AIRCRAFT NAMES

The figures following the names below refer to the number preceding the name in the aircraft listing in Part One. They do not indicate page numbers.

Names other than official names are included where known; however, the list is undoubtedly far from complete. Code names are enclosed in brackets.

Translated names (i.e. those enclosed in parentheses in the main body of the book) are not included, since they are seldom used in designating foreign aircraft.

INDEX OF AIRCRAFT MANUFACTURERS' SYMBOLS

Specific aircraft identification often depends upon a model number preceded by letters identifying the manufacturer. These are sometimes confusing because they are changed after a period of time or because several manufacturers have occasionally been assigned the same symbol. Nevertheless, lacking other information, they can do much to narrow the search for identification.

The organization of this book is alphabetical by manufacturer. If manufacturers' identities can be assigned each symbol, the reader should then have little difficulty in locating a specific aircraft. This eliminates the need for a lengthy index incorporating each and every model number preceded by the same symbol. It must be stressed, however, that this list includes only those symbols that occur more than once; therefore, the Index of Aircraft Symbols should also be checked.

Punctuation in the symbols is variously used in the literature; therefore, do not attach too much importance to it. The punctuation used below was based on what appeared to be predominant usage.

A. W.	Armstrong Whitworth
B	Beagle, Bensen, Berlin, Blackburn, Boeing, Brantly
BD	Berlin Doman
Be	Beriev
Bf.	Messerschmitt
BI	Bereznyak-Isaev
BN	Britten-Norman
Bo	Bölkow
BS	Omega
Bü	Bücker
Bv	Blohm & Voss
B. Z.	Breda-Zappata
C.	Autogiro (Cierva), Caspar, Caudron, Commercial
CA	Commonwealth
Ca	Caproni
C. A. O.	SNCAO
C. C.	Caproni-Campini
CCW	Custer
CL	Canadair
CM	Castel-Mouboussin, Fouga
C. R.	Fiat
CW	Curtiss (Wright)
D.	Dewoitine
DC	Douglas
DGA	Howard
D. H.	De Havilland
D. H. A.	De Havilland Aircraft Pty.
D. H. C.	De Havilland Canada, Huneault
Do.	Dornier
DW	Distributer Wing
E. P.	Edgar Percival
F	Farman
Fa	Focke-Achgelis
FH	Fairchild Hiller
Fi	Fieseler
F. K.	Koolhoven
FL	Flettner
F. N.	Nardi
Fo.	Folland
FW	Focke-Wulf
G.	Fiat, Grumman
GA	General Aviation, Goodyear
G. A. C.	General Aircraft Corporation
G. A. L.	General Aircraft

G. C. A.	Gyrodyne
GL	Gourdou-Leseurre
Go.	Gotha
H	Hanriot, Helio, Hunting Percival
HA	Helwan, Hispano, Hollandair
Ha	Blohm & Voss
HD	Hanriot-Dupont, Hurel-Dubois
He	Heinkel
Ho	Horton
H. P.	Handley Page
Hs	Henschel
I	Polikarpov
I. A.	Dinfia
I. AÉ	Instituto Aerotecnico
IK	Ikarus
IL	Ilyushin
J.	Jona
Ju.	Junkers
K.	Kaman
Ka	Gotha-Kalkert, Kamov
Ki.	Kawasaki, Mitsubishi, Nakajima, Tachikawa
L	Czechoslovakia, Lockheed, Lualdi
La	Lavochkin
LeO	Liore-et-Olivier
LN	Loire-Nieuport
LZ	Doman
M	BFW, Macchi, Mauboussin, Messerschmitt, Miles
MA	Murryair
M. B.	Bloch, Macchi, Marcel Besson, Martin-Baker
M. C.	Macchi, McCulloch, Matra-Cantinieau
Me.	Messerschmitt
M. F.	Maurice Farman
MH	Max Holste
Mi	Mil, WSK
ML	Millet-Lagarde
MV	Monsted-Vincent
N	Northrop
NA	National, North American
NC	Aerocentre, Navy-Curtiss
NH	Nagler
NR	Nagler and Rolz

NW	Navy-Wright
P	Boulton Paul, Hunting Percival, Partenavia, Percival, Piaggio, PZL
PA	Payen, Piper, Pitcairn
PB	Galeao
P. B.	Pemberton-Billing
PC	Pilatus
PD	Piaggio /Douglas
Pe	Petlyakov
P. V.	Port Victoria
R	Renard, Rey, Roussel
Re	Reggiane
Ro.	IMAM
R. S.	Reid & Sigrist
S	De Schelde, Letov, SABCA, Savoia-Marchetti, Short, Siebel, Sikorsky, SIPA, Sparmann, Supermarine
S. A.	Short, Sud Aircraft
S. B.	Short
SC	Short
SE	Sud-Est
SEV	Seversky
Si	Siebel
SK	Skoda Kauba, Somers-Kendall
S. M.	Savoia-Marchetti
SO	Sud-Ouest
S. R.	Saunders-Roe
S. T.	Monospar
Su	Sukhoi
S. V. A.	Ansaldo
S. Z.	Savieliev-Zalewski
T	Thaden
Ta	Focke-Wulf
TS	OKL
Tu	Tupolev
UB	Burnelli
V	Vought, Vultee
VB	Arsenal
VG	Arsenal, Vertigyro
W	Wagner, Waterman, Weick, Weir
WA	Wallis
WM	Weiss Manfred
Z	Cant

INDEX OF AIRCRAFT SYMBOLS

Letters and letter-number combinations are almost always used to designate specific aircraft, even if a name is also assigned to that aircraft. These symbols frequently take the form of a military designation or, as explained in the introduction to the preceding index, a manufacturer's designation.

In the introduction to Part One, Aircraft, paragraph 5c, (page 3) the use of the X and Y prefix for military designations of aircraft that never developed beyond the experimental or prototype stage is explained. By treating the prefix as though it did not exist, like categories of aircraft can be grouped together. For example, the Boeing XB-44 designation is read as B-44. This places it with other bombardment aircraft in the index and assists the uninitiated in finding a designation which may only be known to him as B-44.

Roman numerals always appear before Arabic numerals of similar value.

The citation refers to the identification (i.e., entry) number assigned each aircraft in Part One, not to the page number.

INDEX OF ENGINE NAMES

The figures following the names below refer to the number preceding the name in the engine listing in Part Two. They do not indicate page numbers.

INDEX OF ENGINE SYMBOLS

Symbols associated with specific engines usually consist of letter or letter-number combinations. They often designate a manufacturer's model or a military specification.

Experimental engine designations frequently begin with the letter X. So that similar engines can be categorized together, and to assist the searcher who is unaware of the X prefix (since it is sometimes omitted in the literature), this index treats the X as though it did not exist.

Periods, hyphens, and other marks of punctuation in the symbols are variously used in the periodical literature; therefore, not too much importance should be attached to them.

The figures following the symbols below refer to the number preceding the symbol in the engine listing in Part Two. They do not indicate page numbers.